Scientific method in analysis of sediments

Ager Principles of Paleoecology
Cagniard (Trans. Flinn and Dix) Reflection and Refraction of Progressive Seismic Waves
De Sitter Structural Geology
Ewing, Jardetzky, and Press Elastic Waves in Layered Media
Grant and West Interpretation Theory in Applied Geophysics
Griffiths Scientific Method in Analysis of Sediments
Grim Applied Clay Mineralogy
Grim Clay Mineralogy
Heinrich Mineralogy and Geology of Radioactive Raw Materials
Heiskanen and Vening Meinesz The Earth and Its Gravity Field
Howell Introduction to Geophysics
Jacobs, Russell, and Wilson Physics and Geology
Krauskopf Introduction to Geochemistry
Krumbein and Graybill An Introduction to Statistical Models in Geology
Kuznetsov Ivanov, and Lyalikova (Trans. Broneer) Introduction to Geological Microbiology
Legget Geology and Engineering
Menard Marine Geology of the Pacific
Miller Photogeology
Officer Introduction to the Theory of Sound Transmission
Ramsay Folding and Fracturing of Rocks
Shaw Time in Stratigraphy
Shrock and Twenhofel Principles of Invertebrate Paleontology
Turner and Verhoogen Igneous and Metamorphic Petrology
Turner and Weiss Structural Analysis of Metamorphic Tectonites
White Seismic Waves: Radiation, Transmission, and Attenuation

Scientific method in analysis of sediments

John C. Griffiths

Professor of Petrography,
Department of Geochemistry
and Mineralogy,
Pennsylvania State University

McGraw-Hill
Book Company

New York St. Louis
San Francisco
Toronto London
Sydney

Scientific method in analysis of sediments

Library of Congress Catalog Card Number 67-13901

24802

1 2 3 4 5 6 7 8 9 0 M P 7 4 3 2 1 0 6 9 8 7

Preface

In attempting to present an integrated approach to research problems in the geosciences by means of the scientific method, this book follows the recommendations of Chamberlin's Method of Multiple Working Hypotheses, using a somewhat more modern framework than that available to Chamberlin in 1897. To solve problems in this framework it is essential to define the problem clearly at the outset, state explicitly the possible solutions or hypotheses (or models), and then proceed to test the hypotheses on the basis of experimentally determined observations or data. The arrangement of the chapters in the book follows the scientific method as an organized plan, or algorithm, for problem solving.

The required axioms are stated and an outline of the scientific method as an algorithm is provided in Chap. 1. The elementary aspects of sampling designs are discussed in Chap. 2. A basic model for the measurement of the properties of materials, including a definition of *fundamental* and *derived*, or secondary, *properties*, is presented in Chap. 3. Chapters 4 to 10 discuss the techniques and procedures by which the fundamental properties are measured, and Chap. 11 briefly describes the measurement of some derived properties.

The preliminary steps in recording data, measuring its information content by means of scale theory, and selecting appropriate summary statistics as estimators of the appropriate parameters in the measurement model are described in Chap. 12. This material forms the basis for recommending specific means for data gathering and for rejecting many published procedures which fail to satisfy the requirements of the problem-solving algorithm or the measurement model.

Chapters 13 to 21 describe the statistical analysis of data, and the last chapter presents some general conclusions on the present status of experimental inquiry in the geosciences.

Although illustrative examples are selected from the petrographic analysis of detrital sediments, the approach of the book is quite general and may be applied to any field which permits the isolation and definition of a population and its elements. Similarly, the measurement of properties and its accompanying model is quite general, as are most of the techniques for data gathering; the subsequent analysis of data is by means of conventional statistics with prime emphasis on parametric, as contrasted with nonparametric, analysis. My predilection for parametric analysis is based on the belief that unless the data contain the necessary information for achieving the solution to the problem, there is little point in gathering them, and scale levels of at least interval rank are necessary to attain this desired goal.

The objective, then, is to outline a method, as an algorithm, along with a basis for critical appraisal and testing of the information gathered according to the plan, so that the problem-solving procedure can be improved and some workable solutions achieved. The fact that such solutions are workable is, at present, one of the few criteria which enable an investigator to judge that his work is successful. In the geosciences, in particular, prognostications based on presently existing theories are woefully weak in holding up under the stringent testing of economic demands; the search for natural resources is growing more difficult as the "easily found" targets are used up, and continued success will demand more rigor and greater investment in collecting adequate information to improve our guesses.

The experience summarized in this book was accumulated during teaching and research at a university as well as in daily practice as a petroleum geologist; it was immeasurably augmented by my discussions and clashes with colleagues and, more particularly, with many enthusiastic and critical young minds, both as their teacher and as research adviser. The basis for the book was outlined while I worked with Dr. M. A. Rosenfeld, who contributed materially to the devising of techniques and procedures. A large number of other graduate students will recognize their handiwork incorporated indissolubly in this work; I acknowledge my indebtedness both to their critical comments and to their constructive contributions.

Early drafts of parts of the book have been examined by M. A. Rosenfeld and Professor J. B. Bartoo; many parts have appeared in different form in various books and journals.

John C. Griffiths

Contents

Scientific method in analysis of sediments

1 The scientific method

It follows from these considerations that any definition of "red" which professes to be precise is pretentious and fraudulent. All empirical concepts have this character—not only obviously vague concepts such as "loud" or "hot" but also those that we are most anxious to make precise, such as "centimeter" and "second."

B. Russell, 1948

1.1 Introduction

Analysis is usually associated with laboratory experimentation, and so a course in analysis is usually considered to comprise a description of procedures and instruction in the assembly and handling of equipment; in other words, it is presumed to concern the description of techniques of experimental analysis. This aspect is, however, only a very small part of the subject of *method*. Method deals with the entire aspect of problem solving. It is not sufficient to describe a series of techniques; they must be integrated with the reason and purpose of inquiry. Without this philosophical aspect, analysis is a very limited tool and a rather barren subject.

The description of method concerns the very heart of a science, and it is the theme of the present work that there is one and only one entirely efficient and satisfactory method of approach to scientific inquiry. Initially this method is associated with, and indeed inseparable from, *purpose*. If the purpose of inquiry can be uniquely described, the achievement of the purpose is usually straightforward; if there is any doubt or equivocation, conscious or otherwise, about the purpose, the same doubt and equivocation will be attached to the achievement of that purpose.

Quotation: B. Russell, "Human Knowledge: Its Scope and Limits," p. 277, George Allen & Unwin Ltd., 1948.

The first step in method, then, is to decide the purpose of the investigation; fundamentally, scientific investigation is an organized attempt to answer questions which arise from the curiosity of men's minds concerning the "real world." In simple terms, the first step in method is to ask a question. Let us ask, "How far is the rainbow from the observer?" Some of the early inquiries into this question terminated with the value of minus 93×10^6 miles. This answer was dismissed as meaningless because the distance in front of an observer can hardly be a minus quantity, and no doubt such a question and answer gave rise to prolonged controversy. Eventually it was observed that the rainbow originated from the refraction of light around water droplets. The question became "How far away from the observer is the source of light giving rise to the rainbow?" The answer minus 93×10^6 miles is now resplendent with meaning. From this much oversimplified analogy it can be seen that one of the most important aspects of method is the correct formulation of meaningful questions. Meaningless questions beget meaningless answers!

It should also be clear from the above example that there must have been a technique for estimating distance and there must have been a scale of measurement and a large number of other items which form part of the method of analysis. We may reduce these various steps in the method of problem solving to the following unified plan, which may be considered as a basic formulation of scientific method:

1. Formulate the question.
2. Plan an experiment (or set of experiments) to answer it.
3. State the exact steps involved (the experimental procedure) in obtaining the raw data.
4. Classify and summarize the observations (raw data).
5. Analyze the raw data.
6. Draw conclusions (inferences) from the analysis.
7. Erect a hypothesis on the basis of the inferences.
8. Predict new inferences from the hypothesis.
9. Test the inferences by further experimentation.
10. Accept or reject the hypothesis on the basis of the analysis of the data in step 9 and its agreement or otherwise with prediction.

Let us now suppose that we have formulated a "simple" question such as "What is the weight of this sample of rock?" or "What is the length of this grain of quartz?" It is usually assumed that the answer is some single value of weight or length expressed in a scale of units. In other words, we assume that the object has a specific length or weight and that this property can be defined by a single or unique value; this answer is the *true value*. In simple terms, we search for truth, in this case the true weight or true length of the object. A very large number of other factors are also contained in this assumption, because the weight of an object will not be the same at sea level as it is at 5,000 ft elevation. For this concept of true value to be meaningful, therefore, the conditions under which it is to be determined must be specified.

For the true value to possess the property of being unique it must be a constant under constant conditions.

On this basis, then, we make three fundamental assumptions: first, that a true value exists for every measurement; second, that granting it exists, we can recognize it; and third, that we can specify the conditions under which the true value can be recognized. It is good practice never to lose sight of these three assumptions as some assurance against dogmatism and the acceptance of any answer as completely final.

Although we may feel satisfied in accepting these assumptions, this still does not suffice to enable us to determine a true value; in fact, the determination of a true value for any purpose is exceedingly difficult and is sometimes regarded as impossible. Operationally this is the correct view, but one possible solution to this enigma may be found in the following philosophic approach to the formulation of scientific method:

> But there is a viewpoint, typically modern, that to say an objective is unattainable is not the same thing as saying the objective does not exist. This viewpoint would insist that an objective that may be approached within any given distance, however small, exists, even though it can never be reached. If answers are not attainable in a finite number of steps, they may exist nonetheless in the sense that they are endpoints (limits) of a progress that may eventually approach them within any given degree. Those ends that are unattainable but approachable . . . [are ideals].
>
> That ideals exist is a necessary postulate of all inquiry, and the set of all ideals is the "real world." Statistically this ideal is called the "stochastic limit" . . . the ideal is the "true value," and all truth is ideal.
>
> The true measure of a given distance will be the limit of an infinite set of observations, all in statistical control.[1]

If, then, a true value exists and our experiments suffice to enable us to approach this value by successive approximation, we may reduce the concept of accuracy to a definition, as a basis of judgment of value of any experiment, as follows: *One experiment is more accurate than another when it leads to a closer approximation to a true value.*

This yields a conceptual definition of accuracy; it does not enable us to reach a decision in any specific instance, because we have not yet set up criteria to judge which of two experiments results in a closer approach to the true value. This subject is much too wide in scope to be answered in detail here, but the following illustration may assist us in recognizing the kind of approach we may use.

Suppose three marksmen aim at a target and obtain the results illustrated in Fig. 1.1. Marksman A is a good shot, and his hits cluster closely together although they all miss the bull's-eye; marksman B is a poor shot, and his hits are scattered widely over the target. Marksman C is a novice, and his shots are as often off the target as they are on it, and all are to one side of the bull. We can best characterize these three results by measuring the

[1] C. W. Churchman, "Theory of Experimental Inference," pp. 57–58, The Macmillan Company, 1948.

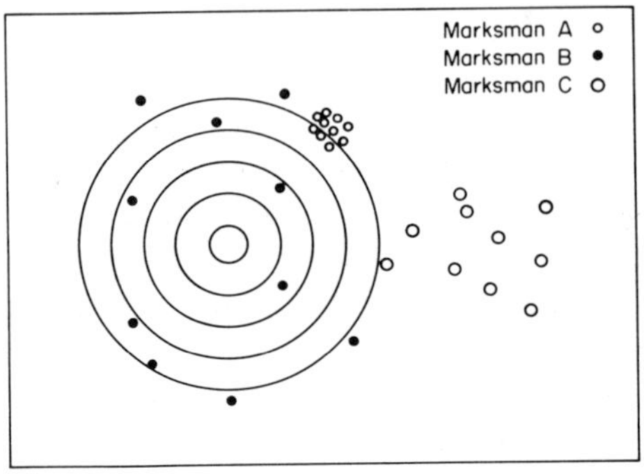

Figure 1.1 Precision and accuracy represented by marksmen shooting at a target.

deviations[1] of each shot in two ways: first, we find the average deviation for each marksman and then measure the deviations of this average from the bull; second, we measure the deviation of each shot from its respective average.

A's shots cluster closely around his average and show an almost constant departure from the bull; i.e., the results contain a systematic bias. B's shots deviate widely from his own average and the bull, but his average is relatively close to the bull. These shots show lack of control. Finally, C's shots show lack of control and possess a bias. The cure is, in each case, to increase control, i.e., reduce the scatter around each average. After the control is increased, the next step is to remove the bias, thus A's control is good, and if he adjusts the sights of the rifle, his shots will approach the bull consistently. B needs practice to increase his control; then his performance will approach that of A. C will have both to increase control and remove bias.

In this experiment the true value is known, and the accuracy of each marksman can be expressed as the deviation of his average from the true value. The precision of each marksman can also be expressed in terms of the deviations from each average. A is precise but inaccurate, B is accurate but imprecise, and C is both inaccurate and imprecise.

It may therefore be deduced that precision and accuracy are not entirely dependent; increase in one does not necessarily imply increase in the other. It does appear, however, that the best approach when the true value is not known is first to improve precision by increasing control and then, by removing bias, to increase accuracy. In most of our experiments we do not know the true value; hence we emphasize increase in precision. Only after this has been successfully accomplished can we attend to accuracy.

On the basis of the foregoing experiment we may now define precision as follows: *One series of experiments is more precise than another when its results show less spread around the average.* This definition may be reduced to an operational procedure by defining the average as the best value or mean of a series of values and measuring departures from this best value by

[1] Where deviation equals the radial distance of each shot from a common center.

the variance of the values, or their spread around the mean. These two statistics may be formally defined as follows:

$$\text{Best value} = \text{mean} = \bar{X} = \sum_{1}^{n} \frac{X_i}{n} \tag{1.1}$$

$$\text{Spread} = \text{variance} = \hat{\sigma}^2 = \sum_{1}^{n} \frac{(X_i - \bar{X})^2}{n - 1} \tag{1.2}$$

where X_i is an individual value of the variable X, $\sum$ is the sign for summation, n is the number of X items, and the sum is to be taken over all n values of the X's. $X_i - \bar{X}$ is an individual deviation from the mean; to measure spread, each is to be squared, and the n squared values are to be summed and divided by one less than the number of items.

If we had been able to use the true value for the mean, we should have used n as the divisor in Eq. (1.2), but since we computed $\bar{X}$ from the sample of X's according to the equation

$$n\bar{X} = \sum_{1}^{n} X_i$$

it can be seen that once given $\bar{X}$ and $(n - 1)X_i$'s, the nth is fixed; hence we lose one degree of freedom when we substitute the best value for the true value.

These two equations are extremely useful, as they enable us to specify the entire set of values by two computed figures: a value for position, the mean; and a value for spread around the best value, the variance. Because the variance is a square unit, to reduce this figure to original linear units we find its square root, which is called the *root mean square*, or standard deviation

$$\hat{\sigma} = \left[\sum_{1}^{n} \frac{(X_i - \bar{X})^2}{n - 1} \right]^{1/2} \tag{1.3}$$

We should also note that it is necessary to square the deviations before summing, as the sum of the deviations around the mean is always zero, a property of the mean. The computation of the mean and standard deviation is illustrated in Table 1.1.

Thus by specifying precision and measuring it we can compare the results of two or more experiments in terms of their relative precision. Furthermore, it can be seen that precision can be increased by reducing the spread, i.e., reducing the standard deviation around the mean, and one way of achieving this objective is to increase n, the number of items or measurements in the experiment. This specification of precision also allows us to decide which of two techniques designed to measure the same property is the more precise. We can, also by this means, compare the relative efficiency of increasing the complexity of the experiment, by introducing more elaborate instrumentation, with the increase of precision gained by adding to the number of experiments. This operational definition of precision is therefore of considerable value in experimental analysis.

It should be emphasized that increase in precision is not necessarily a reflection of increase in accuracy; hence the question remains as to how we can increase accuracy. Without certain assumptions about the nature of scientific inquiry this question admits of no answer; the assumptions here adopted follow from the equating of "true value" to the ideal as contained in the quotation on page 3. The true value, as a stochastic limit, may be approached by successive approximations, and the idealized trend is illustrated in Fig. 1.2.

According to this concept, successive experiments arranged to measure some property show a gradual improvement in precision, i.e., a reduction of spread around a best value; and since each experiment is performed as a series of independent determinations of the same property by different techniques, the bias shown by each best value is gradually reduced. Thus both precision and accuracy are improved, and the mean value of the property at successive intervals of time lies on an asymptotic curve which approaches a constant, the true value, in the limit. This constant value is the true value, and in the example it is the correct value of the property being measured; similarly, in other cases this concept of true value may be looked upon as the correct answer to a question. An illustration of this asymptotic approach to a true value is described by Shewart (1931), using the velocity of light as the property to be determined; this concept of approach to a true value is capable of operational usefulness in many experimental situations (see Chap. 15, for example). Hence on the basis of practical usefulness as well as theoretical

TABLE 1.1

	X_i	$X_i - \bar{X}$	$(X_i - \bar{X})^2$
	5	−3	9
	7	−1	1
	8	0	0
	9	+1	1
	11	+3	9
$\sum_1^n$	40	0	20

$$\frac{40}{n} = \bar{X} = 8 \qquad \frac{20}{n-1} = \hat{\sigma}^2 = 5$$

$$\hat{\sigma} = (\hat{\sigma}^2)^{1/2} = 2.24$$

$$\bar{X} = 8 \qquad \hat{\sigma}^2 = 5$$

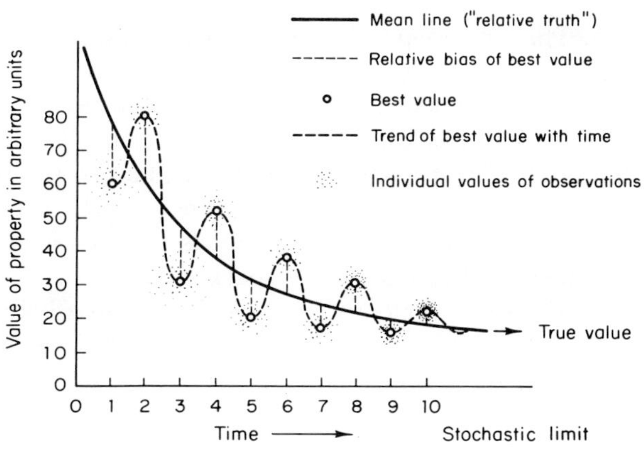

Figure 1.2 Stochastic-limit approach to the "true value."

sufficiency this stochastic-limit theory is attractive; however, the present discussion should not be considered sufficient for its acceptance. Recourse to texts on scientific method wherein such concepts are adequately ventilated e.g., Churchman, 1948, is essential to a thorough understanding of this modern outlook.

We have defined, both conceptually and operationally, a basis for decision in judging accuracy and precision of experimental investigations, and this basis will be used throughout the subsequent discussion; it is necessary to remember that to reach this basis for decision certain assumptions were incorporated in the definitions, and the only way to justify the acceptance of these assumptions is to compare predictions based on the results of experiments with actual happenings in the "real world." Furthermore, by the very nature of the model no experiment ever yields a completely final answer, but different answers may be compared in terms of relative precision and accuracy by means of this model.

1.2 The Nature of the Problem and Solution

The formulation of the general plan of approach to scientific investigation, summed up in the term scientific method, has been outlined in the previous paragraphs; it now follows that we must define the field of inquiry to which we intend applying this plan. Our prime interest is the petrography of sedimentary rocks, and our general objective is to apply the plan of approach to problems in this field. On the other hand, it should perhaps be emphasized that one of the main advantages in using the scientific method is that it is quite general and may be applied to a very wide range of problems, not by any means confined to the field of sedimentary petrography or even to the earth sciences. In fact, the vital steps in reducing a problem to an acceptable solution are to abstract its essential features and to generalize the solution process so that it may be applied in any subsequent similar situation.

The first step, then, is to define the field of sedimentary petrography, and this is best accomplished by describing the materials which form the objects of interest; i.e., the first step will be to describe the properties which enable us to recognize sedimentary rocks. To describe the properties of any material it is necessary to analyze it, and so the immediate problems concern the techniques of analysis of sediments. In attempting to follow this argument it will be found that analysis without a very definite purpose in mind will never achieve a clearly specified answer. At every stage of inquiry it is essential to hold clearly in view the purpose behind the inquiry; we must be able to state why we wish to perform certain experiments, and if the purpose is vague, the accomplishment often becomes impossible. The simplest way to define and maintain a logical purpose is to formulate the objective as a question.

Suppose we wish to know the weight of a certain piece of rock, i.e., how heavy the rock is. It will soon be found that there is no single answer to this simple question. The answer usually given depends on the comparison of certain "known weights" with that of the unknown specimen, but it will be readily understood that the "known weights" are "known" by comparison with some other standard. All our information is of this kind; i.e., no absolute values are known, only values relative to some arbitrary standard.

In most of our experiments we want to make comparisons, and given certain defined conditions, the comparison is simple; but it must be clearly understood that the comparison holds only if the conditions are fulfilled; under other conditions the results of the comparison will be different. This is of fundamental importance in all experimental analysis; it is impossible to determine an absolute value; only relative values can be found. Once this is accepted, experimental analysis becomes much simpler, and the connection between relative and absolute follows from the relationship between the concepts of precision and accuracy defined earlier. The absolute value is the true value; the relative value is useful only when the degree of precision is known. Thus, in any experiment, we attempt to determine the best value together with its penumbra of error expressed as a degree of precision, i.e., the mean and standard deviation. Using this approach allows us to compare the results of two experiments by comparing the mean values; but if the degree of precision associated with each mean is different, the comparison is difficult to evaluate.

Thus in performing experiments aimed at the analysis of materials, we attempt to determine some mean value and the degree of variation associated with this mean under the conditions surrounding the experiment.

It follows from this objective of analysis that by far the simplest kind of information to use is quantitative. Qualitative observations are surrounded by an unknown and usually large degree of uncertainty; i.e., the precision of qualitative observations is usually low and always difficult to specify. In general, it will be found that quantitative analysis is much superior to qualitative, and the basis for this generalization is manifested in the history of scientific development. Three reasonably obvious factors lend encouragement to this conclusion. First, in any scientific field which has advanced beyond the qualitative stage there is a strong tendency to concentrate on the quantitative. Second, those scientific fields which have proved outstandingly productive, measured in terms of scientific or humanistic progress, are without exception those which are most highly quantified. Finally, the only unequivocal definition of truth which escapes the dilemma of modern relativity theory is that based on the stochastic-limit approach, and this demands the use of quantitative data for most efficient operation.

So far, then, we intend to analyze sedimentary rocks by means of experiments which lead to quantitative data, and essentially our results will be expressed in terms of means and standard deviations. As already mentioned, one way of increasing precision is to increase the number of items; hence, any experimental technique which leads to an increase in the number of determinations is generally advantageous. This suggests that experimental techniques should be adopted which allow us to obtain a reasonably large number of determinations, and, for a given degree of precision, the technique which yields most determinations per unit time or cost will generally be the most efficient. Most of our techniques therefore will be those which fulfill these requirements; in general, simple techniques offer some advantage in achieving this end.

The adoption of elaboration in experimental technique with concomitant reduction in the number of experiments and, in contrast, an increase in the number of experiments without elaboration of technique are important alternatives in achieving any given degree of precision. The choice of alternative should be given very careful consideration, and the final decision as to which is the most efficient procedure will be a function of the objective, time, and cost involved in the experimental program.

Two other items are important in considering the general plan of attack. The study of the properties of any material is based on a knowledge of the degree of variation in those properties, and the foremost objective in experimental analysis is to establish the range of variation in each property. If the range is large, differences between materials are likely to be obvious, and less precision is necessary for the determination of valid differences. If the range of variation is small, greater precision is necessary in establishing differences. In differentiating elephants and mice on the basis of weight, qualitative judgment, despite its large degree of imprecision, is adequate; in differentiating mice of the same age, considerably greater precision is necessary in finding valid differences. In general it will be found that we rarely need to compare elephants and mice, and most of our interests lie in ranges of variation which demand more precise evaluation than qualitative estimation can achieve.

Second, the answers to the questions we ask are never final; they are generally stated with some degree of likelihood in mind, and this is generally expressed as a degree of probability. Thus a claim of a difference in color between red and green rocks, even with limited experience, may be accepted with a high degree of probability, whereas a claim of a difference between red and red-brown is less likely to be correct and is received with less confidence. As the precision of the estimates of color difference is increased by suitable experimental procedures and qualitative estimation is largely eliminated, smaller differences are accepted with greater assurance. Thus most of our experimental analyses possess not only degrees of precision but also degrees of confidence. This property of experimental data is usually expressed on a probability scale of zero to unity, with zero indicating complete lack of confidence and one representing complete certainty. Again it should be remarked that most of our analytical results will lie between these extremes and may not approach either very closely.

It will be seen, then, that the object of inquiry is the evaluation of the degree of variation in the properties of sedimentary rocks, and this aim is to be accomplished by means of experimental analysis using simple techniques, as far as possible, which will lead to quantitative data. For precision and

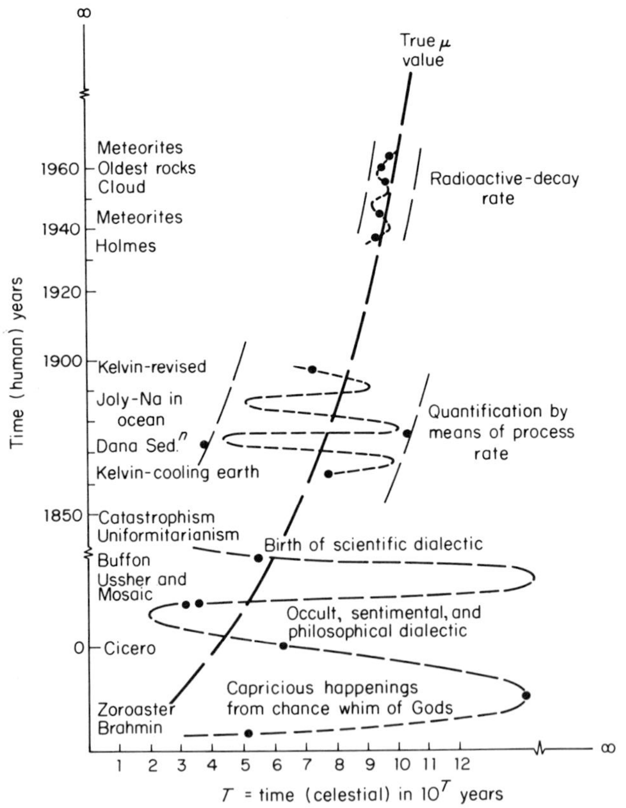

Figure 1.3 Estimation of the age of the earth as an example of stochastic convergence toward some true value.

efficiency we shall rely on relatively large numbers of determinations. The answers to the various questions will be given in the form of summaries of the data as means and standard deviations, the average, and a measure of spread of the data around the average; where possible, these statistics will be accompanied by probability estimates as a basis for specifying the degree of confidence to be placed in the results.

The answers must be considered not as final, absolute answers but as relative to some arbitrarily chosen set of standards. The degree of finality is estimated by the precision and probability associated with the single-valued answer. The wider the spread in precision and the lower the probability values, the less final the answer.

Last, it should be emphasized that increasing knowledge leads to improvement in standards and therefore to improvement in comparisons; and with concomitant increase in precision and certainty the model we have suggested as the basis of operations leads us to expect a successive approach to the final, true value.

As an example of this kind we may cite the change in estimates of that important universe parameter, the age of the earth. The estimates range from approximately zero to infinity, but when these several estimates are plotted graphically against time (Fig. 1.3), it can be seen that the graph illustrates many of the principles stated in the preceding pages.

The estimates have evolved from the qualitative opinions of our early forebears to the quantitative and very precise estimates based on radioactivity decay rates; there seems little question of the improvement in precision of estimation with time. Accuracy, however, still remains elusive and small changes in our point of view toward the model of the universe lead to relatively large changes in the "best value." Apparently we still have not achieved the true value, but there is a definite convergence toward a constant value as our measurement techniques improve and a stochastic convergence as our knowledge grows. Most estimates of universe parameters illustrate this convergence, and the history of science apparently repeats this stochastic convergence quite generally as our overall knowledge of the universe grows (see, for example, estimates of the velocity of light, Shewart, 1931).

We may then expect that the answers to questions concerning the real world will comprise numbers, and these numbers will be expressed as "best values," or averages $\bar{X}$, accompanied by an estimate of precision, the standard deviation $\hat{\sigma}$, and an estimate of assurance, a probability statement which indicates, on the basis of the suppositions (axioms) included in the model, our degree of confidence in the results.

2 Sampling sedimentary rocks

The information as to whether a sample is random is not available through examination of the properties of the sample itself . . . the criterion, therefore, of a random sample has to be sought elsewhere, namely in the process or method of selection. If a random method of selection can be developed, then a random sample can be simply defined as a sample which has been obtained by a random method.

P. O. Johnson, 1949

2.1 Introduction

In all experimental work the analysis is performed on samples; each rock unit is in itself a sample of the earth's crust, and when we examine outcrops, we are sampling these rock units. Subsequently, we collect specimens of 4 by 3 by $1\frac{1}{2}$ in. which are samples of the outcrop. In the laboratory we take these outcrop samples and sample them, choosing a $\frac{3}{4}$-in. square thin section for detailed microscopic description or an aliquot of 100 g or 100 grains for sieving or other analytical purpose. It is quite clear that the part analyzed is a very small sample of the original unit chosen for investigation. When analysis is complete, the description of the samples is applied to the whole unit from which they were taken; and it is necessary, if this description is to be meaningful, to be sure that samples are representative of the unit to which the description is to apply.

Sampling problems are, of course, quite general both in scientific investigations and in our daily lives, and we usually act on the assumption that the sample is representative of the whole unit. Tasting one candy out of a bagful, the 30-day test on cigarettes, testing a few electric lamps out of a

Quotation: P. O. Johnson, "Statistical Methods in Research," p. 199, Prentice-Hall, Inc., 1949.

batch, and describing 10 samples of beach sand or 10,000 ft of well core are all examples of the sampling process. It is, then, important to ensure that the samples are truly representative of the unit we wish to describe.

Sampling problems have been most thoroughly examined by statisticians in connection with large-scale census surveys, and the terminology of the statistician and demographer is in conventional use for this aspect of problems of analysis (Yates, 1949; Cochran, 1953). The large unit chosen for investigation is called the *population*; samples are drawn from this population, the objective being for analysis of the samples to lead to generalizations about the population. Analogous is the logician's argument from the particular to the general, and in both the laboratory and the logical process it is necessary to be sure that the samples of particular items are representative of the population if the generalization is to be valid.

The first step, then, is to define the population; it is necessary, for example, to differentiate between the average height of the American male and the height of the average American male. In the first case we sample a population of heights and find the average; in the second we sample a population of American males, find the average, and measure his height. It is much simpler to handle the first population than the second but much simpler to perform the measurement in the second case than in the first. Similarly, when a petrographer states that he wishes to determine the size of a beach sand, he does not usually mean the area or volume of the sand but the size of the grains of sand making up the beach. Suppose that we wish to determine that a certain granite is uniform in composition; is it correct to rely on samples from quarries and from outcrops, or is it necessary to create additional sampling sites? It should first be remembered that quarries in granite are there for a purpose, and in the case of ornamental and memorial stones one of the desirable qualities is uniformity in the product. Sampling of quarries in such a case may lead to a high degree of uniformity in composition; this would not necessarily be characteristic of the granite population although it is characteristic of the population exposed in quarries. Similarly, many road cuts expose the harder rocks, and fill covers up the softer ones. It is therefore necessary to define the population clearly before arranging a sampling plan.

The next step is to ensure that the samples chosen are representative of the population defined. For example, museum exhibits of rock and mineral specimens are not representative of the materials in the earth's crust because they are usually chosen on the basis of their striking character, rarity, or uniqueness. The terms *type locality*, *typical specimen*, and *holotype* are examples of biased sampling because the item (sample) chosen as standard is usually anything but representative of the general occurrence (population) of the materials. Often the most perfect specimen, the most attractive specimen, or the locality first inspected or "most complete" is the basis of choice. Confusion of such samples with representative samples leads to looseness of definition and continual change in the basis of comparison, a feature all too characteristic of many fields of geological inquiry. Such incorrect sampling introduces a systematic error into the generalization from sample to population; very often the sample description is applied to a different population from that sampled. The effect of systematic errors of this kind is called *bias*, and the samples are biased samples. In general, biased samples lead to biased conclusions.

To be truly representative of a population a sample should be a random sample of that population. When each item of a population of items has an equal chance of occurring in a sample, the sample may be considered a random sample of the population. Very often this definition is taken to imply that a random sample is any sample, but it will be found that ensuring randomness is a very difficult task, and in general it is always necessary to take positive steps to obtain a random sample.

Suppose a population comprises 100 items and a random sample of 10 items is desired; if each item is numbered 1 to 100 and an equivalent set of numbers is entered on cards, the cards may be shuffled and 10 items chosen by successive cutting of the pack. In practice the use of random-number tables is usually the operational basis of random sampling. These tables are convenient, and some guarantee of random sampling is ensured when it is known that the tables have been subjected to rigorous tests for the presence of bias (Fisher and Yates, 1948; Arkin and Colton, 1950).

Ultimately it must be admitted that randomness is one of the most difficult concepts to reduce to operational practice (Johnson, 1949, pp. 187ff.). For most purposes attempts are made to choose a technique for the selection of samples which by its operation appears to avoid bias, and often this is the most that can be achieved. The condition of randomness is essential to most statistical tests; but despite its fundamental importance, in this and other respects, the only positive practice that can be advocated is to be on guard against the introduction of bias during sampling.

2.2 Geological Populations

In sampling populations of interest to the geologist we may define the following "target" populations (Cochran, Mosteller, and Tukey, 1954; Griffiths, 1962a):

1 The *hypothetical population:* that volume of rocks which *occurred* in some given volume of the earth's crust. This population remains a hypothetical population because there is generally some ambiguity concerning those rocks not now represented in the volume of crust defined; i.e., some sediments are believed to have been deposited and then removed, but they may never have been deposited. This material, not present, cannot be sampled; hence the entire population is not equally available to any proposed sampling scheme. Because the absent material is not represented in the samples, there is a possibility of bias.

2 The *existent population:* that volume of rocks existing at present in a defined volume of the earth's crust. This population may be defined unequivocally but is rarely accessible in its entirety for sampling. Hence, while this is the target population for most geological investigations, it cannot be random-sampled; i.e., all parts of the population may be present but are not equally likely to be contained in the sample program.

3 The *available population:* the population represented by the accessible outcrops, road cuts, boreholes, etc. This population should be sampled properly, and conclusions from the samples may be applied to this, the available population, if the sampling program is suitably designed.

For the present, sampling programs should be arranged to represent adequately the available population. In some areas it may be possible to argue from the available population to the existent population; for example, Rosenfeld (1953) in attempting to sample the Oriskany found that the only available sampling sites were outcrops along the flanks and near the tops of anticlines. Hence his samples represent an available population; however, since the Oriskany is a thin formation (averaging around 150 ft thick with an upper limit of 300 ft), Rosenfeld suggested that the Oriskany did not affect the folding of the Appalachians. Since his target population was the size (X_i = long axis) of the quartz grains, he concluded that variation in quartz-grain size was independent of the folding and therefore of the outcrops available. In this case, if Rosenfeld's presumptions are correct, the samples of the available population yield an unbiased estimate of the existent population.

For some populations it may be possible to test such arguments, for example, where exposures are plentiful (e.g., Colorado plateau or Grand Canyon) or where drilling is sufficiently thorough to yield random samples of the entire available population and for all practical purposes the available population coincides with the existent population.

2.3 Definitions and Notation

It is necessary to define elements, samples, populations, and sampling procedures with some rigor to ensure that the inductive argument from samples to populations is justifiable; the notation in Table 2.1 is in general use.

The inductive argument from samples to populations requires that as n increases to N, $\bar{X}$ should converge on μ and $\hat{\sigma}^2$ should approach σ^2. In order to ensure this, the sampling procedure should yield *random* samples of the population.

Test criteria exist for selecting the best statistics from a number which may be used to describe populations by means of random samples. Thus suppose that the variate X_i possesses a normal distribution and that the statistical estimators $\bar{X}$ and $\hat{\sigma}$ are unbiased, consistent, efficient, and sufficient estimators of μ and σ, respectively (see Sec. 12.6; Fisher, 1948, p. 10; Griffiths, 1955a, 1962a).

TABLE 2.1

Characteristic	Sample statistic	Population parameter
Mean	$\bar{X}$	μ
Variance	$\hat{\sigma}^2$	σ^2
Standard deviation	$\hat{\sigma}$	σ
Number of items	n	N

In this table

$$\bar{X} = \sum_{i=1}^{n} \frac{X_i}{n} \qquad \mu = \sum_{i=1}^{N} \frac{X_i}{N} \tag{2.1}$$

$$\hat{\sigma}^2 = \sum_{i=1}^{n} \frac{(X_i - \bar{X})^2}{n - 1} \qquad \sigma^2 = \sum_{i=1}^{N} \frac{(X_i - \mu)^2}{N} \tag{2.2}$$

As an illustration consider the sample variance s^2 defined as

$$s^2 = \sum_{i=1}^{n} \frac{(X_1 - \bar{X})^2}{n} \tag{2.3}$$

This estimator is consistent, converging on a fixed value even in small samples, but it does not converge on σ^2; it is biased, converging on $[n/(n-1)]\sigma^2$. In large samples as n approaches N (infinite), $n-1$ approaches n, and s^2 converges on σ^2. In small samples it is necessary to correct for the bias, and so $\hat{\sigma}^2 = [n/(n-1)]s^2$ is used. It can be seen that multiplying Eq. (2.3) by $n/(n-1)$ yields Eq. (2.2), which is described as the best unbiased estimator $\hat{\sigma}^2$ of σ^2 from small samples (see also Sec. 16.9).

Similarly, the median Md can be used to estimate μ and is both an unbiased and a consistent estimator, but it is less efficient than the mean in a normal distribution (Dixon and Massey, 1951, p. 71).

Sufficiency is a desirable property, but in general it is difficult to find sufficient estimators; the sample mean is a sufficient estimator of the population mean in a normal distribution, no other estimator adding any additional information on the "central tendency" or "parameter of location" of the frequency distribution. Similarly, the sample variance in a normal distribution is a sufficient estimator of the population variance.

It is generally desirable to select statistical estimators which are, at least, unbiased and consistent for inductive argument from sample values to population characteristics.

2.4 Sampling Populations

A random sample is a sample of a population selected in such a manner that each item in the population has equal probability of being selected. The concept of randomness is simple to define, but it is usually very difficult to fulfill the requirements in practice. A considerable diversity of opinion exists on procedures for ensuring randomness (von Mises, 1957; Popper, 1959; Johnson, 1949; McCarthy, 1957); and while it is generally required that a randomization procedure be adopted, this may or may not yield the required sample.

In a geological context if the population is uniformly available, there is usually no problem in random sampling; more often the population is nonuniform, and not all is equally available; then it frequently is impossible to attain a random sample of the population (Griffiths, 1962a; Krumbein, 1960a; Whitten, 1961). It is, of course, not advisable to argue from a sample to a population the sample does not represent.

Consider the usual geological example: the investigator wishes to sample a formation or other geological (mappable) unit. The formation is exposed in outcrops; thus outcrops are the first sampling stage and should be random samples of the population, i.e., formation. Each outcrop is sampled by selecting specimens, and again the specimens should be random samples of the outcrop. The specimen is usually further subdivided in the laboratory into, say, chips for thin sections, and such chips should be random samples of the specimen. From the thin sections a few microscope fields or traverses are selected, and from each field or traverse a few grains are measured or counted.

Random sampling must be the goal at each sampling stage, and this sampling design or pattern is then known as *multistage* or *hierarchical sampling* (Cochran, 1953; Griffiths, 1962a; Krumbein, 1960a), as illustrated in Fig. 2.1. Modifications of this design are in common use in geological investigations (e.g., Olson and Potter, 1954; Potter and Siever, 1956; Krumbein and Slack, 1956).

An interesting variant of this design having important applications in field and laboratory investigations is described by Krumbein (1934a). Each specimen is subdivided into two parts, and one of the two, selected at random, is analyzed individually; the second part from each specimen is homogenized, this bulk is random-sampled, and these samples are analyzed. The second subsample set[1] yields variation which should be a reflection of experimental error associated with the technique, whereas the first subsamples yield variation composed of field sample-to-sample variation plus experimental error. It is then possible to segregate the field sampling variation from the experimental error (Krumbein, 1934a; Griffiths, 1953b). This sampling design is illustrated in Fig. 2.2; it is useful in deciding whether the population sampled is

[1] Actually equivalent to channel samples (see below) of the population.

uniform in the property measured or, equivalently, whether specimen-to-specimen variation is no larger than experimental error.

Many geological populations are not uniform but layered, typically stratified, as in sedimentary rocks; similarly they are layered in igneous rocks (gabbro, ultramafics) and in metamorphic rocks (foliated rocks). From a statistical viewpoint all these populations are stratified. There are two strikingly different sampling patterns. In the first design the sampling is arranged to cross the layers, yielding *channel samples.* Here each channel should be a random sample of the population; if this requirement is fulfilled, each sample is self-weighted and yields unbiased, consistent estimators of the population characteristics. From this sampling design we can estimate the population mean and variance with any degree of precision we wish, e.g., by increasing the number of channel samples.

A second sampling design for layered populations is to sample within the layers; this is *stratified sampling* (Cochran, 1953; Steinmetz, 1956, 1962; Krumbein, 1960a). In this case with appropriate weighting, say by the magnitude of the layers, it is possible to obtain adequate estimation of the population mean and variance; and, from the individual-layer statistical estimators, a knowledge of the within- and between-layer variation can also be derived.

There are, of course, a number of other kinds of variation possible in populations, but nearly all may be encompassed by uniform, layered, and patchy populations; and a patchy arrangement may be treated in the same

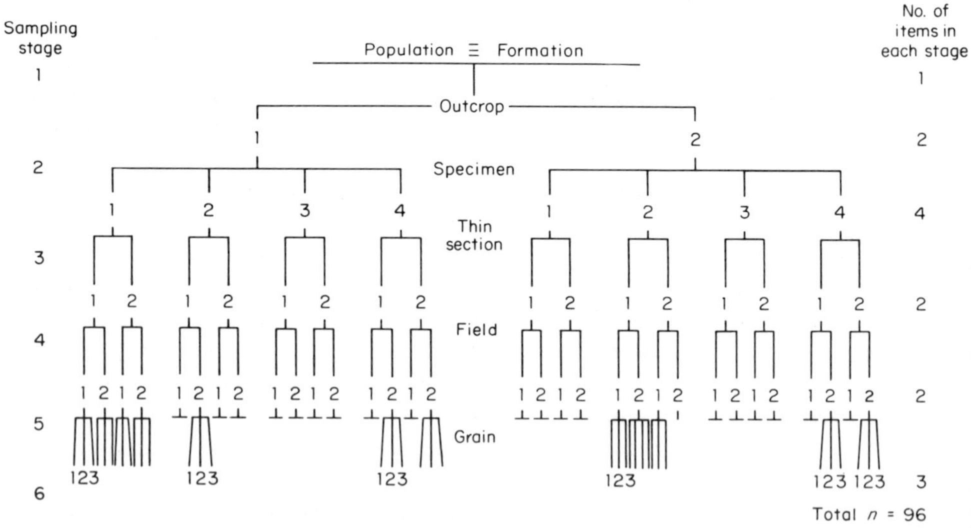

Figure 2.1 Multistage, or hierarchical, sampling design.

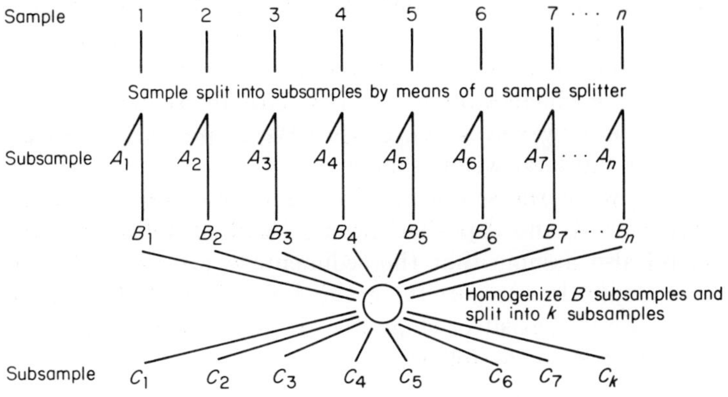

Figure 2.2 Sampling arrangement for comparing sample-to-sample variation with experimental error.

way as a layered population by considering the patch equivalent to a layer. There is then little difficulty in sampling such populations when the variation can be duly diagnosed during the sampling procedure. Suppose, however, that there is a systematic arrangement, layering, or patchiness which cannot be clearly defined before or during sampling; then the specimens are likely to yield biased estimators of population characteristics.

How, then, can we sample a population in such a way that the heterogeneity is avoided or, when desired, exposed? This is particularly appropriate in investigations of sedimentary rocks when we wish to use the sedimentation unit (Apfel, 1938; Otto, 1938) as a basis for sampling the population. Examination of a number of investigations designed to elucidate this problem has led to a sampling model for sedimentary rocks (Griffiths, 1955b; 1962a), and this model has been checked in a number of cases (Hulbe, 1957; Steinmetz, 1956; Cochran, 1960; Griffiths, 1955b, 1959b, 1960a, 1960d; Wood and Griffiths, 1963).

2.5 A Model for Sampling Sedimentary Populations

Suppose a source material made up of broken rocks and their weathered products on top of a hill (Fig. 2.3) moves by gravity down the slope to the deposit of stage 1. It is our wish to determine the size of the quartz grains in the stage 1 deposit; the population consists of the long a axis of quartz grains in the deposit. We take samples of, say, 50 grains from the deposit with one restriction: that each sample be large enough to include the largest quartz grain that occurs in the deposit—not a very serious practical limitation. We measure the length of the long a axis on each grain in each sample and calculate the mean and standard deviation (or variance) of each sample, as well as the variance of the sample means. As grains of quartz of all sizes are randomly distributed through the deposit, each sample yields a mean and a variance which estimates the population variance. Variation among sample means is, however, small, so we have a size variation that is small between samples but large within samples.

Now let the sediment of stage 1, or of the source material, be eroded, transported, and deposited several hundred miles away from the source (we could also assume that the sedimentary processes of selective sorting have had ample time to sort the grains into relatively homogeneous sets and form the layered deposit of stage 3). Here we again take samples according to two restrictions as follows:

1 Each sample must be large enough to contain the largest quartz grain in the layer.
2 Each sample must be confined to a single layer [or sedimentation unit (Otto, 1938) or phase (Apfel, 1938)].

Then each of the long *a* axes of the quartz grains is measured; in this case the range in length of *a* is very much reduced in each layer, so that the mean size is not a good estimate of the population mean, and the variance of each sample is small. Differences among sample means from layer to layer are large; i.e., variation is small within samples and large among sample means.

Clearly, we need to sample these two deposits differently; at stage 1 we need few samples of many grains; at stage 3, many samples of few grains. There are, of course, all intermediate stages (stage 2). In fact, we must run preliminary analyses to determine which is the most effective sampling pattern.

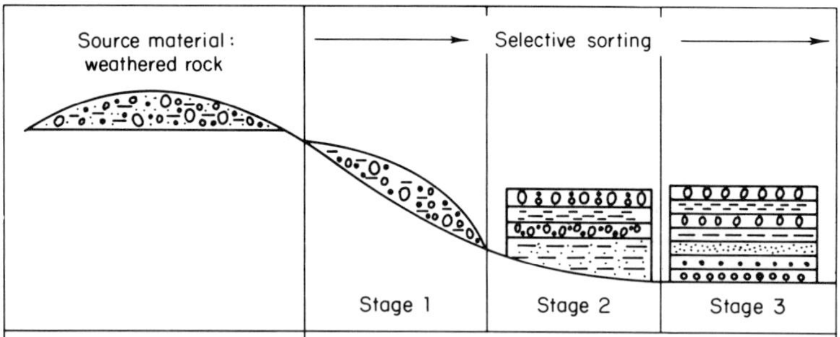

Source of variation	Relative magnitude of variation		
Between samples	Small	Moderate	Large
Within samples	Large	Moderate	Small

Figure 2.3 Processes of sedimentation and the basis of sampling. *(From H. B. Milner, "Sedimentary Petrography," vol. 1, "Methods in Sedimentary Petrography," 4th ed., fig. 93, p. 607, George Allen & Unwin Ltd., 1962.)*

In an investigation of the size and shape of quartzose pebbles from glacial, fluvial, and beach gravels of New Jersey, Steinmetz (1956, 1962) has shown that a channel sample is quite efficient in estimating the population mean and variance in all cases (although not the most efficient sample program in layered deposits). However, we cannot decide on the basis of a single channel sample whether the deposit is layered or not, and it is therefore necessary to decide whether it is worthwhile investigating this aspect. From a geological point of view it is certainly necessary, because the population mean μ and the variance σ^2 of the long axes of quartz grains from glacial, fluvial, and beach deposits may be similar, and the fundamental difference may well be that in a till the layering is absent whereas in a beach sand it is perfect.

If this is the case, it is necessary to assume that the population is layered and sample the "layers" as recommended above; comparison of the variance within and between samples will then determine the presence or absence of layering. No ambiguity is introduced by layered sampling of a massive (homogeneous) deposit, but lack of stratified sampling of a stratified deposit may well give rise to biased conclusions. In other words, the size and shape of the sampling unit are not critical in a homogeneous population, but a sampling unit which ignores the strata in a stratified population is liable to yield biased conclusions and is certain to lose valuable information.

Channel samples across the whole deposit are equivalent to single random samples. They yield an estimate of the population mean and variance but do not offer any information on the internal arrangement of the elements of the population; the internal arrangement may well be vital in geological investigations.

No satisfactory definition of the size and shape of the primary sampling unit (sedimentation unit of Otto or phase of Apfel) has yet been offered except in the case of the 1955–1956 investigation of the Homewood quartzite (Griffiths, 1962a). In some sedimentary rocks having layers or lenses that are very small, i.e., microscopic, the definition of a primary unit is extremely difficult.

In conclusion then, it is necessary to arrange the sampling program in small strata, i.e., to use layers of contiguous grains; the size and shape of these strata (units) can be decided only by trial and error for the present. To solve the sampling problem it is necessary to run a preliminary set of experiments to decide the size and shape of the primary units.

2.6 Determining the Number of Samples

It may be demonstrated under very general conditions that with a small number of specimens each composed of a few measurements, the means of the samples show less variability than the individual measurements. In Table 2.2 the individual observations range from 1 to 110 units, whereas the means vary from 5.25 to 37.25 units. This relationship can be expressed in a useful form as[1]

$$\sigma_{\bar{X}} = \sigma/\sqrt{n}$$

$\sigma_{\bar{X}}^2$ is the variance among means and σ^2 is the variance of the individual observations. This relationship may be shown to hold under very wide conditions (Hoel, 1947, p. 69; Kemeny et al., 1959, p. 167). Now let us suppose we have a good estimate of σ; we can then reduce $\sigma_{\bar{X}}$ by increasing n, as illustrated in Fig. 2.4. Increasing n from 1 to 4 reduces $\sigma_{\bar{X}}$ to half its value, and increasing n to 16 reduces the value of $\sigma_{\bar{X}}$ to a quarter of its original value. This becomes an expensive way of increasing precision in many experiments because of the rapid increase in n. Nevertheless, it suggests two guiding rules for sampling design. First, the larger the numbers of measurements, the greater the precision; or, written in another way, the numbers of samples chosen should be proportional to the magnitude of the variability. If one material varies twice as much as another in some property under investigation, then, for equal precision in mean values, there should be four times the number of measurements in the more variable material than in the less variable.

Second, in increasing the precision of experimental analysis a careful balance should be maintained between the following alternatives:

1 Increasing elaboration of experimental procedure and its concomitant reduction in the number of experiments performed per unit time (and/or cost).
2 Maintaining a simple and less exact procedure but increasing precision by repeating (replicating) the experiment on each sample (with, of course, an accompanying increase in cost or time).

[1] $\sigma_{\bar{X}}$ is often called the standard error of the mean.

TABLE 2.2
Range in Mean Compared with Range in Individual Observations

		Sample No.			
	Individual operator	1	2	3	4
	1	110	50	3	15
	2	20	30	5	40
	3	15	1	6	81
	4	4	2	7	8
	$\sum_{i=1}^{n=4} X_i$	149	83	21	144
$\bar{X} =$	$\sum_{i=1}^{n=4} \frac{X_i}{n}$	37.25	20.75	5.25	36.00

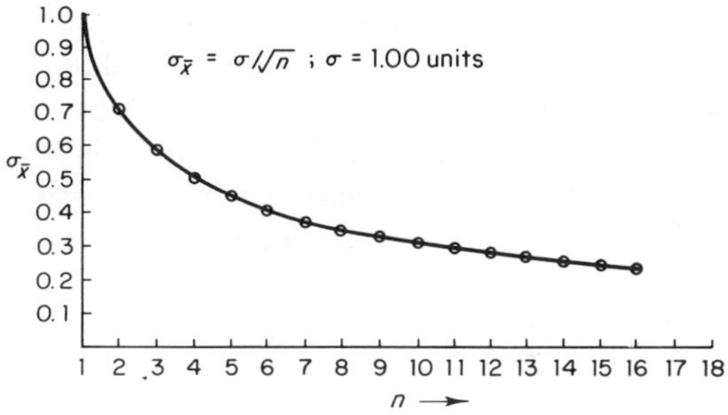

n	$\sigma_{\bar{x}}$
1	1.0
2	0.707
3	0.577
4	0.500
5	0.447
6	0.408
7	0.378
8	0.353
9	0.333
10	0.316
11	0.301
12	0.288
13	0.277
14	0.267
15	0.258
16	0.250

Figure 2.4 Relationship between standard error and sample size in simple random sampling.

2.7 Sampling Practices in Geology and Allied Fields

Now that general rules and guiding principles for the selection of samples suitable for analysis have been outlined, it is necessary to examine the kinds of samples in use and to decide whether these samples fulfill the requirements for efficient representative sampling.

In reconnaissance investigation single samples of large deposits are analyzed to estimate the characteristics of a rock. An example of this type is a *spot* sample containing many grains taken at a point in a rock unit. It should, of course, be a random sample, and it cannot yield a very precise estimate of a population parameter unless the rock unit is very uniform. To offset some of the disadvantages of a spot sample a number of spot samples can be combined and subsequently split to a convenient size in order to obtain greater representation. This *composite*, aggregate, or compound sample will again yield limited information, essentially an estimate of the population mean and variance, about the rock unit sampled. Each spot sample should be randomly selected, and great care must be taken in homogenizing and subdividing the composite (Otto, 1937).

The problem of sampling by means of spot samples has been discussed by Apfel (1938) and Otto (1938), who propose taking a spot sample from a unit layer of sediment. The unit layer is called a *phase* by the former and a *sedimentation unit* by the latter, who defines the sedimentation unit as "that thickness of sediment deposited under essentially constant physical conditions" (Otto, 1938). This kind of sampling presupposes that homogeneous units can be selected in the field, and so there is set up a kind of stratification, random spot samples being selected from the strata. Two difficulties arise. First, there is a practical difficulty in sampling the sedimentation unit or phase when it is very small, as indeed it must be in many sediments if the definition is to be fulfilled.

The second obstacle is the more serious because choice of a homogeneous unit presupposes that the variability can be estimated before analysis. In many cases the only guarantee of "constant conditions" would imply that the sampling unit is a single grain.

It must be clear that any sampling plan which includes the selection of a single sample will yield only a single mean value, which may or may not be representative of the population sampled. Furthermore, the only estimate of variability obtainable from a single sample is a measure of within-sample variability; and if greater variability exists between samples than within samples, this estimate will be a very poor approximation to the population value.

It is necessary therefore to obtain a series of samples, and once this procedure is decided upon, the arrangement of the series becomes an important aspect of the sampling problem. A number of plans have been specifically designed to fit certain environmental patterns such as a beach sand, alluvial fan, or delta (Krumbein and Pettijohn, 1938, p. 15); but such plans, by the very process of fitting them to the environment, are likely to induce bias in the sampling (see, for example, Fisher, 1942, pp. 62, 74). Hence it is more advantageous to choose a sampling plan which is symmetrical but not related to any imagined gradient in the population to be sampled if unrestricted random sampling is to be the goal. On the other hand, there is no reason to neglect more efficient arrangements such as stratified random sampling or cluster sampling (Cochran, 1953; Yates, 1949) if the basis for stratification or choice of cluster can be shown to improve the sampling program by reducing the variability without the introduction of bias. Single-phase, or sedimentation-unit, sampling is exactly of this type in its intentions, but in practice the intensions can rarely be attained without reconnaissance investigation to determine the levels of variability.

Geological sampling may be subdivided into two main groups: surface samples, i.e., samples taken within 100 ft of the earth's surface, and subsurface samples, obtained from deeper horizons.

The most common surface samples are outcrops, which may be natural or artificial; quarries, sandpits, road and railroad cuts, and riverbanks are typical examples of locations where outcropping rock makes the site suitable for sampling the earth's crust. In general, outcrops always yield biased samples of rock units. The bias in most cases is fairly obvious, because the majority of outcrops are artificial and exist for a very definite purpose, often yielding material with desirable properties, e.g., claypits giving clay for refractory bricks or sandpits giving sand for glass manufacture. In such cases, of course, the outcrops represent only a small and unusual portion of the total population. On the other hand, road and railroad cuts expose hard rocks, and fill usually obscures the softer ones. In using outcrops as a basis of a sampling plan it is therefore necessary to realize the nature of the bias; sometimes this bias can be estimated and the sample statistics corrected for bias. Generally this is not possible, and it is then necessary to decide the nature of the population sampled (the available population) and avoid applying the results from the samples to parts of the rock unit not represented in the outcrops.

There is no doubt that as our knowledge of the variability of the properties of sediments increases, corrections for sampling bias will become possible, but until this desirable condition can be achieved, it is better to restrict the analytical results to the population sampled than to risk applying them to parts of the rock unit not available for analysis.

Auger holes are common sampling sites in tropical countries, where natural and artificial outcrops are few and far between and the rocks are loosely consolidated. The arrangement of auger holes in a random-sampling grid is reasonably simple but is rarely attempted because the geologist has still to learn the dangers of bias existing in systematic sampling plans. Frequently when the geological structure is unknown, a symmetrical grid becomes, by favorable accident, a random-sampling grid, but this is the exception rather than the rule. Generally, during augering, a bias enters the sampling because of technique limitations; it is usually more difficult to recover sands by this procedure than clays.

Artificial pits are sometimes used to expose soft rocks, but because of their cost they almost always yield biased samples, the pit site being chosen to reveal some special contact or structure of the rock formation.

Subsurface samples are taken from mines, shafts, and boreholes. Mines and shafts are really artificial outcrops, and they are generally constructed in areas of specific commercial interest; there is no doubt of bias in these cases, but again if the population is correctly defined, the samples may be randomly selected for certain specific purposes (Hazen, 1958, 1961; Sichel, 1947).

Boreholes drilled in search of water, oil, uranium, and various other ores also yield samples of sediments, and much of our information on sedimentary rocks has been compiled on the basis of samples from cores taken during drilling. Quite frequently, during exploration drilling, the cores are random samples of the formations present in an area; but more generally the borehole sites are chosen on certain specific structures such as anticlines or near faults, and the sampling may then be biased to an unknown extent. Here again it is possible to define a population of which the samples are representative, and this is the only recourse in many cases.

In oil-development work two kinds of drilling are common, rotary and cable-tool, or percussion, drilling. Rotary drilling consists in boring a hole into the earth by means of a rotating bit; grinding, cutting, and abrading are the principal features accompanying this procedure. To obtain samples of rocks drilled, several different techniques are in common use. First, coring yields the most attractive type of rock samples, and various kinds of coring yield different results. Diamond coring is most expensive but usually leads to recovery in the range of 90 to 99 percent. Conventional cores are taken in a barrel attached to the end of the drill pipe, and the bit on the end of the barrel has a hollow center. The drilled rock passes up into the core barrel, and when the barrel is filled, the whole string of drill pipe is removed from the hole and the core barrel detached. This is a very expensive procedure on a deep hole and introduces hazards which may lead to loss of hole.

A more recent innovation is wireline coring, wherein the barrel may be removed without pulling the entire string of drill pipe. Core recovery varies from 60 to 90 percent and cannot but introduce bias in most cases, as the softer sands and clays are usually lost. In the extreme case the producing sand, which is often most friable, is completely lost, and only the barren sediments are represented in the cores.

During rotary drilling the drilling mud, carrying broken-down fragments of the formation, is passed over a screen before pouring into the mudpit, and these screen samples are often used for analysis. It should be remembered that any rock which breaks down and disperses readily in the mud stream will not be retained by the screen and will not therefore be represented in the sample collection.

Cable-tool drilling uses a different principle from rotary drilling, the action of the bit being chiefly crushing and grinding, hence the name percussion drilling. Cores are rarely taken in this process, but chips may be obtained from the hole during cleanouts and, of course, are quite suitable samples for analysis. Assuming that contamination from higher horizons is negligible, chip samples will, when combined, be a composite sample of the horizon being penetrated. Soft, easily broken down rocks are of course not represented; for example, samples of producing Oriskany gas sand from the Pennsylvania area are almost nonexistent because they are friable and readily break down during percussion drilling.

In all these cases bias is present, but the degree of bias varies considerably, from almost complete recovery by diamond coring with little bias due to loss, to cable-tool drilling, wherein losses are often large. In addition, drilling sites are not selected randomly but with very definite purposes in mind; hence the magnitude of the bias may be high. In every case the investigator has the responsibility either of attempting to estimate and correct for bias or of defining rigorously the exact population represented by the samples.

Most geological sampling is accompanied by bias, and during an investigation of the Oriskany orthoquartzite Rosenfield (1953) found that, despite a randomly arranged grid, the unequal availability of outcrops led to a very marked bias because in the folded Appalachians the flanks of the anticlines, and sometimes their tops, are exposed but synclinal outcrops rarely occur. The alternatives are (1) to describe the population as that exposed in the available outcrops and not attempt to extrapolate to the Oriskany formation as a whole and (2) to show that analysis of outcrop samples does not lead to biased estimates of the population parameters of the property being studied. For example, if the size of the quartz grains is independent of the folding of the Oriskany sandstones, the anticlinal outcrops will be representative of the entire Oriskany. The Oriskany is a thin formation and is therefore unlikely to have affected the general folding of the Appalachians; if this is true, the available outcrops determined by the folding and subsequent erosion are independent of the grain size and are representative of the entire Oriskany.

At least for the present such arguments will be necessary to establish that geological samples are representative and approach the requirement of randomness; perhaps when enough is known about the variation in the properties of sediments, the bias can be evaluated and due allowance made for it.

2.8 Testing for Randomness

It seems clear from the above discussion that random sampling is difficult to achieve in geological investigations unless we are prepared either to create new sampling sites which will yield random sampling or to change our target population from the hypothetical and existent to the available type of population. When it is finally decided that it is necessary to describe existent populations rather than available ones, the necessary investment will have to be made to achieve random sampling. In effect, when we say that it is impossible to obtain random samples from existent geological populations, we are implying that the investment required to achieve this sampling goal is not worthwhile. Geological prediction will suffer from serious shortcomings until this attitude is modified; and, of course, this will be particularly true where geological predictions are applied to already costly exploration for natural resources. Major improvement in predictive success would follow upon fulfillment of the requirements of most efficient random-sampling designs. It is almost impossible to make firm estimates of the characteristics of populations from biased samples unless the bias is known.

When a sampling design has been arranged and requisite samples taken, it then becomes necessary to demonstrate that random sampling has been achieved. Unfortunately, no tests exist to demonstrate randomness; the most we can achieve is to make a thorough examination for bias. Random-number tables, which are generally used as the basis for random sampling, are very rigorously tested against various kinds of ordering, and a number of ideas have been expressed in defining what tests a random selection should fulfill (Fisher and Yates, 1948; Arkin and Colton, 1950; Tippett, 1952; RAND Corporation, 1955). However, as long as problems exist with the conceptual definition of randomness, it will be exceedingly difficult to establish beyond doubt what a truly random sampling entails in practice (Fisher, 1942, 1956; Jeffreys, 1948; von Mises, 1957; Popper, 1959).

A simple test which is of wide application follows from the relationship

$$\sigma_{\bar{X}} = \sigma / \sqrt{n}$$

and from the central-limit theorem, which states that means of samples of size n tend to normality as n grows large (see Secs. 13.7ff.). Then, since the interval $\bar{X} \pm 3\sigma_{\bar{X}}$ includes 99.7 percent of the area under a normal curve, we may claim that if random samples of size n are taken, we expect 3 means in every 1,000 to fall outside these limits.

As an example may be cited an investigation of the lengths of long a axes of quartzite pebbles from the Montoursville gravel undertaken as a class exercise in 1960–1961. Two series of samples were taken, the first comprising 5 channel samples of 105 pebbles each and the second, 15 layers of 30 pebbles each, arranged in a stratified pattern. The grand mean and standard deviation are $\bar{X} = -5.358$ phi and $\hat{\sigma} = 0.827$ phi; the respective standard errors for means of samples of size $n = 105$ are $\hat{\sigma}/\sqrt{105} = 0.081$; and for $n = 30$, $\hat{\sigma}/\sqrt{30} = 0.151$, and the corresponding intervals (confidence limits, Sec. 13.8) are:

n	$\hat{\sigma}$	$\bar{X} \pm 3\hat{\sigma}_{\bar{X}}$
105	0.0808	$-5.358 \pm 0.242 = -5.116$ to -5.600
30	0.151	$-5.358 \pm 0.453 = -4.905$ to -5.811

It can be seen from Fig. 2.5 that all five channel samples fall within the confidence limits; i.e., there is no evidence from this test to deny that the means of the channel samples are random samples from a population with $\mu = -5.358$ phi and $\sigma = 0.827$ phi. In the case of the stratified samples 2 out of 15 fall outside these limits, which exceeds the 3 in 1,000 by a considerable margin; we deduce that either a very unfortunate accident has occurred, yielding 2 of the 3 expected in the first 15, or else these sample means vary more than would be expected from means of random samples taken from a population with mean $\mu = -5.358$ phi and $\sigma = 0.827$ phi.

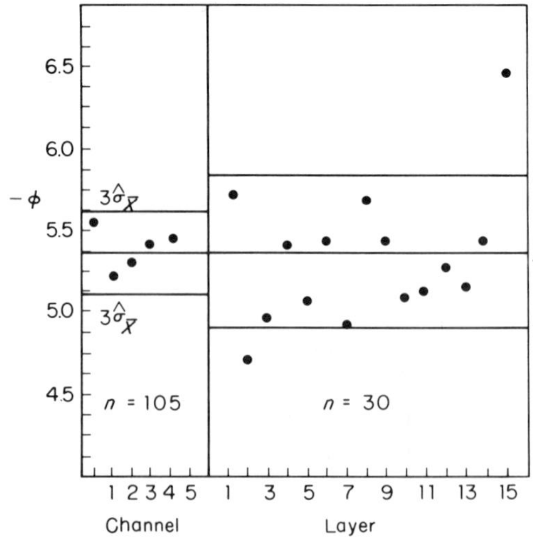

Figure 2.5 Testing for random sampling using confidence limits; population = pit face in Montoursville gravel (*Griffiths, 1959b*); element = long a axis, in phi units.

Since the processes which lead to variation in grain size tend systematically (selectively) to sort out specific sizes and accumulate them into layers, we expect, for geological reasons, that this will be the outcome; in fact, if the variation were random throughout this population as a whole, strata would not exist in terms of grain size. It can be seen from Fig. 2.5 that strata means vary much more than channel means, although they are not strictly comparable without adjustment for differing sample sizes.

A number of other tests for the presence of systematic patterns of variation, i.e., bias of various kinds, are described in succeeding chapters. It is evident that to employ a test, the nature of the bias expected should be specified and the outcome then tested against the specific departures from randomness. There is no omnibus test against all possible alternatives to randomness.

2.9 Some Recommended Procedures

Just as there is no possibility of designing an omnibus test for all possible (unspecified) departures from randomness, it is equally impossible to recommend a simple comprehensive procedure which will be suitable for all occasions. Experimental designs and their accompanying sampling programs vary widely and are especially designed for specific types of problems. It is necessary to be familiar with the various experimental designs available for the majority of situations; fortunately, some excellent textbooks exist as reference guides (Cochran and Cox, 1957; Cochran, 1953; Yates, 1949; Hansen, Hurwitz, and Madow, 1953; and see subsequent chapters in the present book). In general, the less that is known about the population under investigation, the simpler the design and sampling pattern will be. Increasing knowledge about possible outcomes leads to more sophisticated and more efficient sampling programs.

Geological problems require, at present, an initial reconnaissance investigation to determine the magnitude of variation before any but the simplest patterns of random sampling may be used. A model has already been proposed for sedimentary populations (Fig. 2.3), but even with this model, exact recommendations of size of sample and minimum number of samples for a specific problem in estimation require a fairly precise knowledge of the relative magnitudes of variation at different levels, and this, in turn, necessitates a preliminary experiment.

It is always necessary to specify the objective, and the more precisely this can be achieved, the clearer the formulation of the problem will be—including the sampling pattern. If the problem requires estimates of parameters of a population, it is necessary to obtain an initial estimate of the variance before fixing sample size and number of samples. It is also advisable to choose a probability level and make initial calculations to determine the approximate sample size and number which the available resources will permit. These requirements were met in agronomic field experiments by quite elaborate and extensive uniformity trials devoted solely to this purpose. On the basis of these evaluations the size and shape of sampling units, known as blocks and plots in agronomy, are arranged as squares and rectangles, respectively, the latter crossing the entire block. The plots are equivalent to channel samples, and each extends over the range of soil heterogeneity in a block. Similar experiments will be necessary to evaluate the most desirable sample sizes and shapes in studying geological populations. In fact, while necessary reconnaissance experiments have been performed for this purpose, it is rare indeed for an experiment to be exclusively devoted to the solution of a sampling problem, and this will be necessary before firm recommendations can be achieved.

Discussions of the problem of sampling (geological populations) are given by Krumbein (1953a, 1954a, 1960a, 1960b), and there are a number of examples of the use of various sampling designs in the literature (Chayes, 1956a; Whitten, 1961; Krumbein and Slack, 1956; Olson and Potter, 1954; Potter and Olson, 1954; Krumbein and Miller, 1953; Griffiths, 1953b). Some experiments aimed at studies of specific problems in sampling design are discussed in describing designs for homogeneous, i.e., massive, and layered populations based on experiments performed by graduate students (Griffiths, 1959a). Steinmetz (1956) first tested a variety of sampling designs for patterns from gravels, and Cochran (1960) investigated variation in bulk density as a sampling problem. In the field of mining, Hazen (1958, 1961; Becker and Hazen, 1961) has investigated representative sampling in assays; and Sichel (1947, 1952), Krige (1951), and Ross (1950) have discussed various aspects of sampling the auriferous veins in the Rand goldfields, South Africa. Somewhat more general discussions of sampling are included in statistical texts (Cochran, 1953; Yates, 1949; Hansen, Hurwitz, and Madow, 1953; Deming, 1950). A very instructive discussion of sampling with quite general implications (Cochran, Mosteller, and Tukey, 1954) concerns the possible bias due to "nonresponse," which may have been included in an investigation into the sex behavior of the human male (Kinsey, Pomeroy, and Martin, 1948).

Extensive information on sampling exists, and the literature on sampling applied to geological populations is steadily growing, but until specific experiments aimed primarily at solutions to geological sampling problems are completed, no very exact guide of general use can be expected.

3 The properties of sedimentary rocks

To exhibit the structure of an object is to mention its parts and ways in which they are inter-related.

B. Russell, 1948

3.1 Introduction: Concept of Properties

The definition of an object usually entails a description of its properties and the way in which they are interrelated. In the case of the sedimentary rocks there are an infinite number of properties which could be used, but some appear to be more fundamental than others. As a basis for judging which properties are fundamental, two criteria are used: (1) the properties should be as generally applicable as possible; i.e., they can be used to characterize any and all objects; (2) they are the simplest set of properties which can achieve this end. The criteria are perhaps best summarized in a mathematical sense as the necessary and sufficient conditions for a complete and unique description of an object.

A rock may be axiomatically defined as an aggregate of minerals, where the elements, which are undefined, are the minerals, and the aggregate, or collection, is the set of elements (minerals) of which the rock or population is composed. An equivalent generalized definition is therefore: *A population is an aggregate of elements.*

The characteristics of the population comprise the combined properties of the elements and the characteristics of their association. As a first step

Quotation: B. Russell, "Human Knowledge: Its Scope and Limits," p. 267, George Allen & Unwin Ltd., 1948.

the elements may be classified, and for this purpose a set of criteria is necessary which will assign each and every element to a single class. The classes are mutually exclusive (disjoint) and exhaustive; i.e., the subsets (classes) range from the empty set (no elements) to the entire population. The equivalent procedure in describing a rock is to identify the minerals (subsets or classes of elements), and the diagnostic properties of different minerals are those which lead to enumeration of the subsets.

It is then necessary to count the number of elements which fall in each class or to estimate the proportions of elements (or minerals) in the aggregate, i.e., population, or rock.

After the elements (minerals) are classified and their proportion in each class is determined, each element is characterized by its properties (or range of properties); e.g., the size and shape of each element are determined. Of course, in practice, this reduces to specifying the frequency (probability) distributions of the variates, size and shape, of the elements; these distributions may be confined to the elements of a single class (monodisperse) or may combine elements from all classes (polydisperse). In the latter case adequate weighting of the frequencies by the proportions of elements in each class is advisable.

These four steps encompass the properties, mineral composition, estimates of relative proportion, and size and shape. If now the position of each element in R space (where $R = 3$ for Euclidean space) can be uniquely defined, the aggregate is completely defined. Unfortunately, in practice it is not feasible to define the position of each element, and this characteristic is described by two substitute measures, the orientation of each element and its position with regard to the neighboring elements (called *packing*).

3.2 The Defining Equation for the Population

In practice, then, the population is characterized by the kinds and proportions of its elements, their sizes and shapes, and their orientation and packing. We may symbolize this characterization as

$$P = f(m,s,sh,o,p) \tag{3.1}$$

where P, an index characterizing the population (rock), is a function f of the properties of its elements (mineral particles); and the properties are the kinds and proportions of elements (m_1, m_2, m_3, . . . , m_k for k classes of elements), their sizes s, shapes sh, orientation o, and packing p. The different classes of elements are characterized in terms of size, shape, etc., by frequency distributions of various kinds, and the function contains the parameters θ_i of these distributions or, in practice, suitable statistical estimators $\hat{\theta}_i$ of these parameters.

The classification of the elements and the statement of their relative proportions may be represented as a binomial distribution if there are two classes ($k = 2$) or as a multinomial distribution if there are more than two classes ($k > 2$); the details of this procedure are described in Chaps. 9, 10, and 14. Suitable statistical estimators for substitution in Eq. (3.1) are the means $\bar{X}$. Classification requires that criteria be established which lead to mutually exclusive and exhaustive classes, and for mineralogical analysis this is usually taken to imply that each and every mineral is defined by suitable diagnostic criteria and that subsequently their relative proportions are determined by counting the frequency in each class in some sample of the population. In fact, the number of classes varies with the objective of the analysis and always represents some compromise between the ideal of "complete" identification and some degree of "incompleteness," in which at least one class contains representatives of a number of different minerals. For example, suppose it is decided to identify quartz (A), feldspar (B), and mica (C) and ignore all others; then ideally this is equivalent to using four classes A, B, C, and others (not A or B or C). If the diagnostic criteria used to assign elements to classes are not entirely effective, various other classes may occur, such as AB, AC, or BC; i.e., the original classes are not mutually exclusive. It is always necessary to test the effectiveness of the diagnostic criteria and to show that the classification is consistent (see Chaps. 9 and 10). This requirement of consistency is even more critical when counts of relative proportions are involved; indeed, misclassifications are much easier to recognize when the data are quantitative.

If the elements have been satisfactorily classified, the sizes of the elements are measured; if a single class is selected for this purpose (say quartz) and a suitable measurement procedure is chosen (see Chaps. 4 and 5), the frequency distribution of the variate generated by the procedure will be, at least approximately, logarithmic-normal (or phi-normal). The appropriate estimators are then the mean $\bar{X}$ and standard deviation $\hat{\sigma}$ or variance $\hat{\sigma}^2$. If more than one class is included in the size measurement, the frequency distribution will almost certainly be polymodal, and *order statistics* such as the median Md_ϕ and percentile deviation PD_ϕ are the appropriate measures (again on a logarithmic phi scale). It has not proved possible, as yet, to find suitable weighting factors for the size distribution per class to simplify these frequency distributions.

Similarly, if shape is measured on the elements from a single class and the measures are combined into variates such as sphericity (see Sec. 6.1.1) or are used as linear compounds of the logarithms of the variates (see Sec. 6.3.2), then the distributions are again approximately normal, and the mean $\bar{X}$ and standard deviation $\hat{\sigma}$ are suitable estimators. There are few proposals to measure shape across different classes of elements.

Measurement of orientation of the elements of a single class has not been routinized, and the procedure described (Chap. 7) leads to an arbitrary mean; the only estimator for inclusion in the defining equation (3.1) is the standard deviation as a measure of degree of parallelism, or preferred orientation, of the elements.

Suitable procedures for the measurement of packing p exist and, under suitable conditions, lead to binomial or multinomial frequency distributions; here again the mean $\bar{X}$ is a suitable summary statistic.

There are, of course, many other properties which may be used to characterize aggregates (populations or rocks), e.g., pore space, hardness, etc., but these are behavioral or derived properties in the sense that variation in each is really dependent on the variation in the five properties listed in the defining equation. These five are therefore the *fundamental* properties of the population, and, at least in theory, they can be used to predict the behavioral or derived properties. These five properties are therefore adequate to characterize the aggregate and are the necessary and sufficient conditions to lead to a unique index P.

3.3 Interrelationships among the Properties

When the fundamental properties have been chosen, the next step is to devise a set of procedures for the specification of each property. This amounts in essence to the development of a set of techniques which will lead to a set of measurements, i.e., a set of numbers on some scale or scales, which may be substituted in Eq. (3.1) and which, when suitably combined, will yield a value of P which represents a unique description of a rock specimen. This approach (Emery, 1954; Emery and Griffiths, 1954a; Griffiths, 1958a) appears to have promise of wide application. The greatest difficulty it presents is the lack of understanding of the interrelationships among the properties measured by the several techniques.

Many techniques exist for the measurement of grain size, but upon adequate testing most of them result in only partial achievement of the objective of the experimental technique. In general, concepts are confused with operations and vice versa; it is necessary, therefore, to evaluate the interrelationships among the properties and to decide in each case whether a technique is, in fact, achieving the objective for which it is devised.

We shall consider the well-known property of grain size and, by treating this procedure in detail, evaluate others by analogy. It is necessary to recognize two features at the outset: (1) the objective of the analysis must be defined; (2) the analytical technique is to be applied to a sedimentary rock. The resulting size measurement P_s will be some function of the properties of the rock, so that we can describe the experiment symbolically as

$$P_s = f(m,s,sh,o,p) \tag{3.2}$$

Now let us define our objective. In general we wish to determine grain size for two purposes: (1) as a basis of comparison between rock specimens and/or (2) as a basis for interpretation of the manner in which the rock was formed. In the second case we evidently wish to know the grain size of the material as it was deposited, and we are willing to assume that the size of the material as it now exists in the rock has not changed significantly since deposition (Miller and Olson, 1955). This automatically eliminates such materials as clay minerals[1] and, in general, most matrix minerals (Krynine, 1948) and the cementitious materials which have recrystallized. It follows, therefore, that these materials should be removed before analysis if objective 2 is to be achieved. This, incidentally, considerably simplifies disaggregation procedure. It seems, therefore, that we actually wish to determine the grain size of only a part of the assemblage of minerals in the rock.

Now, let us consider a typical procedure used in grain-size analysis, namely, the combination of sieving and pipette sedimentation. The first step is to reduce the rock specimen to its individual components without, of course, affecting their size. It is clear that there are considerable difficulties involved, but let us assume that this objective is successfully attained. In terms of Eq. (3.2) we have, by disaggregation, destroyed the original fabric of the rock, i.e., destroyed original orientation and packing[2] but without affecting any of the other properties. Hence this size measure P_{s1} can now be written

$$P_{s1} = f(m,s,sh) \tag{3.3}$$

It will be noted that the size measure P_{s1} obtained by this technique is a function of the three variables composition, size, and shape of the aggregate of grains analyzed. It is necessary to establish that P_{s1} is really a measure of size and is not affected by composition or shape; or, if it is affected, suitable correction factors can be applied. Suppose the aggregate is composed of quartz grains and mica plates; then there is a difference in composition, and variation in amount of mica will affect the size measure P_{s1}. It is therefore necessary to estimate the amount of mica or quartz and apply a correction factor. Of course there is a difference in shape between quartz and mica components, but as shape is here *completely dependent* on composition, the one correction factor will suffice to remove the effect of mica and the effect of shape difference.

This hypothetical example illustrates that the interrelationships among the properties are important in evaluating the size measure obtained from the experiment. Let us consider one other hypothetical example, namely, a case where more than one mineral occurs in the aggregate analyzed for grain size but where the shape of the different components is *independent* of composition. By independence is meant that each of the several components shows similar,

[1] Actually, it appears to be impossible to determine the grain size of clay materials as deposited even in recent sediments, and, indeed, in general we cannot even achieve the more limited objective of determining the grain size of this material as it now exists in a rock specimen.

[2] Note that disaggregation does not remove fabric but changes it from some systematic pattern, such as a layered arrangement, to a random arrangement; this means that the properties which are embraced in the term "fabric" no longer interact with the properties of interest and are unaffected by, and do not affect, the measurement procedure.

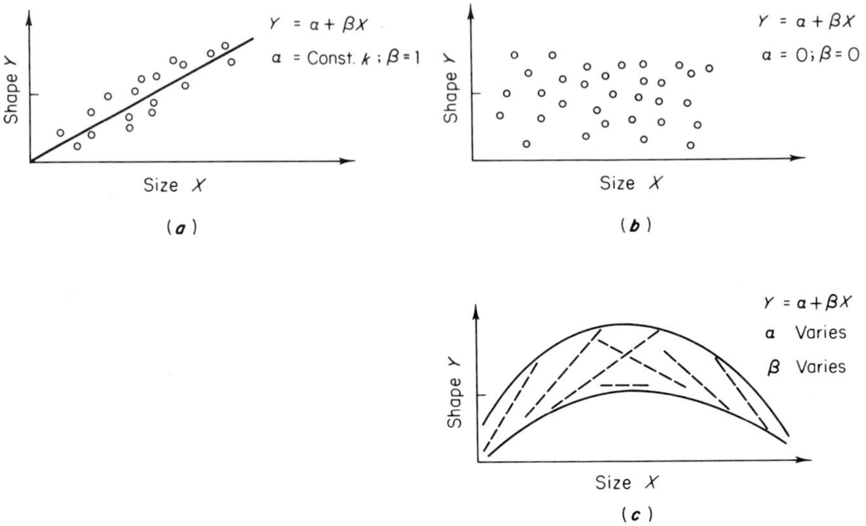

Figure 3.1 Interrelationships among properties of sediments. (*a*) Dependence; (*b*) independence; (*c*) interdependence. [*Reprinted from Measurement of the Properties of Sediments by J. C. Griffiths, J. Geol.,* **69**:*492 (1961), by permission of The University of Chicago Press. Copyright 1961 by the University of Chicago.*]

perhaps wide, ranges in shape. In such a case no correction factor is necessary; all components will behave similarly in sieving, and the measure of size, P_{s1}, will vary with size changes and will not be affected by changes in amounts of components. It may, however, be necessary to correct for variation in amounts of different minerals, as density differences will materially change the size measure obtained by pipette sedimentation.

Now let us consider sedimentary rocks rather than hypothetical examples and attempt to decide whether their properties are dependent or independent. Quartzites and many beach sands may be considered monomineralic for the present purpose; i.e., their major components will be quartz (usually over 70 percent) and rock fragments of chert and quartzite (10 to 15 percent). In this case no problem should arise, because the aggregate is a close approximation of a homogeneous one-component system. Most sediments, however, are graywackes or arkoses (some 80 percent by volume), and these are polycomponent systems. The bulk components are quartz (30 to 70 percent), feldspar (10 to 50 percent), and rock fragments (10 to 50 percent); to simplify the approach, we may consider quartz and feldspar together on the basis that they resemble each other in shape and therefore follow the hypothetical example of independence earlier mentioned.

Rock fragments of quartzite and chert do not materially differ from the quartz grains in shape, but in sediments with rock fragments that tend to be platy, such as fragments of shale, slate, phyllite, schist, and gneiss, the shapes of the different components may vary widely from that of quartz grains; in this case the group called rock fragments includes some grains which are similar to quartz grains in shape, e.g., quartzite and chert, and some grains which are more like micas. Here shape is not dependent on composition; neither is it independent. These rocks (most graywackes and some arkoses will be included in this group) illustrate a new relationship of partial dependence and partial independence or *interdependence*. To illustrate the meaning of this interrelationship among the properties, it is necessary to consider some of the fundamentals of experimental design.

First, we may represent the physical meaning of dependence, independence, and interdependence by reference to a two-variable system (see Fig. 3.1). In Fig. 3.1*a*, size changes with shape, showing a direct linear dependence; in Fig. 3.1*b*, size is represented as independent of shape; and in Fig. 3.1*c*, size is related to shape, but the size-shape relationship is not the same from one size range to the next. In terms of experimental design there is an interaction between size and shape in Fig. 3.1*c* (Fisher, 1942, 1948; Brownlee, 1948; Snedecor, 1956). In the first case (Fig. 3.1*a*), if each size were corrected for shape, the size measure would be independent of shape. In the second case (Fig. 3.1*b*), no correction factor is necessary; size is independent of shape, and hence the measure of size is a real measure of size. In the last case (Fig. 3.1*c*), a different correction factor is necessary for each limited size range.

A three-variable system, such as the size measure P_{s1} arising from sieving a polycomponent system, is even more complex; the sources of variation are summarized in Table 3.1 and are pictorially represented in Fig. 3.2. If the

TABLE 3.1
Three-factor Experimental Design to Illustrate Sources of Variation

Size Shape Orientation	Three main effects
Size shape Size orientation Shape orientation	Three first-order interactions
Size shape orientation	One second-order interaction

Source: Reprinted from Measurement of the Properties of Sediments by J. C. Griffiths, *J. Geol.*, **69**: 491 (1961), by permission of The University of Chicago Press. Copyright 1961 by the University of Chicago.

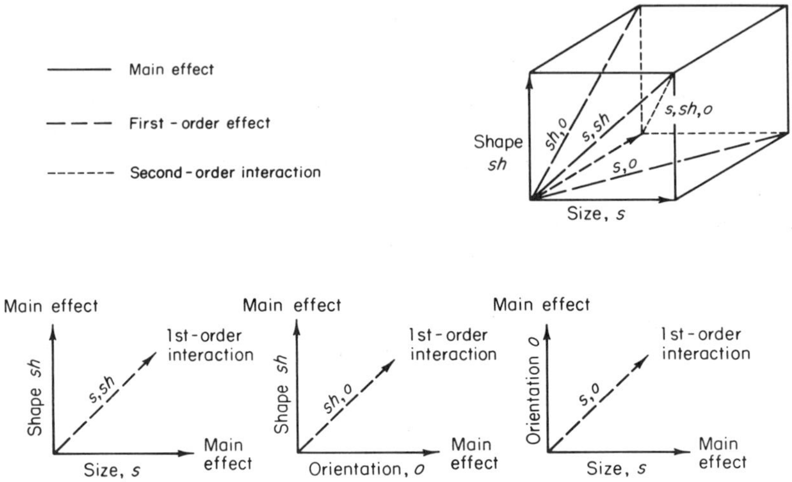

Figure 3.2 Geometrical illustration of three-factor experiment and its interactions. [*Reprinted from Measurement of the Properties of Sediments by J. C. Griffiths, J. Geol.,* **69**:*493 (1961), by permission of The University of Chicago Press. Copyright 1961 by the University of Chicago.*]

three variables are interdependent, it means that the interactions are contributing materially to P_{s1}, the variation measured.[1]

In such a case no simple correction factor can be used, but each variable must be separately and simultaneously evaluated by means of a suitably designed experiment. At least for the present such a course is fraught with difficulty, and some simpler means of size measurement is a more practicable alternative.

In many cases the sample of sedimentary rock cannot be disaggregated without altering the sizes of the constituents to be measured, and then it is customary to use thin-section size analysis instead (Krumbein, 1935). It is conventional to choose quartz as the mineral of interest, and so composition becomes a fixed constant and can be ignored. The size measure adopted may be represented symbolically by P_{s2}, where

$$P_{s2} = f(s,sh,o,p) \tag{3.4}$$

Again it is clear that in this multivariate system the properties are interdependent and interacting; the length of intercept measured across the quartz grain is a function of size of grain, shape of grain, direction of section, and orientation of grain. The density of grains, i.e., packing, will also affect their orientation (Cailleux, 1938). Here we have a system comprising four main

[1] It should be noted that the two-dimensional pictures in Fig. 3.2 do not contain the second-order interaction, so that studying the system by means of isolated pairs is liable to be misleading if the second-order interaction happens to be an important contributor.

effects, six first-order interactions (of the form s, sh; s, o; etc.), four second-order interactions (of the form s, sh, o; s, sh, p; etc.), and one third-order interaction (s, sh, o, p). Quite evidently if there is a change in fabric with no change in size of grain, the measure P_{s2} will change; it is not, therefore, solely a measure of size but is indirectly dependent on a number of other variables; i.e., size variation is *confounded* with variation from other sources in this "size measure" P_{s2}.

Many attempts have been made to calibrate these size measures by comparison of one with another, and this amounts to studying the relationship between P_{s1} and P_{s2} or between

$$f_1(m,s,sh) = f_2(s,sh,o,p) \tag{3.5}$$

If composition is constant in the set of samples studied, that is, $m = k$ as in "pure" quartzites, the equation reduces to

$$kf_1(s,sh) = f_2(s,sh,o,p)$$

and the calibration comprises an attempt to adjust for the random fabric on the left against the systematic fabric on the right-hand side of the equation. This can usually be accomplished in sands which approximate a monomineralic type, but the correction factor changes as composition changes (Rosenfeld, Jacobsen, and Ferm, 1953; Friedman, 1958). Here again it is clear that the presence or absence of interaction is the important feature involved in evaluation of the two techniques.[1]

Finally, we may return to the simplest and most direct measuring technique—a procedure exceedingly popular in the physical sciences—and measure the size of quartz grains independently of their arrangement in a rock. In other words, we disaggregate the rock specimen (the only precaution being not to affect the size and shape of the quartz grains) by destroying, if necessary, all other constituents, and we measure the size and shape of the quartz grains. This may be represented symbolically as P_{s3}, where

$$P_{s3} = f(s,sh) \tag{3.6}$$

This is a single two-dimensional system, and there are two main effects (s and sh) and one first-order interaction (s, sh). If, however, we measure both main effects simultaneously on the same grain, we can resolve this measure into a single independent measure of size. A technique which achieves this purpose has been devised by Hulbe (1955). Can we, in fact, segregate size from shape? Imagine a person addressing a group in a room from the doorway and saying, "The cord in my hand is attached to a spherical object. Shall I bring the object into the room?" The audience cannot answer without

[1] Incidentally, nearly all attempts to rationalize thin-section size measurement have utilized Euclidean geometry and therefore comprehend, at most, a three-dimensional metric of flat space, whereas the presence of interactions clearly implies that the space involved in performing this measurement is curved.

some information about the magnitude of the object. Similarly, if the statement is "The cord in my hand is attached to a 2-meter object," the audience cannot tell whether the object can pass through the doorway. Such considerations indicate that size and shape are inseparable in a physical object, and generally both are necessary if the object is to be satisfactorily described.

If, then, we wish to measure the grain size of a sedimentary rock and the objective is to use this measure of size to determine the history of the rock, it is necessary to measure the size directly; on the other hand, if the objective is to compare the "size" of one specimen with another or to find a difference between two specimens in "size," then as long as technique is standardized and its precision (reproducibility) determined, this objective can be achieved by almost any technique. However, such a difference, if found, can be interpreted only with the greatest difficulty. Most indirect measures are of this kind.

There is a plethora of techniques purporting to measure size, shape (Krumbein and Pettijohn, 1938), orientation (Griffiths and Rosenfeld, 1950), packing (Emery and Griffiths, 1954a; Kahn, 1956), and composition (Chayes, 1956a), but they all suffer from the same drawbacks: they confound variation from a number of properties in a single measure, and in order to define a rock specimen there is no alternative other than to evaluate each property of the aggregate simultaneously and then attempt to decide whether the set of measurements contains any redundant information by multivariate analysis (Kendall, 1957).

3.4 Interrelationships between Fundamental and Derived Properties

The preceding analysis leads to a general conclusion, namely, that to evaluate a meaningful index which is uniquely characteristic of a rock it is necessary to measure each of the five variables and interrelate them. This objective has rarely been attempted (see, however, Griffiths, 1962b, 1963b), but suitable statistical techniques exist and have frequently been used in genetics and psychology.

The interdependence of the five fundamental properties also plays an important role in the more complex properties of sedimentary rocks. Thus porosity is a function of these five properties, and an estimate of "absolute porosity" may be expressed in terms of Eq. (3.1) by considering the index P the porosity measure. Then porosity, or the volume of voids, may be defined as

$$P = f(m,s,sh,o,p) \tag{3.7}$$

Porosity, as generally measured, assumes that the fluid filling the pores is a noninteracting variable, and if the voids could be evacuated completely, this view would be correct; but in most studies of porosity some fluid is introduced into the pores, and there is an interaction between the fluid and the components of the rock. One reason for the use of gases in the determination of porosity is the supposedly negligible interaction and hence the easier manipulation and reproducilibity of the porosity index measured. It should be fully realized that to determine the porosity of a rock to water it is necessary to take into account the interaction; attempts to establish the reservoir capacity of a rock by using fluids which are presumed not to interact cannot result in a measure which can be compared with the porosity of an interacting fluid. Again the concept of interdependence of the interacting variables appears to be most fruitful, and the porosity index measured may be represented as

$$P_0 = f(r,l) \tag{3.8}$$

where P_0 = porosity index
r = rock
l = fluid

It is not possible to evaluate r and l separately, because if the two variables are independent, as this procedure presupposes, they will not interact. The absence of interaction in the experimental procedure and its presence under field conditions may largely account for the discrepancy between reservoir behavior as measured in the laboratory and in the field.

Similarly, attempts to introduce a correction factor for r are unlikely to achieve more than limited success unless the factor is capable of encompassing the variation of the five interdependent interacting variables. An empirically determined correction factor can be used but only in the case where the values of the five variables entering r (and therefore the correction factor) show limited variation. In other words, empirically derived correction factors will be useful only over the range of variables included in the experimental determinations upon which the correction factor is based; i.e., correction factors can be used only for a single homogeneous set of rocks.

Bulk density is a function of grain density and porosity and is a somewhat more complex variable than porosity. Grain density is frequently assumed to be a linear function of the density of the rock components, the minerals. As can be seen from the above discussion, the grain density of a rock is a function of the five variables, and this function is hardly likely to be linear except under very limited conditions. Indeed, any measurement of rocks which assumes linear (or, as it is sometimes written, additive) relationships of the components should be regarded with the greatest suspicion unless linearity has been demonstrated by adequate experimental analysis.

Permeability is a vector or directional property of fluids and rocks and may be looked upon as the dynamic aspect of porosity. In nearly all experimental determinations of permeability the equations evaluating flow rate contain a constant, which is in fact a correction factor believed to account for the rock variable (Pirson, 1950; Muskat, 1937). Once again it should be clear that a constant which embraces the interacting variations of five interdependent variables plus the interaction between these five and the fluid used to measure the flow rate is something of a miracle. Discrepancies between laboratory measurements of permeability and reservoir behavior in the field will remain as long as this oversimplification is tacitly accepted. There is no doubt that experimental determinations of permeability under conditions simulating those in the field will be difficult and perhaps can only be approached by successive approximations; nevertheless, the alternative of accepting discrepancies of varying and considerable magnitude is at least equally unsatisfactory.

For our present purpose, therefore, in analyzing the characteristics of sedimentary rocks it is necessary to evaluate the variation in the five fundamental properties, composition, size, shape, orientation, and packing, and, as far as possible, to interrelate these variables by expressing their functional relationships. In the final analysis the characteristics of a rock may be expressed as an index which will be a complex function of the five interdependent variables.

Measurement of the more complex properties can be evaluated only by relating them to these five variables and simultaneously accounting for other interacting variables introduced during the experiment. It seems worthwhile to emphasize that any conversion or correction factor introduced to account for the five or more interacting variables in the experimental system is unlikely to have more than very limited application, and the limits within which the factor is useful will be determined by the magnitude of variation in one or all of the variables supposedly accounted for by the correction factor. Any correction factor determined largely by "rational analysis" and including assumptions of linearity with no experimental confirmation is likely to be more misleading than helpful.

It seems clear that to be efficient and meaningful a study of such natural systems as rocks must include all the necessary variables and cannot afford to neglect or oversimplify the interrelationships between these variables. Experimental arrangements which accomplish this purpose are an integral part of the scientific-method approach; hence it is necessary to be familiar with this approach to obtain meaningful answers to questions which concern natural systems. Specifically, interacting variables, which are characteristic of interdependent relationships, can be taken into account only when experimental designs are used which permit estimates of the relative importance of the interactions together with those of the main effects or main variables during one and the same experiment.

4 Measurement of grain size

In *principle* the operations by which length is measured should be *uniquely* specified. If we have more than one set of operations, we have more than one concept, and strictly there should be a separate name to correspond to each different set of operations.

P. W. Bridgman, 1946

4.1 Some General Considerations

It is customary, after deciding what properties are to be studied, to choose some technique which appears "suitable" and proceed with the measurement. Such practice usually leads to a mass of data which is interpretable with difficulty if at all. It is essential at this stage in experimental analysis to realize that there is a series of steps between deciding which property to measure and choosing the procedure by which the measurement is to be performed. The first step is to examine the *concept* designated by the property; thus the concepts of size and shape are related to volume, and very considerable difficulties arise when these two properties are measured independently and an attempt is made to determine what the measurements mean. It has already been mentioned (page 40) that a statement of size without shape or vice versa is, on theoretical grounds, difficult to evaluate.

Conceptually, there is an intuitive belief that size refers to bigness, but no clear definition of size has been given or can be given except by comparison with some reference standard. Quartz grains are small when compared with chairs and tables, large when compared with atoms and molecules. What,

Quotation: P. W. Bridgman, "The Logic of Modern Physics," p. 10, The Macmillan Company, 1946.

then, is meant by size? The only answer that can be given is in terms of a scale in which the universe lies at one extreme and subatomic particles at the other. The size scale generally chosen is a linear scale, and the range for particles in sediments encompasses fractions of a micron to meters, a range greater than $1:10^6$. By size of particle or grain in a sediment is meant a number on this numerical scale. How do we obtain this number? At this stage we descend from conceptual considerations of the property of size to the operational definition of a performance which will lead to a number on the chosen numerical scale (see Fig. 4.1, upper half).

When this performance is attempted, it becomes obvious that there are many ways of obtaining a number on this numerical scale, and each way may result in a different number. The question arises, "What is the basis of choice of procedure?" The most comprehensive answer is that no one way is better than another. To choose a procedure it is necessary to "know" something of the property to be measured, and the only way we can "know" this is to perform experiments. Initially the simplest way is to resort to the expedient of deciding that the best procedure is the one yielding the most precise result; then we can operationally define the best procedure on the basis of the best value, the mean, and the spread of values around the mean estimated by the standard deviation. Here again we meet the two concepts precision and accuracy, and once more we see that it is simple enough to be precise because we possess a unique operational definition of precision. Accuracy is still a difficult concept to specify, and, ultimately, the only way we can solve this dilemma is to follow the recommendations of the earlier discussion, i.e., to use a number of different procedures. Each procedure yields a best value and a measure of precision; and by refining the procedures and increasing their

Evolution	Concept Conceptual definition Operational definition Procedure Technique Initial measurement	
Precision	Choice of technique	
Accuracy and meaning of measurement	Measured data Technique Procedure Operational definition Conceptual definition Concept	Judgment of accuracy and meaning (Operational definition, Conceptual definition)

Figure 4.1 The problem of measurement of properties.

precision if the best values tend to lie along an asymptotic curve (see Fig. 1.2), we may presume that we are improving accuracy as well as precision.

In practice two characteristics of experiments and measurements are used to assist us in obtaining the information we need. First, we use precision by assuming that of two techniques which achieve the same purpose, the one with the highest precision, i.e., the best reproducibility, is preferable. The purpose must be decided beforehand and the techniques and procedures designed to achieve it.

After choosing a property to be described it is necessary to define it conceptually, reduce the conceptual definition to an operational definition, and select an operation or technique which will best fulfill the operational definition. Subsequently when the results of the operation, i.e., experiment, are analyzed, an attempt is made to retrace our steps from operation to conception (Fig. 4.1). Initially the choice of technique depends upon the degree of precision, but the technique finally chosen is the most meaningful or the technique which makes it simplest to proceed from operation to conception. This is akin to saying that the experimental technique desired is of known precision and leads to meaningful measurements. It should be noted that the term meaningful is the obstacle here, and unless we can decide what this implies, the experiments will be very unsatisfactory; once, however, we choose our objective, the term meaningful becomes equivalent to accuracy and is a useful practical guide to the choice of procedure.

In summary, then, we start with a concept and reduce it to an operation which will yield information in the form of measured data of known precision; then a technique is selected which best fulfills the operational definition and yields the highest precision, and the data collected by means of this technique are translated through the operational definition to the concept. If the experiment is successful, the data will be meaningful in terms of the concept and will ultimately lead to increasing accuracy in approaching a true value by successive approximation. Figure 4.1 portrays the steps in the reasoning which lead to meaningful measurements.

4.2 The Concept and Measurement of Size

The concept of size is connected with volume, and in order to understand the meaning of size we must consider regular bodies which can be simply defined in terms of volume. Thus for a cube or sphere the volume may be defined as the cube of a single linear intercept uniquely related to the shape of the regular body; i.e., for a cube the volume is defined as the cube of an edge, whereas in a sphere it is the cube of a radius multiplied by a constant ($\frac{4}{3}\pi$). The next step is to consider a less regular body, such as a rectangular polyhedron which possesses a volume defined by the product of its three unequal edges; these definitions can be extended to increasingly less regular bodies, and in such cases a greater number of dimensions must be specified to achieve a satisfactory value; ultimately, for an irregular body such as a grain of quartz in a sediment, n dimensions become necessary for the complete description of its volume (Griffiths and Smith, 1964).

From the volume the size of a body may be obtained by choosing a unique axis and comparing the different bodies on the basis of this common axis; no difficulty exists in choosing a unique axis in the case of a cube or sphere, but some arbitrariness of choice is introduced in the case of the rectangular body, and the arbitrariness increases as the irregularity increases. Nevertheless, one means of defining size is to apply the principle of choosing a single uniquely defined axis as the basis of a comparative scale. One definition of size is, then, the measure of the length of some uniquely defined axis.

Another expedient, frequently adopted in practice, is to measure volume "directly" by displacing a volume of fluid, and this leads to another definition of size. Thus we may measure the volume of an irregular body by displacement in water and compute, say, the side of a cube or the radius or diameter of a sphere which possesses the same volume. Here again we obtain as the result of an operation a comparative measure of size of a body, and this procedure is applicable to all bodies whose volume can be individually determined by displacement.

Finally, we can adopt a different approach, which is less direct, i.e., measuring some property which is related to size and using this measure as the basis of comparison. The determination of settling velocities of particles in water and the reaction of a particle to a sieve mesh are procedures of this kind.

More involved techniques lead to measurements which become more and more difficult to relate to the original concept. Ultimately the measuring technique yields a measure which, while it may still be used for the comparison of one result with another, has no traceable relationship to the original concept. It is therefore necessary to decide whether the concept is of interest or whether the purpose of the experiment is to find a difference which may or may not be interpretable. The second objective is much simpler to achieve, but the results are, of course, very much less meaningful.

4.3 Measurement of Grain Size in Sediments

The purpose of measuring grain size in sediments is to understand the processes which result in the formation of sedimentary rocks, and frequent attempts have been made to reconstruct the environment of sediment formation from measurement of grain size of the constituent particles. The practical difficulties of defining size measurement of irregular particles have led to a pronounced emphasis on indirect measuring techniques; hence most procedures of size analysis represent attempts to study behavior of particles under conditions presumed to simulate natural processes. The manner in which quartz grains settle in a fluid, such as water or air, has been the basis of many recommended procedures. Usually the experiments are designed to measure the rate of settling of particles and, by means of a model, e.g., Stokes' law, to translate the settling velocity to some size scale. Such a procedure generally involves two stages in obtaining a measurement: first, the rock sample (or population) must be disintegrated into its component particles (or elements), and then the sizes of the particles (elements) so obtained are measured. It will be convenient to consider these two stages separately, although the effects of both are contained in the measurement of grain size and are rarely independently evaluated.

4.4 Disintegration of Rocks into Their Constituent Particles

Sedimentary rocks vary in their degree of consolidation from loose sands and clays through friable and loosely adherent materials to compact and very hard sandstones, shales, and clay stones. No one technique can encompass the entire range, and so disintegration procedures are rarely standardized. Two features must be borne in mind when attempting to find a suitable procedure for disintegrating any rock.

First, the principal objective of the analysis is to define the size of the grains present in the sediment as and when it was formed. This objective is rarely attainable, and most investigators would be satisfied to achieve a more limited objective and measure the size of the individual grains as they exist in the rock sample. Unfortunately, even this aim is difficult to achieve, because measurement of the size of grains usually contains in varying degree the effect of the technique used to break the sample down to its mineral components. The third objective, and the one most commonly achieved, is a measure of the size of the grains or elements as a function of the technique-sample interaction; attempts to interpret this result as "size" are generally abortive. Nevertheless, if the objective is limited to comparison of two or more specimens (samples) without an attempt to interpret any differences which emerge, this procedure is quite adequate.

The second important feature, which has also never been systematized, is that to be effective, a disintegration technique must be specifically designed to suit the composition and texture of the rock to be analyzed. Thus there is no problem involved in breaking down a loose, loosely consolidated, or friable rock, and the ideal of complete disintegration without affecting the size of the individual particles can be achieved very simply. Increasing degrees of consolidation demand increasingly harsh treatment of the rock, and it is advisable to arrange a systematic program of increasing severity as follows:[1]

1 Crushing with the fingers.
2 Stirring gently in water (warm the water if necessary).

[1] Details of these techniques are described in most textbooks on the analysis of sediments, e.g., Krumbein and Pettijohn, 1938 and Milner, 1962.

3 Prolonged shaking in a bottle in water (with or without boiling).
4 Crushing gently with a rubber pestle in a porcelain mortar.
5 Crushing in an iron mortar with an iron pestle.

These techniques represent physical procedures for the disaggregation of rocks, and increasing violence will, of course, increase the risk of affecting the grain size of the constituent particles. The aim is, therefore, to use the least severe treatment which will suffice to disintegrate the rock.

Consolidated rocks are, of course, much more difficult to disintegrate without affecting the size of the grains present in the rock. The types of consolidation arise from two sets of components in sedimentary rocks, and unfortunately the effects of these components are rarely independent.

One set of components is represented by the cements, and the disintegration of cemented rocks usually demands solution and destruction of the cement. It is, of course, obvious that the dissolved components do not enter the subsequent measurement of size, but it should also be realized that the cement, generally crystalline, was originally in a different state when the rock was formed. Strictly, therefore, if the objective is to estimate the size of grains as deposited, the cement can be ignored and its destruction presents no serious problem in achieving the objective. On the other hand, if the objective is to measure the size of the grains present in the rock as collected, then some estimate of the distribution and quantity of cement must be obtained. This objective cannot be achieved by removal and destruction of the cement.

A number of different cements are common in sedimentary rocks, the most common being carbonates and silica; such sulfides as pyrite and pyrrhotite are also frequent, and phosphate, sulfates, and organic materials are not rare. The removal of these cements is generally achieved by solution. Most carbonates are soluble in dilute acids, and dolomite, a mixed magnesium and calcium carbonate, is removed fairly readily in warm dilute hydrochloric acid. Sometimes weaker acids are used, e.g., dilute acetic acid, when it is suspected that fragile components such as diatoms, radiolarians, and siliceous foraminiferans are present and may be destroyed by violent effervescence.

In most graywackes and perhaps in many other sediments the chief cement is a mixed carbonate of ankeritic type, in which calcium, magnesium, iron, manganese, and perhaps other ions are present. The most suitable solvent in this case is warm oxalic acid, which serves the double purpose of dissolving carbonate and reducing the ferric to the ferrous ion to keep it soluble. Many refractory rocks yield to this acid when the chief cementing agent is a mixed carbonate.

Sulfides, generally pyrite, are destroyed by warm dilute nitric acid, and pyrrhotite can be removed with dilute hydrochloric acid. Sulfates of calcium can be dissolved in hydrochloric acid, gypsum in warm concentrated acid and anhydrite with milder treatment in dilute acid. Barium sulfate, barite, demands warm concentrated sulfuric acid. Phosphates are dissolved in warm dilute nitric acid, and organic cementing materials can be removed by dissolving in carbon tetrachloride, acetone, or alcohol.

The most troublesome cement is, of course, silica, and it seems likely that silica cement in small quantity is present in most consolidated rocks. Hot strong alkalis have been suggested for the removal of silica cement, but such a treatment almost certainly affects the detrital grains in the rock and is hardly to be recommended except as a last resort. Various techniques in which the rock is saturated with a liquid which then is forced to crystallize have been proposed to break up refractory sedimentary rocks but have not been very successful. Recommended techniques include freezing and thawing water; saturation with, and subsequent crystallization of, sodium hyposulfite; and saturating with calcium bicarbonate and adding dilute acid, the resulting effervescence being presumed to disintegrate the rock.

The second set of components is the matrix, which is common in arkoses and graywackes.[1] Matrix is largely composed of fine-grained minerals such as quartz, micas, chlorites, and members of the clay mineral groups, principally kaolinite, illite, and montmorillonoids. Bauxite and iron hydroxides are also common constituents of the matrix of sediments. Here again a plethora of treatments exists for the disintegration of the different mineral constituents, and this problem is the same as that met with in attempting to reduce a clay or mud to its component particles. There can be no doubt in this case that the particle size measured will be a function of the treatment, and grain-size distribution of the finer-grained sediments is in reality a measure of the efficiency of the dispersion effected by the treatment.

The necessity for the disintegration of the matrix or of fine-grained sedimentary rocks is, then, an effective obstacle to the complete analysis of grain size. The objective in this case must be something other than the grain size of sediments as deposited and usually something other than the grain size as it exists in the rock sample. Nevertheless, it is necessary to disintegrate matrix and the finer-grained sediments, in the former case in order to estimate the size of the other detrital particles and in the latter in order to fractionate the sediments for mineralogical identification procedures. In describing the techniques and procedures used in this phase of the analysis of sediments no specific objective is stated. Two attitudes may be taken. The most common approach is to standardize the techniques and compare the results, i.e., compare the settling velocities of the dispersed material or the effective grain sizes calculated from the settling velocities. Any differences which arise in the final data are, of course, of dubious significance, because differences in settling velocity will arise from differences in original mineral composition, either in kind or degree, differences in size, or differences in chemical environment of

[1] All petrographic terms used in the sense defined by Krynine (1948).

the sediment, such as pH, or base-exchange capacity, or kind of base. Indeed, standardization of technique is in this case a very serious problem, as the material is very sensitive to its environment.

The second approach is to attempt to reduce the fine-grained material to its smallest particle size, i.e., to use the most efficient dispersant. If this is taken literally, a very harsh treatment may be the most effective; but under normal conditions it is almost impossible to decide what particle size means in such a case.

For the present purpose, therefore, two approaches will be adopted. First, we can treat matrix as we did the cement and merely endeavor to disintegrate the rock, irrespective of the effect of technique on the matrix or fine-grained sediment, and measure the size of the particles, e.g., quartz, not affected by the technique. Second, when it is essential to consider the finer-grained minerals, particularly of the clay group, the objective will be to compare the results of the experiments, using carefully standardized treatments for disintegrating and dispersing the rock samples.

When a clay is suspended in water, it tends to ionize into micelles—cations and anions. The reactions can be represented as follows:

$$\text{H clay OH} \rightleftharpoons \text{H}^+ + (\text{clay})\text{OH}^-$$

$$\text{H clay OH} \rightleftharpoons \text{H}(\text{clay})^+ + \text{OH}^-$$

Any treatment which will assist the ionization may be considered as an aid to dispersion. Cation hydrolysis is the more important volumetrically, and anion hydrolysis is not well understood.

To increase cation hydrolysis it has been found that a sodium salt is advantageous, and sodium increases the amount of dispersion more efficiently than most other common ions. This may be attributed to the following idealized happenings.

Upon the addition of sodium salt in a solution of suitable concentration the clay becomes a sodium base-saturated clay,

$$\text{NaOH} + \text{H clay OH} \rightleftharpoons \text{Na}(\text{clay})\text{OH} + \text{HOH}$$

the sodium clay ionizing more readily than the original hydrogen clay. In general, a sodium clay breaks down into smaller particles than any other type of base-saturated clay.[1]

This example is very much idealized; it has been found that NaOH, a strongly ionizing base, is not the *most* efficient peptizer, a combination of a strong base with a weak acid radical being much more efficient. This performance, probably reflecting the effect of anion exchange, has led to the use of sodium carbonate and sodium oxalate as peptizers for dispersing sediments. In general, the use of a strong base (usually sodium) and a weak acid (usually

[1] Rarer ions such as lithium or chromium can be used, but their concentrations become more critical and the reaction more difficult to control.

organic) is the most efficient treatment for the dispersion of natural clay materials.

This again is only part of the truth because in practice, particularly in the treatment of oil-well drilling muds composed of clay and water, sodium phosphate and sodium tannate are generally far more efficient than most other peptizers. We may add, therefore, that a strong base, a weak acid, and a large acidic anion are the most effective dispersants. The reasons for this are somewhat obscure but may be largely accounted for by considering both cation and anion exchange in the average case; in certain special cases, e.g., clays saturated with calcium and magnesium, the complex salts of these acids form large anion complexes which isolate the calcium ion from reaction with the clay-water system. A calcium base-saturated clay is flocculated by contact with fresh water, and some such treatment is necessary to achieve efficient dispersion.

Dispersion agents are generally known as peptizers, whereas the opposite effect, flocculation, is brought about by coagulants. It must be remembered that the difference is one of degree, because any dispersion agent will act as a coagulant if the concentration is suitably high. In practice it is customary to suspend the clay in water and add a known volume of a known concentration of peptizer. The suspension and dispersion are effectively improved by placing the clay-water suspension containing the peptizer on a mechanical shaker for periods of from hours to days.

Ammonium hydroxide and sodium oxalate are commonly used peptizers in grain-size analysis. The former is moderately efficient as a peptizer and can be evaporated during the subsequent drying of the sample. When sodium oxalate is used, the weight of the peptizer in the aliquot taken for size analysis must be allowed for during the computation of weight percent of the material (see Table 4.1).

Finally, it must be remarked that most sedimentary rocks owe their consolidation to more than one factor; in most cases both cement and matrix are present, and a very small amount of silica cement in a fine-grained matrix of clay is probably as effective a binder as large amounts of silica in an almost pure quartz sandstone. Diagenetic effects, which generally increase with age and/or with proximity to folding, usually result in a redistribution and recrystallization of these mixed cements and perhaps also of the matrix material; in the absence of positive information on the exact composition of the binding materials no standardized procedure can be recommended. In general it is best to examine the rock in thin section under the microscope before attempting any disintegration procedure, and if the composition of the binding material can be decided, a technique for disaggregation can be designed to meet the requirements.

In the absence of knowledge of the factor or factors underlying the consolidation the only recourse is to attempt each recommended procedure, starting with the least violent and increasing the severity of the treatment in

TABLE 4.1
Data Sheet for Computing Size Distribution in Weight Percent from Pipette Sedimentation

(*a*) Weight of material into graduate ________ g. T ______°C. Sample No. __________

Finer than, phi units	Depth, cm.	Settling time for ______ °C, sp. gr. ___ Hr.	Min.	Sec.	Date and time due off	Beaker No.	Weight dry beaker	Weight beaker + sample to constant weight	Weight sample	Minus peptizer (*b*)	(*b*) ×50	$\frac{(b)}{(a)}$ cumulative % finer than
4.0	20											
4.5	20											
5.0	10											
5.5	10											
6.0	10											
6.5	10											
7.0	10											
7.5	10											
8.0	10											
8.5	10											
9.0	10											
9.5	10											
10.0	10											
10.5	5											
11.0	5											

Notes: (Dispersing agent, type and amount; dispersing procedure; etc.) ____________________

____

____

____

____

Analyst____________________ Date____________________

Source: These laboratory instructions are extracts from a syllabus, "Handbook of Laboratory Instructions," edited by M. A. Rosenfeld and used in graduate courses in sedimentology in The Pennsylvania State University.

steps. It is preferable to use physical disaggregation when possible because the effects can be more readily evaluated, but in most cases of rocks older than the Mesozoic a combination of physical and chemical procedures, generally repeated successively, is necessary. Many moderately well consolidated graywackes and arkoses can be disintegrated by means of gentle crushing after treatment with warm dilute oxalic acid.

It must be remembered that the technique chosen for disintegration will almost certainly affect the size of the grains, and perhaps the solution to the problem lies in using an experimental design which will segregate the effects of different disintegration techniques. Very few experiments of this type have been attempted, but it certainly represents a worthwhile approach (see Chap. 19); perhaps more attractive is the use of thin-section size analysis, wherein disintegration of the rock is unnecessary. In some cases, particularly in silica-cemented quartzites, this may, indeed, be our only choice, but this technique also possesses disadvantages and is not by any means a panacea (see Sec. 4.9).

4.5 Procedure for the Measurement of Grain Size

Many practical procedures purport to yield a measure of the size of particles of disaggregated materials; each technique has certain specific advantages and usually a set of limiting conditions within which the technique operates satisfactorily. Unfortunately, since few natural materials meet the conditions required by these techniques, all procedures in practice are, in varying degree, compromises between theoretical demands and practical difficulties in meeting these demands. Details of many of these techniques have been incorporated in readily available texts (e.g., Krumbein and Pettijohn, 1938; Dalla Valle, 1943; Milner, 1962), and there is little purpose in recapitulating them in their entirety.

The measurement of grain size of sedimentary rocks is not usually an end in itself but is one means to an end. Grain size is considered to offer information which may be interpreted in terms of processes of sedimentation; i.e., it is a reflection of the manner in which the sediment is formed. As most sediments are formed by sedimentation in water, procedures which reflect the behavior of particles in water have been used as one approach to the measurement of grain size. The advantages and disadvantages of the several procedures used for this kind of size measurement will be exemplified by a description of one commonly used technique, pipette sedimentation. Since most of the difficulties which beset this procedure are common to similar techniques, its description will therefore apply generally to most "wet" analytical procedures.

Pipette sedimentation is limited to grain sizes of equivalent diameters less than 50 microns. For larger sizes, sieving is generally adopted. In some cases, particularly when sedimentary rocks cannot be disaggregated without breaking their component grains, measurement of grain size by some technique which does not demand disaggregation is convenient; for this purpose measurement of grain diameters in thin sections of rocks is a possible recourse. The final technique, measuring diameters of loose grains, has not been practiced so much as might be expected from its apparent theoretical and practical advantages. Discussion of these four procedures occupies the balance of this chapter.

It should be abundantly clear that the measurement of size of the constituents of polydisperse heterogeneous component systems represents a very difficult problem, and it seems unlikely that any one technique will prove entirely satisfactory in achieving a complete definition; it is therefore inevitable that more than one technique will be necessary, each procedure supplying part of the information required for a complete description of the size of these complex natural materials. Failure to recognize the complexity of the material has encouraged the adoption of techniques essentially designed to attack simple systems composed of few components, and this is one of the greatest stumbling blocks in attempting an interpretation of the measurement of grain size of the mineral constituents of sedimentary rocks.

4.6 Measurement of Grain Size by Pipette Sedimentation

Pipette sedimentation is one of many techniques which attempt to measure the grain size of the dispersed components of sedimentary rocks by settling in water. The procedure is comparatively simple, the apparatus readily available in most laboratories, and experimental control, at least over a limited size range, easily attainable, so that the analysis is adaptable to mass production by moderately skilled personnel. Furthermore, since the measurement of settling velocity of the grains in water is the aim of the experiment, it appears, at least upon superficial examination, that this technique closely simulates natural conditions of sedimentation. These factors have led to the common adoption of pipette sedimentation as a measurement of grain size.

It is necessary to disperse and disaggregate the rock before beginning the measurement of size by pipette sedimentation, but as this procedure has already been described, it will be assumed that the rock has been reduced to its individual components. It must be emphasized that the term individual components is a function of the technique used to disaggregate and disperse the rocks; variation in size arising from the use of different dispersing procedures will be included in the final measurement of settling velocity determined by pipette sedimentation.

In practice 20 g of disaggregated sample is added to distilled water in a liter flask and the suspension made up to 1 liter in volume. Peptizers may be introduced at this stage to maintain the dispersed state, and the suspension is then poured into a liter graduate (Fig. 4.2).

The suspended material is stirred until homogeneous, and the experiment starts by allowing the material to settle. Pipette samples of 20 ml volume are taken at fixed times and depths from the suspension and evaporated to dryness, the weight of the sediment after drying representing the concentration of material in a 20-ml aliquot. The concentration of the original suspension is 20 g in 1,000 ml; hence the weights in successive 20-ml aliquots represent the amount, in weight percent, present at fixed depth after a fixed time. Sample time and data sheets covering the necessary computations are given in Tables 4.1 and 4.2. Ideally all the particles should begin settling at the same moment, but in practice it has been found that ensuring a homogeneous suspension is all that is necessary (Rittenhouse, 1939).

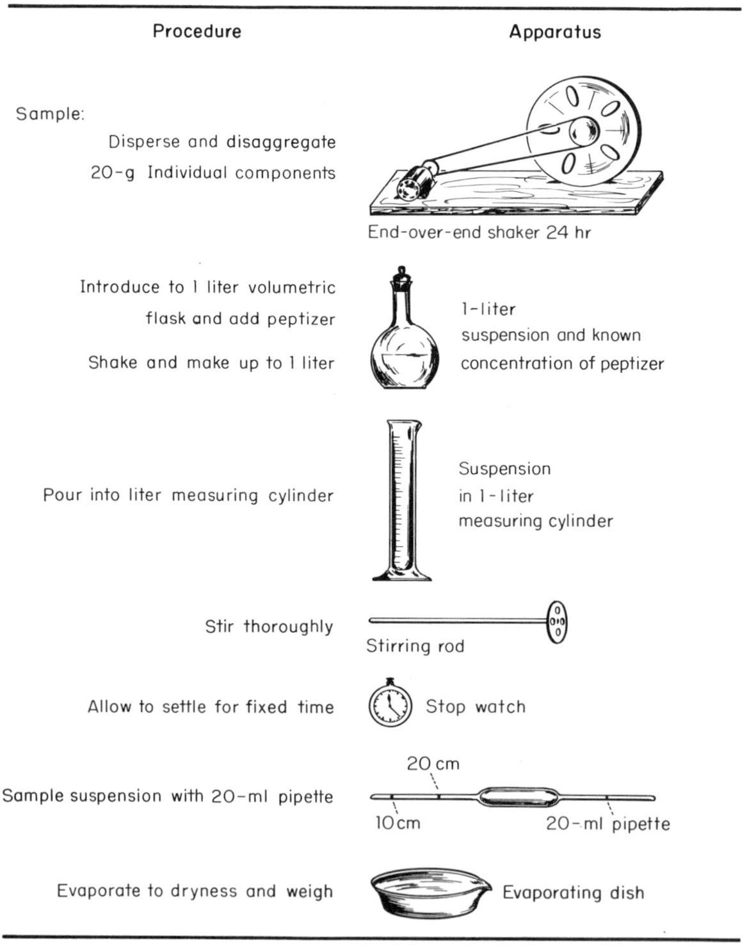

Figure 4.2 Flow sheet and equipment for determining settling velocity of particles by pipette sedimentation. (*After M. A. Rosenfeld, "Handbook of Laboratory Instructions."*

TABLE 4.2
Specimen Timetable for Settling Times and Depths of Sampling during Pipette Sedimentation[a]

Finer than, phi units	Microns	Depth, cm	Settling time for 20°C sp. gr. 2.65		
			Hr	Min	Sec
4.0	62.5	20			58
4.5		20		1	56
5.0	31.2	10		1	56
5.5		10		3	52
6.0	15.6	10		7	42
6.5		10		15	
7.0	7.8	10		31	
7.5		10	1	1	
8.0	3.9	10	2	3	
8.5		10	4	5	
9.0	1.95	10	8	10	
9.5		10	16	21	
10.0	0.98	10	32	42	
10.5		5	32	42	
11.0	0.49	5	65	25	

[a] These times and depths are for the conditions stated; 20°C, atmospheric pressure, and specific gravity of 2.65. See Eq. (4.1).
Source: After M. A. Rosenfeld, "Handbook of Laboratory Instructions."

The principle relied on in this technique is that if two particles start falling at the same time, the larger will settle faster than the smaller; hence pipette samples taken at constant depth will contain fewer and fewer coarse particles at successive intervals of time. Each sample is then a proportion, weight percent, of material less than some definite size (Table 4.2).

Settling velocity is transformed to a size scale by means of the model incorporated in Stokes' law; a description of the derivation of this law is given in Krumbein and Pettijohn (1938) and Dalla Valle (1943), and the operating formula is

$$V = \frac{2}{9}\frac{(d_1 - d_2)gr^2}{\eta} \tag{4.1}$$

where V = velocity or rate of settling cm/sec
d_1 = density of particle, g/cm^3
d_2 = density of fluid, g/cm^3
g = acceleration of gravity, cm/sec^2
r = radius of particle, cm
η = viscosity of fluid, poises

In practice it is generally assumed that quartz is the main constituent of the sedimenting system, and d_1 then becomes 2.65. Water is generally used as the suspending medium; hence, $d_2 = 1.00$. For any one experiment the temperature, pressure, and locality are constant; hence η is fixed at 0.01 poise (at 20°C and atmospheric pressure). Gravitational acceleration is usually taken to be 980 g/cm². Hence, the relationship becomes

$$V = \frac{2}{9} \frac{(2.65 - 1.00)980r^2}{0.01} \tag{4.2}$$

so that

$$V = Cr^2$$

where $C = 3.59 \times 10^4$ at 20°C and atmospheric pressure.

In practice, the velocity is separated into its two components, distance and time, and it is conventional to use a fixed distance of, say, 20 or 10 cm, marked on the stem of the pipette, and variable time intervals. It is convenient to substitute in the equation required sizes for r and so obtain a set of times at which to take samples; e.g., a quartz grain with a diameter of 31.2 microns (5 phi) will pass the 10-cm mark in 1 min, 56 sec, so that a sample taken at this time and depth will theoretically contain only particles finer in size than 31.2 microns diameter. Table 4.2 gives sizes and times for quartz particles, but it is advisable to determine the conditions, and particularly the daily variation in temperature, in the laboratory and compute a suitable set of values for oneself. It is advantageous to use a constant-temperature bath if available, since in most laboratories the temperature gradient can be very steep throughout the day and for fine grain sizes of around 1 micron diameter the time at 20 cm approaches 8 hr.

The final result is then a cumulative frequency distribution of weight percent of material finer than certain grain sizes, and, of course, this may be simply expressed as "coarser than" by subtracting each weight percent from 100.

4.7 The Limitations and Assumptions Involved in the Use of Pipette Sedimentation

Clearly, this procedure entails a number of assumptions and restrictions, and we must evaluate some of them in order to establish the meaning of the size measurement obtained. It has already been mentioned that the efficiency of the dispersion and disaggregation will affect the result, and in the case of fine-grained sediments containing large proportions of clay minerals, for example, dispersion is an extremely important point, so much so that for shales and clays the size measurement is largely a reflection of the effect of the dispersion technique used and rarely, if ever, reflects the size of the particles, either as originally deposited to form the sediment or as they exist in the specimen before treatment. Nevertheless, if the technique is carefully standardized, the experimental results may be used to achieve the third objective, i.e., to compare grain sizes of different samples of rock. It should, however, be fully realized that size differences in this case may be a reflection of differences in original mineral composition and the differential response of different minerals to dispersion techniques and will not be solely a reflection of differences in original particle size.

If we assume that dispersion and disaggregation effects are negligible, there are many other conditions entailed in the use of the Stokes' law model. Of immediate importance are the size limits over which the technique applies. For example, one limit occurs in the fine sizes when Brownian motion of the molecules begins to affect the sedimenting particles, very fine particles remaining indefinitely in suspension. For practical purposes 0.5 micron may be taken as the limiting diameter for freely sedimenting systems. Since the settling velocity of particles of this size is extremely slow, the experiment may take several days to complete. In such cases, it is essential to control the temperature within narrow limits in order to avoid convection currents in the suspension. The time can be decreased in these cases by reducing the distances of travel, i.e., taking the pipette sample at, say, 5 cm depth.

Coarse particles of several millimeters diameter create turbulence during their settling, and their settling velocities become very high; hence, there is an upper limit which theoretically falls in the range of 50 to 70 microns (Krumbein and Pettijohn, 1938, p. 100). In practice a diameter of 50 to 55 microns is generally taken as the upper limit. The theoretical limitations and the comparison of computed diameters with empirically determined diameters is discussed by Nevin (1946).

Even within the size limits of 0.5 to 50 microns a large number of other restrictions also exist if the Stokes' law model is applied, and analysis of the problems connected with the fulfillment of these requirements has received considerable attention. A number of alternatives have been proposed to avoid some of the more obvious difficulties, and for simple systems of particles it has been demonstrated that improvements can be achieved by the use of these alternative models. An excellent analysis of the various suggestions and their accompanying advantages and disadvantages has been given by Krumbein (Krumbein and Pettijohn, 1938), and an attempt to study a wider range of conditions is described by Nevin (1946). It should be realized that there are two extreme views involved in this approach: the first aims at the derivation of a theoretical model which when applied to a set of data will yield a rigorous measure of size; the second viewpoint ignores this aspect and bases the approach on standardizing the technique and comparing the results. A procedure proposed by Doeglas (1946), for example, ignores the effect of size on the settling velocity of particles and uses a very large settling column

(6 ft), weighing the sediment deposited at the base of the column at successive time intervals and calibrating the settling velocity irrespective of the departure from the ideal model. Results from this procedure are not strictly size measurements, and their relationship to the concept of size is uninterpretable. They do, however, form a valid basis for comparison of results of one sample with another. The exact physical meaning of a difference is obscure in this case.

Nearly all attempts to evaluate a theoretical model have failed, and, as is generally the case, the more rigorous the mathematical approach, the greater the departure of the resulting model from reality. In attempting to take the effect of shape into account, for example, rigorous analysis has been attempted and confirmation of the results of this analysis sought by empirical determination of the effects of differently shaped particles, such as spheres, rods, plates, etc. Factors have been devised to allow for shape differences and particularly for departures from the ideal sphere. In sediments particles of quartz occur which depart widely from a spherical form; indeed, the range in sphericity to be expected when measured by one technique is from about 0.5 to 1.0 unit (Curray and Griffiths, 1955). Similarly, micas are known to be platy in shape. It is possible in this case to apply a correction factor which will offset the effect of the proportion of mica plates present if the proportion of mica is known. This has, of course, rarely been determined, and it is doubtful whether, say, 5 percent of mica in a rock sample demands an elaborate correction factor. Nevertheless, because the shapes of the two minerals are here mutually exclusive and dependent on composition, the use of a correction factor is valid and would result in some small improvement. Another, much larger proportion of many sedimentary rocks comprises fragments of shaly rocks. Their shape varies from plates like micas to ellipsoids of varying sphericity like quartz, and at the same time their size and composition are not constant. In this case shape is not dependent on composition, as in the case of quartz and micas, and allowing for all three constituents, it can be seen that shape, size, and composition are neither dependent nor independent, as already stressed (Chap. 3). Interdependent factors of this kind give rise to complex interactions, and no constant correction factor can hope to encompass the range of variation involved. This is the most important restriction which is not fulfilled by any of the approaches in use; no amount of rational analysis can remove its effect, and, so far at least, no experiment has been suitably designed to evaluate it.

The main stumbling block, then, is that variations in "size" measured by pipette sedimentation are really reflections of complex interactions between three interdependent properties, that is, $P = f(m,s,sh)$.

All "wet" techniques suffer from this same difficulty, and in effect no one is better than another. Which is chosen becomes therefore a matter of convenience; in the present case one of the techniques easiest to control, i.e., yielding the highest precision, has been selected.

4.8 Measurement of Grain Size by Sieving

Settling in water has generally been considered unsatisfactory for sedimentary particles coarser than 50 microns diameter. Nevertheless, settling velocities have been used to much coarser sizes (Doeglas, 1946); and for comparison the settling velocities may be transformed to a size scale by means of a model such as Stokes' law, Wadell's modified formula, the impact law, or an empirically determined calibration. Such an expression may have no direct physical meaning in terms of the concept of size but can be used as a basis for comparison.

An alternative technique, and one which is probably most popular, is to sieve the disaggregated sample. In this procedure a weighed disaggregated aliquot, usually 100 to 150 g, is introduced into a set of sieves arranged in descending order of decreasing coarseness (Krumbein and Pettijohn, 1938, fig. 33, p. 139); the set of sieves is shaken, and the residues in each sieve are weighed and recorded (Table 4.3). This analysis yields results in terms of weight percent between two sieve sizes.

Sieves or screens are usually made of brass with a wire mesh cloth spread across the base. Wire mesh cloth with square holes of specified diameters is a standard product, and high precision is generally available. The U.S. standard sieve sizes extend from cloth rated at 325 mesh (44-micron-diameter holes) to 1-in. mesh (25.4 mm diameter) in steps arranged on a geometric scale with an interval of $4\sqrt{2}$. Sieves within the interval 62.5 microns to 2 mm are generally used for sedimentary rocks.

It has been demonstrated (Swineford and Swineford, 1946) that shaking the screens is best performed by means of a mechanical shaker, such as a Ro-tap, and this has become standard equipment in most laboratories. It has been found that 15 to 25 min is the most efficient shaking interval; this conclusion is based on a compromise aimed at avoiding wear of sieve mesh and particles, which tends to take place beyond 20 min, and completeness of separation, which follows the "law" of diminishing returns. It is, of course, necessary to avoid using so much material that some grains fail to make contact with the sieve mesh, and this has resulted in the recommendation of 100 to 150 g, for the average sedimentary sample.

In practice, with most rocks it is necessary to remove the fine particles, which tend to agglomerate onto the coarser sieves and choke the meshes of the finer sieves. For sedimentary rocks, and dry sieving, the limiting size beyond which fine material becomes troublesome lies in the neighborhood of 105 microns, and considerable error generally arises in the fine-sized sieves. With industrial powders of homogeneous composition successful separations down to 5 microns and below have been obtained.[1]

It has been generally considered that for most particles, particularly

[1] Electroformed micromesh sieves are available to diameters of 5 microns from Buckbee Mears Co., St. Paul, Minn.

TABLE 4.3
Data Sheet for Computing and Recording Grain-size Analysis by Sieving

Analyst ______________

Date ______________

Screen Analysis

If a coarse gravel:

(*a*) Wt. original sample ______________ $100\,\frac{b}{a}$ = ______________ % material 4 mm (*c*)

(*b*) Wt. material 4 mm (−2 phi) ______

If acidized:

(*d*) Wt. sample before acid ______________ $100\,\frac{e}{d}$ = ______________ % insoluble material (*f*)

(*e*) Wt. residue ______________

If decanted:

(*ff*) Wt. sample decantation ______________ $100\,\frac{g}{ff} - m$ ______________ % material 0.044 mm (4.5 phi) (*h*)

(*g*) Wt. residue after ______________

Wt. into sieves ______________ (*j*)

Mesh	(*m*) raw wt.	(*n*) est. % aggr.	(*p*) corrected wt. $(m - mn)$	(*r*) $\frac{p}{q}$ %	% of insoluble (rch)	% of entire sample (rcfh)	Cumulative %
5							
10							
18							
20							
25							
30							
35							
40							
45							
50							
60							
70							
80							
100							
120							
140							
170							
200							
230							
270							
325							
Pan							

Total ______________ (*k*)

Loss ______________ $(j - k)$ ______________ (*q*) ______________ ______________ ______________

TABLE 4.3 *(continued)*

Remarks (disaggregation procedure, unusual difficulties, other pertinent comments):

Source: M. A. Rosenfeld, "Handbook of Laboratory Instructions."

quartz grains, the behavior on a sieve is determined largely by the intermediate diameter, and many discussions of this topic exist. Rittenhouse (1941, 1943a) has demonstrated that the diameters of particles on any sieve show a considerable range in intermediate diameter. In this investigation many of the difficulties which arise during sieving are discussed, and the dominant error appears to be introduced by oversized holes in the sieve mesh. Reproducibility of sieving sand has been investigated by Walker (1941), and the variation in repeated analyses of the same samples in the same and different laboratories is shown to be far from negligible (see also Ames, Irani, and Callis, 1959).

It seems clear that sieving, like pipette sedimentation, represents a compromise between an ideal aim and a practical achievement, and a very large number of errors enter into the experimental procedure. Many of these errors are common to all laboratory procedures, and their effects can be evaluated and removed by suitable experimental design. Certain difficulties, however, are inherent in the technique and cannot be overcome by any practicable means at our disposal.

First, as would be expected, when a single sample is analyzed by pipette analysis and by sieving, covering the range in size from 1 to 50 microns and 62.5 to 2,000 microns, respectively, there is always a discrepancy in the results between the two techniques. A device for producing a smooth cumulative curve despite the anomaly is described by Griffiths (1952a). It should be realized that this is merely a device for producing a smooth graphic expression of the analysis, and it does not remove the physical discrepancy or difference due to errors in the techniques.

Second, the features affecting such wet techniques as pipette analysis also affect sieving procedures. The "size" measurement obtained comprises variation in three interdependent interacting variables $P = f(m,s,sh)$, which cannot be evaluated independently to arrive at a correction factor. Many attempts, both by rational analysis and empirical determination, have been made to correct for variation in shape, but the fact that the three factors interact and are interdependent automatically foredooms such attempts.

Correction factors can be applied only to simple one-component systems in which shape and size variation is limited.

It must be concluded that size measurement by sieving yields measures which may be compared when analytical technique is suitably standardized, but, once more, attempts to interpret the physical meaning of the measure in terms of size are unlikely to succeed with complex polycomponent mixtures such as disaggregated samples of sedimentary rocks.

4.9 Measurement of Grain Size in Thin Section

It has already been mentioned (page 54) that some rocks cannot be disaggregated without breaking the component grains, and in such cases size measurement by pipette sedimentation and/or sieving of the disaggregated residue is really a measure of the efficiency of the disaggregation procedure. Krumbein (1935) has proposed measuring the size of the grain of such rocks in thin section. Such a procedure has much to recommend it for all sediments, and adopting a single technique, if it were possible, would in itself be of very considerable advantage.

In this procedure it is customary to select a specific mineral, usually quartz, and measure some diameter or intercept across the grain, either directly under the microscope by means of a micrometer eyepiece or indirectly on the projected image of the grain. Microprojector or camera lucida drawings of the grain images may be used for this purpose.

The measurements of lengths of intercept are usually recorded in millimeter or micron units, and they may be grouped into a frequency distribution directly. The size frequency distribution in this case is a number frequency and not weight percent.

The outstanding difficulty in this case is to decide which intercept to measure, and it is necessary, for reasons previously outlined, to choose a definition which leads to a unique axis (page 46). Frequently the longest intercept through the grain is chosen, but a number of definitions exist which achieve the same purpose; one depends on the selection of the shortest intercept through the projected image and the subsequent measurement of the longest intercept perpendicular to the shortest (Griffiths and Rosenfeld, 1950). There appear to be no a priori reasons why one intercept should be better than another as long as the definition results in the selection of a unique axis (see, however, Griffiths and Smith, 1964). The analysis of different samples may be compared in any case. Once more, however, any attempt to interpret the exact meaning of the measurement in terms of absolute size is frustrated.

In this procedure, since the rock is not disaggregated, the original fabric is undisturbed, and the measurement can be expressed as

$$P = f(m,s,sh,o,p) \tag{4.3}$$

As a single mineral is selected for measurement, there is no variation in composition m, and m becomes a constant; it can therefore be omitted from further consideration, and the expression becomes

$$P = f(s,sh,o,p) \tag{4.4}$$

Many ingenious attempts to evaluate the meaning of P have been made, again largely by rational analysis but sometimes empirically. For sedimentary rocks, at least, the rational analysis has been much too naïve, and it seems obvious that attempts to evaluate the different factors as independent variables are unlikely to offer any solution to the general problem. The only published analysis performed by this means is Krumbein's (1935) original single sample of St. Peter sandstone. This is unfortunate, not only because it is a single sample, but also because it happens that the St. Peter sandstone is a quartzite and is therefore unlikely to be representative of any but the simplest sediment in terms of composition.

It must be realized that whatever unique axis is chosen as a basis for measurement, the length of the axis actually measured, and therefore the size measure obtained, is a function of the size and shape of the grain, the orientation of the grain and/or of the direction in which the section is cut, and of the density of the grains in the section, i.e., packing. Again, as size and shape interact with orientation and packing, no correction factor can be used to adjust separately for these variables. It is necessary to realize that to evaluate the exact effect of size divorced from the effects of the other factors would necessitate an experiment designed to evaluate the effects of all four factors, the six first-order interactions, the four second-order interactions, and the single third-order interaction.

The literature discloses numerous attempts to calibrate size measured in thin section with size measured by sieving and/or other techniques, but except for the simplest possible idealized cases no success has been achieved. An empirical study aimed at evaluation of this aspect has been described (Rosenfeld, Jacobsen, and Ferm, 1953); and from the results of these analyses of 32 samples, including three different rock types, it seems apparent that rocks of different mineral composition cannot be treated alike. In particular, discrepancies in rocks of simple composition, such as quartzites, are small (Friedman, 1958), but as complexity of composition increases, the discrepancy between the size measurement obtained from the two different techniques becomes very large.

In conclusion, then, it is well to remember that if the objective is a comparison of several sets of samples to define a difference, any technique suitably standardized is likely to be satisfactory. Each technique supplies, however, different information; e.g., pipette sedimentation yields proportion by weight in certain "size" classes, and thin-section size analysis yields proportion by number in certain size classes. In addition, if the important variable is the size of the quartz grains, then pipette sedimentation and sieving will not yield a meaningful measure unless the sediment is of very simple composition, i.e., a quartzite. For this purpose thin-section analysis would be preferable; on the other hand, if we wish to determine the amount of silt irrespective of composition, pipette sedimentation is the better analytical technique.

In no case can the above measures be interpreted into absolute size measures, as each technique yields a measure of a variate which is a complex function of three or more interacting variables. Similarly, if it is desired to compare the results of thin-section size analysis with that determined by sieving, it is advisable to use the empirical approach for most sedimentary rocks. Rational analysis has proved insidiously misleading in attempts to predict the result of analysis of sedimentary rocks.

4.10 Measurement of Grain Size in Loose Grains

Mature consideration of the size-measurement techniques practiced suggests a set of requirements which any improved technique should meet; it seems obvious that the complexity of sedimentary rocks precludes any single practicable procedure from achieving the specific objective of determining the grain size of mineral constituents of a rock as deposited. First, a technique should be capable of successful application to rocks of widely varying degrees of consolidation and as far as possible must avoid elaborate disaggregation procedures. Second, because mineral composition is a complex variable in sediments, results of techniques which attempt to treat the rock irrespective of kinds and proportion of constituents must fail because they can yield only the average size of a gram of sand; and if the gram of sand is not constant in composition, this average will possess dubious meaning. Sieving, pipette sedimentation, etc., fall at this hurdle. One requirement is, then, to simplify mineral composition, i.e., choose a technique which can be applied to a single mineral species.

To this end thin-section size analysis meets the requirements of avoiding disaggregation procedure, and in this technique it is possible to concentrate on a single mineral species. Unfortunately, the aggregate is still a complex, and the presence of other minerals tends to interfere with the size measurement; in particular, the fabric attributes of orientation and packing are the most troublesome features. Nevertheless, this technique is probably simpler to evaluate by a correctly designed experiment than any other so far described. It seems likely, however, that if the quartz grains can be separated from the other constituents without breakage, a still better measure of size may be attainable.

One of the main difficulties in disaggregation is that different minerals are affected differently, and 100 or even 20 g represents a large amount of rock to produce without elaborate precautions. If, however, 100 grains of a mineral least likely to be affected by disaggregation procedures could be used, many of these disadvantages would be avoided. For this purpose the ubiquitous quartz grain seems admirably suited. Furthermore, if other minerals are ignored, their destruction is not important, and so most chemical techniques, which hardly affect the quartz grains, can be used with impunity. On the basis of procedure, then, it seems likely that the measurement of loose grains of quartz would offer a number of definite advantages.

In addition, number frequency is much more suitable for statistical analysis than weight percent, and 50 to 100 grains should be ample for most measuring procedures. Actually, from a small rock sample 50 grains can easily be loosened, and only the most refractory consolidated quartzite would offer much trouble in achieving this kind of sample.

Furthermore, a homogeneous material is some assurance of a homogeneous size frequency distribution, a requirement not achieved by any technique which ignores the complexity of mineral composition. There appears, then, to be ample justification for this approach on the basis of technique difficulties and analytical requirements (Ames, Irani, and Callis, 1959; Irani, 1961).

Is there any justification on theoretical grounds? It seems clear from the theoretical treatment of the problem in Chap. 3 that the use of loose quartz grains leads to a unique measurement and one that is simplest to interpret. In this case the measurement is a function of two interacting variables, and size or size and shape simultaneously can be measured by this technique. Furthermore, the long axis is in this case likely to approach very closely the true longest axis of the grain if it can be assumed that the grain exposes its maximum projection area. Adjustments of procedure during mounting can be made to achieve this in most cases. Theoretically, then, this is the best "size" measurement we can achieve.

Finally, a model may be proposed which, with a few restrictive assumptions, will enable this size measurement to be interpreted in terms of sedimentational environment (Miller and Olson, 1955).

It will be advantageous here to return to the objective of grain-size analysis. The geological aim is to estimate the size of the particle as it was sedimented and from it to deduce the conditions of deposition. Let us propose the hypothesis that any unit particle settling in a fluid will be accompanied by another only if they are of equivalent hydraulic size. This postulate has been very effectively utilized by Rubey (1933) and Rittenhouse (1943a) in attempting to explain variation in proportion of accessory heavy minerals in sediments.

If the postulate is accepted, when a mineral particle is found closely associated with, say, a quartz particle in a sedimentary rock sample of detrital origin, their association will be considered indicative of original hydraulic equivalence. In this way the grain-size distribution of the quartz grains reflects the size distribution of the material as deposited; and if the effective density of any material and the grain-size distribution of the associated quartz particles are given, the grain-size distribution of the nonquartz detritus can also be determined.

Thus, when in a thin section of sedimentary rock a quartz particle, rock fragment of granite, clay particles, and, say, a shell fragment are found associated *as detritus*, it will be presumed on the basis of this model that all these materials were hydraulically equivalent at the site and time of deposition. This would imply that their *effective* densities were similar and that the rock fragment, if denser than quartz, would be smaller; the shell fragment of carbonate, if larger than quartz, was therefore less dense; the clay presumably settled as flocs. It is obvious that if this sediment is disaggregated and dispersed, the "size distribution" determined by sieving or pipette sedimentation will yield a completely misleading picture of the conditions of deposition because it will separate particles which behaved in a similar manner. If the quartz grains can be separated, however, the size distribution of this mineral will reflect the conditions at the site of deposition; indeed, if the experiment could be performed on a homogeneous disaggregated quartz aggregate, all the techniques would be equally suitable and would differ only in magnitude of experimental error.

In operating with this model there are, of course, a number of assumptions involved; first, the use of the model presupposes that the quartz grains settle freely. It seems likely that in sediments with much clay, such as low-rank graywackes and perhaps some arkoses and high-rank graywackes, the clays are flocculated and include in the floc at least silt-size quartz grains. What volume of sediments is affected by failure to meet this requirement can be determined only by trial; in extreme cases, such as turbidity currents and various kinds of mudflows, hydraulic equivalence may still be operative, but conditions of the sedimenting systems are vastly different from those of "normal" detritus. Under such conditions no grain-size technique presently practiced can yield grain-size distributions which can be used to interpret conditions of sedimentation.

A second assumption is that there is equal availability of all sizes of material; this requirement is probably met by the ubiquitous quartz particle but is less likely to be fulfilled by other materials which occur as detritus in sedimentary rocks. One advantage of the hydraulic-equivalence principle is that given the grain size of quartz and the effective density of another component, the hydraulically equivalent grain-size distribution of the second component can be calculated; if the size of the second component is then measured, departures of the observed from the calculated may be interpretable. Here again, empirical determination of quartz-grain size is the only certain way of judging the degree of fulfillment of this requirement.

Measurement of the grain size of the quartz grains and translation of the measurement into conditions of deposition also entail the assumption that the size of the quartz grain has not been affected since deposition in the rock, i.e., has passed through the time-diagenetic barrier without change (Miller and Olson, 1955). Here the physical and chemical stability of quartz is some assurance of fulfillment of the assumption, but other detritus may depart widely from this requirement. Once more the departure from the size distribution, computed via hydraulic equivalence, may be of use in deciding stability of the material, always supposing other factors, such as equal availability, may be eliminated.

One final requirement applies solely to the comparison of grain-size distributions of quartz and some other component. In this case, if the interpretation is to be meaningful, it must be shown that the technique does not differentiate between quartz and the other component; i.e., there must be no interaction between technique and particles.

The determination of grain size of loose quartz grains is attractive on a theoretical basis because this technique is the only one which yields the basis for a definition of a unique size which is at the same time a true measurement; i.e., the longest axis is indeed the longest axis of the grain and not the apparent long axis.

Third, the material requirement of a few, say 50, grains of a stable mineral such as quartz avoids the difficulty of elaborate disaggregation techniques and their effect on the size measurement. Largely because of this, but also because the mineral composition has been simplified and the fabric of the rock destroyed, the resulting measurement is the simplest to evaluate and involves the fewest interactions $[P = f(s,sh)]$.

Finally, the grain size of quartz grains measured as individuals is likely to yield the best basis for geological interpretation in terms of environmental conditions of sediment formation.

The restrictions or assumptions of the model proposed as a basis for interpretation here need adequate ventilation and elucidation by experimental testing, and it seems likely that a suitably designed experiment using sediments formed under known sedimentational conditions and subjected to varying degrees of subsequent change would help to evaluate the effect not only of sedimenting conditions but also of diagenesis on detrital rocks. This may then be applied to rocks of unknown origin.

The comparison of size distributions through the hydraulic-equivalence model between minerals susceptible to change with those not susceptible under definitely specified conditions would be a powerful tool in interpreting environments from the petrography of the older sediments.

4.11 Procedures for the Determination of Size in Loose Quartz Grains

A procedure for the measurement of grain size in loose grains of quartz has been described by Pye (1943), and a new procedure has been developed by Hulbe (1955). In the latter case a sample of from 20 to 50 grains is obtained from the rock to be analyzed and mounted in a rectangular plastic rod with square ends. The grains are projected through a microprojector, and the measurement of size is performed upon the projected image.

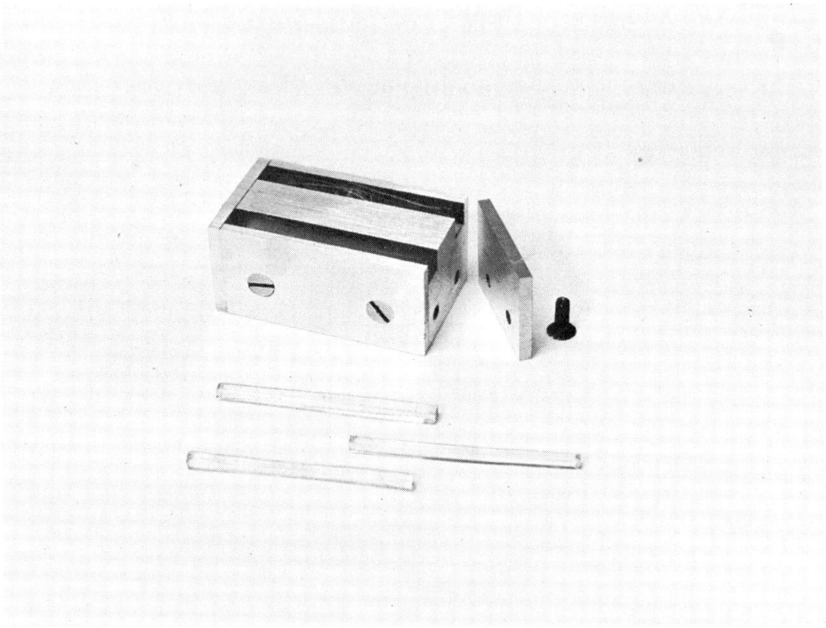

Figure 4.3 Aluminum mold and rods of Castrolite holding quartz grains; rods are 2 in. long. (*After Hulbe, 1957.*)

Castolite[1] is used as the thermoplastic, and the rod shape is obtained by means of a polished aluminum mold (Fig. 4.3). The mold is painted with releasing compound and half filled with plastic; the plastic is hardened by adding promoter and warming gently for a few minutes. The 20 to 50 quartz grains are then mounted and aligned along the hardened plastic; this generally ensures a stable position and exposure of the maximum projection area. Subsequently, more Castolite plastic is added, together with the requisite amount of promoter, and the mold is warmed in an oven at 100 to 110°C until the plastic achieves its permanent set (about 10 min). The rod is then removed from the mold and buffed until transparent. A finished rod with 50 grains is illustrated in Fig. 4.3.

The permanent mount is easily stored, and the grains can be individually numbered. The square section of the rod enables the operator to examine both maximum and minimum projection area and so permits measurement of three-dimensional sphericity. By varying the size of the mold, grains from silt to gravel sizes can be treated by the same technique and direct size and

[1] Castolite promoter and mold-releasing compound may be purchased from Ward's Natural Science, Rochester, N.Y.; a polishing wheel and motor are also obtainable from the same source.

sphericity measurements obtained from the projected image. Roundness can also be estimated from the image.

One disadvantage of the technique is the similar refractive indices of quartz and Castolite plastic, but sufficient contrast can be obtained in most cases, and some ambiguity of grain margin is always present because the grains tend to wedge out toward the margin. Increasing contrast will not remove this difficulty, but by using different operators some measure of the variation due to this ambiguity can be obtained.

5 Expression of measured data

The uniform pattern of static variability is revealed by arranging the variable in a frequency distribution or frequency series, which is simply a summary of an array of the variable arranged from smallest to largest. . . . When graphed the figure is called a "histogram," "frequency polygon," or "frequency curve," depending upon the manner of graphing.

J. G. Smith and A. J. Duncan, 1945

5.1 Introduction

When an experiment is performed and, by means of a suitable technique, a number of measurements are obtained, the next step is to examine and analyze the data. When the number of measurements is small, this is a fairly simple process, and the whole set of data is easily comprehended; but usually the set of data is large, and then it is difficult to comprehend unless ordered in some fashion. Generally, this ordering is achieved by means of a classification; thus, measurements are usually numbers on a scale, and for the present purpose we may subdivide scales into two kinds, continuous and discontinuous. If a scale is continuous, any number or fraction of a number can occur in the resultant set of measurements. On the other hand, if a scale is discontinuous, only certain values can occur in the measured data. In measuring grain size in millimeters, for example, the sizes obtained as measured data are not restricted to certain values, but all intermediate fractions of a millimeter may occur; hence, the scale is a continuous one. On the other hand, in counting the number of garnets in 300 grains of heavy residue, the smallest number is zero, followed by one, two, etc. There is in

Quotation: From "Sampling Statistics and Application," by J. G. Smith and A. J. Duncan, p. 1. Copyright 1945, McGraw-Hill Book Company. Used by permission.

this case no such thing as $1\frac{1}{2}$ garnets; a garnet is either present or not present. The scale proceeds by integers and is therefore discontinuous.

It should be emphasized that this subdivision into continuous and discontinuous is largely a matter of convenience, because it is possible to convert the number of garnets to a percentage figure, when the scale automatically becomes continuous. Similarly, a continuous scale is really a conceptual ideal and rarely can be achieved in practice, because the measuring technique limits the precision of the result; i.e., a millimeter scale really permits measurement in millimeters only, and there are no parts of millimeters involved so the scale is discontinuous. In practice all scales are discrete at some finite level of subdivision. Nevertheless, it is usually convenient to differentiate between continuous and discontinuous scales (see also Chap. 12).

The first step is to subdivide the scale, and generally a continuous scale is subdivided into classes at some convenient figure. Thus, suppose a set of measurements extends from 1 to 100 mm, all possible numbers within this range being permissible. When a large number of measurements is obtained as numbers in this range, it is convenient to group the set into classes, i.e., subdivide the scale into convenient intervals of, say, 5 or 10 units per group. Generally, the intervals are equal, but this is not always essential although it is most convenient for purposes of subsequent computation. The first requirement is to decide how many classes or groups to use; conventionally, some 10 to 20 classes are chosen, the precise number being dependent on how large a set of data is available. Fewer classes accompany fewer data to avoid too many empty cells.[1]

When the classification has been decided upon, the set of measurements is grouped by counting the number of measurements which fall in each class; the data are then recorded in a table which, it should be remembered, is a summary of the original set of measurements. This tabular arrangement, consisting of the number of measurements occurring within each class or each unit interval on the scale, is called a *frequency table*. By grouping the data into classes the original continuous range in values of the data is lost, and the set of data in the class is considered to be centered around the class midpoint. An example is given in Table 5.1.

This set of data has been classified on the basis of two variables: the class interval and the number or frequency of measurements in each interval. The choice of interval is arbitrary, but once chosen, it will affect the frequency values of the data in each class; thus, if a scale is divided into 20 class intervals which are subsequently condensed to 10, the frequencies of the first classification will be about doubled in the second. In other words, the class interval chosen on the measurement scale is the independent variable, and the frequencies depend on the class interval chosen. It is customary to record the two variables as independent and dependent, respectively.

[1] Considerations of the loss due to grouping of data are mentioned by Fisher (1948, p. 50), and appropriate adjustments are suggested.

TABLE 5.1
Frequency Distribution of Grain-size Measurements

Class interval mm	Class interval ϕ	Midpoint mm	Midpoint ϕ	Frequency, %	Cumulative frequency, %
0.5–0.25	1–2	0.375	1.5	1	1
0.25–0.125	2–3	0.187	2.5	2	3
0.125–0.0625	3–4	0.094	3.5	6	9
0.0625–0.0312	4–5	0.047	4.5	21	30
0.0312–0.0156	5–6	0.023	5.5	40	70
0.0156–0.0078	6–7	0.012	6.5	25	95
0.0078–0.0039	7–8	0.006	7.5	4	99
<0.0039	8+		?	1	100

Size is usually measured on some linear scale such as millimeters or inches, and this scale is then subdivided to obtain the class intervals for grouping of the data. A very large number of subdivisions exist; in grain-size investigations the system of subdivision used is called a *grade scale.* In order to select a suitable grade scale, it is necessary to be aware of the purpose and objective for which it is to be used—a suitable grade scale for size is not necessarily the same as that for shape. In general, a grade scale should form a useful basis for description and comparison of results, it should not be affected by the technique of measurement, and preferably it should have some physical justification. On the basis of these criteria few of the grade scales proposed are justified; by far the most useful and acceptable scale is that proposed by Wentworth (1922).

The size scale is subdivided into class intervals which are not equal. Successive class intervals are related to each other by the ratio of $1:\sqrt{2}$. Each class limit is twice the preceding interval and half the succeeding one. Each class interval on this grade scale also has a descriptive title, such as coarse sand, fine sand, coarse silt, etc., yielding a basis for the standardized description of the grain size of sedimentary rocks.

The Wentworth grade scale was subsequently modified by the Committee on Sedimentation of the National Research Council (Trask, 1942), and this scale is now the basis of most of the grain-size analyses performed in the United States. It has several advantages.

This grade scale forms a useful basis of comparison for sedimentary rocks in terms of grain size, and the associated descriptive terms represent a standardized form for the specification of sediment grain size. Fortunately, a very large number of analyses have been published using this scale, and so a useful reference background of data exists expressed in terms of the Wentworth grade scale. The scale is independent of measuring technique. It has been suggested that 2 mm represents the upper capillary limit and 62.5 microns approaches the size limit of suspended detritus under normal sedimenting conditions. On the basis of utility a set of sieves, the U.S. standard series, yields a very close sizing coinciding with this grade scale; in the range from 2,000 to 62.5 microns there are 21 sieves on the $4\sqrt{2}$ interval, yielding 4 sieves per Wentworth grade. Finally, Krumbein has proposed a logarithmic transformation the phi scale (ϕ) to conform to the Wentworth grade scale, thus increasing the utility of this scale by simplifying computation.

The modified Wentworth grade scale is illustrated in Table 5.2, along with sieve-mesh equivalents of the U.S. standard sieve series, the phi equivalents, and the appropriate descriptive terms.

Most sediments comprise material of many grain sizes, and several grade names must be applied to encompass the range of size variation. A number of alternatives have been proposed, and triangular diagrams have been designed to simplify the description and comparison of sediments of different grain sizes. The best description of the grain-size distribution is by the use of the frequency table already mentioned, and attempts to replace this frequency distribution by less precise descriptive terms can only lead to loss of information.

If it is essential to summarize the size frequency distribution, we can portray the distribution either graphically or by summary descriptive statistics. The main difficulty met with in using these procedures to compare size distributions arises because the frequency distributions are generally incomplete (possess open ends) or because they are too complex, e.g., polymodal with modes of different magnitude. These difficulties arise from the nature of the frequency distribution, which in part, is a function of the technique used to measure grain size and, in part, arises from the polycomponent character of the bulk of the sediments. If a technique is adopted which leads to a uniform symmetrical curve such as a normal distribution (see Chap. 13), then the graphic expression and summary statistics are materially simplified; with normal or near-normal frequency distributions complicated descriptive terms become unnecessary, and the comparison of size distributions from different sediments is simplified and becomes more rigorous. If these simpler distributions can be obtained, the use of relatively vague terms to describe the results of detailed analysis not only is pointless but replaces a precise definition by a less precise description. Indeed, if the experimentally determined grain-size distribution of a sediment is used merely to obtain a vague descriptive title, it is largely a waste of time.

On this basis, as mentioned earlier (page 58), the techniques of sieving and pipette sedimentation do not lead to the most efficient or most meaningful measure of size. Measurement of a single component, such as quartz, in loose-grain mounts is infinitely superior and leads to size distributions which are relatively simple frequency curves and which can be rigorously described

TABLE 5.2
Grain-size Grade Scale[a]

Mesh[b]	mm	Phi	Grain type[c]	Rock type
	2,048–4,096	−11 to −12	VL boulder	Gravel or conglomerate
	1,024–2,048	−10 to −11	L boulder	
	512–1,024	−9 to −10	M boulder	
	256–512	−8 to −9	S boulder	
	128–256	−7 to −8	L cobble	
	64–128	−6 to −7	S cobble	
	32–64	−5 to −6	L pebble	
	16–32	−4 to −5	M pebble	
	8–16	−3 to −4	S pebble	
5	4–8	−2 to −3	VS pebble	
6	3.36–4.00	−1.75 to −2	Granule	
7	2.83–3.36	−1.5 to −1.75		
8	2.38–2.83	−1.25 to −1.5		
10	2.00–2.38	−1 to −1.25		
12	1.68–2.00	−0.75 to −1	VC sand	Sand or sandstone
14	1.41–1.68	−0.5 to −0.75		
16	1.19–1.41	−0.25 to −0.5		
18	1.00–1.19	0 to −0.25		
20	0.84–1.00	0.25–0	C sand	
25	0.71–0.84	0.5–0.25		
30	0.59–0.71	0.75–0.5		
35	0.50–0.59	1–0.75		
40	0.42–0.50	1.25–1	M sand	
45	0.35–0.42	1.5–1.25		
50	0.297–0.35	1.75–1.5		
60	0.250–0.297	2–1.75		
70	0.210–0.250	2.25–2	F sand	
80	0.177–0.210	2.5–2.25		
100	0.149–0.177	2.75–2.5		
120	0.125–0.149	3–2.75		
140	0.105–0.125	3.25–3	VF sand	
170	0.088–0.105	3.5–3.25		
	Microns			
200	74–88	3.75–3.5		
230	62.5–74	4–3.75		
270	53–62.5	4.25–4	C silt	Silt or siltstone
325	44–53	4.5–4.25		

TABLE 5.2 (*continued*)

Mesh[b]	mm	Phi	Grain type[c]	Rock type
	31.2–44	5–4.5		
	15.6–31.2	6–5	M silt	
	7.8–15.6	7–6	F silt	
	3.9–7.8	8–7	VF silt	
	1.95–3.9	9–8	C clay	Clay or shale
	0.975–1.95	10–9	M clay	
	0.487–0.975	11–10	F clay	
	0.243–0.487	12–11	VF clay	

[a] Modified from *Nat. Res. Council Rept. Comm. Sedimentation, 1941* (essentially Wentworth's scale, *J. Geol.*, **30:** 377–392 (1922).

[b] U.S. standard mesh retained on.

[c] VL, very large; L, large; M, medium, S, small; VS, very small; VC, very coarse; C, coarse; F, fine; VF, very fine.

by a few representative statistics. On the basis of elegance of treatment and meaningfulness of the measurements, i.e., their ease of interpretation, this experimental technique is far superior to indirect measurement.

It has generally been found that whatever technique is used to measure grain size, the resultant frequency distributions are highly skewed if the measuring scale is linear or based on equal intervals. One of the advantages of Wentworth's scale is that it is a ratio scale, and if each interval is considered equal, the skewness of many size distributions is materially reduced. This is equivalent to drawing a graph of the frequency distribution using a logarithmic scale for the size variable. However, statistical computation is difficult and laborious when using Wentworth's scale (Wentworth, 1929), and it is some advantage to convert the measurements from the arithmetic scale in, say, millimeters to a logarithmic scale in which the equal ratios of Wentworth's scale become equal arithmetic intervals. Logarithms to any convenient base would suffice, but Krumbein has designed a log transformation especially adapted to the Wentworth grade scale, and this, the phi scale, has been adopted for grain-size frequency distributions (Krumbein, 1938).

Krumbein made use of the $\sqrt{2}$ ratio of the Wentworth grade scale as follows:

$$d = 2^{-\phi} \tag{5.1}$$

where d is the diameter in millimeters; then

$$\log d = -\phi \log 2$$

$$d = -0.30103\phi \tag{5.2}$$

The diameters of the grains in most sediments are less than 1 mm; hence, the use of the minus sign permits fractions of a millimeter to be expressed as positive integers on the phi scale. As an example, suppose we wish to express 0.25 mm in phi units; then

$$\log 0.25 = \bar{1}.39794 = -0.30103\phi$$

$$\phi = \frac{-(\bar{1}.39794)}{0.30103} = \frac{-(-0.60206)}{0.30103}$$

$$= 2$$

A summary of the phi units equivalent to millimeter and sieve-mesh diameter values is given in Tables 5.2 and 5.3, and a more extended table is given in Griffiths and McIntyre (1958).

The phi scale will be used for size frequency distributions in the subsequent account so that the reader can become familiar with this scale; and, in general, comparison of size distributions by both graphic and statistical analysis will be performed on the basis of the phi scale. Increasing familiarity will enable the analyst to think in terms of the phi scale and will facilitate mathematical manipulation of size frequency distributions and the interpretation of results in terms of geological significance.

5.2 Graphical Expression of Measured Data

The tabular classification of frequency distributions does not readily lend itself to visual comparison of grain-size distributions, and the next step is to represent the frequency distribution as a graph. The basic two-dimensional form of the coordinates for this graph is derived from the frequency table (Table 5.1); the independent variable X, grain size in this case, is plotted along the horizontal scale, or abscissa; and the dependent variable Y, the frequency, is plotted on the vertical, or ordinate, axis. In most cases graphic scales are arranged to increase away from zero near an origin at the intersection of the axes, but if millimeters are used, this convention is reversed in graphing grain-size distributions, the millimeter values increasing toward the left, or origin, side of the abscissa scale. The phi scale, on the other hand, increases numerically away from the origin, but it should be remembered that because of the minus sign in the transformation equation (5.2) a numerical increase equals a decrease in grain size.

A large number of procedures have been proposed for the graphic expression of grain-size distributions, and new ones are still suggested from time to time (e.g., Doeglas, 1946). For most purposes the histogram is the simplest and most useful form of expression. In this case the class intervals are plotted along the abscissa, and the frequency in each class interval is represented by a column of appropriate height measured along the ordinate

TABLE 5.3
Transformation of Phi to Millimeters and Microns

Phi	0.000	0.125	0.250	0.375	0.500	0.625	0.750	0.875		Phi
−4	16.000	17.43	19.01	20.73	22.62	24.68	26.9	29.35		
−3	8.000	8.73	9.51	10.73	11.31	12.34	13.46	14.68		
−2	4.000	4.36	4.75	5.18	5.66	6.17	6.73	7.34	Millimeters	
−1	2.000	2.181	2.378	2.594	2.828	3.084	3.363	3.668		
−0	1.000	1.090	1.189	1.297	1.414	1.542	1.681	1.883		
	Microns	545	595	648	707	772	841	917	1,000	+0
		273	297	324	354	386	421	458	500	1
		136	149	162	177	193	210	229	250	2
		68.1	74.4	81.2	88.4	96.5	105.1	114.6	125	3
		34.1	37.2	40.6	44.2	48.3	52.6	57.3	62.5	4
		17.0	18.6	20.3	22.1	24.1	26.3	28.6	31.25	5
		8.53	9.23	10.1	11.0	12.0	13.1	14.3	15.67	6
		4.26	4.64	5.97	5.53	6.03	6.57	7.16	7.84	7
		2.13	2.32	2.53	2.76	3.01	3.28	3.58	3.92	8
		1.07	1.16	1.27	1.38	1.51	1.64	1.79	1.960	9
		0.522	0.580	0.634	0.691	0.754	0.821	0.895	0.980	10
		0.266	0.290	0.317	0.345	0.377	0.410	0.447	0.490	11
		0.133	0.145	0.158	0.173	0.188	0.205	0.224	0.245	12
		0.066	0.073	0.079	0.086	0.094	0.103	0.112	0.122	13
		0.033	0.036	0.039	0.043	0.047	0.051	0.056	0.061	14
Phi		0.875	0.750	0.625	0.500	0.375	0.250	0.125	0.000	Phi

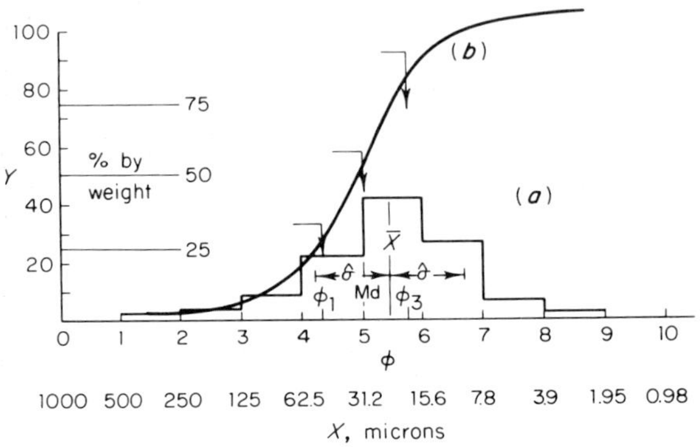

Figure 5.1 Frequency histogram (*a*) and cumulative frequency curve (*b*) of data in Table 5.1.

scale. It is conventional to make the areas of the columns proportional to the frequency in each class. It is necessary to follow a standardized procedure in preparing these graphs, as variations in scale will alter the appearance of the graph, and comparison of different grain-size distributions is then difficult. The procedure adopted by the author is to use standard graph paper[1] having 20 divisions to the inch on both ordinate and abscissa scales. The data from sieving and pipette sedimentation are expressed as weight per cent coarser than certain grain sizes, and the weight percent is plotted along the ordinate axes with 1 in. = 20 percent. Along the abscissa scale 1 phi unit equals 1 in., and this graph paper will then include 10 phi-unit intervals. The grain-size distribution of Table 5.1 is plotted as a histogram in Fig. 5.1*a* as an example.

The frequency polygon obtained by joining the midpoints of the histogram columns has not been used very often in sedimentary petrography. It represents the size frequency distribution by ordinates at the midpoints of the class intervals, and it possesses all the disadvantages of the histogram without any extra advantages.

If the class interval is changed, the appearance of the histogram is also changed; and as no standardized class intervals have been adopted for grain-size analysis, different authors have used different class intervals. This makes the visual comparison of histograms very difficult. In order to overcome this disadvantage the frequencies in succeeding classes are accumulated (final column, Table 5.1), and the cumulative histogram can then be plotted. This manner of graphic presentation is independent of changing class intervals, and in practice, the cumulative-frequency polygon based on the cumulative histogram and rounded to a smooth curve is now the most popular graphic device for representing grain-size distributions. As an example, the data of Table 5.1 are represented as a frequency histogram and cumulative curve in Fig. 5.1*a* and *b*.

Interpretation and comparison of cumulative size frequency distributions are relatively simple; thus, if the curve maintains its form but is translated

[1] Codex No. 318A, supplied by the Codex Book Company, Inc., Norwood, Mass.

along the abscissa, the translation reflects a change in grain size (curve 3 represents a finer-grained sediment than curve 2 in Figs. 5.2*a* and 5.3*a*). Similarly, a change in slope or steepness of the cumulative curve represents a change in "sorting" or uniformity of grain size (Figs. 5.2*b* and 5.3*b*; curve 3 represents a more poorly sorted or less uniform sediment than curve 1 or 2).

Despite this apparent simplicity of interpretation, attempts have been made to return to the frequency curve but to obtain a frequency distribution which is independent of class interval on the abscissa scale. Such attempts are based on the relationship between the frequency and cumulative-frequency curve: the latter may be looked upon as the integral of the former. Hence, to derive a frequency curve from the cumulative curve it is necessary to find its derivative by differentiation. Krumbein describes a graphic procedure by means of tangents to achieve this derivative curve, which is then called the *unique frequency curve* (Krumbein, 1934b). Brotherhood and Griffiths (1947) propose a mathematical procedure which achieves the same end.

The unique frequency curve differs from other graphic procedures principally in its ordinate-axis scale; the abscissa scale remains grain size, but the ordinate scale is now a measure of change in slope of the cumulative curve and not frequency (see Brotherhood and Griffiths, 1947). In very steep cumulative curves, i.e., very well sorted sediments, values of over 100 percent on the ordinate scale are common. The main advantage of the unique frequency curve is that the most frequent diameter, the modal grain size, is readily obtained by reading the diameter value corresponding to the maximum height of the curve. This, the commonest diameter, reflects the commonest current velocity; but its exact implications are vague because the frequency curve must be symmetrical for this diameter to be most meaningful, and grain-size distributions achieve symmetry only on a logarithmic scale. The complexity of procedure necessary to obtain the unique frequency curve militates against its general adoption, and the difficulty of returning to the physical significance of characteristics of the unique frequency curve suggests that it has very limited value in grain-size investigations.

When graphs have been used to represent grain-size distributions, the aim, intentional or unintentional, has been to obtain a reasonably simple curve, preferably unimodal and ideally a well-known frequency distribution such as the normal curve. The tendency for many size distributions of natural sediments to approach this ideal when size is plotted on a logarithmic scale is probably the basis for this concept. As a direct result a graphic procedure has been proposed for testing the normality of size distributions. In this case the abscissa scale may be logarithmic or, more conveniently, arithmetic, using the phi transformation, whereas the frequency is represented as a probability scale along the ordinate axes. Codex graph paper may be used with arithmetic probability scales (Codex No. 3127) or logarithmic probability scales (Codex No. 3128). In either case if the size distribution is normal, the graphic plot will be a straight line. Generally, the experimentally deter-

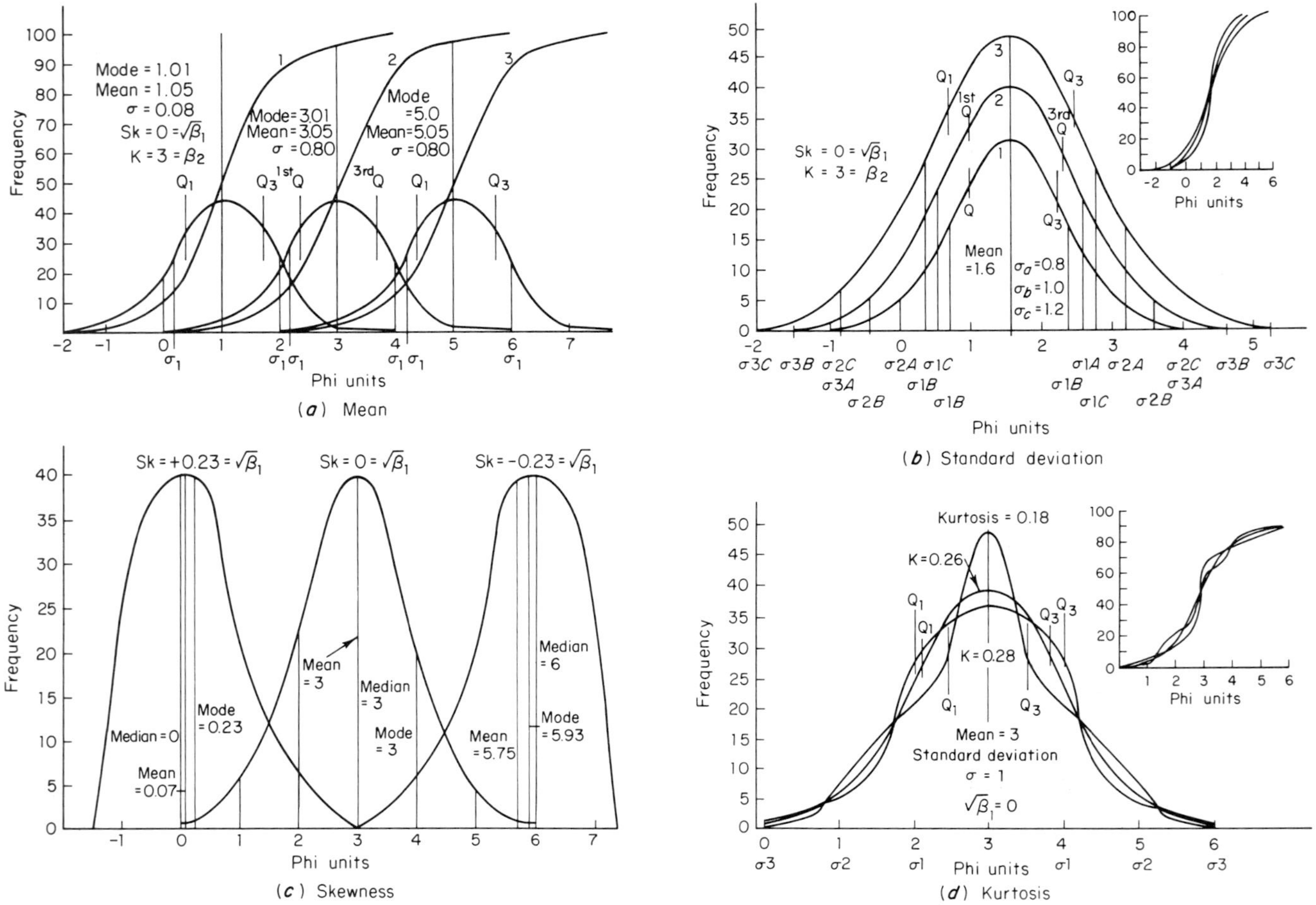

Figure 5.2 Frequency and cumulative frequency curves.

mined points do not lie exactly on a straight line, and the question of how much smoothing may be considered legitimate is left to the intuition of the analyst. As an illustration, the curves from Fig. 5.2 are plotted on arithmetic probability (phi) paper in Fig. 5.3.

A somewhat more rigorous approach has been described by Otto (1939), who attempts to reduce the finding of statistics from the size distributions plotted on arithmetic probability graph paper to an entirely graphic procedure. Doeglas (1946) also uses a probability frequency scale but with an arithmetic

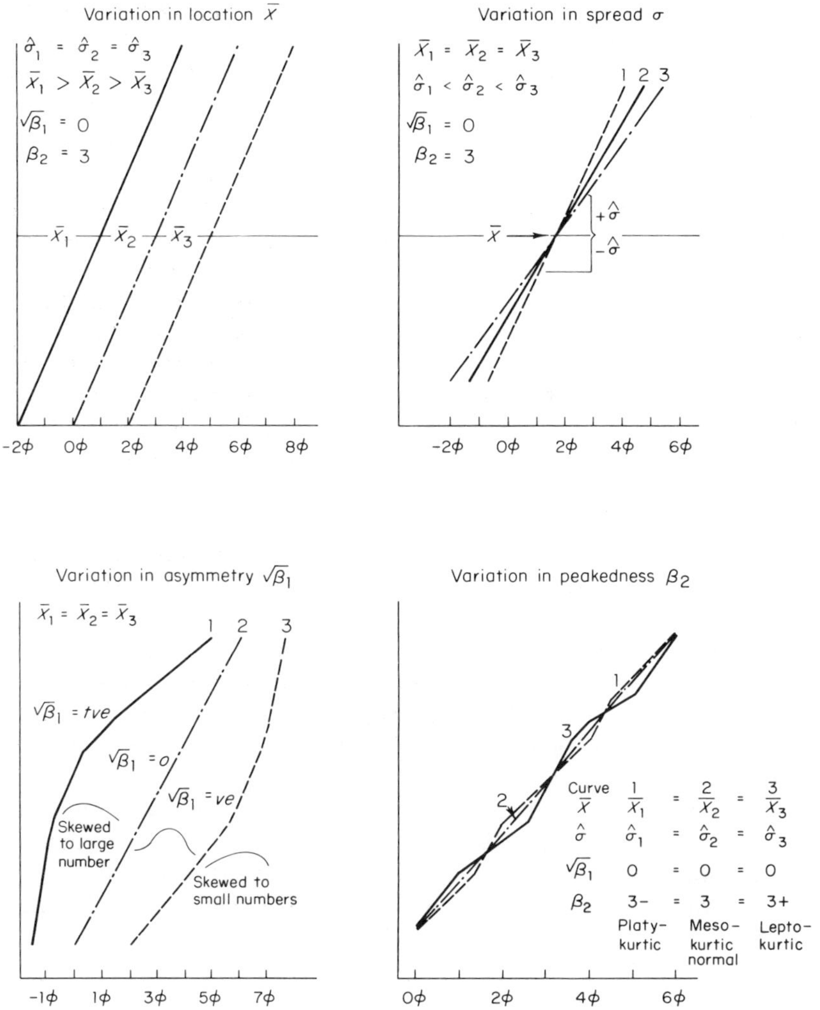

Figure 5.3 Cumulative frequency distributions on arithmetic-probability graph paper.

millimeter size scale to illustrate another approach to the interpretation of grain-size distributions of sediments.

One may conclude that there are a very large number of procedures advocated for expressing the grain-size distribution of sedimentary rocks, and, at least upon superficial examination, it would be difficult to recommend one in place of the other; indeed, each manner of expression adds more information, suggesting that it may be necessary to use all these approaches to interpret size distributions so that information of geological value results. Much of the difficulty has arisen because grain-size distributions of sediments are not simple unimodal distributions; e.g., when pipette sedimentation and sieving are used, the cumulative curves possess open ends. In most sediments the cumulative curve cuts the 10-phi ordinate at 90 percent or less, implying that the sample analyzed has appreciable quantities of material much finer-sized than 1 micron equivalent diameter. Attempts have been made to close this curve at its fine end; e.g., by using an arithmetic millimeter size scale along the abscissa, Doeglas (1946) reduces the fine end to obscurity. Griffiths (1952b) suggests closing the open end on the fine side by adjusting for the composition of the clay minerals present in the sediment. Unfortunately, no suggestion has achieved general adoption, so that the problem of the open end is still unsolved. For the present, therefore, the interpretation of size frequency distribution expressed as cumulative curves, as unique frequency curves, or as probability curves is likely to remain qualitative and largely subjective unless an appropriate change in analytical technique is adopted which results in a simple curve (see, for example, Griffiths 1962a).

5.3 Statistical Expression of Measured Data

A number of attempts have been made to introduce a rigorous and objective approach to the comparison and interpretation of grain-size frequency distribution by means of statistics. The main function of statistics at this stage in experimental analysis is to summarize the characteristics of the size frequency distributions; the statistics, to be of maximum use, should contain as much information as the graph or frequency table which they are intended to summarize. The development of statistics has, therefore, been compelled to follow and depend upon the selected graphic procedures.

Van Orstrand (1925) was among the first to attempt to reduce size distribution to frequency curves suitable for such analysis; he recommended the use of moment statistics. Later Wentworth (1929) proposed the adoption of moment measures as descriptive statistics for size frequency distributions, but both these approaches are cumbersome because they are based on the use of logarithmic class intervals. Hatch and Choate (1929) applied a statistical approach to the description of size distributions of artificially produced materials and were able to develop a relatively simple transformation from percent by weight to number of frequency. The curves were approximately normal in form. Trask (1932) applied quartile statistics to the reduction of data from size analysis and proposed the use of the median as an average diameter and the well-known Trask sorting coefficient *So* as a measure of uniformity of grain size. He also found it necessary to use a measure of skewness or asymmetry of the size frequency distributions (see Table 5.7). In this case both the logarithmic size scale and the open-endedness of the distributions were adequately overcome. Finally, the statistical approach was broadened and unified by the detailed presentation of Krumbein (1936a, 1939b) and Krumbein and Pettijohn (1938). By means of the phi transformation Krumbein (1936c) was able to simplify the computation of the quartile statistics and adopt conventional measures for the median, quartile deviation, skewness, and kurtosis. Subsequently, and again largely through the use of the phi scale, Krumbein (1936a) was able to apply moment measures to size frequency distributions in those cases where the curves were not open-ended. During and after this period of evaluation of statistical devices a number of authors attempted to invent statistics for specific purposes which would avoid some of the disadvantages contained in the more conventional approach; needless to say, these "statistics" have not proved very popular (Baker, 1920; Niggli, 1935; Griffiths, 1951a).

Once more we may note the existence of a plethora of statistical devices which are available to analysts of grain-size frequency distributions. A summary of those most likely to be met in the literature concerned with grain-size analysis is given in Table 5.7.

In general, the incomplete data which lead to open-ended size frequency distributions compel us to use quartile statistics, but these measures serve to characterize only the central part of the distributions. When it is required to extend the statistics to cover a greater portion of the curve, it becomes necessary to use percentiles (Griffiths, 1951a; Inman, 1952; Folk and Ward, 1957). Ultimately, if the graphs imply a very irregular size distribution, no statistical summary is sufficient, and the best that can be achieved is some graphic plot. Inman (1952) has attempted to clarify the mathematical background of quartile and percentile statistics and relate them to the corresponding moment measures. He proposes the use of five diameters as a basis for the characteristics of size frequency distribution determined by sieving and pipette sedimentation.

It seems clear that additional manipulation of the statistics now in use in an attempt to obtain more satisfactory summaries of size frequency data will suffer from the law of diminishing returns. It can be deduced from the very detailed presentations of Krumbein and more recent evaluations by Inman (1952), Folk and Ward (1957), McCammon (1962a, b), Friedman (1962), and Middleton (1962) that the difficulties in developing adequate summary statistics arise from the complex nature of size frequency distributions. This, in turn, is largely a function of the technique of analysis; the open-ended character and skewness of many size distributions, for example, arises largely from the presence of a clay fraction, and the "size" of this clay fraction is determined by the dispersion and disaggregation procedure used to reduce the rock sample to its individual components. In fact, the "size" measure obtained from sieving and pipette sedimentation [$P = f(m, s, sh)$; see Chap. 3] is largely responsible for the difficulties arising in the graphic presentation and statistical reduction of these data. Two features of experimental analysis appear to have been obscured in this approach: first, the size measure obtained cannot be related to the concept of size and only empirically to the geological objective for which the analysis is performed; second, the complex nature of the size distribution arises because the material is complex in composition. The techniques fail principally because they are not designed to embrace polycomponent mixtures.

On the other hand, if the objective of the analysis is to be achieved and homogeneous size frequency distributions are to be obtained, the material analyzed must be homogeneous to start with, and the technique adopted should be designed to characterize this homogeneous distribution. Screen analysis of crushed slate, limestone, and various other one-component systems produces a close approximation to a normal distribution (Hatch and Choate, 1929). Sieving a disaggregated quartzite also achieves this desirable end (Krumbein, 1935; Rosenfeld, Jacobsen, and Ferm, 1953), but sieving a polycomponent material cannot but result in a nonuniform size distribution. Once this is realized, the advantage of size analysis by means of loose grains is readily understood (page 66).

A homogeneous material leads to a homogeneous size distribution if the technique is satisfactory; e.g., size analysis of quartz grains in thin section tends to produce a normal frequency distribution, but the failure to measure fine sizes below 31.2 microns (6 phi units) in diameter leads to a cutoff in the frequency distribution and therefore to a skewness which is characteristic of the technique, not the material. On this basis, as argued in earlier chapters, we are led to the use of loose-grain size analysis. The size frequency distributions obtained from this procedure are homogeneous; they can be truly represented by a histogram and can be adequately characterized by conventional moment statistics. Furthermore, these statistics possess the qualities most desirable in analytical tools (Fisher, 1948; Griffiths, 1955a); they are consistent and unbiased in the sense that as the number of measurements increases, the statistics approach their respective population parameters. They are also more efficient in that they are more precise for any given number of measurements than any other statistics which may be used. Finally, if the frequency distribution is a close approximation to a normal distribution, they are also sufficient statistics, i.e., no other statistics are necessary to characterize the frequency distribution. If these criteria of consistency, efficiency, and sufficiency had been applied to most of the statistics so far proposed for the characterization of size frequency distributions, the statistics would have been rejected at the outset.

If we now adopt a technique for grain-size analysis which yields homogeneous size distributions and subsequently characterize these distributions by means of moment statistics which are consistent, efficient, and sufficient, we may exercise an additional privilege by testing these statistics for sampling error. Thus, random samples from some population of sedimentary grains in which the size distribution is normal will yield nonnormal size distributions due to accidents of sampling. It is necessary, therefore, before attempting to offer physical interpretation of a statistic, to establish that its magnitude significantly exceeds the value which may arise from errors of sampling (Griffiths, 1955a, 1959b; Milner, 1962).

5.4 Computation of Statistics Based on Moment Measures

The development of statistics based on moment measures has an elaborate and thoroughly ventilated mathematical background (Fisher, 1948, pp. 70ff.). Three objectives are involved. First, the procedure should result in a small set of characteristics or statistics which adequately describe the frequency distribution. The fulfillment of the criteria of consistency, efficiency, and sufficiency ensure that this objective is attained. Second, a means must be developed for testing the significance of the statistics to ensure that their magnitude is not merely a result of accidents of sampling; a modification of the procedure used to compute statistics from the moments has been evolved by Fisher (1948, pp. 73ff.) to fulfill this requirement. The exact steps are described in a subsequent chapter after description of the normal distribution and introduction of tests of significance (Chap. 13).

Finally, although we are immediately concerned with size frequency distributions, it is some advantage to adopt a procedure which is generally applicable to all frequency distributions and is independent of the property measured. The succeeding procedure achieves all three of these objectives.

The mean in a symmetrical distribution is the position of the center of gravity of the distribution, and the mean is the first moment; each succeeding moment can be assigned an analogous physical significance.

Moments may be defined as (Rider, 1939)

$$m_k = \frac{1}{n}\Sigma(X_i - \bar{X})^k \tag{5.3}$$

and as the first[1] moment m_1 is the mean,

$$m_1 = \frac{1}{n}\Sigma(X_i - \bar{X}) = 0$$

[1] The zero moment $m_0 = (1/N)\Sigma(X_i - \bar{X})^0$ is the area of the curve.

In practice it is convenient to compute the moments around an arbitrary origin, usually a class midpoint selected in the neighborhood of the mean (Griffiths, 1962a). It is then necessary to transform from moments n computed around the assumed mean to moments m around the true mean[1]

$$m_1 = 0 \tag{5.4}$$

$$m_2 = n_2 - n_1^2 \tag{5.5}$$

$$m_3 = n_3 - 3n_2n_1 + 2n_1^3 \tag{5.6}$$

$$m_4 = n_4 - 4n_3n_1 + 6n_2n_1^2 - 3n_1^4 \tag{5.7}$$

A further adjustment is necessary if, as is usual, the grouped data from a frequency table are used as a basis for moment calculation. In this case it is necessary to introduce a correction for class interval c, and the formulas become[2]

$$m_1 = \bar{X} = \bar{X}' + cn_1 \tag{5.8}$$

$$m_2 = s^2 = c^2(n_2 - n_1^2) \tag{5.9}$$

$$m_3 = c^3(n_3 - 3n_2n_1 + 2n_1^3) \tag{5.10}$$

$$m_4 = c^4(n_4 - 4n_3n_1 + 6n_1^2n_2 - 3n_1^4) \tag{5.11}$$

The mean and variance are equal to the first and second moments, respectively, but it is conventional to reduce the third and fourth moments to obtain measures of skewness and kurtosis. One commonly used approach is to transform the moments from powers to linear units by dividing by the second moment or variance raised to an appropriate power as follows:[3]

$$\sqrt{\beta_1} = \frac{m_3}{m_2^{\frac{3}{2}}} = \frac{m_3}{\sigma^3} \tag{5.12}$$

$$\sqrt{\beta_2} = \frac{m_4}{m_2^2} = \frac{m_4}{\sigma^4} \tag{5.13}$$

In a normal distribution $\sqrt{\beta_1} = 0$ and $\beta_2 = 3$, and following Krumbein and Pettijohn (1938, p. 251), the skewness Sk and kurtosis K may then be

[1] The derivation of these formulas can be found in most statistics textbooks, e.g., Rider (1939), Johnson (1949), and Hoel (1947).

[2] Krumbein avoids this correction by using a class interval of 1 phi unit (Krumbein and Pettijohn, 1938, table 29, p. 251), but for other class-interval units it is necessary to include this correction.

[3] The sample estimates are designated as b_i, and they estimate the appropriate population parameters β_i.

expressed as

$$Sk = \frac{\sqrt{\beta_1}}{2} = \frac{m_3}{2\sigma^3} \tag{5.14}$$

$$K = \beta_2 - 3 = \frac{m_4}{\sigma^4} \tag{5.15}$$

The statistics $\sqrt{b_1}$ and b_2 may be used in a test for significance (Johnson, 1949, p. 151), or as an alternative, Fisher's g statistics may be derived and tested for significance (see Griffiths, 1955a; Milner, 1962). The g statistics may also be used as measures of skewness g_1 and kurtosis g_2, and the only apparent disadvantage is the lack of suitable reference background of values expressed in terms of g statistics (see Table 5.5).

The actual procedure for computing the moment statistics is best performed in an organized manner, and a tabular guide, together with a worked example, is included in Table 5.4.

The first step is to compute the powers using an assumed midpoint (see totals of Table 5.4). The last column [entitled $f(d - 1)^4$] is included as a check on the calculations (see step 2 of Table 5.4).[1] The moments around the assumed mean n_{1-4} are then found (step 3, Table 5.4) and corrected to moments m_{1-4} around the true mean (step 4, Table 5.4). The subsequent steps leading to the mean $\bar{X}$, the variance $\hat{\sigma}^2$, standard deviation $\hat{\sigma}$, skewness Sk, and kurtosis K are sufficiently obvious to need no further explanation.

5.5 Examples of Grain-size Frequency Distributions

The size of particles in an aggregate, such as a sedimentary rock, is a measurement consisting of a number on some scale, and, as already mentioned, different measurement procedures lead to different numbers and even different scales (see Sec. 12.5). It is therefore necessary at the outset to realize that the level of information contained in the measurement is not always the same, some techniques yielding more precise and more meaningful measurements than others. As a preliminary statement it seems likely that size measurement by indirect procedures, when obtained from a polycomponent aggregate (population), achieves at best an ordinal[2] level, and order statistics are advisable, if not mandatory. On the other hand, size measurement performed in thin section may achieve interval[2] level; and when direct measurement is used, the measurement may achieve a ratio[2] level.

The frequency model generally used for grain-size frequency distributions is the logarithmic-normal distribution (Krumbein, 1938); and, while considerable controversy exists concerning the applicability of this model, it may be

[1] Generally known as the Charlier check (Yule and Kendall, 1948, p. 156).
[2] These terms are defined in Sec. 12.5.

TABLE 5.4
Computation of Statistics from the Moments of a Frequency Distribution

X = midpoint of each class $\quad f^a$ = frequency

$\bar{X}$ = arithmetic mean $\quad \bar{X}'$ = assumed mean = 5.5

c = class interval = 1 phi $\quad d = \dfrac{X - \bar{X}'}{c}$

Class limits, phi units	X_ϕ	f	d	fd	fd^2	fd^3	fd^4	$f(d-1)^4$
1–2	1.5	1	−4	−4	16	−64	256	625
2–3	2.5	2	−3	−6	18	−54	162	512
3–4	3.5	6	−2	−12	24	−48	96	486
4–5	4.5	21	−1	−21	21	−21	21	336
5–6	5.5	40	0					40
6–7	6.5	25	1	25	25	25	25	0
7–8	7.5	4	2	8	16	32	64	4
8–9	8.5	1	3	3	9	27	81	16
1 Total		100		−7	129	−103	705	2,019

2 Charlier check on totals

$$\Sigma f(d-1)^4 = \Sigma fd^4 - 4\Sigma fd^3 + 6\Sigma fd^2 - 4\Sigma fd + \Sigma f$$

$$2{,}019 = 705 + 412 + 774 + 28 + 100$$

3 Moments around the arbitrary origin

$$n_1 = \frac{\Sigma fd}{\Sigma f} = -0.07 \qquad n_2 = \frac{\Sigma fd^2}{\Sigma f} = 1.29$$

$$n_3 = \frac{\Sigma fd^3}{\Sigma f} = -1.03 \qquad n_4 = \frac{\Sigma fd^4}{\Sigma f} = 7.05$$

4 Moments around the true mean corrected for grouped data

$$m_1 = \bar{X} = \bar{X}' + cn_1 = 5.5 - 0.07 = 5.43$$

$$m_2 = \hat{\sigma}^2 = c^2(n_2 - n_1^2) = 1.29 - 0.0049 = 1.2851$$

$$\hat{\sigma} = \sqrt{\hat{\sigma}^2} = \sqrt{1.2851} = 1.133$$

$$m_3 = c^3(n_3 - 3n_2n_1 + 2n_1^3) = -1.03 + 3 \times 1.29 \times 0.07 - 2 \times 0.000343$$

$$= -0.7598$$

$$\sqrt{b_1} = \frac{m_3}{\hat{\sigma}^{3/2}} = -\frac{0.7598}{1.4544} = -0.5224$$

$$Sk = \frac{\sqrt{b_1}}{2} = -0.262$$

TABLE 5.4 (*continued*)

Class limits, phi units	X_ϕ	f	d	fd	fd^2	fd^3	fd^4	$f(d-1)^4$

$$m_4 = c^4(n_4 - 4n_3n_1 + 6n_1^2n_2 - 3n_1^4)$$

$$= 7.05 - 4 \times 1.03 \times 0.07 + 6 \times 0.0049 \times 1.29 - 3 \times 0.00002401$$

$$= 6.7994$$

$$b_2 = \frac{m_4}{\hat{\sigma}^4} = \frac{6.7994}{1.6512} = 4.117$$

$$K = b_2 - 3 = 4.117 - 3 = 1.117$$

5 Summary statistics

$$\bar{X}_\phi = 5.43 \qquad \hat{\sigma}_\phi = 1.133$$

$$\sqrt{b_1} = -0.522 \qquad b_2 = 4.117$$

[a] Data from size frequency distribution of Table 5.1.

argued on theoretical grounds and on the basis of empirical data that this is the best approximation to size frequency distribution (see, for example, Sec. 13.12). To achieve a normal distribution it is customary to transform the variate, grain size, to the phi scale, and it is also necessary to sample the population correctly and to perform the measurement in a rigorously prescribed manner (see Griffiths, 1962a). Any departure from these recommendations, introduced either explicitly or implicitly, will lead to departures from the model; e.g., the logarithm of radii of oranges of a single kind may well be log-normal, but if oranges of different kinds are mixed in different proportions and sampled, the resulting frequency distribution will almost certainly be polymodal. If, in addition, a few bananas are added, the frequency distribution of samples will vary almost uncontrollably; here, size, shape, and composition are varying, and this variation is confounded in the so-called "size measurement." The case of sedimentary rocks, or indeed any polycomponent aggregate, is strictly analogous. In the succeeding account, therefore, the log-normal model is used as a basis, and deviations from the model are related to variation in measurement procedure and to departures from the ideal of random sampling. Statistics, to be meaningful, must be tested for significance; if they cannot be so tested, attempting a physical interpretation of their variation is to attempt the impossible; only very vague and ill-defined conclusions can be based on such statistics. It will be observed that in most attempts to relate such statistics as order statistics, medians, quartiles, and percentiles, the basic model against which variation in the order statistics is compared is the normal model (Inman, 1952; Folk and Ward, 1957; McCammon, 1962a, b); it is not, in general, possible to make statements about statistics which cannot be related to any known frequency-distribution model.

Adopting the normal model for size measured in phi units permits the use of the moment measures as relevant summary statistics; and, in the following accounts, the four moment measures will be expressed in terms of the parameters arithmetic average μ, variance σ^2 or standard deviation, σ and the Pearsonian beta statistics: $\sqrt{\beta_1}$ as a measure of asymmetry, or skewness, and β_2 as a measure of peakedness, or kurtosis. The statistical estimators of these parameters are $\bar{X}$, $\hat{\sigma}^2$, $\hat{\sigma}$, $\sqrt{b_1}$, and b_2 respectively; as a first step, the estimates of the beta statistics will be tested for significance, and it is known that in a normal distribution $\sqrt{\beta_1} = 0$ and $\beta_2 = 3$. Departures from these parametric values arising from errors of random sampling are specified in tables (e.g., Pearson and Hartley, 1954, tables 34b, c, pp. 183–184), and the significance of the estimates b_i of the parameters β_i are readily determined for any given sample size by means of these tables. An analogous procedure using the Fisher g_i statistics as estimators of the γ_i parameters is described in Griffiths (1955a) and in Milner (1962, p. 578).

5.6 Size Frequency by Indirect Measurement

When sieving and/or pipette sedimentation technique is used to obtain a size measurement, the independent variable X is the logarithm (or phi equivalent) of the measure obtained, and the dependent variable or frequency Y is usually weight percent. The measurement is expressed as $P = f(m, s, sh)$, and the average grain size represents the average grain size of a gram of sediment. Quite obviously, the measure varies with variation in the gram of sediment, and one of the main difficulties in using this as a measure of grain size is that the sample size is artificial, a percent by weight; in such a case with sample size n unspecified, the degrees-of-freedom concept is difficult to evaluate, and so tests of significance are impossible to perform. The "statistical estimators" obtained are therefore obscure, and while some attempts have been made to interpret variation in skewness and kurtosis of the sample size distributions empirically, the results are rather vague and hardly justify the experimental costs. This criticism will apply to any size measurement aimed at relating "size" to geological or genetic features and performed by indirect procedures on polycomponent mixtures.

This is, nevertheless, the most common practice, and in an extensive study of size and sorting in sediments an attempt was made to resolve the problem by using the "large-sample" approach; the medians of 6,976 grain-size analyses taken from the literature and the writer's files yield the frequency histogram of Fig. 5.4. The frequency distribution appears to be unimodal and, at least approximately, symmetrical. By using a class interval of 0.5

phi unit, moment analysis yields $\bar{X} = 3.339$ phi, $\hat{\sigma} = 2.422$ phi, $\sqrt{b_1} = 0.0168$, and $b_2 = 5.241$.

As a first step the b_i may be tested for significance by comparison with the Pearson and Hartley tables (1954); the degrees of freedom here can be obtained from the number of observations in the sample, $n = 6.976$. The value of $\sqrt{b_1}$ is not significantly different from zero; hence $\sqrt{\beta_1} = 0$ is an acceptable value, and the frequency curve is symmetrical. The peakedness estimated as $b_2 = 5.241$ is, however, highly significant, and the curve is leptokurtic; i.e., $\beta_2 \neq 3$, which rejects the normal model. It is now necessary to examine the nature of the deviations from the model; for this purpose the normal curve with parameters $\mu = 3.339$ and $\sigma = 2.442$ was calculated (as described in Sec. 13.5), and this normal curve is superimposed on the histogram in Fig. 5.4. The class-by-class deviations of the observed histogram from the theoretical normal curve can then be tested by chi square (see Sec. 17.4), and, of course, they are highly significant. There is an excess of observed over expected in the size ranges beyond 10, beyond -2, and between 1 to 4 phi units; there is a deficiency of observed compared with expected in the size ranges 1 to -2 and 4 to 10 phi units.

There are a number of possible explanations for these discrepancies; e.g., it has been suggested that there is a paucity of fine gravel in the population of

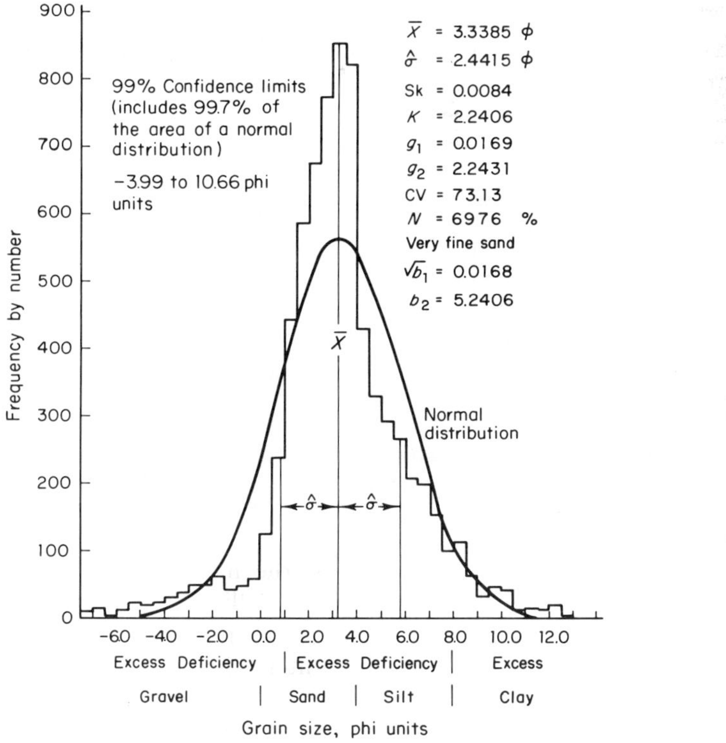

Figure 5.4 Size frequency distribution by sieving and pipette sedimentation; medians of various detrital sediments.

particles in sedimentary rocks (Pettijohn, 1957, p. 47), which may account for the deficiency in the range 1 to -2 phi units. However, it is most remarkable that there is a deficiency in the silt sizes (4 to 8 phi, or 62.5 to 3.9 microns), because there is certainly no lack of silt sizes in the population of particles in sediments. It appears more likely that the excesses in the clay, sand, and coarse-gravel sizes reflect the emphasis of petrographic investigation using x-ray analysis, the optical microscope, and individual megascopic analysis, respectively; whereas, where no suitable technique exists, in the size ranges between those suited to x-ray and microscopic and between microscopic and individual handling procedures, the existing sediments have been neglected. If this explanation is correct, the sampling bias reflected in the peakedness is to be attributed to operator selectivity and is no reflection of nonnormality in the size frequency distributions; at least for the present, this is the most convenient explanation. It is also clear that if this explanation is correct, there is little point to adding further analyses to this collection because all that will result is a greater deviation in the peakedness and greater emphasis on the bias.

On the basis of this moderately large sample, therefore, we may conclude that the size frequency distribution of medians of sediment samples, where size is determined by indirect procedures, is effectively log-normal. The mean $\bar{X}$ is therefore an unbiased estimator of the population mean μ of all detrital sediments, and μ is likely therefore to be closely approximated by 3.339 phi units, about 100 microns. The absence of asymmetry suggests that increasing sample size will not materially change this value. The standard deviation, on the other hand, is likely to be an underestimate, because "errors of estimation" in the variance are strongly affected by deviations in the fourth moment. Nevertheless, if we accept this value as a basis for argument, it can be stated that $\mu \pm 3\sigma$ yields a range in size from -3.99 to 10.66 phi units (approximately 8 mm to 0.5 micron), and it may be stated that we expect about 3 medians in every 1,000 to lie beyond this range (99.7 percent of the area of a normal curve is included in the $\pm 3\sigma$ range about the mean; see Sec. 13.7). This then presents a possible approximation to the size frequency distribution of sediments when size is measured by sieving and pipette sedimentation.

5.7 Size Frequency Distribution by Measurement in Thin Section

Measurement of size frequency distribution in thin section differs quite markedly from measurement by indirect technique, both in the nature of the independent variable, which is *apparent length* (or intercept), and in the nature of the dependent variable, which is generally a number frequency. The independent variable, apparent length, may be defined as $P = f(s, sh, o, p)$, and then it will be seen that this measure contains confounded information from variation in size, shape, orientation, and packing as well as from their various interactions. Here, again, the measure is difficult to specify in terms of the concept of size; and while it appears to be somewhat more directly related to the concept than the variable obtained from indirect measurement, its level of information is unlikely to be better than interval scale level. On the other hand, the sample size n is specified, and the concept of degrees of freedom for significance testing is easily applied.

Using the empirical approach, we fortunately have a moderately large number of analyses to which we can refer; and subject to minor variations, largely attributable to differences in the exact procedural steps, the results of thin-section size measurement are quite consistent. The size limitations are quite serious, and, on the fine side, measurement of grains of diameters less than 5 phi units (31.2 microns) is exceedingly difficult, generally because the presence of matrix in the finer-grained sediments tends to obscure the grain boundaries. On the coarse side, the grains may well exceed the microscope field, and here again there is some difficulty in measurement. For quartz grains in the sand sizes the technique is simple and effective. A number of other limitations have been described (e.g., Krumbein, 1935; Chayes, 1950a, b, 1956a, b; Elliott, 1952; Friedman, 1958).

A large-sample approach may again be attempted. By using a single sedimentary rock, the Salt Wash member of the Morrison formation from the Colorado plateau (Griffiths et al., 1956), as parent population, some 15,000 measured axes of quartz grains were collected by three operators (actually three teams of five operators each) from 50 thin sections, using the sampling arrangement of 300 grain axes per thin section, patterned as 10 grains in each of 10 fields by three operators. The resulting frequency histogram is shown in Fig. 5.5, and the relevant statistics are

$$\bar{X} = 2.758; \qquad \hat{\sigma} = 0.662; \qquad \sqrt{b_1} = -0.076; \qquad b_2 = 3.007$$

Upon testing the b_i for significance, we find that $\sqrt{b_1}$ is significant and b_2 is not; here, then, $\sqrt{\beta_1} = 0$ is rejected, but $\beta_2 = 3$ is accepted. Again, assuming that the parent population is phi-normal, we can calculate the expected normal frequency distribution with $\mu = 2.758$ and $\sigma = 0.662$, and the relevant histogram is superimposed on the observed histogram in Fig. 5.5. The deviations of observed from expected are also inserted; no large deviations, and therefore no consistent bias, are evident. It appears likely, therefore, that this is again the result of sampling errors and, in particular, reflects the difficulty entailed in sampling from a stratified population without adequate definition of the strata. If we accept that the parent population is phi-normal, then to achieve random sampling there is no specific recommendation other than to take more samples of the same kind; i.e., if more were taken, it would be expected, when appropriate weighting of the strata is attained, that $\sqrt{b_1}$ would converge on $\sqrt{\beta_1} = 0$. The value of 0.076 is not large in an absolute sense, so that perhaps not many additional samples would be necessary; but few investigators would contemplate sample sizes in excess of 15,000. It is obvious, therefore, that if the population distribution is normal, a very carefully designed sampling plan is necessary to obtain unbiased sample estimators for small samples of the size $n = 100$ to 300.

A number of measurements of the long a axis of quartz grains of various sandstones measured in thin sections are presented in Table 5.5 with their accompanying sampling arrangements; nearly all the examples show evidence of biased sampling in possessing significant asymmetry $\sqrt{b_1}$ or peakedness b_2, and these several biases are largely to be attributed to difficulties in achieving random samples from stratified populations wherein the strata limits are not defined. The ideal sampling pattern appears to require measurements on small sets of contiguous quartz grains, say 4 to 6, taken as, say, 10 locations in each thin section, and a large number, around 30 to 50 in the average case, of thin sections; fulfillment of these requirements tends to reduce the bias arising from incorrect weighting of the strata (see also Sec. 17.3).

The mean size varies from 0 to 4 phi units, as would be expected of sand-size quartz grains; the standard deviation varies from 0.3 to 1.5 phi units: well-sorted quartz-grain long axes have values near 0.3 unit and the most poorly sorted sandstones have standard deviations in excess of 1.0 phi

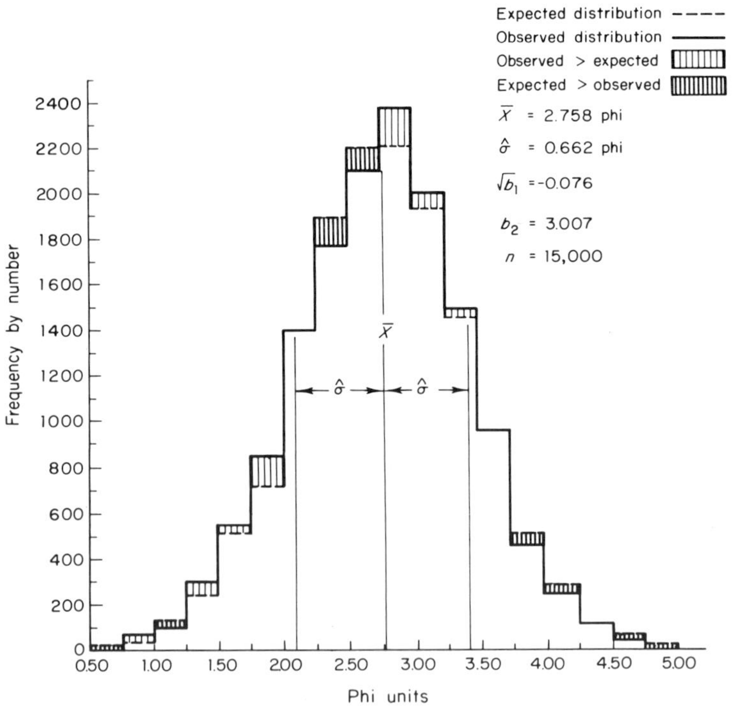

Figure 5.5 Size distribution in thin section; quartz grain axes from Salt Wash sandstone, Morrison formation, Colorado plateau. (*After Griffiths et al., 1956.*)

unit. It should also be observed that local sampling of quartzites, such as the Oriskany, yields standard deviations ranging from 0.3 to 0.8 phi unit; i.e., the quartzites are well to moderately sorted, whereas the formation as a whole is poorly sorted ($\hat{\sigma} = 1.0$ phi unit), and this is true of quartz grains from most formations. The range of quartz-grain size may therefore be estimated as some three standard deviations on either side of the mean, or from 0 to 6 phi units around a mean of 3 phi. The sampling arrangement required to estimate the standard deviation within specified confidence limits is calculated in Sec. 17.3, using the range 0.3 to 1.5 phi units.

Another feature of considerable interest in sampling sediments for the measurement of grain size is that in most cases the mean is related to the dispersion when indirect measurement is used (Griffiths, 1951b; Inman, 1949; and Chap. 15) but is independent in the majority of examples when the lengths of quartz grains are measured in thin section (e.g., Griffiths, 1958b). In random samples from a normal distribution the mean is independent of the variance and standard deviation, and this is a requirement which should be achieved in sampling sediments if both estimators are to be used to characterize the size properties of grains in sedimentary rocks.

5.8 Size Frequency Distribution by Direct Measurement

There is an intuitive appeal in a simple direct-measurement procedure, and this has ample justification from developments in many physical sciences; rods and clocks yield basic measurement data which are precise and generally accepted as accurate. Size, by direct measurement of axes of isolated individual grains, offers a possibility of relating the measure directly to the concept as $P = f(s,sh)$; and, as mentioned earlier, this measurement procedure yields data which attain a ratio level of information. In addition, it is likely that the variable is in direct one-to-one relationship with current velocity (see Chap. 13) so that it may permit interpretation in terms of energy of environmental processes and so prove to be a powerful tool in the petrologic interpretation of grain-size data. This procedure has been developed to the stage where all sizes, from at least coarse silt to boulders, can be analyzed by the same technique, which is a very considerable advantage. If we confine attention to quartz grains, by far the largest part of the grain-size range is likely to be comprehended by this technique.

Pye (1943) was among the earliest to recommend this procedure, and many workers (for references see Pettijohn, 1957, pp. 51–54) have analyzed gravelly deposits in this way. Hulbe (1955) developed an alternative technique which permits measurement of shape and size simultaneously, thus leading to the evaluation of $P = f(s,sh)$ as a "clean" measurement; i.e., it is possible to segregate variation of shape and size by mathematical analysis (Griffiths and Smith, 1964). Measurement of simultaneous variation in size and shape supplies a very sensitive set of data for differentiating environments (Hulbe, 1957; Biederman, 1959; McIntyre, 1961).

TABLE 5.5
Long *a* Axis of Quartz Grains in Sandstones; Measured in Thin Section

Horizon	outcrop	$\bar{X}\phi$	$\hat{\sigma}\phi$	$\sqrt{b_1}$	b_2	Quartz, %	Specimens	Grains	Total	Locality	Rock type
Tuscarora[a]	T_1	1.948	0.532	0.148	3.457**	77.9	20	25	500	Pa.	Quartzite
	T_2	1.836	0.454	0.008	3.764**	77.9	20	25	500	Pa.	Quartzite
Tuscarora[b]		1.823	0.966	−0.279**	2.546†		46	25	1,175	Pa.	Quartzite
Homewood[a]	H_1	1.080	0.673	−0.244*	3.203	70.1	20	25	500	Pa.	Quartzite
	H_2	1.535	0.558	0.109	3.117	70.1	20	25	500	Pa.	Quartzite
Oriskany[a]	O_1	1.586	0.367	0.068	3.886**	94.1	20	25	500	Pa.	Quartzite
	O_2	1.381	0.500	0.433	3.487*	94.1	20	25	500	Pa.	Quartzite
Oriskany[c]		1.659	1.040	−0.075**	3.650**	87.0	87		9,743	N.Y.–Ky.	Quartzite
St. Peter[d]		1.833	0.906	0.291**	2.717		13	40	520	Ottawa, Ill.	Quartzite
Baraboo[d]		1.798	0.824	−0.527	3.804		13	40	520	Wis.	Quartzite
Sharon[d]		0.565	0.861	−0.406**	4.119**		17	20	340	Ohio	Quartzite
Cow Run sand[e]		1.240	0.703	0.196**	3.494**	39.7	38	40	1,520	W.Va.	Low-rank graywacke
Maxton sand[d]		1.461	0.631	−0.076†	3.612**	51.1	33	40	1,320	W.Va.	Low-rank graywacke
Weir sand[d]		3.650	0.37	0.301**	2.920**	51.1	19	40	760	Ky.	Low-rank graywacke
Chipmunk sand[f]		2.760	0.660	−0.076**	3.007	53.5	71	35	2,485	N.Y.–Pa.	Low-rank graywacke
Chipmunk sand[g]		2.990	0.557	0.099	3.382	55.3	48	20	960	N.Y.–Pa.	Low-rank graywacke

Middle Kittanning[h]	1.975	0.837	1.228**	2.333**	43.7	75	20	1,500	Pa.	Low-rank graywacke
Pocono[i]	2.870	1.070	0.068	2.555**	62.2	16	50	800	Pa.	Low-rank graywacke
Berea[i]	3.370	0.720	0.402**	3.700**	63.2	16	50	800	Ohio	Low-rank graywacke
Pottsville[b]	0.892	0.836	−0.225**	3.231†		54	25	1,348	E. Pa.	Low-rank graywacke
Salt Wash[j]	2.758	0.662	−0.075**	3.007	61.0	50	300	15,000	Colo. plateau	Arkosic quartzite
Lyons[k]	2.359	0.811			48±	32	36	1,152	Colo.	Arkose
Fountain[k]	2.307	0.914			45±	49	36	1,764	Colo.	Arkose
Waynesburg[h]	1.452	1.024	1.228	2.333	49.8	75	20	1,500	Pa.	Arkose
Paleozoic "Grits"[d]	1.888	1.111	0.223	3.201	27.6	7	40	280	United Kingdom	High-rank graywacke

Note: Throughout the book the following convention is used: * means significant at the 5 percent probability level; ** means significant at the 1 percent probability level; *** means significant at the 0.1 percent probability level; † means not significant.

[a] Wood, 1960. [b] Landy, 1956. [c] Rosenfeld, 1953. [d] Griffiths, unpublished. [e] Griffiths, 1958b. [f] Stanonis, 1956. [g] Packman, 1961. [h] Siegrist, 1961. [i] Emery, 1954. [j] Griffiths et al., 1956. [k] Hubert, 1958.

Here, again, the sample size n is directly determinable, and so the concept of degrees of freedom becomes operational, and the statistical estimators based on moment analysis can be tested for significance. The log-normal or phi-normal model will again be used as the fundamental population frequency distribution; indeed, in performing detailed studies of lengths of axes of quartz and quartzite rock fragments, this model, accompanied by a carefully specified sampling arrangement, suffices to yield sample frequency distributions which are not significantly different from random samples from a normal distribution (Griffiths, 1955b, 1958b, 1959a, b, 1962a).

It has been found that if the population is completely available, such as a completely exposed face in a gravel pit (Griffiths, 1959b) or a scree (Griffiths, 1959a), channel samples which are self-weighting, taken across the entire available, i.e., existent, population, yield statistical estimators which are not significantly different from those expected in random sampling from a normal distribution; when stratified random samples are taken, the number of ele-

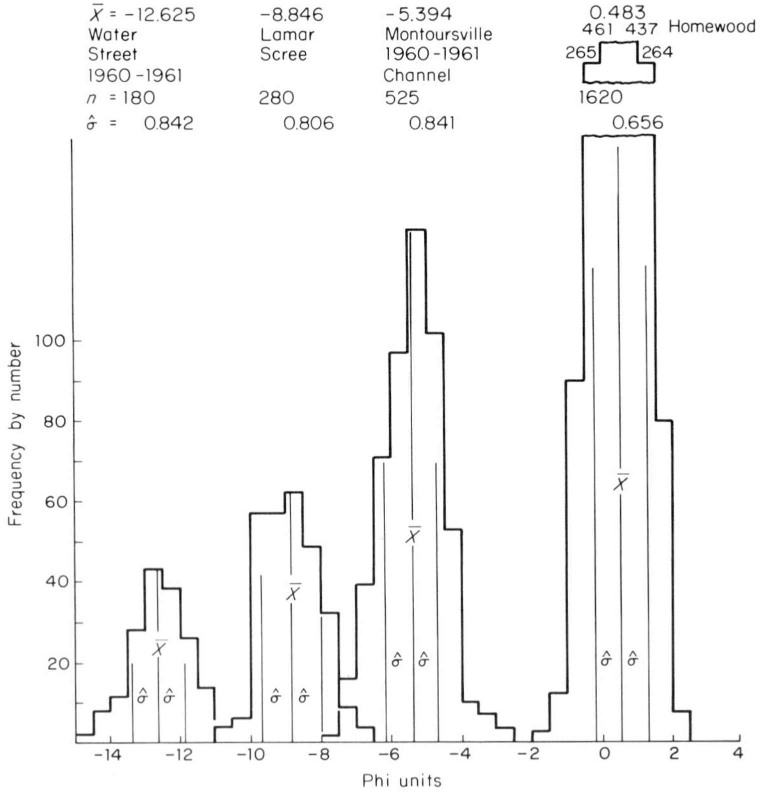

Figure 5.6 Frequency histograms of long a phi axes of particles in some detrital sediments.

ments per sample and the number of layers sampled are quite critical in achieving unbiased estimators. As the sampling becomes balanced, the estimators converge on normality, as would be expected. On the other hand, spot samples, which in layered deposits include parts of the population and duplicate some layers and omit others, yield statistics which converge only slowly toward unbiased estimators (see, for example, Steinmetz, 1956, 1962; Griffiths, 1959a, b). Since in sampling thin sections the samples tend to be spot samples, the significant asymmetry and peakedness exhibited in the data of Table 5.5 are probably to be attributed to this feature, which induces sampling biases of various kinds. In general, it has been found that 30 to 50 elements per sample, when the sample is confined to a single homogeneous population, i.e., layer or sedimentation unit, and some 16 to 20 layers, is a desirable sampling arrangement (Steinmetz, 1956, 1962; Griffiths, 1959a, b, 1962a). Extensive tests of the sampling model proposed in Fig. 2.3 have demonstrated that achievement of unbiased estimators characteristic of random samples demands very careful sampling designs, and even small departures from an appropriately balanced design introduce significant bias. In order to achieve random sampling of log-normal or phi-normal populations, therefore, one must have a procedure permitting a precise control of the sampling program; direct measurement performed on individual elements (grains or pebbles) is the simplest means of attaining these specifications. Neither thin-section measurement nor indirect techniques (such as sieving and pipette sedimentation) are able to achieve this sampling control without very elaborate reconnaissance investigations. In the former case a well-defined balanced sampling program is very difficult to specify without a preliminary reconnaissance devoted to the sampling problem. In the latter, the sample sizes, 20 or 100 to 150 g, result in spot sampling; and convergence on unbiased estimators requires prohibitively large sample sizes.

A number of isolated examples of direct-measurement data are illustrated in Fig. 5.6 and Table 5.6; they were compiled during sampling experiments repeated successively on the same populations in order to resolve the sampling problem[1] (Griffiths, 1959a, b, 1962a). They represent the extremes of the sampling model (Fig. 2.3), the Tuscarora scree being a population approximating stage 1 of the model and the Montoursville gravel and Homewood (Pottsville) quartzite being approximates to stage 3 of the model. It is recommended that in any initial foray the sampling plan designed for a stage 3 stratified population be used. Then, if the population is a stage 1 type, no loss will be suffered, although the sample size is somewhat larger than absolutely necessary (see, for example, Steinmetz, 1956, 1962); whereas if the population belongs in stage 2 or 3 the estimates will tend to be unbiased, or at least they will supply data upon which subsequent recommendations for achieving random samples can be readily and precisely defined. If a stage 1

[1] These experiments were carried out under the author's direction in graduate classes at The Pennsylvania State University.

TABLE 5.6
Summary Statistics for Long *a* Axis of Particles from Sediments; Direct Measurement

	Pedro, California[a]			Pennsylvania[b]			
						Tuscarora Scree	
Statistic	Beach, plan 1	Dune	Beach, plan 2	Homewood, 1955–1956	Montoursville gravel channel samples	Water Street, 1962	Lamar
$\bar{X}$	1.218	1.206	1.165	0.483	−5.394	2.678	−8.846
$\hat{\sigma}$	0.399	0.340	0.328	0.656	0.841	0.833	0.806
$\sqrt{b_1}$	−0.704**	−0.076	−0.186	−0.017	0.035	0.016	−0.151
b_2	5.475**	3.224	3.397†	2.929	2.860	2.765	2.673
N	250	250	600	1,620	525	436	280
χ^2	17.637	5.532	23.639	6.212	13.39	5.74	16.10
d.f.	7	5	6	6	8	7	6
P	$0.02 > P > 0.01$	$50 > P > 30$	<0.001	$50 > P > 30$	$10 > P > 5$	$50 > P > 10$	$0.02 > P > 0.01$

[a] Hulbe, 1957.
[b] Griffiths, 1959a, b, and unpublished work.

program is applied to a stage 3 population, either the estimators will be biased, or if the estimators are unbiased, the samples will approximate channel samples, and only the population mean and variance will emerge. In this case the patterned variation within and among the layers will be lost, and this may well be the critical information in interpreting environmental processes. For example, the range of quartz-grain size in a glacial and in a beach deposit may well be the same, and even the mean sizes may be similar, but the glacial till is relatively unstratified, i.e., approaches a stage 1 type of population, whereas the beach deposit will almost certainly be a stage 3 population (Steinmetz, 1956, 1962). Hence, specification of the within- and among-sample variation is quite critical in differentiating degrees of selective sorting characteristics of different environmental processes.

There are too few samples in Table 5.6 to generalize about the range in mean grain size, which extends from 1 phi unit (500 microns) to about −13 phi units. The deposits, however, are representative of a wide range in sorting, and the standard deviation extends from 0.3 to 0.84 phi unit. The best-sorted sediments are, as expected, beach and dune sands with standard deviations around 0.3 phi (compare with the sorting of quartz-grain long axes measured in thin section), whereas the scree, which may be expected to contain a wide range in grain size, has standard deviations of about 0.8 phi unit. It will be observed that the range in length of long axes of quartzite pebbles from the Montoursville gravel approximates that of boulders in the scree. The difference between these deposits is contained in the "sorting" within and between samples, with the scree resembling a stage 1 population and the gravel representing an approach to a stage 3 type of population.

The two investigations of quartz-grain long axes from Pedro beach, San Mateo County, California (Hulbe, 1957), illustrate the effect of different sampling arrangements; sampling plan 1 for the beach consists of five (spot) samples of 50 grains each, whereas sampling plan 2 for the beach represents an attempt to achieve stratified random sampling by selecting 6 grains from small samples of about 20 to 50 grains at each of 100 intersections of a grid. The grid consisted of five vertical lines 2 in. apart and 20 horizontal lines spaced $\frac{1}{2}$ in. apart. The mean and standard deviation of the 600-grain (plan 2) sample are similar to those of the 250-grain (plan 1) sample; the asymmetry and peakedness are, however, much reduced in the plan 2 sampling as compared with plan 1, suggesting that stratified sampling, when it achieves randomness, will yield unbiased sample statistics. Neither of the beach analyses yields normal frequency distributions on the basis of testing by the chi-square criterion; the departures occur both in the center and on the tails in plan 1, whereas the main deviations occur in the two central classes in plan 2. The only specific recommendations for improving the sampling arrangement using plan 1 are to increase both sample size and number; on the other hand, the expected phi-normal model requires more grain measurements in the center of the distribution in plan 2.

5.9 Applications of Size Frequency Measurements

A very large number of applications of grain-size data are recorded in the literature (for references see Pettijohn, 1957, p. 54), but despite a long period of investigation and an extensive background of data the results are still largely empirical, qualitative, and open to alternative interpretations (Pettijohn, 1957). It is quite evident that despite a large investment in time and money the size data generally collected admit of very little unequivocal interpretive petrology. Some of the many reasons for this disappointing outcome have been discussed; largely the information level achieved is inadequate for the interpretation attempted, and this may be due to inadequate measurement procedure, incorrect sampling (and therefore biased estimators), and a general lack of rigor in the approach. Some possible alternatives are recommended which only further experience can affirm or deny; to date, the main advantage of size frequency data is their use as a descriptive, petrographic, rather than an analytic, petrologic, tool.

Attempts to differentiate environments by means of size frequency data are common, and Pettijohn gives an extensive discussion together with the reservations which the results seem to require. Nevertheless, in specific and particulate investigations size data are useful; for example, qualitative specification of environments is certainly feasible (Krumbein and Aberdeen, 1937; Doeglas, 1946; Griffiths, 1951b, 1952a; Folk and Ward, 1957; Friedman, 1962). A discussion of the relationship of size and sorting is included in Chap. 15. Again for specific engineering questions in recommending gravel-pack sizes, grain-size data are essential (Griffiths, 1952a).

It can also be shown, at least in some cases, that grain size may be related to oil-reservoir behavior (Illing, 1939; Cloud, 1941; Krumbein and Monk, 1943) and that variation in grain size is a critical factor in determining the incidence of oil reservoirs (Griffiths, 1952a, 1958b, 1960b, 1963b; Stanonis, 1956; Pachman, 1961; Smith, 1964).

It is also obvious that grain size is one of the fundamental properties characterizing an aggregate, but to measure the level of information achieved it will almost certainly be necessary to analyze the accumulated data by some form of multivariate analysis (e.g., Griffiths, 1962a, 1963b; Miller and Kahn, 1962, pp. 315ff.; Griffiths and Smith, 1964).

TABLE 5.7
Summary Table of Statistics for Size-frequency Distribution[a]

Name of statistic	Symbol	How derived	Comment
Summary of Measures of Central Tendency			
Millimeter, median	Md	Graphically: the 50% diameter on cumulative-frequency curve	An average of position only. Not susceptible to mathematical manipulation
Phi median	Md_ϕ	Where $\mathrm{Md}_\phi = -\log_2 \mathrm{Md}$	
Millimeter mode	Mo	Graphically: the highest point on the frequency curve	Not mathematically manipulatable. Represents the predominant size present in the distribution
Phi mode	Mo_ϕ		
Millimeter arithmetic mean	$\bar{X}$ M_a	For ungrouped values: $\dfrac{\Sigma X_i}{N}$ For grouped values: $\dfrac{\Sigma(f\cdot\text{class midpoint})}{f}$	May be manipulated mathematically. Represents the center of gravity of the distribution, that is, the "weight × distance" midpoint
Millimeter geometric mean	M_g	$\sqrt[N]{X_1X_2X_3\cdots X_N}$ $= \text{antilog}\,\dfrac{\Sigma(f\cdot\log_{10}\text{class midpoints})}{\Sigma f}$ By direct conversion from the arithmetic mean of the phi distribution	May be manipulated mathematically. Arithmetic mean of a log distribution equals geometric mean of corresponding arithmetic distribution
Phi arithmetic mean or Logarithmic mean (Krumbein)	$\bar{X}_\phi$	$\dfrac{\Sigma(f\cdot\phi\text{ class midpoint})}{\Sigma f}$ Derived as the first moment (see calculation table for moments)	Most commonly used and most significant representation of center of gravity of distribution. May be mathematically manipulated
Inman phi mean	M_ϕ	$\dfrac{P_{16}+P_{84}}{2}$	Inman, 1952
Folk mean	M_z	$\dfrac{P_{16}+P_{50}+P_{84}}{3}$	Folk, 1957

TABLE 5.7 (*continued*)

Name of statistic	Symbol	How derived	Comment
Summary of Measures of Dispersion			
Millimeter quartile deviation	QD_a	$\frac{Q_3 - Q_1}{2}$	Measures half the spread of middle 50 percent of distribution; linear measure only; cannot be mathematically manipulated
Phi quartile deviation	QD_ϕ	$\frac{Q_{3\phi} - Q_{1\phi}}{2}$ = $\log_2 So$ (So = Trask's sorting coefficient)	Measures half the spread of middle 50 percent of distribution; linear measure only; cannot be mathematically manipulated
Phi percentile deviation	PD_ϕ	$\frac{P_{90\phi} - P_{10\phi}}{2}$	Measures half the spread of middle 80 percent of distribution. Linear measure only, cannot be mathematically manipulated
Trask's sorting coefficient	So	$\sqrt{\frac{Q_3}{Q_1}}$ = antilog to base 2 of QD_ϕ	Values are not readily visualized in terms of spread because the scale is geometric
Arithmetic standard deviation, mm	σ_ϕ	$\sqrt{\frac{\Sigma[f(\text{class midpoint} - \bar{X})^2]}{\Sigma f}}$	Measures average spread of curve about M_a. Not used on open-ended curves
Phi arithmetic standard deviation or logarithmic standard deviation (Krumbein)	σ_ϕ	Computed from second moment (see calculation sheet for moments) If a straight line on ϕ probability curve $\frac{P_{84\cdot2} - P_{15\cdot8}}{2}$	Measures average spread of curve about M_ϕ. Most commonly used and most significant measure of dispersion. In phi units
Geometric millimeter standard deviation	σ_g	Conversion to millimeters from σ_ϕ	
Inman phi deviation	σ_ϕ	$\frac{P_{84} - P_{16}}{2}$	Inman, 1952
Inclusive graphic standard	σ_1	$\frac{P_{84} - P_{16}}{4} + \frac{P_{95} - P_5}{6.6}$	Folk and Ward, 1957

TABLE 5.7 (*continued*)

Name of statistic	Symbol	How derived	Comment
Summary of Measures of Asymmetry			
Millimeter quartile skewness (arithmetic)	Sk_a	$\dfrac{Q_1 + Q_3 - 2\text{Md}}{2}$	An arithmetic linear measurement of the distance between the arithmetic mean of the two quartiles and the median
Geometric millimeter quartile skewness	Sk_g	$\sqrt{\dfrac{Q_1/Q_3}{\text{Md}^2}}$ $\log Sk_g = \dfrac{\log Q_1 + \log Q_3 - 2 \log \text{Md}}{2}$	A geometric form of the above. Bears a relationship to the above analogous to the relationship of Trask's sorting coefficient to quartile deviation
Trask's log of skewness, mm	$\log Sk$	$\log \dfrac{Q_1Q_3}{\text{Md}^2}$	This is the log of the square of Sk_g. Q_1, Q_3, Md in millimeters
Phi quartile skewness	$Sk_{q\phi}$	$\dfrac{Q_{1\phi} + Q_{3\phi} - \text{Md}_\phi}{2}$ $\dfrac{-\log Sk \text{ (Trask)}}{0.602}$	An arithmetic measurement of a log distance
Phi skewness	Sk_ϕ	Computed from third moment	Not usable in open-ended curves
Phi skewness	α_ϕ	$\dfrac{M_\phi - \text{Md}_\phi}{\sigma_\phi}$ or $\dfrac{P_{16} + P_{84} - 2P_{50}}{P_{84} - P_{16}}$	Inman, 1952
Second phi skewness	$\alpha_{2\phi}$	$\dfrac{(P_5 + P_{95})/2 - \text{Md}_\phi}{\sigma_\phi}$ or $\dfrac{P_5 + P_{95} - 2P_{50}}{P_{84} - P_{16}}$	
Inclusive graphic skewness	Sk_I	$\dfrac{P_{16} + P_{84} - 2P_{50}}{2(P_{84} - P_{16})} + \dfrac{P_5 + P_{95} - 2P_{50}}{2(P_{95} - P_5)}$	Folk and Ward, 1957

TABLE 5.7 *(continued)*

Name of statistic	Symbol	How derived	Comment
Summary of Measures of Peakedness			
Arithmetic quartile kurtosis, mm	K_{qa}	$\dfrac{P_{90} - P_{10}}{Q_3 - Q_1}$	(P_{90}, P_{10}, Q_3, Q_1 in millimeters.) Ratio of PD to QD. Reciprocal of Krumbein's
Phi quartile kurtosis	$K_{q\phi}$	$\dfrac{P_{90} - P_{10}}{Q_3 - Q_1}$	Same using phi units
Phi kurtosis	K_ϕ	Computed from fourth moment	Extremely sensitive to tails of distribution
Phi kurtosis measure	β_ϕ	$\dfrac{(P_{95} - P_5)/2 - \sigma_\phi}{\sigma_\phi}$ or $\dfrac{P_{95} - P_5 - P_{84} + P_{16}}{P_{84} - P_{16}}$ or $\dfrac{P_{95} - P_5}{P_{84} - P_{16} - 1.0}$	Inman, 1952
Graphic kurtosis		$\dfrac{P_{95} - P_5}{2.44(P_{75} - P_{25})}$	Folk and Ward, 1957

[a] Where $Q_3 = 25\%$ mm $= 75\%$ phi and $Q_1 = 75\%$ mm $= 25\%$ phi.

Note: Moment measures and mathematically calculated means cannot be computed for open-ended distributions.

P_5 = diameter in phi units at 5th percentile

. . . .

P_{25} = diameter in phi units at 25th percentile = 1st quartile Q_1
P_{50} = diameter in phi units at 50th percentile = median Md
P_{75} = diameter in phi units at 75th percentile = 3rd quartile Q_3

Source: Modified from M. A. Rosenfeld, "Handbook of Laboratory Instructions."

6 Measurement of grain shape

There are however manifolds in which the fixing of position requires not a finite number but either an infinite series or a continuous manifold of determination of quantity. Such manifolds are constituted for example by the possible determinations of a function for a given domain, the possible shapes of a figure in space, et cetera.

G. F. B. Riemann, 1868

6.1 Introduction

The general pattern of discussion of each fundamental property follows along the usual lines. Starting with the concept, attempts are made to reduce the concept to an operational definition; this is followed by development and testing of procedures for collecting data; and eventually an attempt is made to collect some data. The form of the frequency function, the sampling program, and, ultimately, a summary of the possible interpretation of the assembled observations lead us back to a statement of meaning in terms of the concept. The properties, it should be remembered, are measured as a means to an end, so that success may be ultimately measured in terms of information which advances us toward some specified objective.

Shape is probably the most fundamental element property of any particle and is certainly one of the most difficult to specify for any but the simplest shapes; nevertheless, since it implicitly, in terms of habit, forms the basis of identification of many mineral constituents and, explicitly, is considered to reflect provenance and process of formation, it is one of the most important properties of the particles of sedimentary rocks.

Quotation: G. F. B. Riemann, in D. E. Smith (ed), "A Source Book in Mathematics," p. 415, Harvard University Press, 1929.

Shape is conceptually and tautologically defined as form or as the expression of external morphology. Theoretically, shape may certainly be considered as independent of magnitude, and topological form illustrates this independence; in practice, however, it is almost impossible, or at least impracticable, to separate size and shape, and it is doubtful whether they can be separated operationally in examining the properties of the particles of sediments. One means of measuring shape is to determine curvature by estimating radii (e.g., Le Corbeiller, 1954); this procedure is quite general and envisages both positive and negative curvature. It resembles the recommendations for measurement of "roundness" put forward by Wadell (1932, 1933, 1936). Such a procedure requires a set of measurements, and difficulties arise in combining the set; the meaning of average "shape" expressed in this manner is difficult to imagine. Out of a large number of other recommendations for reducing shape to an operational definition, Wadell's contributions and their subsequent modifications are probably the most satisfactory.

In general, shape measurement may be developed from considerations of regular solids; thus the shape of a sphere may be specified by the length of a single diameter, that of a cube by a single uniquely defined axis. A rectangular block requires three axial measurements (any three perpendicular edges); and as the body becomes less regular, or if the shape is not specified beforehand, a larger and larger number of measurements becomes necessary to specify the complete shape. In fact, if we choose arbitrarily a center point within the particle, say the center of the largest inscribed circle, as recommended by Wadell (1932), and then measure the lengths of a number of radii joining the center to the margin as in Fig. 6.1, we can reduce the specification to a sampling problem. For a smoothly convex surface the number of axes necessary to specify the shape completely would be few (fewest, say two, for a sphere); and the number of axes required increases as the particle becomes less spherical. If the surface of the particle becomes corrugated, then, by judicious arrangement of the axes in clusters, the details of roughness can be specified to any required degree of refinement. Analysis of variance between and within sets of systematically spaced radii would lead to definition of how

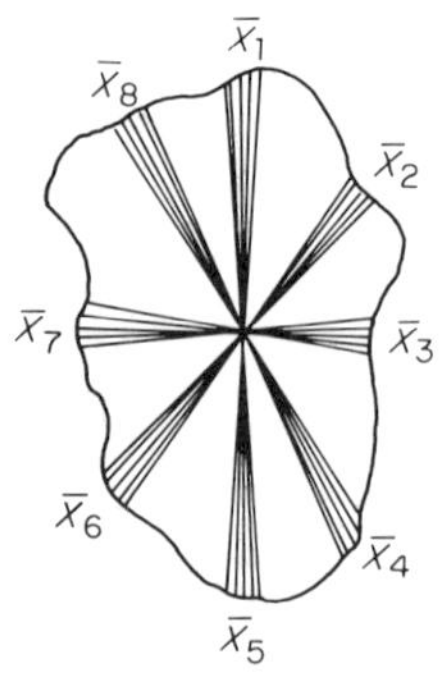

Figure 6.1 Measurement of shape of irregular particles. Compare variation in X_i (roundness) with variation in $\bar{X}$ (sphericity) by analysis of variance.

many measurements would be required and how they should be arranged to achieve a mean "shape" with any required degree of precision.

The development of shape measurement clearly implies increasing the number of measurements as the shape increases in irregularity; thus Wadell recommended sphericity and roundness measurements to specify those particles which varied from spherical to less regular, and subsequently surface texture was added (Krumbein and Pettijohn, 1938, pp. 303ff.). If this development is followed, there appears no limit to the degree of detail it is possible to envisage; i.e., spheres need few measures, ellipsoids more—say three, angular ellipsoids more than three; it is always possible to choose two particles equally spherical, equally round, with the same surface textures but of different "shapes." Apart from the sampling solution mentioned above, there seems to be no ultimate solution along these lines.

Concepts of the shape of particles in sediments have arisen from genetic considerations, overall shape being related implicitly if not explicitly to the original form of the particle, roundness being a function of wear, surface texture of certain environmental differences, and so on. Such conjectures rarely offer adequate solutions to measurement problems because they are vaguely formulated and are generally oversimplified. It seems more promising to attempt to define a more specific objective and then evaluate the measurement technique in terms of the fulfillment of that objective. Thus in estimating volume of quartzite pebbles it can be shown that any three orthogonal axes will predict volume with some 98 percent precision (Griffiths and Smith, 1964); additional axial measurements appear to add little information. We are quite generally forced to admit that to specify the shape of an irregular particle we require n measurements. It seems necessary therefore to select some few among these measurements in some way, and the decision of which to select becomes a function of the precision required in achieving some specified objective. Ultimately, as in the use of size measurement, the requirement of shape measurement will terminate with a number on some scale, and the value of this scale in terms of information concerning some specified objective will determine whether the number has any meaning. If it has clear meaning, collecting this information will automatically become a routine; if the meaning is not clear and the data collected do not contribute to our understanding of the aggregate of particles under investigation, whether they are sedimentary rocks or galactic nebulae, the measurement will generally be ignored.

6.1.1 Operational Procedures. The large number of recommendations devised for the measurement of some aspect of shape can be subdivided into indirect and direct measurement procedures. The indirect measurement of size usually requires the assumption of a shape factor; similarly, any attempt to measure shape by indirect means—settling in air or water, rolling down slopes, etc.—will require some restrictions on size or assumptions about size. It seems strange that in contrast to size measurement so few procedures exist for indirect measurement of shape, but apparently most of the procedures have been designed to consider the shape of each particle. Thus in evaluating shape measurement we can simplify the defining equation directly and consider two of the common procedures in use. First, in thin section it is customary to select one mineral, such as quartz, and measure its shape. Second, we can disaggregate the assemblage and concentrate on individual grains. Several different procedures can be used to perform the actual measurement, but nearly all those so far advocated require one of these two preparatory steps. In evaluating these techniques we shall examine the simplest routine first, namely, the visual estimation of various aspects of shape by comparison with a calibrated set of standards.

6.2 Determination of Shape by Visual Comparison with a Set of Standards

Visual comparison with standards is attractive because of its simplicity; it is a widespread technique in routine observations both in the field and in the laboratory but possesses very insidious disadvantages because it is so deceptively simple. First, it is a psychophysical procedure of *subjective assessment* and almost always suffers from personal prejudice, so that different observers achieve different results. Furthermore, the results for one observer vary from time to time. It requires, therefore, very elaborate precautions to

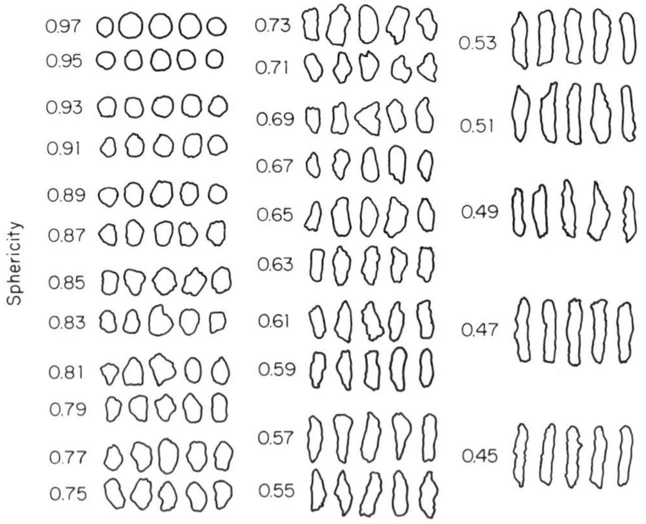

Figure 6.2 Images for estimating visual sphericity. [*After G. Rittenhouse, A Visual Method of Estimating Two Dimensional Sphericity, J. Sediment. Petrol.,* **13:***80–81 (1943), Society of Economic Palaeontologists and Mineralogists.*]

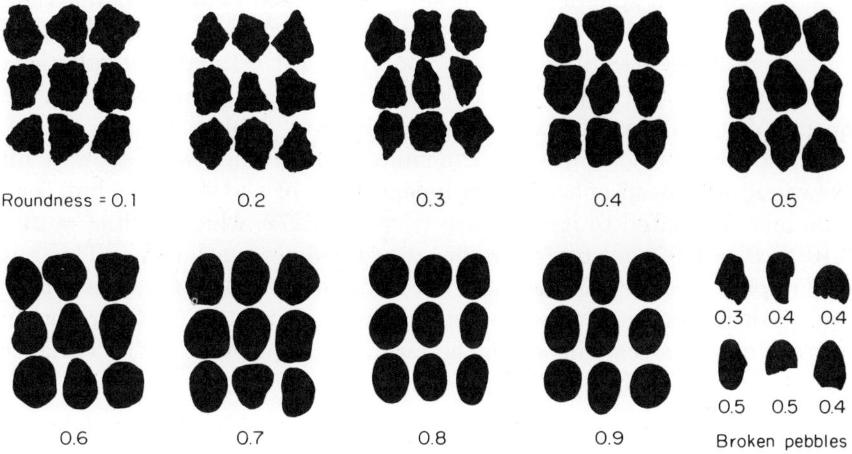

Figure 6.3 Images for estimating visual roundness. [*After W. C. Krumbein, Measurement and Geological Significance of Shape and Roundness of Sedimentary Particles, J. Sediment. Petrol.,* **11:***68 (1941), Society of Economic Palaeontologists and Mineralogists.*]

ensure even reasonable reproducibility. Identification of minerals, fossils, and rocks in hand specimen, in the field, and under the microscope relies heavily on subjective assessment and is very directly a function of the experience of the observer. Estimates of color and various other attributes have been the subject of prolonged investigation, with the result that an extensive body of literature has accumulated. Examples of such investigations are Yule's early work on grayness (1911), Pearson's work on various attributes of animals (K. Pearson, 1895, 1902; E. S. Pearson, 1948), and more recent investigation into attributes depending on sensory perception (Stevens, 1955, 1956; Thurstone, 1927; Coombs, 1952, 1959). The following summary of an attempt to evaluate visual comparison in the estimation of shape may serve to emphasize some of the pitfalls.

The objective is to estimate the sphericity and roundness of 21 selected quartz grains taken from a beach sand by comparison with charted standards, as shown in Figs. 6.2 and 6.3. We begin with the defining equation

$$P = f(m,s,sh,o,p)$$

m becomes a constant and because the grains are mounted loose on a glass slide, the effects of o and p are removed. The size range is not large and may be neglected for the present purpose.[1] Thus, the measure P is presumed to be some measure of "shape."

[1] The effect of size on shape measurement may be very important when the size range is large.

The range in sphericity of these 21 grains is 0.71 to 0.93, and the range in roundness is 0.10 to 0.80. Seven operators examined each of the 21 grains on each of 3 days, leading to $3 \times 7 \times 21 = 441$ estimates of sphericity and roundness, respectively. The relevant statistical summaries for the frequency distributions of these data are given in Table 6.1.

The distributions are quite definitely nonnormal and show a marked skewness; the mean sphericity is independent of the variance, but the mean roundness is related to its variance ($r = -0.4272$, which with $n = 21$ yields $0.10 > P > 0.05$; not quite significant at the 5 percent level). This relationship implies a larger variance at lower roundness values; it was used as a basis for a transformation of the raw data following Bartlett's (1947) recommendations to obtain a scale in which the mean is independent of the variance (see Rosenfeld and Griffiths, 1953, p. 565).

6.2.1 Experimental Design. The experiment contained three specified sources of variation, namely, differences among grains, operators, and days; the resulting design is a three-way classification, as shown in Table 6.2.

The 21 grains were a small sample from an infinite population of grains in the beach sand, the 7 operators a small sample from an infinite population of operators, and the 3 days a small sample from an infinite population of days; and so we may use the infinite-population model (Bennett and Franklin, 1954, p. 348; Eisenhart, 1947, p. 15; Chap. 20).

This is an unfortunate sampling design for this kind of experiment because, if the first-order interactions are not zero, there is no straightforward basis for testing the main effects (see Chap. 20); furthermore, there is no basis for testing the second-order interaction, which is almost certainly important. (See Table 6.3.)

TABLE 6.1
Summary Statistics of Sphericity and Roundness of 21 Quartz Grains from a Beach Sand

Statistic	Sphericity	Roundness
$\bar{X}$	0.817	0.424
$\hat{\sigma}$	0.045	0.131
$\sqrt{b_1}$	0.203*	−0.329***
b_2	2.656*	2.766
n	441	441
χ^2	87.07	41.84
d.f.	6	4
P	<0.001	<0.001

TABLE 6.2
Experimental Design for Three-way Classification[a]

Source of variation	d.f.	Expected mean squares
Among grain means	$g - 1 = 20$	$\sigma_\epsilon^2 + d\sigma_{pg}^2 + p\sigma_{dg}^2 + pd\sigma_g^2$
Among operator means	$p - 1 = 6$	$\sigma_\epsilon^2 + d\sigma_{pg}^2 + g\sigma_{pd}^2 + dg\sigma_p^2$
Among day means	$d - 1 = 2$	$\sigma_\epsilon^2 + p\sigma_{dg}^2 + g\sigma_{dp}^2 + pg\sigma_d^2$
Grains × operators	$(g - 1)(p - 1) = 120$	$\sigma_\epsilon^2 + d\sigma_{pg}^2$
Grains × days	$(g - 1)(d - 1) = 40$	$\sigma_\epsilon^2 + p\sigma_{gd}^2$
Operators × days	$(p - 1)(d - 1) = 12$	$\sigma_\epsilon^2 + g\sigma_{pd}^2$
Grains × operators × days plus all unassigned sources	$(g - 1)(p - 1)(d - 1) = 240$	σ_ϵ^2
Total	$gpd - 1 = 440$	

[a] See Sec. 20.2.

About 25 percent of the variation in magnitude is "error" or unassigned variation, a very large ignorance factor. Replicates would have helped to evaluate whether this was a reflection of second-order interaction or simply lack of control in the experiment. Before proceeding with the analysis it is necessary to point out that the source of variation labeled days is really a reflection of changes in order of estimating the "shape" of the grains; the first day the operators ran the grains from 1 to 21, the second 21 to 1, and the third 1 to 21. It would be difficult to imagine a physical meaning for a grain's changing shape from day to day, but estimates of shape could well vary as the order changed. In fact, it was found that an operator tended to judge the first grain by comparison with the chart and the second from memory of the chart and the first grain. Memory played an increasing role as the experiment proceeded. Thus there are good procedural reasons, as well as design requirements, for randomizing the order of treatment of the grains in future experiments. From hindsight it seems evident that this procedure probably contributed more to upsetting the experiment than any other single cause; however, one learns best from experience (particularly mistaken experience).

The first-order interaction days (order) by grains proved nonsignificant; the interactions operator by days and operator by grains were however, highly significant. Apparently, then, operators were inconsistent from day (order) to day (order) and from grain to grain Thus no general recommendations are really possible from these results except that if this technique is to be used, experimental control must be considerably improved in some way to reduce these inconsistencies.

There are, despite these inconsistencies, differences due to change in order among operators and among grains which are significantly larger than the respective interaction terms. It is perhaps of interest to note that because the test suggests that the grains-by-days variance is effectively zero ($\sigma_{gd}^2=0$), the main effect, designated grain differences, can be tested against the interaction grains by operators and is highly significant, i.e., exceeds variation due to interaction. This is, of course, necessary if the procedure is to be capable of supplying information on variation in grain shape. The term for testing differences among operators is not so obvious; and, while operator differences are much larger than any interaction, or combination of interactions, the test procedure is not clear-cut. It is sufficient to suggest that operators are different enough to require segregating of their effect if grain differences are to be compared. These operator differences are illustrated graphically as the solid line in Fig. 6.4. The interaction operator by days (order) is also illustrated; and while the similar order led to similar estimates, the reversal of order led to marked inconsistency. This could be reduced by randomizing the order of treating the grains on each occasion.

The mean differences among grains are illustrated as the solid line in Fig. 6.5; by means of the least significant differences (LSD) it can be seen that at the 5 percent probability level differences of two units in the second place cannot be distinguished; i.e., the scale is effectively 0.80 to 0.82, 0.82 to 0.84, etc. The range of insensitivity increases to 0.04 unit at the 0.001 probability level. This suggests that from two to five classes are all that can be effectively used in the range 0.8 to 0.9 sphericity unit.

TABLE 6.3
Condensed Analysis of Variance Table for Sphericity of Quartz Grains in Beach Sand[a]

Source of variation	d.f.	Mean square	Variance ratio F
Grains	20	160.2198	27.137***
Operators	6	88.795	6.575**
Days	2	12.992	3.88*
Grains × operators	120	5.904	1.7735***
Operator × days	12	13.504	4.056***
Grains × days	40	3.4532	1.04
Error	240	3.308	
Total	440	12.6492	

[a] Data transformed to arcsin values.
Source: M. A. Rosenfeld and J. C. Griffiths, An Experimental Test of Visual Comparison Techniques in Estimating Two Dimensional Sphericity and Roundness of Quartz Grains, *Am. J. Sci.*, **251:** 567 (1953).

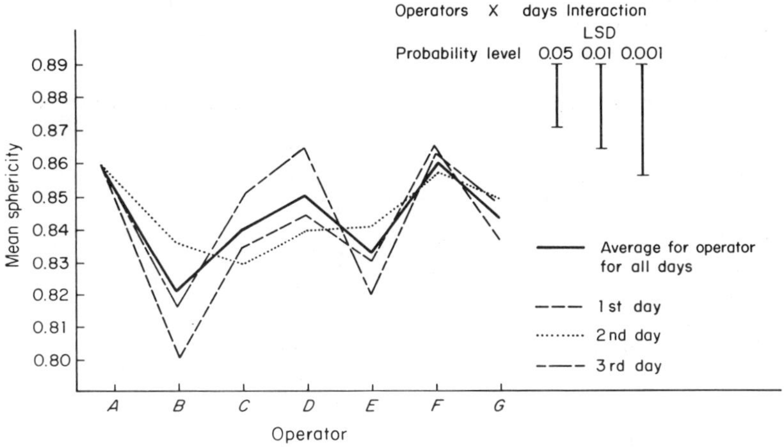

Figure 6.4 Variation in sphericity for each operator and for each operator on each day. [*After M. A. Rosenfeld and J. C. Griffiths, An Experimental Test of Visual Comparison Technique in Estimating Two Dimensional Sphericity and Roundness of Quartz Grains, Am. J. Sci.,* **251:***568 (1953).*]

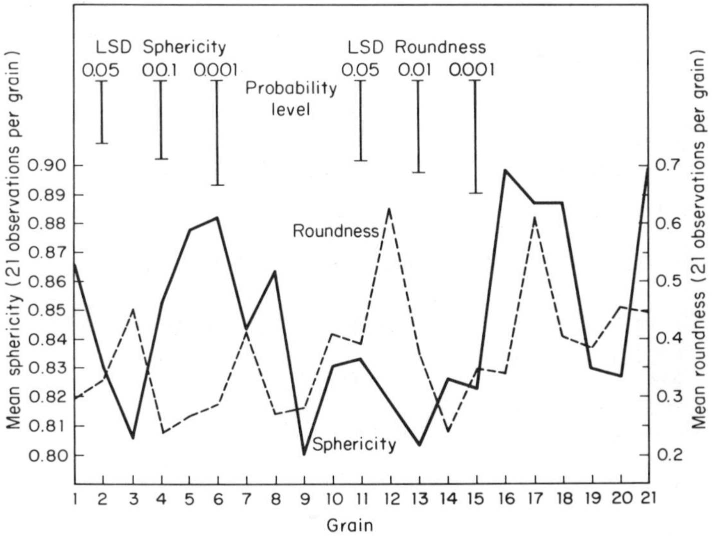

Figure 6.5 Mean sphericity and roundness for each of the 21 grains. [*After M. A. Rosenfeld and J. C. Griffiths, An Experimental Test of Visual Comparison Technique in Estimating Two Dimensional Sphericity and Roundness of Quartz Grains, Am. J. Sci.,* **251:***571 (1953).*]

TABLE 6.4
Condensed Analysis of Variance Table for Roundness of Quartz Grains in Beach Sand

Source of variation	d.f.	Mean square ×10⁴	Variance ratio F
Grains	20	14.914	28.96***
Operators	6	7.686	14.96***
Days	2	0.245	<1.0
Grains × operators	120	0.515	1.936***
Operators × days	12	0.334	1.255
Days × grains	40	0.2274	0.855
Error	240	0.2655	
Total	440	1.0989	

Source: M. A. Rosenfeld and J. C. Griffiths, An Experimental Test of Visual Comparison Techniques in Estimating Two Dimensional Sphericity and Roundness of Quartz Grains, *Am. J. Sci.*, **251:** 575 (1953).

The roundness data are tested in the same manner, with the results summarized in Table 6.4. Again the error factor is large; two first-order interactions are zero, and only the grains-by-operators effect is significant. Also the differences among grain means and operator means are large and significant.

TABLE 6.5
Components of Variance: Standard Deviation

Source of variation	Sphericity	Roundness
Grains	0.033	0.103
Operators	0.014	0.042
Days	0.002	0.001
Operators × grains	0.0101	0.038
Operators × days	0.007	0.006
Days × grains	0.0	0.005
Error	0.024	0.062

Source: M. A. Rosenfeld and J. C. Griffiths, An Experimental Test of Visual Comparison Techniques in Estimating Two Dimensional Sphericity and Roundness of Quartz Grains, *Am. J. Sci.*, **251:** 573, 580 (1953).

The LSD is 0.2 unit at the 5 percent probability level; in other words, the scale between 0.2 and 0.7 unit consists of only three effective classes (see Fig. 6.5).

The magnitude of variation expressed as the standard deviation is illustrated in Table 6.5; and while the variation in roundness appears about three times larger than that in sphericity, in terms of grain differences the next highest contributor is error, and this is also three times larger in roundness than in sphericity. The results are therefore quite comparable in a relative sense throughout the table for both variables; i.e., variation from different sources is similar in relative magnitude for both estimating techniques.

6.2.2 Relationship between Sphericity and Roundness. It is generally believed that sphericity and roundness are related, but on the basis of these 21 estimates they are not (Rosenfeld and Griffiths, 1953, p. 581). However, when operator mean roundness and sphericity are plotted in place of grain mean roundness and sphericity, the relationship is clear, as shown in Fig. 6.6.

This graph is quite revealing; it shows that when operators estimate high sphericity, they also estimate high roundness, and the order of treatment of the grains clearly affects the estimates (see central inset). In nearly every case the relationship between sphericity and roundness using these techniques is consistently a reflection of variation among operators and not necessarily a characteristic of grains.

6.2.3 Remarks on Technique. It seems clear that it is not possible to estimate sphericity and roundness independently by visual comparison, and when this technique is used, these two procedures represent two estimates of the same shape feature P. Hence if independent assessment of sphericity

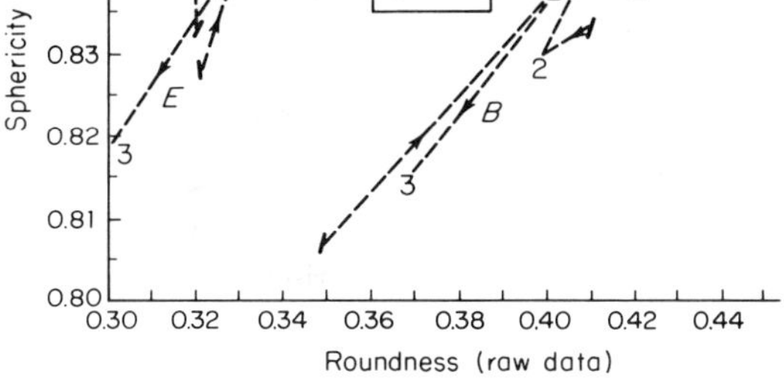

Figure 6.6 Operator means by days for sphericity versus roundness. 1, first day; 2, second day; 3, third day. Letters A to G refer to operators. [*After M. A. Rosenfeld and J. C. Griffiths, An Experimental Test of Visual Comparison Technique in Estimating Two Dimensional Sphericity and Roundness of Quartz Grains, Am. J. Sci.,* **251:***582 (1953).*]

and roundness is required, some other technique must be adopted for at least one of the properties.

The scale of both these estimates is probably not better than ordinal on the basis of this experiment, and to use analysis of variance on these data is questionable. The fact that the analyses give equivalent results irrespective of the transformation used is a reflection of the fact that the measurement level, to a first approximation, is invariant to such transformations and is therefore unlikely to have achieved interval rank (see Sec. 12.5).

On the basis of this experiment it seems evident that information on shape would best be searched for other than by these techniques. If it is decided to use visual comparison, the experimental design must be made more elaborate, and the procedure must be demonstrated to be in control. Lack of adequate precautions implies, in effect, that one is merely comparing variation in one's prejudices confounded with a large contribution from unassigned sources of variation, i.e., error. Most procedures which rely on subjective assessment using comparison standards suffer from the same disadvantages.

6.3 Axial-ratio Shape Measurement in Thin Section

Another possible approach is measurement of shape performed on the two-dimensional image of grains in thin sections. This is particularly necessary when it is impossible to disaggregate the rock specimen without affecting the shape of the grains. The procedure may be analytically defined as

$$P_{sh\,2} = (m,s,sh,o,p)$$

and by choosing a single mineral constituent, say quartz, composition m becomes a constant, but no further simplification is possible. When all properties vary simultaneously, the variation in $P_{sh\,2}$ receives contributions from the four main effects size s, shape sh, orientation o, and packing p, together with six first-order interactions of the form s, sh; s, o; etc.; four second-order interactions of the form s, sh, o; s, sh, p., etc.; and one third-order interaction s, sh, o, p. In other words, the measure obtained can be written

$$P_{sh\,2} = f(s,sh,o,p)$$

The subsequent results illustrate the difficulty in evaluating $P_{sh\,2}$ and suggest that the values obtained contain confounded information which is complexly and only remotely related to the concept of shape.

6.3.1 Procedural Details. Upon choosing the constituent it is necessary to define the selection procedure for the axes with some rigor, and as a measure of circularity, i.e., two-dimensional sphericity, the two axes may be selected in the manner described by Griffiths and Rosenfeld (1950, p. 237). The procedure comprises obtaining the projected image of the grain, either by drawing the projected outline by using a microprojector or from a photomicrograph, and then finding the shortest dimension (b axis) with parallel rule by trial and error. After determining the short dimension, the rectangle tangent to the grain-image outline is completed and the long a axis measured perpendicular to b, as in Fig. 6.7. If a millimeter rule is attached to the parallel rule, the measurements can be read directly as the axes are defined. The same procedure can be performed under the microscope by using an eyepiece containing a graduated square mesh grid. The order of selection, short b axis first, is followed because it was found, in initial experiments, that if operators are permitted to select the long axis, a very considerable contribution to variation is introduced by differences in choice, and this source of variation is much better controlled by the procedure described.

The data resulting from these determinations can be expressed in a number of different ways. If the millimeter values are plotted on a scatter diagram, the observations tend to show heteroscedasticity; i.e., the variability increases as the magnitude of the a and b axes increases (Fig. 6.8a). By using logarithmic (phi) scales the spread of the values tends to be homoscedastic (Fig. 6.8b). It is customary in such cases to use the ratio of the two measurements if millimeter scales are used, and both ratios b/a and a/b have been proposed (Krumbein and Pettijohn, 1938, p. 288; Bokman, 1952). Since the information and its interpretation remain the same in either case, the b/a axial ratio will be adopted for the present discussion. A histogram exemplifying the frequency of axial ratios of quartz grains in Bradford sand is given in Fig. 6.9 (see also Griffiths and Rosenfeld, 1953, fig. 6, p. 206; Rosenfeld, 1953, fig. V.27; Griffiths, 1958b, fig. 6). The normal model may be used to graduate the curve if the number of measurements is fairly large, although cutoffs tend to arise because the b/a scale is limited to the range 0 to 1.

In cases where there are many circular grains there is a marked asymmetry because the frequencies accumulate at unity.

A large number of results are recorded by Bokman (1957), and the relevant statistical summary is given in Table 6.6; it is noteworthy that only four of the examples show significant asymmetry but nearly all are platykurtic. Kurtosis is affected by deviations from normality ($\beta_2 = 3$)

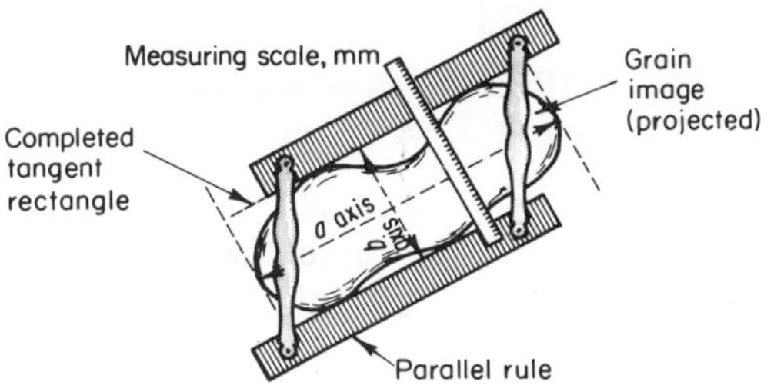

Figure 6.7 Measuring axial-ratio shape in thin section.

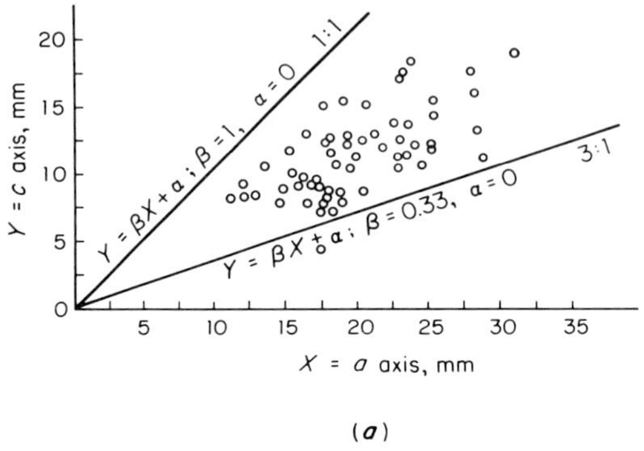

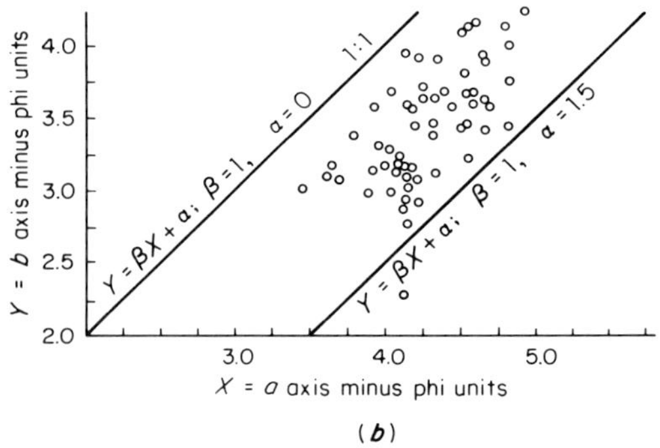

Figure 6.8 Long *a* axis vs. short *b* axis of pebble images. (*a*) Heteroscedastic in millimeters. Variation increases with increasing size. (*b*) Homoscedastic in phi (logarithmic) units. Variation independent of value of variate.

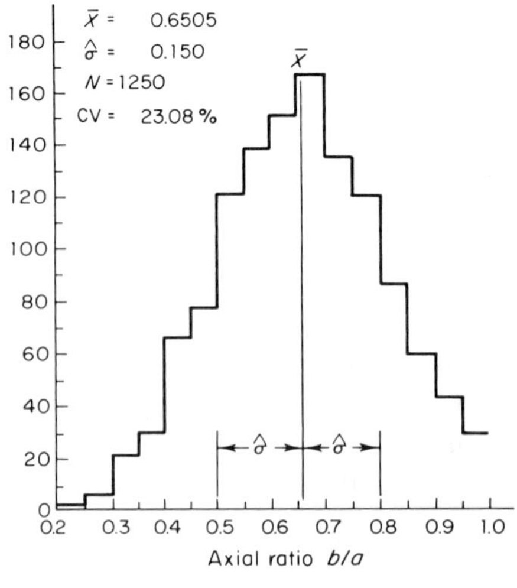

Figure 6.9 Frequency distributions of axial ratio b/a of quartz grains from Bradford sand. [*After J. C. Griffiths, and M. A. Rosenfeld, Further Test of Dimensional Orientation of Quartz Grains, in Bradford Sand, Am. J. Sci.,* **251:***206 (1953).*]

TABLE 6.6
Summary Statistics of Axial-ratio Frequency Distribution of Quartz Grains

Formation	$\bar{X}$	$\hat{\sigma}$	$\sqrt{b_1}$	b_2	n
Atoka	0.696	0.140	−0.078	2.555**	795
Hardinsburg	0.703	0.143	0.036	2.647*	750
Aux Vases	0.679	0.148	0.0003	2.658*	754
Cypress	0.707	0.138	−0.115	2.755	744
Misener	0.702	0.142	0.171*	2.489**	746
Wilcox	0.670	0.159	0.012	2.534**	755
Graneros	0.713	0.141	0.098	2.540**	721
Berea	0.686	0.173	−0.028	2.516**	752
Dakota	0.729	0.140	−0.318**	2.879	694
Morrison	0.688	0.157	−0.102	2.574**	784
Recent Mustang I., Tex.	0.649	0.152	−0.094	2.896	451
Recent Chenier, La.	0.670	0.141	−0.134	2.740*	854
Belle Canyon	0.663	0.150	−0.037	2.624**	791
Recent Holly Beach, La.	0.700	0.159	−0.345**	3.014	801
Tertiary, S. America	0.638	0.169	0.034	2.522**	413
Recent Topsail Beach, N.C.	0.608	0.160	−0.107	2.007**	277
Recent Grade Isle, La.	0.651	0.170	+0.120	2.475**	400
Recent Mississippi River	0.678	0.161	−0.152*	2.562**	802
Recent Padre I., Tex.	0.661	0.143	−0.125	2.569*	453
Total	0.685	0.150	0.0990**	2.684**	12,737

Source: Recalculated from data by Bokman, 1957.

both in the center and on the tails, and the departure measured by kurtosis is much more sensitive to deviation in the tails. Hence small excesses in the tails are probably the source of the platykurtosis of these distributions.

A much more striking feature is the limited range of mean value of axial ratio, which in Bokman's data extends from 0.608 to 0.729 for 19 different sedimentary rocks; similarly, the standard deviation varies only from 0.138 to 0.173, and as these statistics are mostly calculated from frequencies in the range $n = 277$ to 854, they should be reasonably stable estimates.

A set of similar measurements based on analyses of seven different sandstones by an entirely independent group of analysts is summarized in Table 6.7, and once again the range in mean axial ratio, despite wide variation in sampling design, is very small: from 0.649 to 0.682, with standard deviations from 0.142 to 0.173. This would suggest that the range of variation in axial ratio is very limited in all sediments; in Table 6.7 it is also obvious that the range 0.655 to 0.682 may be found in one and the same rock specimen from thin sections cut in different directions. This implies that orientation of

TABLE 6.7
Comparison of Axial Ratio of Quartz Grains in Various Sediments, Thin-section Measurement

			Axial ratio b/a		
Sediment	Samples	Grains	Mean	Standard deviation	Rock type
Pocono	16	800	0.6609	0.1739	Graywacke[a]
Berea	16	800	0.6691	0.1730	Graywacke[a]
Bradford Sand 5820	1	1,250	0.6506	0.150	Quartzose graywacke[b]
P4-1					
Parallel to bedding	1	400	0.665	0.148	Quartzose graywacke[b]
Parallel to maximum permeability	1	1,000	0.681	0.150	Quartzose graywacke[b]
Parallel to minimum permeability	1	1,000	0.682	0.144	Quartzose graywacke[b]
Inclined to maximum permeability	1	1,000	0.655	0.142	Quartzose graywacke[b]
Oriskany	87	9,981	0.677	0.142	Quartzite[c]
Salt Wash	50	15,000	0.649		Arkosic quartzite[d]
Cow Run sand	38	1,520	0.671	0.168	Graywacke[e]
Oswego sandstone	13	650	0.656		Graywacke[f]

[a] Emery, 1954. [b] Griffiths and Rosenfeld, 1953. [c] Rosenfeld, 1953. [d] Griffiths et al., 1956. [e] Griffiths, 1958b. [f] Griffiths, unpublished work.

section contributes as much to variation in axial ratio as different rocks and confirms that contributions to variation in P_{sh_2} from at least one first-order interaction, namely, shape by orientation *sh, o*, is extremely important.

If we now examine an analysis of variance of the results for the thin sections cut from the same rock specimen (Table 6.8), we find certain characteristic features which appear to apply to shape measurements of all kinds. First, the major proportion of the variation, about 94 percent in this case, occurs at the error level and suggests that the main (and perhaps the only important) source of variation is contained in axial-ratio differences among small sets of grains. There is no inconsistency of operators over fields or of operators over thin sections, and this feature is also general (see, for example, Griffiths and Rosenfeld, 1953), but there is a marked difference among operators. From the point of view of future experiments along these lines, then, it is essential to arrange the experiment to remove operator differences, and then operator inconsistency will also be in control. There are significant differences in mean axial ratio among means of both fields, within thin sections, and thin sections. Apparently axial ratio varies from area to area within one thin section (perhaps within laminae) and from thin section to thin section, depending on the direction in which the section is cut. In general, this suggests that mean axial-ratio shape of quartz grains tends to be approximately constant in many different kinds of sedimentary rocks.

6.3.2 ***Bivariate Scatter Diagrams of Axes.*** An alternative mode of expressing these data, which appears capable of generalization to more than two dimensions and permits a more exact treatment, is represented by bivariate analysis of the axes measurements (Griffiths et al., 1956, pp. 40–45; and Chap. 22). In this case a model approach can be used if the axes are plotted either in millimeters on log-log graph paper or, after the data are transformed to phi units, arithmetic graph paper. In the former case when the grain section or image is circular, the *a* and *b* axes are equal, and by using identical scales on the coordinate axes the points plot on a straight line which bisects the angle between the axes. Similarly, if axial-ratio shape is constant, independent of size, the points representing the grains will fall on a straight line parallel to this bisecting axis on the graph (see Figs. 6.9 and 6.10). These features are best represented in analytical form by using the standard equation for a straight line (see Chap. 22 for detailed discussion),

$$Y = mX + k$$

where Y represents the b axis, or the logarithm of b if arithmetic graph paper is used, X represents the a axis (or its corresponding logarithm), and m and

TABLE 6.8
Analysis of Variance of Axial-ratio Measurements from Bradford Sand Specimen P4-1

Source of variation	d.f.	Sum of squares	Mean square	Variance ratio F
Thin sections	3	0.901	0.300	13.99***
Operators	3	0.738	0.246	11.48***
Operators × thin sections	9	0.234	0.0260	1.216
Fields (within thin sections)	36	1.163	0.0323	1.506*
Operators × fields	108	2.574	0.0238	1.115
Grains within fields (error)	3,840	82.061	0.0214	
Total	3,999	87.672	0.0219	

Source: After Griffiths and Rosenfeld, 1950, table 7.

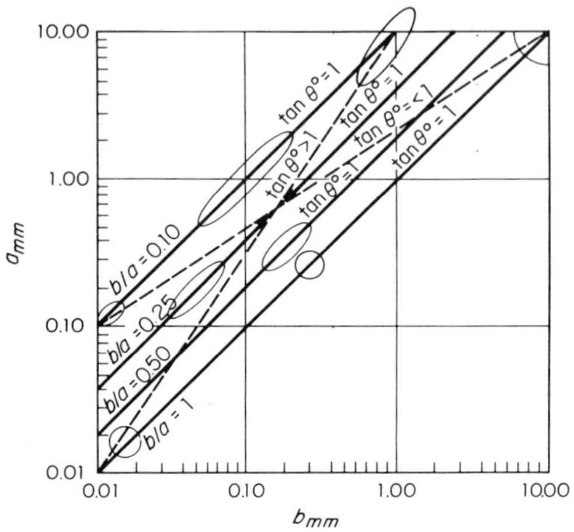

Figure 6.10 Graphic representation of model hypotheses for axial-ratio shape.

$\tan \theta° = 1$	(shape is constant over size)
$\tan \theta° < 1$	(elongation decreases with increasing size)
$\tan \theta° > 1$	(elongation increases with increasing size)

k are the coefficients defining its slope and position, respectively.[1] For circular grains $m = 1$ and $k = 0$, and the line bisects the angle between the coordinate axes Y and X ($\tan \theta = 1$, where $\theta = 45°$). For any constant shape $m = 1$ and k equals some positive real number.[2] If the grain sections become more circular with increasing size, the line will converge on the bisecting diagonal at higher values of a and b, and m equals a positive fraction while k equals a positive value. On the other hand, if elongation increases with increasing size, the line will converge on the bisecting diagonal for low values of a and b. It is theoretically possible that the scatter will not fall on a straight line or indeed not show any very simple functional relationship, but in practice the grain measurements concentrate around a straight line fairly closely, and mean values fall on, or very close indeed to, a straight line.

We now fit lines to the data by the method of least squares (see Chap. 22), and the degree of concentration around a straight line is measured by the correlation coefficient. In practice, since $b < a$ in millimeters, the scatter is unlikely to be bivariate normal but will show varying degrees of cutoff (see Fig. 6.8*b*) and hence asymmetry; but as a first approximation it is convenient to use simple linear regression and correlation as summary statistics describing the bivariate scatter diagram. If tests of significance are performed, the resulting probability statements will not be exact; but as the number of observations increases, the bivariate frequency distributions will tend to normality, and for large numbers of measurements (say over 100) the results should not be seriously in error. We may summarize and compare the arithmetic and logarithmic representation as follows.

[1] Because the grain axes are defined so that $a > b$, only the top half of the graph will be used; if the data are expressed in phi units, which is the negative logarithm of the millimeter value to the base 2, only the bottom half will be used.
[2] k is always positive because $a_\varphi < b_\varphi$.

Summary of Representation of Axial-ratio Shape of Quartz Grains Measured in Thin Section

AXIAL-RATIO SHAPE

Define a = long axis, then $\log a = X$.
Define b = short axis, then $\log b = Y$.
Shape is expressed in

	Millimeters	Logarithms
	$b = ka^m$	$Y = k' + mX$
Case 1: If $m = 1$ (Shape is constant over size)		
	$b = ka$	$Y = k' + 1X$
Case 2: If $m > 1$, say $m = 2$ (Shape changes; elongation increases with increase in size)		
	$b = ka^2$	$Y = k' + 2X$
Case 3: If $m < 1$, say $m = \frac{1}{2}$ (Shape changes; circularity increases with increasing size)		
	$b = ka^{1/2} = k\sqrt{a}$	$Y = k' + \frac{1}{2}X$

We may now set up models in the form of hypotheses for testing as follows:

Hypothesis: H_0; shape in phi units	$Y = \beta x + \alpha$		
For constant circular shape	$\beta = 1$	and	$\alpha = 0$
For constant shape	$\beta = 1$	and	α = positive value
For circularity increasing with size	$\beta < 1$		
For elongation increasing with size	$\beta > 1$		

The test statistic is $(\beta - m)/\hat{\sigma}_m$, and it follows Student's t with degrees of freedom $n - 2$.

As an example consider the scatter diagram of Fig. 6.11 showing the plot of mean values of b and a axes of quartz grains expressed in phi units of 50 specimens of Salt Wash sandstones (Griffiths et al., 1956, pp. 40ff., and fig. 11.8). It is quite clear that the data approximate a straight-line relationship and the scatter is small, as shown by the correlation coefficient $r = 0.99$. The relationship for 25 ore-bearing specimens is

$$Y = 1.0535X + 0.4685$$

and the hypothesis of $\beta = 1$ is not rejected; hence this relationship is approximately

$$Y = X + 0.4685$$

In the original report the value for b/a in millimeters is calculated to be 0.649. It is also of interest to observe the manner in which the relationship between the axes stabilizes as means are taken; in Table 6.9 the degrees of association between the grain axes are expressed as coefficients of correlation. For both populations, ore-bearing and barren, the relationships are similar, about 54 percent (r^2) in common to both axes. At the level of the individual grains, however, the axes in ore-bearing sandstones possess nearly 60 percent in common, whereas those in the barren sandstones possess less than 30 percent in common. This difference rapidly disappears when field means are considered (approximately similar with 65 percent in common), and the specimen means show almost identical association, almost 99 percent in common. Evidently unless individual grain axes are considered, the variation in axial shape rapidly disappears, and all sediments will tend to yield the same results.

Similar features are common to shape variation measured by many other techniques, both two- and three-dimensional, and indeed to other properties such as grain size and orientation, so that the main source of variation in the measured properties of sediments appears to reside in variation among indi-

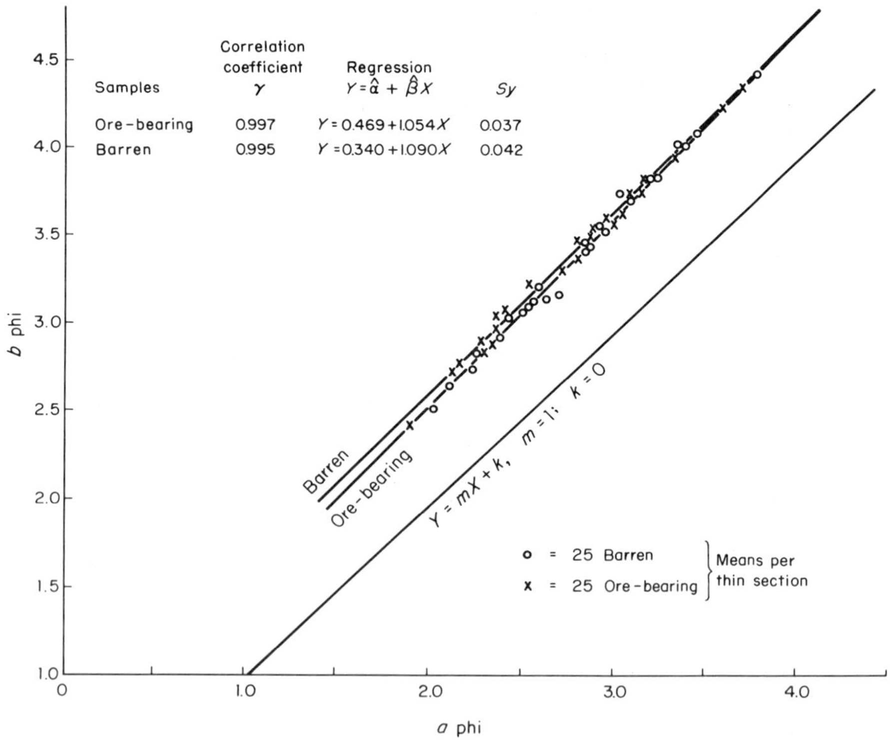

Figure 6.11 Association between long *a* and short *b* axes in quartz grains from Salt Wash sandstones. (*After Griffiths et al., 1956.*)

TABLE 6.9
Association between a and b Axes of Quartz Grains Measured in Thin Sections of Salt Wash Sandstones

		Correlation coefficient	
No. of items	Source of variation	Ore bearing	Barren
25	Specimen means	0.9931	0.9927
10	Field means	0.8049	0.8171
30	Grain means	0.7697	0.5369
7,500	Total population	0.7490	0.7325

vidual grains; and the relationships among means of sets of grains can be completely misleading.

Since the axial measurements obtained from grains in thin sections, or in their projected images, are representations of apparent axes and variation in their measurements is essentially "apparent variation," the evaluation of P_{sh_2} is quite evidently a complex, actually multivariate, problem which requires solution of the equation

$$P_{sh_2} = f(s,sh,o,p)$$

to be adequately interpreted.

With regard to the specific objectives, if it is required to compare two specimens (or generally two populations) without attempting to place a physical interpretation on the results of the comparison, this technique is quite satisfactory. If it is necessary to interpret the physical meaning of the results of the experiment, e.g., relate some difference found to a difference in source area or processes of sedimentation, this technique will not at present yield suitable results.

6.4 Three-dimensional Measurement of Shape

The next step in the measurement of grain shape is to attempt some new approach; we have tested visual comparison and measurement in thin section without very encouraging results. From experience with grain-size measurement the final step is to reduce the complexity of the measurement procedure to its fundamental level; i.e., using the defining equation as a guide, we must devise a procedure which will permit more or less direct measurement of shape. In other words, starting with the five fundamental properties, we shall attempt to isolate that property about which we require information. We remove the effects of most variables by disaggregating the rock specimen, i.e., reducing the population to its elements. Then we select one specific kind of element, particularly that one unaffected by disaggregation technique which is considered to reflect the information on geological processes in which we are interested, and measure the property of interest on that set of elements. We are now forewarned about some of the difficulties we are likely to meet, e.g., operator variation and, particularly, the necessity for ensuring control to remove operator inconsistency. We must not overlook the possibility of losing most of the required information by taking means of compounded sets. There are still some involved problems to be met, such as how many measurements we should make and in what manner.

An experiment was devised as a reconnaissance test to compare three different techniques, visual comparison by means of Rittenhouse's chart (1943*a*), measurement of two axes as recommended by Riley (1941), and measurement of three axes as recommended by Krumbein (1941*b*). The design was not the most efficacious for our purpose, but it suffices to add to the basis for recommendations for future work. In this experiment 13 operators[1] collected 13 specimens from the Ordovician Oswego low-rank graywacke at an outcrop at Skytop on Route 322 in Centre County, Pennsylvania. The specimens were disaggregated, and 50 quartz grains selected from each specimen were mounted on the edge of a glass slide, following the routine suggested by Plumley (1948). On the maximum projection area the long a and intermediate b axes were measured; and by turning the mount on edge, the short c axis was measured in the minimum projection area. Visual sphericity was estimated by comparison of the maximum projection image with the standard chart (Fig. 6.2).

These procedures yield three estimates of sphericity on the same grains, namely, the visual estimate according to Rittenhouse ψ_R; approximation to Riley's sphericity ψ_{Ri} using the long and intermediate axes as follows:

$$\psi_{Ri} = \left(\frac{b}{a}\right)^{1/2} \tag{6.1}$$

and Krumbein's sphericity calculated as

$$\psi_K = \left(\frac{bc}{a^2}\right)^{1/3} \tag{6.2}$$

As the experiment was intended to illustrate a simple experimental design, each operator analyzed a single specimen; hence operator difference and specimen difference are confounded. The frequency distributions may be represented by their summary statistics (Table 6.10).

It seems clear that the three-dimensional measurement yields the lowest sphericity value, as would be expected, and the coefficient of variation is

[1] Students in a graduate class learning how to perform experiments.

TABLE 6.10
Summary Statistics of the Frequency Distribution of Sphericity of Quartz Grains from Oswego Graywacke Measured by Several Techniques

Technique	Mean $\bar{X}\psi$	Standard deviation $\hat{\sigma}\psi$	Coefficient of variation	$\sqrt{b_1}$	b_2	No. of grains
Krumbein	0.754	0.089	11.68	−0.318**	2.873	650
Riley	0.810	0.063	7.78	−0.618**	3.80**	650
Rittenhouse	0.818	0.072	8.80	−0.346**	3.016	650
Curray[a]	0.753	0.077	10.53	+0.030	2.695	300

[a] Curray, 1949; Curray and Griffiths, 1955.

highest in this procedure; all three techniques lead to significant asymmetry, and the Riley values are also significantly leptokurtic. The fourth set of results (Curray) represents an independent set of measurements of 300 quartz grains from two specimens of Oswego graywacke from the same outcrop measured in the same manner and expressed as Krumbein sphericity. The mean sphericity is gratifyingly close to the set of 13 specimens, and the standard deviation is larger than that in either of the two-dimensional techniques. The statistics are not significantly different from normal.

The results of the different measurement techniques can be compared in pairs by means of scatter diagrams (Fig. 6.12 *a* to *c*); if the two techniques yield identical results, they will fall on a straight line bisecting the angle between the coordinate axes. If they differ by a constant factor, the lines will be parallel; if they show inconsistencies, the lines will converge and perhaps cross. For the purpose of discussion it will be assumed that the Krumbein technique is most informative and reliable and the visual technique is least reliable for sphericity estimation. We first compare the two-dimensional estimations, the visual and Riley sphericities (Fig. 6.12*a*). Twelve of the thirteen lines bunch closely around the 1:1 line, and the thirteenth shows marked departure. Since specimen differences and operator differences are confounded, this aberrant result may be due to either, but it is most likely due to operator difference. Not only is this line different from the others, but it crosses them at a very low angle, indicating that the operator was overestimating the visual sphericity in the low values (and underestimating it in the higher values) compared with the Riley measurement. Many of the other lines converge on the 1:1 line, and very few are parallel. This indicates not only that operators differ but that they show inconsistent differences. This demands objective measurement as a basis for estimating sphericity and

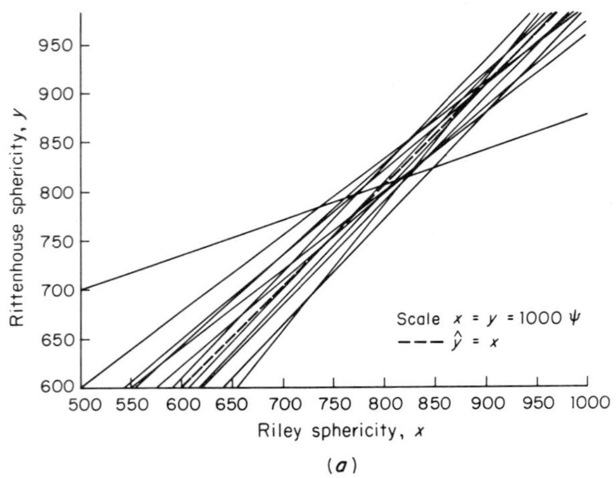

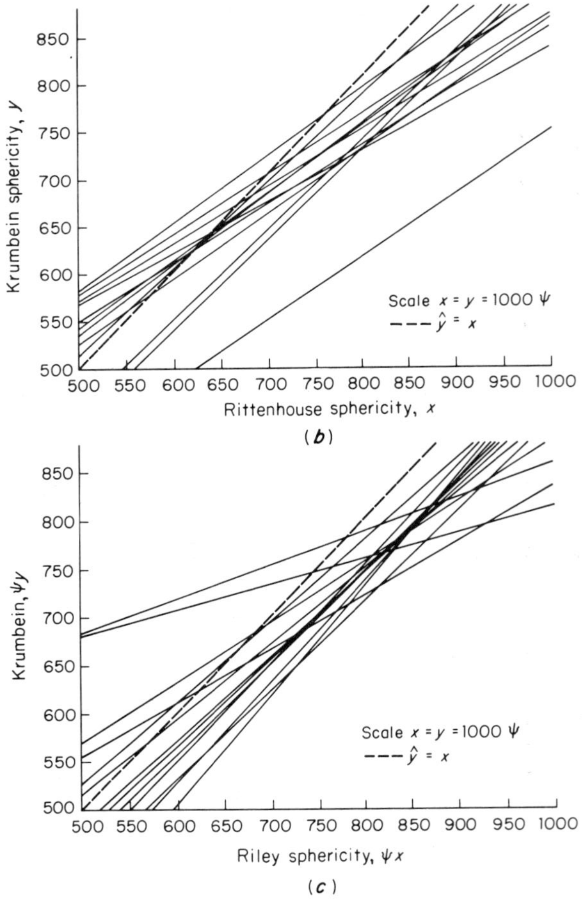

Figure 6.12 Relationships among three techniques for sphericity measurement. (*a*) Regression of Rittenhouse sphericity Y on Riley sphericity X; (*b*) regression of Krumbein sphericity Y on Rittenhouse sphericity X; (*c*) regression of Krumbein sphericity Y on Riley sphericity X.

a design which effectively removes the effect of different operators and reduces their inconsistency.

A comparison of the visual estimation with the Krumbein measurement technique (Fig. 6.12*b*) shows somewhat larger discrepancies among the operators; at least three depart from the other ten. Furthermore, all the lines cross the 1:1, indicating that sphericity is visually underestimated in grains with relatively low sphericity and overestimated in grains with high sphericity.

Finally, a comparison of the two measurement techniques also implies marked discrepancy in which possibly four operators are markedly different (Fig. 6.12*c*). The remaining nine are not parallel to the 1:1 line but are more nearly parallel than in the two previous comparisons Measurement is definitely helpful, but experimental control is not something to be taken for granted.

As a result of these several experiments it is clear that if shape is to be estimated at all, the expenditure of time and energy required by the experimental design certainly necessitates measurement, and it may as well be in three dimensions.[1]

6.5 Sphericity and Roundness of Quartz Grains in Sediments

Curray (1949) investigated the nature of variation in the sphericity and roundness of quartz grains in sediments, using three representatives of each of the petrographic groups arkose, low-rank graywacke, and quartzite (rock types as defined by Krynine, 1948). There were no fixed ideas on the number of specimens or outcrops, etc., required because the magnitude of variation at different levels was unknown; so, as a first approximation, two specimens per formation were arbitrarily selected. Since it was believed on the basis of the relevant literature that shape varied with size, three sieve sizes were included in the experiment, and 50 grains in each set were measured. The sampling design is summarized in Table 6.11.

In measuring roundness it was observed that the estimations were differentially affected by different degrees of magnification, a feature common to all attempts to measure shape and one which again emphasizes the importance of interactions among the petrographic properties in nearly all operational procedures. For roundness only one size was used, the 120-mesh, or 125-micron, sieve size. There were 2,700 sphericity measures and 900 roundness values.

6.5.1 Procedure. The 50 grains of quartz were mounted on the edge of a slide by the technique of Plumley (1948, p. 562), and the axes were

[1] As an expression of opinion it may well be practicable to use visual estimation of roundness, but the relatively small range of variation in sphericity will necessitate measurement if the data are to be meaningful.

TABLE 6.11
Sampling Plan for Estimation of Sphericity of Quartz Grains in Sediments

			Mesh[a]			
Rock type	Formation	Sample number	35 (500)	60 (250)	80 (177)	120 (125)
Arkose	Roslyn, Eocene,	R_1		×	×	×
	Washington	R_2	×	×		×
	Sespe, Oligocene,	S_1	×	×		×
	California	S_2	×	×		×
	Fountain, Pennsyl-	F_1	×	×		×
	vanian, Colorado	F_2	×	×		×
Low-rank gray-wacke	Miocene, Trinidad	M_1		×	×	×
		M_2		×	×	×
	Bradford Sand, Devo-	B_1		×	×	×
	nian, Pennsylvania	B_2		×	×	×
	Oswego, Ordovician,	T_1	×	×		×
	Pennsylvania	T_2	×	×		×
Orthoquartzite	Oriskany, Devonian,	O_1	×	×		×
	Pennsylvania	O_2	×	×		×
	Homewood, Pennsyl-	H_1		×	×	×
	vanian, Pennsyl-vania	H_2		×	×	×
	Gatesburg, Cambrian,	G_1	×	×		×
	Pennsylvania	G_2	×	×		×
Total 3	9	18	11	18	7	18

[a] The equivalent in microns is given in parentheses.
Source: After J. R. Curray and J. C. Griffiths, Sphericity and Roundness of Quartz Grains in Sediments, *Bull. Geol. Soc. Am.*, **66(9)**: 1077 (1955) with permission of the Geological Society of America.

selected from the projected images by the procedure used by Griffiths and Rosenfeld (1950, p. 205). The sphericity was expressed as the Krumbein intercept sphericity, the actual value being obtained by graphic determination from a Zingg chart (Krumbein, 1941b, pl. 1, p. 168).

6.5.2 Frequency Distributions of Sphericity by Rock Type and Grain Size. The histograms of sphericity value are given for rock types, for grain sizes, and for the whole set of measurements in the original publication (Curray and Griffiths, 1955, pp. 1078, 1080–1081). The relevant summary statistics are included in Table 6.12. All the frequencies are unimodal but show significant departure from normality in one or other of the β statistics. Nevertheless, if these frequency distributions are compared with normal distributions possessing the same mean and variance, a chi-square test of goodness of fit (see Sec. 17.4) shows no significant departure from normality. An example may clarify these different results. The frequency distribution of sphericity values of 900 quartz grains from three quartzites is shown in Fig. 6.13, and the expected normal distribution with the same mean and variance is included as a continuous curve. A class-by-class comparison using the chi-square criterion shows no evidence of departure from normality, but the moment measure for kurtosis is significantly different from normal at the 5 percent probability level. The main contributions to chi square are indicated in the diagram, and the excesses and deficiencies are seen to be relatively small. They imply that there are deficiencies of observed compared with expected results at the tails and in the shoulders, which presumably accounts for the platykurtosis. This is partly due to the fact that a sphericity in excess of unity is impossible, whereas in the expected normal curve there is no such cutoff. As the spherical grains increase in frequency, this discrepancy will become larger, and a skewness toward low sphericity values will develop. This is a disadvantage attributable to the range in the scale values from zero to unity. Another possible contributor to departures from normality is the possible nonrepresentativeness of the sampling arrangement; for example, a comparison of the 2 specimens of 50 grains each measured

TABLE 6.12
Summary Statistics for Frequency Distributions of Sphericity in Quartz Grains in Sizes and Rock Types

Size, microns	$\bar{X}$	$\hat{\sigma}$	$\sqrt{b_1}$	b_2	n
Sizes over All Rock Types					
500	0.765	0.077	−0.206*	2.981†	550
250	0.747	0.082	−0.081†	2.852†	900
177	0.758	0.085	−0.097†	2.489**	350
125	0.747	0.087	0.168*	2.884†	900
All sizes	0.752	0.083	−0.151**	2.855†	2,700
Rock Types over All Sizes					
Arkose	0.733	0.086	−0.335**	2.870†	900
Graywacke	0.750	0.081	−0.154*	2.846†	900
Quartzite	0.772	0.078	−0.109†	2.734*	900
Total	0.752	0.083	−0.151**	2.855†	2,700

by Curray with the 13 specimens of 50 grains each measured in the class experiment shows an increase in asymmetry ($\sqrt{b_1}$ significant), and it seems likely that the sampling pattern is a very important feature in developing such departures from normality. At least for the purposes of discussing the results of this investigation, the normal model appears an adequate first approximation; the departures from normality are negligible in the total population of 2,700 grains testing by the chi-square criterion, and the magnitude of the asymmetry, which is significantly different from zero at the 1 percent probability level, is not very large. Such departures are then to be attributed to cutoffs at unity arising from the scale limits and possibly nonrepresentative sampling patterns.

6.5.3 Analysis of the Data. As a first step it seems clear that differences in sphericity across grain sizes are inconsistent, and on the basis of these data there is no obvious relationship between grain size and sphericity within the range 125 to 500 microns. It should be emphasized that since sieving depends on the size-shape interaction, this is what one would expect, and, if the relationship between size and sphericity is to be evaluated, sieving should be avoided as a preparation procedure.

Perhaps the most striking feature of Curray's investigation, shown in almost every case (e.g., Curray and Griffiths, 1955, figs. 4 and 5) is that the variation in sphericity of quartz grains is different for different rock types. This heterogeneous variance clearly implies that comparisons of mean sphericity cannot be undertaken by using this sampling pattern. The variation in mean sphericity per sample of the quartzites is larger than that of the graywackes; and those, in turn, show greater variability than the arkoses (Curray and Griffiths, 1955, fig. 5, p. 1087).

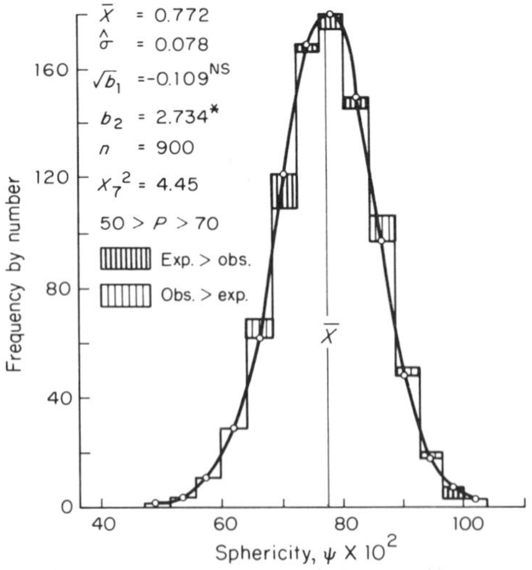

Figure 6.13 Sphericity of quartz grains in quartzites. *(Data from Curray, 1949.)*

TABLE 6.13
Summary Statistics of the Frequency Distributions of Roundness of Quartz Grains in Three Rock types

Rock type	$\bar{X}$	$\hat{\sigma}$	$\sqrt{b_1}$	b_2	n	$P\chi^2$
Arkose	0.378	0.074	−0.295*	3.010	300	0.01–0.001
Graywacke (low rank)	0.418	0.085	0.464**	4.400**	300	0.02–0.01
Quartzite	0.488	0.087	0.690**	3.772*	300	<0.001
Total	0.378	0.094	0.408**	3.974**	900	<0.001

The most important conclusions concern the variability, and until the variabilities are stabilized, a comparison of means is apt to be more misleading than helpful. The mean is, of course, independent of the variance in random samples from a normal distribution (Chaps. 13 and 15); and as we may assume normality here with little loss of generality in the conclusions, our question will essentially concern the randomness of the samples. Since size and sphericity appear to be independent in these data, we may use all three sizes as three replicates of the same specimen, thus yielding three sizes per two specimens per three formations for the three rock types, i.e., 54 samples. A plot of mean against variance for these 54 sample sets of 50 grains each is given in Fig. 6.14, and there appears little evidence of any simple relationship between means and variance for the entire population. We may therefore conclude that these could, indeed, be random samples from some normal parent population. We may therefore state that the population mean is 0.750 for the sphericity of quartz grains in sediments, and the standard deviation is closely approximated by the large-sample value 0.083 sphericity unit. We expect the range in sphericity to be largely encompassed in $\pm 3\sigma$ or $0.750 \pm 0.249 = 0.500 - 1.000$, and if we assume random sample sets of size 50 we expect 3 in 1,000 to be outside the limits $\pm 3\sigma_{\bar{X}}$ or $0.750 \pm 0.012 = 0.738 - 0.762$. As only one of the formation means (arkose) and none of the means for different grain sizes lie outside these limits, we have no evidence except for the arkose that any of these classes differ significantly, and all may be random samples from some normal population.

On the other hand, if we subdivide the population into the three rock types, arkose, graywacke, and quartzite, the scatter of mean against variance is not random; arkose samples tend to vary widely in variance but less in mean value. Quartzite samples tend to show the opposite trend, less variability in variance but a wide scatter in means, and the graywackes (low rank) show relationships between these two. Evidently these samples are

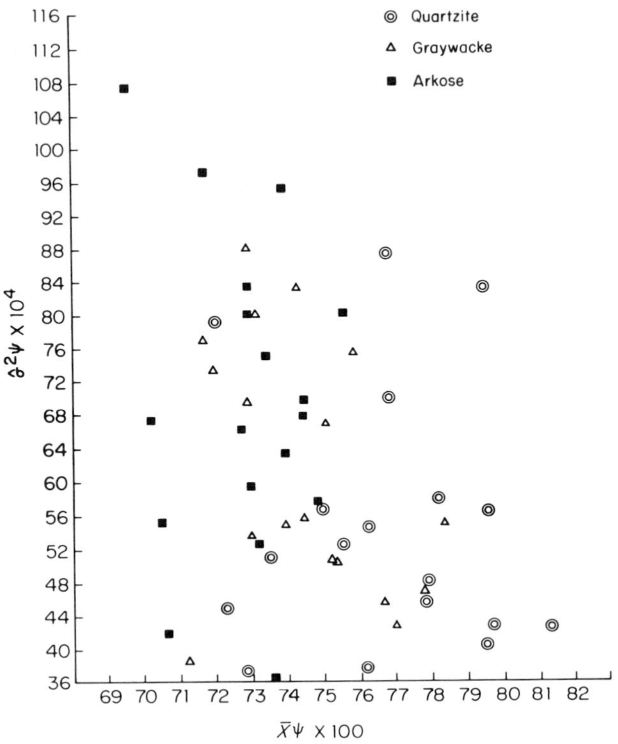

Figure 6.14 Mean and variance of sphericity measurements. (*Data from Curray, 1949.*)

not likely to be random samples from some normal population, and the probable departure is essentially in nonrandomness. Of course, if this did not appear, the grouping into rock types would have little meaning in terms of sphericity.

The frequency distribution of roundness is definitely nonnormal, and there are very large departures from normality in all the frequency distributions both when tested by chi square and when tested by moments. The marked skewness and leptokurtic character of these curves is quite obvious from the frequency distributions (Curray and Griffiths, 1955, figs. 6, 7, pp. 1088, 1089). Presumably, then, the sampling arrangement is very far from random, or the frequency distribution of roundness of quartz grains is nonnormal in sediments, or both may contribute to these results. The use of one grain size may be partly responsible if size and roundness are not independent.

In the published article the analyses of the contributions to the variability in roundness of the quartz grains in the different rock types are treated separately and yield very different results. They may be usefully summarized by studying relationships between mean and variance for each 50-grain sample, of which there are two specimens in each of three formations, or six per rock type. Again when the total population is considered, no trend appears, but when each rock type is considered separately, a markedly non-random arrangement is displayed (Fig. 6.15). Once again the arkoses show a wide variability in variance for a very small range in mean, whereas the graywacke and quartzite samples show a trend in which the variance increases with increase in mean roundness. The scatter in the quartzite is quite large and implies that these formations differ in both sphericity and roundness more than we should expect if the grouping were to lead to homogeneous classes in sphericity and roundness of quartz grains. Once again the heterogeneity of variance prevents straightforward comparison of means.

These diagrams were idealized as a basis for the sampling pattern in sediments and form the basis of the model presented in Fig. 2.3. First of all, these different relationships are due not necessarily to difference in rock types but rather to degree of selective sorting of the grains on the basis of shape. The arkoses are all near-source sediments, the graywackes vary, and the quartzites are of several different kinds. Ideally we can expect near-source sediments, which have suffered relatively little selective sorting, to show wide variability within samples and little among sample means. On the other hand, far-traveled sediments will tend to show wide variability in sample means and less variability within samples. Since quartzites are generally thought of as far-traveled stable detritus, they would be expected to illustrate the latter case, and so some of the "formations" included in this experiment are not true quartzites in terms of genesis although they may be in terms of composition (composed largely of quartz). In fact, since the Gatesburg quartzite contains some 10 percent of microcline, it is likely to be more like an arkose in the pattern of its variability in measured petrographic properties.

In future studies, then, it will be necessary to arrange the sampling pattern differently in rocks which have suffered different degrees of selective sorting; and unless this sampling is appropriately arranged, it will not be possible to find differences among mean shapes (and probably sizes and other measured properties will behave in a similar manner).

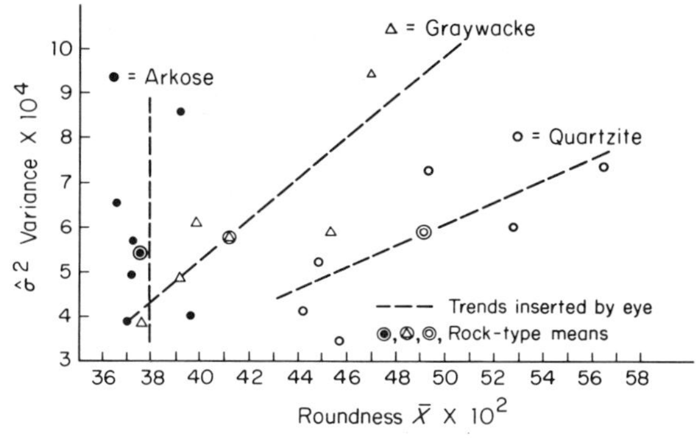

Figure 6.15 Mean and variance of roundness measurements. *(Data from Curray, 1949.)*

6.6 Conclusions on the Measurement of Shapes of Grains in Sediments

This series of investigations aimed at elucidation of the most efficient procedure for studying changes in shape of grains in sediments has led to some firm conclusions with regard to the nature of the experimental design; and it surely indicates that unless these factors are duly accounted for, the information on the petrography and petrology of sediments will be at best clouded and at worst definitely misleading.

First, it is necessary to use a procedure which actually contains information on the concept of the property of interest. Second, it is necessary to be able to sample the population in an appropriate and quite definite manner; i.e., the sampling arrangement must be fitted to the variability pattern of the population. In the case of sphericity it is therefore necessary to measure the property directly and in at least three dimensions. Perhaps roundness can successfully be estimated by visual comparison, but in both cases operator variation must be segregated, and the experimental determination must be demonstrably in control if the results are to be meaningful. This is particularly necessary in the case of subjective assessment procedures such as visual comparison with standards, although it cannot be neglected with impunity even in measurement procedure.

Finally, it is unnecessary and may add to loss of information if ratios are used to express the multidimensional measure; it is advantageous to use the measurement on logarithmic scale and study these as multivariate scatter diagrams accompanied by the appropriate statistical measures. In several recent investigations along these lines it has been found that the short c axis contains the most information, the long a axis values are next in importance, and frequently the intermediate b axis may be omitted with small loss of information (Hulbe, 1957; Biederman, 1959; Griffiths and Smith, 1964).

If this procedure is adopted, each axis measure is an estimate of "size," and their mutual interrelationships express the simultaneous variation in size and "shape," hence fulfilling exactly the requirements theoretically deduced from the defining equation. This indicates that the measure obtained, P_{sh}, is a function of size, shape, and their mutual interaction, that is, $P_{sh} = f(s, sh)$. Only when size is shown to be independent of shape can we separate one from the other and talk meaningfully about a univariate measurement.

6.7 Variation in Grain Shape and Its Interpretation

A number of investigations of variation in the shape of particles in sedimentary rocks have been performed in both the laboratory and the field, and attempts have been made to interpret their meaning in a geological context (see, for example, Pettijohn, 1957, pp. 544ff.). In the laboratory, tumbling-barrel experiments have been repeated many times since Daubrée's (1879)

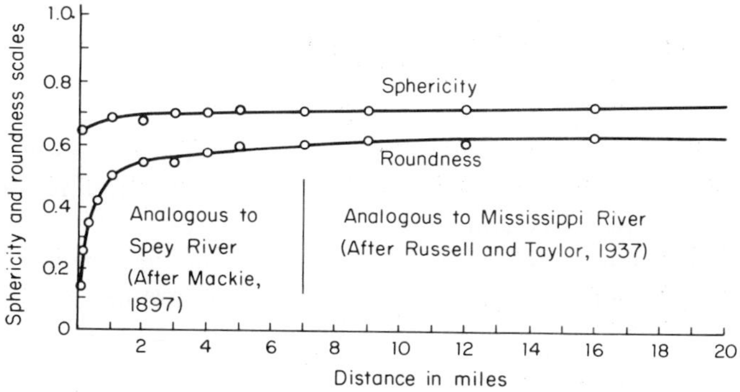

Figure 6.16 Form of the function representing the relationship of change in particle shape with distance (or time). [*Based on data in The Effect of Abrasion on the Size, Shape and Roundness of Rock-fragments, by W. C. Krumbein, J. Geol.,* **49**:*491, table 1 (1941), by permission of The University of Chicago Press. Copyright 1941 by the University of Chicago.*]

initial experiments (e.g., Krumbein, 1941a; Thiel, 1940). The principal factor involved in these experiments is wear. An analogous study in the field has been described by Mackie (1897), who examined the change in shape of particles along the Spey, a relatively short, fast river in Scotland; wear was again the dominant factor. The form of the function representing the relationship of change of shape with time or distance is similar in these investigations (Fig. 6.16); sphericity starts around 0.5 to 0.6 unit, changes rapidly in the first few miles, and then changes exceedingly slowly. Similarly, roundness, which starts around 0.2 unit in laboratory experiments, changes rapidly in the early stages and then reaches what is apparently an equilibrium value. It seems that change of shape of particles arising from abrasion or wear may take place quite rapidly in the early stages of sedimentation, near the source area, and subsequently becomes approximately constant, or, in other words, rapidly approaches an equilibrium value.

An investigation of the change in shape of quartz grains with distance over a range of 1,100 miles in the Mississippi River sands was described by Russell and Taylor (1937); in this case, there was little absolute shape change over the entire distance.[1] It seems evident that the samples in this study represent a part of the curve in Fig. 6.16 where shape changes are very slow. It may be concluded from these two series of investigations that, in the bulk of the detrital sediments, wear plays little role except in those few near-source sediments. For the greater part, the process of selective sorting is by far

[1] The results are difficult to evaluate unequivocally because, from the experiments on measurement earlier described, the interaction of operator with measurement procedure could have encompassed the entire range of variation described by Russell and Taylor (1937).

the dominant factor. If this interpretation is correct, a model may be proposed which starts with a population of particles, preferably ubiquitous quartz grains, in which both sphericity and roundness show a wide range of variation and in which a small random sample of, say, 50 grains will include representatives of the entire range; differences among sample means will be small. This is a stage 1 population of the sampling model (Fig. 2.3). Subsequently, after transport of several miles to several hundreds of miles and deposition, accompanied by selective sorting, in layers (strata or sedimentation units), a small sample of, say, 50 grains taken within a layer or unit will show a restricted range in sphericity and roundness; there will be considerable differences among sample means representing different layers or units. This is a stage 3 population from the sampling model (Fig. 2.3). Thus any population of quartz grains will display the same range and mean, but the relationship between the means and variances of small random samples will reflect the degree of selective sorting and will in effect be an index of maturity (Folk, 1951). This model embraces the results from laboratory investigations, field studies, and the measurements made on sediments by Curray (1949; Curray and Griffiths, 1955). It will be observed that the sampling is again a critical feature in evaluating the relationship, and it appears advisable to apply stratified random sampling, appropriately arranged, to all populations, since the relationship of means to variances among sedimentation units is the feature which enables one to differentiate between near-source and far-traveled deposits. It may be added that implicit in the model is the assumption that most source areas have quartz grains which range widely in both sphericity and roundness and that, to a first approximation, all shapes are equally available; if this assumption is valid, studies of grain shape will not supply information on differences in source area but only on processes of erosion, transportation, and deposition.

A feature rarely considered in shape studies, and one which is quite evident in examining sedimentary rocks, is that extremely angular particles or particles of low sphericity are rare and occur in sediments dominated by chemical rather than detrital processes. For example, in the Salt Wash sandstones on the Colorado plateau the quartz grains vary widely in shape but appear similar to those in the typical detrital sediments (Griffiths et al., 1956; and Table 6.7). In specific loci in some specimens, i.e., thin sections, the quartz grains are extremely angular and irregular, and generally these sediments show strongly disturbed textures due to slumping; similarly, in some sediments the quartz grains and quartzose rock fragments are both angular and irregular in shape in restricted areas where carbonate cement is common (Griffiths, 1958b, 1960d). Kuenen and Perdok (1962) investigated the effect of chemical solution, and their results appear to confirm these observations.

It may be concluded, therefore, that in studying variation in the shape of quartz grains, the main feature is the relationship of the means to the variances of small random samples of homogeneous sedimentation units. The information accruing from these studies reflects degree and intensity of the processes of selective sorting and may be used to differentiate depositional environments (Hulbe, 1957; Biederman, 1959; McIntyre, 1961); it will not usually suffice to elucidate problems of provenance.

In those cases where an unusual degree of angularity and unusually low sphericity occur, the shape of the quartz grains is largely a reflection of chemical solution accompanying unusual episodes in early or late diagenesis.

Studies in progress on pebbles of quartzite from the Tuscarora scree (Griffiths, 1959a); the Montoursville gravel (Griffiths, 1959b); some glacial, fluvial, and beach gravels (Steinmetz, 1956, 1962); and the Olean conglomerate (Griffiths and Smith, 1964) suggest that these elements also exhibit similar patterns of variation. If these investigations confirm their early indications, the relationships among means and variances within and among samples of sedimentation units may encompass a wide range in grain size; this will imply that shape, measured directly on individual elements, varies independently of size when variation in composition is limited and the effects of fabric are removed.

7 Measurement of grain fabric

This is the arrangement typically exhibited in thermally metamorphosed rocks, and the microstructure described, a criss-cross or *decussate* structure, is for such rocks highly characteristic. The component crystals lie in all directions; *not* at random, by the operation of a mathematical law of chance, but as part of a definite mechanical expedient for minimizing internal stress.

A. Harker, 1939

7.1 Introduction

When the kind of particle m and its size s and shape sh have been specified, the particle is identified; a rock, however, is an aggregate of particles, and many characteristics of rocks depend on the position of their particles in space. There is no known way of reducing this measurement to a single procedure, and in general there are two main concepts, which embrace (1) the position of an individual particle and (2) its mutual relationship to other particles; the first of these is called orientation o and the second packing p. In approaching both of these properties first the concept and then the operational procedures will be described; no clear means for deriving the concepts exist at present, and there are few evaluations of the present procedures (cf. Potter and Pettijohn, 1963). It is quite clear that the measures used are interdependent both among themselves and in regard to the other three properties (e.g., Pettijohn, 1957, p. 74).

Quotation: A. Harker, "Metamorphism," p. 35, Methuen & Co., Ltd., and E. P. Dutton & Co., Inc., 1939.

7.2 Measurement of Grain Orientation: Initial Considerations

Ideally the position of a particle in space should be uniquely defined with reference to some standard axes. In a few unusual cases this can be achieved when the particle is identified and its arrangement specified after extraction from the aggregate (Schwarzacher, 1951; Holmes, 1941; Krumbein, 1939a); some pebbles in gravel deposits and tills can be handled in this manner. In general, it is not possible to follow this practice on grains smaller than, say, 10 mm diameter (cf. Schwarzacher, 1951); hence the fine gravels, sands, silts, and clays must be approached in a different way. First, since the arrangement of the particles in the aggregate must be preserved, thin sections, polished surfaces, and fracture surfaces are generally used. Since a surface is a two-dimensional plane across a three-dimensional body, it represents a sampling of the aggregate; it is necessary to show that the surface sampled is representative of the aggregate from which it was taken. The presence of a unique plane is very difficult to determine and has only rarely been attempted, and a unique plane, e.g., the bedding plane, is not necessarily representative of the arrangement in the aggregate; it is common practice to select a surface perpendicular to bedding. In some cases where directional properties are known, the surface of maximum or minimum permeability may be used (Griffiths and Rosenfeld, 1950, 1953; Griffiths, 1952b; Hutta, 1956). Generally the surface is presumed to be a random sample of the aggregate perpendicular to bedding.

Orientation of a particle may ideally be regarded as the inclination of a fixed direction (axis or plane) within the grain to a fixed direction (axis or plane) outside it. In practice there appears to be no way of finding a unique direction within the grain except by using a crystallographic axis such as the *c* axis of quartz grains (e.g., Fairbairn, 1949). Attempts to relate "true" long axes of quartz grains to the *c* axis have not so far led to unique specification of long axis in thin section (Ingerson and Ramisch, 1942; Rowland, 1946). Attempts to distinguish one end of a quartz grain from another have been made (Dapples and Rominger, 1945), but they were not found practicable by Griffiths and Rosenfeld (1950; see also Schwarzacher, 1951). Hence, in practice the measurement of orientation has been reduced from the ideal concept to an attempt to specify degree of preferred orientation or parallelism of *apparent* long axes of grains, e.g., quartz, measured in thin sections cut with reference to the bedding plane where this is clearly recognizable.

When this operational definition is adopted, the measurement procedure is aimed at establishing presence or absence of preferred orientation, and the extremes are random orientation, wherein equal numbers of grain axes point in each direction, and perfect preferred orientation, in which all the axes point in one direction. In practice the extreme of perfect random orientation is rarely found, and most aggregates possess some degree of preferred orientation. Since several tens to hundreds of grains are measured, the specification of degree of preferred orientation must take the numbers measured into account.

7.3 Experimental Procedure

In practice it is necessary to select a single mineral entity such as quartz (m = constant) and measure its orientation. It seems likely that the selection of mica (megamica; see Sec. 9.3) would be more efficient and perhaps more meaningful because its disk shape leads to more consistent orientation (sh = constant, and range in size reduced), but difficulties arise as a result of bending and twisting of mica flakes in many sediments.[1] If quartz is chosen, a uniquely defined axis must be selected; the long axis defined and measured as described on page 120 and in Fig. 6.7 is one such procedure. Next, it is necessary to select a fixed reference direction and then measure the angle of inclination of long axis to the fixed reference direction. Since there appears no way of distinguishing the ends of the axes, the inclination measurements extend over 180 rather than 360°. Again, in practice, if there is a preferred orientation and measurements are performed by using the mean as the fixed direction, the long axes tend to concentrate in two modes at small angles to the reference direction (Fig. 7.1*a*), leading to a U-shaped frequency distribution (Fig. 7.1*b*). Since the reference direction may be rotated without loss of generality, it is more convenient to measure inclination as deviations from the reference direction (Fig. 7.1*c*), leading to a range of measurements from +90 to −90° and a unimodal frequency distribution (Fig. 7.1*d*). The data so collected may be recorded on data cards simultaneously with size and shape measurements (see Fig. 12.1).

A number of difficulties arise in the use of this measurement procedure, some of which are common to all procedures, such as the problem of sampling discussed in Sec. 7.4. The interdependence of the measurements may be illustrated by the difficulty which arises when the grain image in section approaches circularity; in such circumstances choice of the long axis becomes ambiguous, and so variation in degree of preferred orientation is a function of shape. Similarly, size affects the measurement, the smaller grains (less than, say, 50 microns) being more difficult to measure.

It has been shown (Griffiths and Rosenfeld, 1950, fig. 6, p. 7) that the variance decreases with increase of elongation, and in many cases the more circular grains should be rejected from measures of the degree of preferred orientation.

The measurement procedure leads to consideration of the variance (standard deviation), as the mean is largely arbitrary in most cases (Jizba,

[1] When inclination of quartz, measured as described in the succeeding procedure, is plotted against axial ratio, it is found that variance of inclination decreases sharply below an axial ratio of 0.5 (see Griffiths and Rosenfeld, 1950, fig. 6, and the example of Peru Sand described in Sec. 7.7).

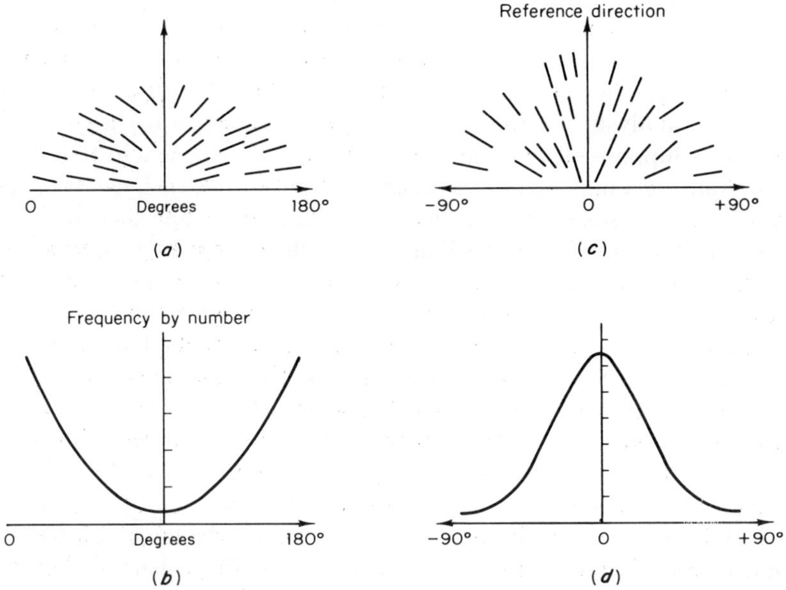

Figure 7.1 Unimodality produced by rotation of reference directions. (*a*) Orientation of long *a* axes; reference direction bedding; (*b*) U-shaped frequency distribution from (*a*); (*c*) orientation of long *a* axes measured as deviations from bedding; reference direction rotated 90°; (*d*) unimodal frequency distribution from (*c*). (*After Griffiths et al., 1956.*)

1953); degree of preferred orientation is then estimated by the variance or standard deviation, and the variance-ratio test is a suitable procedure for testing significance of orientation (see also Potter and Mast, 1963). The smaller the variance (or standard deviation), the higher the degree of preferred orientation. This procedure also leads to well-known frequency distributions for the measurements when the fabric is simple.

7.4 Sampling Procedure

A thin section is a sample of the specimen, which, in turn, is a sample of the aggregate; since not all the (quartz) grains in a thin section are measured, this represents another subsampling stage. In practice a number of microscope fields may be selected, say 10, and a few (4 or 5) grains measured in each field. In practice the fields are distributed as a random 3-4-3 grid of fields over the area of section, and the grains are randomly selected in each field.

If the fabric is simple, this will generally suffice to estimate the degree of preferred orientation, but it is not generally known at the outset whether the population is simple or complex in fabric. This leads to a very troublesome sampling problem for which no completely satisfactory solution has been found. The analysis of the fabric in terms of degree of preferred orientation measured by the variance of inclination takes the form of a variance comparison (analysis of variance; see Chaps. 18 et seq.) at the different sampling levels. Suppose all the grains are pointing more or less in the same direction, representing a high degree of preferred orientation; then each field is a random sample of the *same homogeneous* population, and each field will yield an estimate of the population variance, i.e., the degree of preferred orientation. Variance among mean inclinations per field will be small. On the other hand, suppose there is no preferred orientation, and approximately equal numbers of grain axes point in each direction. Again each field (set of grain-axis inclinations) will be a random sample of the *same homogeneous* population, and variance within the fields will estimate the variance of the population; variance among field means will be small. Therefore the difference between a small and a large variance among grains within fields is all that differentiates a high from a low degree of preferred orientation.

It is, of course, well known that sediments are layered and that direction and degree of preferred orientation may vary from layer to layer. Suppose the microscope fields fall within the layers and the degree of preferred orientation within layers is high but differs from layer to layer in direction. Then variance within fields will be small and between fields will be large. Indeed, total variance for the thin section may be very misleading in this case.

A more realistic model arises when, as in practice, the microscope fields overlap the layers. In this case, by using the last model, variance within fields becomes large and between field means averages out, leading to results similar to that from random orientation.

These different cases are illustrated diagrammatically in Fig. 7.2, where case 1 represents random orientation; case 2, a high degree of preferred orientation common to a number of thin sections. Case 3 illustrates differences among thin sections but a high degree of preferred orientation within each thin section. Case 4 represents a layered sediment where each field is confined within a layer; in case 5 this sampling restriction fails. It seems evident that this sampling problem must be solved if realistic estimates of degree of preferred orientation are to be obtained. It should also be evident that these possibilities arise wherever a fabric or structure is analyzed by a measure of orientation, e.g., in petrofabrics, in studies of paleocurrent directions and paleomagnetism, and in many studies of structural geology. It is, of course, conceptually simple to decide that measures should be confined to homogeneous populations, but it is exceedingly difficult to achieve this in practice.

Attempts have been made (e.g., Emery, 1954) to show that the terms massive and bedded really lead to disparate results in terms of fabric (degree of preferred orientation); differences do appear, but there is also considerable

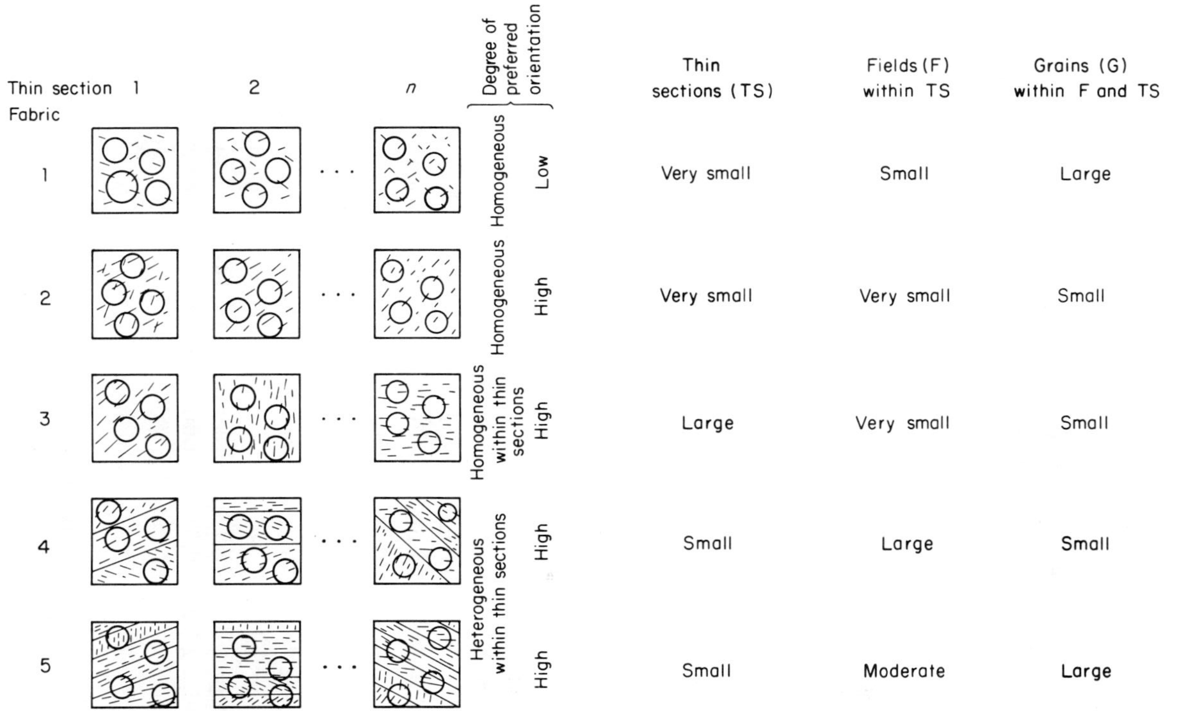

Figure 7.2 Sources of sampling variation in measurement of preferred orientation of long axis of quartz grains in thin section.

overlap (see also Hutta, 1956; Griffiths et al., 1956). The results of applying these procedures to 50 specimens of Salt Wash sediments is described by Griffiths et al. (1956, pt. 2, pp. 47ff.), and the sampling problem was first discussed in this report.

7.5 Frequency Distributions of Long-axis Orientation

It is extremely useful to know the form of the frequency distributions likely to arise when a measurement procedure is applied to the elements of a population, and in the case of measurement of preferred orientation by the described procedure the range may be deduced fairly readily under some restrictions; in particular, it is assumed that the fabric is homogeneous over the population sampled (say a thin section).

Suppose as a first extreme that there is no preferred orientation or that the orientation is random, i.e., isotropic or massive. Then an equal number of inclinations will occur in each class over the range ($2a$) of $+90°$ through 0 to $-90°$. This results in a rectangular frequency distribution, and it has been shown that in such a case the mean is $\mu = a/2$ and the variance is $\sigma^2 = 2{,}700$, or the standard deviation is 52° (Griffiths and Rosenfeld, 1953). This is the worst possible case for a homogeneous fabric but not of course for a heterogeneous one.

On the other hand, if all the grains point in the same direction, the variance will be zero; suppose, as a more realistic model, that the inclinations yield a normal frequency distribution, the 6σ, which is more or less equal to the range 180°, or $\sigma = 30°$, and $\sigma^2 = 900$.

We expect in practice to encounter frequency distributions varying between rectangular and normal, and in general this is what we find; Rosenfeld (1953, pp. 116ff.) attempted to fit a parabolic frequency function to his observations from measurements of the preferred orientation in 87 specimens of Oriskany orthoquartzite, but a combination of rectangular and normal would appear to fit his data. Emery (1954) found a range of standard deviations from 52 to 21° in specimens from the Berea and Pocono low-rank graywackes; and Hutta, in an extensive investigation of orientation, by the procedure herein described found a range of 20 to 58.5° in 322 thin sections (Hutta, 1956, fig. 98, p. 86). A number of examples of heterogeneous fabric are illustrated by Griffiths et al. (1956, pp. 47ff.).

7.6 Data Analysis

In order to ensure that a suitable estimate of the variance is obtained, it is advisable to compute the minimum variance by following the recommendation of Chayes (1954a; see also Griffiths et al., 1956, p. 78). A tabular sheet convenient for this purpose is shown in Table 7.1, although if large numbers of these computations are to be performed, a more efficient procedure

TABLE 7.1
Minimum Variance Origin

Sample #										
		N =						$R = 180$		
(1)	(2)	(3)	(4)	(5)	(6)	(7)	(8)	(9)	(10)	(11)
Lower class limit X_a	Frequency F	Cumulative frequency $S(F)$	Cumulative product $S(F)(X_a)$	$\bar{X}_a$	$\bar{X}_a - \bar{X}_1$	$\frac{N_a}{N}$ K_a	$1 - K_a$	$\frac{1}{2}R(1 - K_a)$	$\bar{X}_a - \bar{X}_1 + \frac{1}{2}R(1 - K_a)$	$K_a \times$ col. 10
−80										
−70										
−60										
−50										
−40										
−30										
−20										
−10										
0										
+10										
+20										
+30										
+40										
+50										
+60										
+70										
+80										
+90										
					0.00	1.00	0.00	0.00	0.00	0.00

Note: Minimum variance origin is maximum negative value of $I_v = K_a[(\bar{X}_a - \bar{X}_1) + 90(1 - K_a)]$. Maximum variance origin is maximum positive value of $I_v = K_a[(\bar{X}_a - \bar{X}_1) + 90(1 - K_a)]$.

TABLE 7.2
Value of Variance Ratio F for Number of Grains Measured and Corresponding Minimum Standard Deviations Necessary to Determine the Presence of a Significant Degree of Preferred Orientation

		Probability level		
		<0.05	<0.01	<0.001
$\nu_1 = \nu_2 = 800$	F	1.123	1.180	1.241
	$\hat{\sigma}_2^2$	2,404.2	2,415.0	2,168.6
	$\hat{\sigma}_2$	49.03	49.14	46.57
$\nu_1 = \nu_2 = 200$	F	1.26	1.39	1.551
	$\hat{\sigma}_2^2$	2,142.3	1,942.4	1,740.8
	$\hat{\sigma}_2$	46.29	44.07	41.72
$\nu_1 = \nu_2 = 150$	F	1.315	1.458	1.657
	$\hat{\sigma}_2^2$	2,053.2	1,836.0	1,624.5
	$\hat{\sigma}_2$	45.31	42.25	40.30
$\nu_1 = \nu_2 = 100$	F	1.39	1.59	
	$\hat{\sigma}_2^2$	1,942.45	1,698.113	
	$\hat{\sigma}_2$	44.07	41.21	

Source: Modified from Griffiths, 1952b, table 8, p. 53.
Note: For example, when 100 axial inclinations are measured, a standard deviation less than 41.2 is required to establish a degree of preferred (nonrandom) orientation at the 1 percent (P_{01}) probability level.

is to program the technique for an electronic computer.[1] The frequency histogram is also useful in displaying the form of the data and may indicate their unimodality or otherwise and whether the application of minimum-variance procedure is necessary.

With the minimum variance given, the presence of some degree of preferred orientation can be decided by a number of tests. Chayes used both chi square and the correlation coefficient in testing petrofabric data (Fairbairn, 1949, chaps. 22 and 23); Eisenhart (mimeographed notes, 1954) and Tukey (in Griffiths et al., 1956) described tests of such data (see also Rusnak, 1957a, b; Potter and Pettijohn, 1963). Watson and Irving (1957) describe tests for paleomagnetic data, and Griffiths and Rosenfeld (1953) and Griffiths (1953a) describe the variance-ratio test. Since the Tukey isotropicity test yields results substantially similar to those of the variance-ratio test in 50

[1] A computer program has been prepared for the IBM 1620 electronic computer, and the write-up may be obtained from the Department of Geochemistry and Mineralogy, The Pennsylvania State University, University Park, Pennsylvania.

specimens of Salt Wash sandstones (Griffiths et al., 1956, fig. 11-18, p. 77), the latter test is certainly more convenient and easily applied in practice.

This significance test is based on the fact that in a homogeneous fabric the largest variance will be found in a rectangular frequency distribution representing random orientation. The variance ratio (see Sec. 17.9) is usually expressed as $F_{\alpha,\nu_1,\nu_2} = \sigma_1^2/\sigma_2^2$, where F is the Snedecor's F distribution (see Fisher and Yates, 1948, pp. 39ff.; Arkin and Colton, 1950, p. 117) with ν_1 being the degrees of freedom for the larger variance σ_1^2 and ν_2 the degrees of freedom for the smaller variance σ_2^2. Since the largest variance in a rectangular distribution with range $\pm 90°$ is 2,700, then $\sigma_1^2 = 2{,}700$, and the number of degrees of freedom for this theoretical distribution is strictly infinite. If the number of grains measured is n, then $\nu_2 = n - 1$; and by selecting an appropriate probability level (P_{05}), the F value for a significant departure from randomness can be determined, Table 7.2. It is recommended that $\nu_1 = \nu_2 =$ number of measurements (when $n > 100$) to reduce the sensitivity of the test (Griffiths and Rosenfeld, 1953, p. 213; Griffiths, 1952b, p. 53).

7.7 An Illustrative Example: Grain Orientation in the Peru Sand

The Peru sand[1] is a fine-grained, silty low-rank graywacke of Pennsylvanian age from Oklahoma. Two thin sections were cut from the core, both

[1] Dept. of Mineralogy Pennsylvania State Cat. No. 5857. The core sample was supplied by W. E. Johnson of the Penn Grade Crude Oil Laboratories, Bradford, Pennsylvania. He and J. N. Breston, of the Pennsylvania Oil Industry, Bradford, Pennsylvania, supplied the permeability plot.

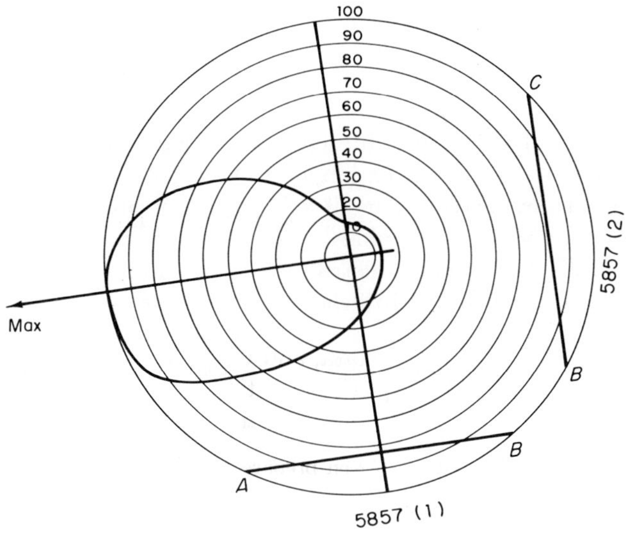

Figure 7.3 Directional permeability plot for Peru sand core, PSU 5857.

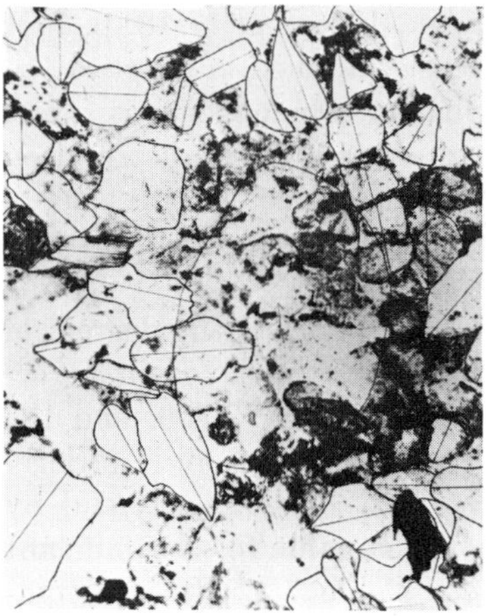

Figure 7.4 Microscope field of thin section showing quartz-grain images; long *a* axes and permeability plot of a Bradford sand, PNY 22(1) (see Griffiths and Rosenfeld, 1950, pp. 250ff.); magnification 200×.

perpendicular to bedding. One was parallel and the other perpendicular to maximum permeability directions (see Fig. 7.3). The permeability plot is markedly asymmetric, suggesting a possible imbrication in the fabric. Each thin section was sampled by 10 fields arranged in a 3-4-3 grid over the area of rock section. Twenty-five grains were randomly selected from each field and the long *a* axis, short *b* axis, and inclination of long axis were measured on the projected images by a single operator (L. M. Banks). A typical field with grain images and long axes marked is shown in Fig. 7.4. The data consist of 250 measurements of each of the three variates on each thin section.

7.8 Analysis of the Data

Quartz grains are ellipsoidal in shape; if they show a strong preferred orientation, it is to be expected that the shape, expressed as the axial ratio b/a, will differ in different directions. A test of axial-ratio variation is therefore some indication of the presence of a preferred orientation. The frequency distributions of axial ratios of quartz grains from each thin section of core 5857 are summarized as relevant statistics in Table 7.3.

The frequency distributions are unimodal and symmetrical ($\sqrt{b_1}$ not significantly different from zero). They differ from a normal distribution in peakedness. Both sets of data are platykurtic; i.e., there is a deficiency in the central classes and an excess at the tails of the frequency distributions. These deviations are not large, because when the observed frequencies are compared with expected normal frequency distributions with the same means and standard deviations, the chi-square (χ^2) values show no evidence of departures from normality (see Table 7.3). Hence the frequency distributions are approximately normal.

TABLE 7.3
Summary Statistics for Frequency Distributions of Axial Ratios b/a of Quartz Grains from Specimens 5857(1) and 5857(2) of Peru Sand

Statistic	5857 (1)	5857 (2)
$\bar{X}$	0.623	0.639
$\hat{\sigma}$	0.158	0.168
$\sqrt{b_1}$	0.0001	0.079
b_2	2.496*	2.293**
n	250	248[a]
χ^2	10.41	16.09
P	$70 > P > 80$	$30 > P > 20$

[a] Measurements of axial ratio on two grains not recordered.

If the two thin sections are now compared in terms of axial ratio of the quartz grains by analysis-of-variance procedure,[1] differences among means per thin section are not significantly larger than variation among field means within thin sections (Table 7.4). If the two thin sections are separately analyzed (Table 7.5), it can be seen that axial ratios of quartz grains are quite homogeneous in the section parallel to maximum permeability, but in the thin section parallel to minimum permeability differences among field means are significantly (at the 1 percent probability level) larger than vari-

[1] Analysis of variance is described in Chaps. 18 et seq.

TABLE 7.4
Analysis of Variance of Axial Ratio of Quartz Grains in Core 5857 of Peru Sand

Source of variation	d.f.	Sum of square	Mean squares	Variance ratio F
Between thin sections	1	26,484.64	26,484.64	1
Among fields	18	968,716.20	53,817.57	2.117*
Among grains	487[a]	12,152,271.36	25,423.16	
Total	497[a]	13,147,472.20		

[a] Degrees of freedom reduced by 2 to account for missing values.

TABLE 7.5
Analysis of Variance of Axial Ratios b/a of Quartz in Individual Thin Sections of Peru Sand

Sources of variation	d.f.	Sum of squares	Mean square	Variance ratio F
(a) Thin Section 5857(1) Parallel to Maximum Permeability				
Among fields	9	249,067.56	27,674.17	1.104
Among grains (error)	240	6,014,417.12	25,960.07	
Total	249	6,263,484.68	25,154.56	
(b) Thin Section 5857(2) Parallel to Minimum Permeability				
Among fields	9	719,648.68	79,960.96	3.101**
Among grains (error)	238	6,137,854.22	25,789.30	
Total	247	6,857,502.90	27,540.17	

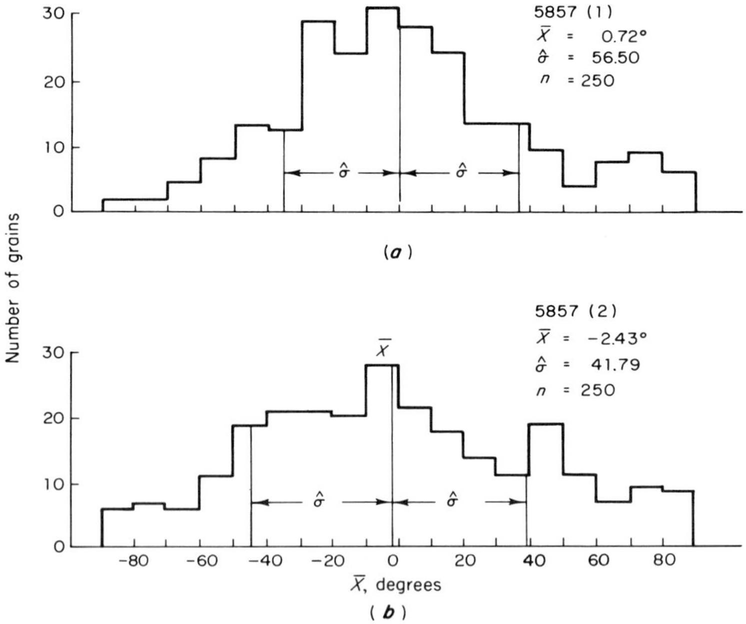

Figure 7.5 Frequency histograms of quartz-grain orientation in thin sections of Peru sand. (*a*) Section parallel to minimum permeability; (*b*) section parallel to maximum permeability.

ation among grains. It seems evident that axial ratio varies in a patchy manner in the section perpendicular to maximum permeability. This, in turn, suggests that while orientation in the maximum permeability direction may be homogeneous, it is possible that in the section perpendicular to that direction [5857 (2)] orientation is not homogeneous. It would certainly be prudent to treat the thin sections individually in the initial analysis.

The frequency distributions of long-axis inclination are illustrated in Fig. 7.5*a* and *b*, and the pronounced mode in the section parallel to maximum permeability [5857(1), Fig. 7.5*a*] is in marked contrast to the less well defined mode in the section perpendicular to maximum permeability [5857(2), Fig. 7.5*b*]. The relevant statistics are included in Table 7.6.

The analysis of variance for the inclination of long *a* axis of quartz grains in these two thin sections is summarized in Table 7.7, and the results are similar to those for the axial-ratio shape of the quartz grains.

While field means contribute no more to variation than grains within fields, the sections show somewhat divergent features; in the direction parallel to maximum permeability the field mean variance is less than the grains variance, whereas in the section perpendicular to maximum permeability field mean variation is larger than that among grains within fields. There is little point in comparing mean inclination per thin section, because we wish to compare degree of preferred orientation, i.e., the variances. Since both thin sections are homogeneous, the total variance may be used for the comparison, yielding a variance ratio $F = 1{,}700.29/1{,}307.81 = 1.300$, which, with degrees of freedom $\nu_1 = 249$, $\nu_2 = 248$, is significant at the 5 percent level. Therefore in this example the highest degree of preferred orientation occurs in the direction parallel to maximum permeability.

TABLE 7.6
Summary Statistics for Inclination of Long *a* Axis of Quartz Grains from Peru Sand

	Thin section	
Statistic	5857 (1)	5857 (2)
$\bar{X}$	0.72	−2.43
$\hat{\sigma}$	36.50	41.80
n	250	249[a]

[a] One measurement of grain inclination not recorded.

TABLE 7.7
Analysis of Variance of Inclination of Long *a* Axis of Quartz Grains in Individual Thin Sections of Peru Sand

Sources of variation	d.f.	Sum of squares	Mean square	Variance ratio F
(a) Thin Section 5857(1) Parallel to Maximum Permeability				
Among field means	9	5,378.76	597.64	1†
Among grains (error)	240	320,266.08	1,334.442	
Total	249	325,644.84	1,307.811	
(b) Thin Section 5857(2) Parallel to Minimum Permeability				
Among field means	9	22,946.18	2,549.58	1.528†
Among grains (error)	239	398,726.89	1,668.31	
Total	248	421,673.07	1,700.29	

7.9 Some General Conclusions from Analysis of Orientation Data

From analysis of this kind, in which care is taken to segregate interfering sources of variation, such as differences among operators and differences among field means, it is generally possible to decide the degree of preferred orientation, and the highest degree of preferred orientation is generally found in sections parallel to maximum permeability (Griffiths and Rosenfeld, 1950, 1953; Griffiths, 1953a; Hutta, 1956). One of the most important sources of variation, when a number of operators is employed to perform the rather extensive number of measurements, is the interaction operators by thin sections. If this interaction is significant, operators differ inconsistently from thin section to thin section, and the experiment breaks down; comparisons at higher levels are then impossible to interpret. Similarly, if variation among fields is significantly larger than that among grains, it implies a difference in inclination in different areas of the slide. Averaging inclination, using the total variance, in such cases may be unjustifiable, and its meaning is questionable. The sampling problem described earlier arises when variation is not homogeneous over a thin section or other population of interest. Whenever measurements of inclination are performed, it is necessary to obtain homogeneity before comparing means (average direction or orientation) or variances (degree of preferred orientation). If there is no preferred orientation, i.e., if orientation is random, then the average is hardly representative of any specific direction assignable to some physical cause such as current direction, magnetic bearing, etc. Furthermore, if there is a high degree of preferred orientation but means vary from sampling level to sampling level, combination of means without appropriate weighting is not very meaningful. At least for the present it is necessary to demonstrate homogeneous variance before attempting to assign physical significance to orientation.

When a number of thin sections from a single formation are measured, widely varying results are generally obtained, some specimens showing a high degree of preferred orientation, others not differing from random orientation. An example showing the varying degree of preferred orientation in 33 specimens of Mississippian Maxton oil sand (a low-rank graywacke) from West Virginia is illustrated in Fig. 7.6. Only 40 grains were measured per thin section, and this would suffice to define the degree of preferred orientation only if the orientation was very closely parallel; nevertheless, the results are similar to those of 50 specimens from Morrison Salt Wash sandstones of the Colorado plateau (Griffiths et al., 1956, fig. II-11, p. 55), where 300 grain measurements per thin section were collected. In Fig. 7.6 the significance levels of $P = 0.1$, 1, 5, and 10 percent are inserted; and because the number of grains measured is 40, the 10 percent level is the more realistic. The range of standard deviation is from 27.4 to 59.1, and at the 10 percent probability level 17 of the 33 specimens show no significant departure from randomness. It would, of course, be of dubious value to average the 33 specimens, and in 17 specimens the average inclination is of dubious physical significance.

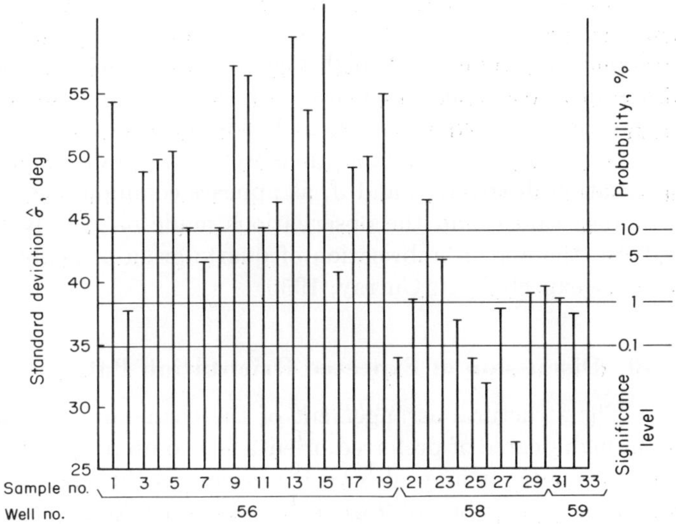

Figure 7.6 Perfection of parallelism in quartz-grain orientation, Maxton sandstone, W.Va.

TABLE 7.8
Bedding versus Preferred Orientation in Some Mississippian Sediments

$N = 32$

	Standard deviation of inclination, degrees							
	Significant					Not significant		
	25	25–30	30–35	35–40	40–46	46–50	>50	Total
Cross-bedded	1	1						2
Bedded		9	3					12
Faintly bedded		1	2	1	4		1	9
Massive			2	2	3	1	1	9
Total	1	11	7	3	7	1	2	32

An attempt to relate this measurement to sedimentary structure was made by Emery (1954), who analyzed 16 specimens from the Mississippian Berea oil sands of eastern Ohio and 16 specimens of Mississippian Pocono sandstones from Pennsylvania, all low-rank graywackes. He also classified the microscopic character of the thin sections into those which showed obvious bedding (bedded), those which were isotropic (massive), and those in which the presence of structure could not be decided (dubious). The results are summarized in Table 7.8. There is quite clearly an association between well-defined bedding and high degree of preferred orientation, but there are also some obvious discrepancies. Emery found the highest degree of preferred orientation ($\hat{\sigma} < 30°$) in current-bedded sandstones.

In no case where thin sections were cut parallel to bedding was preferred orientation discovered; and if, as appears common, e.g., the Peru sand, some Bradford sands, and the observations made on pebbly deposits by Cailleux (1938), there is an imbrication of quartz grain or pebble long axes, then this is to be expected (cf. Curray, 1956).

7.10 Discussion of Expected Orientation Patterns

The structural arrangement of the elements in an aggregate may range from high degree of preferred orientation to no preferred orientation (random arrangement); it may also be homogeneous throughout the population analyzed, or it may be heterogeneous. At least four different types exist, and, of course, there are transitions between them. First, there is a sediment (or specimen) with homogeneous high degree of preferred orientation; second, there are sediments with high degree of preferred orientation in which layers (laminae) differ both in direction of orientation and in perfection; third, there are heterogeneous sediments in which the preferred orientation is patchy (disturbed texture); and finally, there are sediments with no degree of preferred orientation (massive). Sediments which illustrate these patterns certainly exist and yield typical frequency distributions; the first generally shows a unimodal symmetrical frequency histogram. The second and third both exhibit polymodal distributions, and the fourth yields a rectangular frequency histogram. While this is quite clear in a conceptual sense, it is very often difficult to decide in practice between the types of fabric unless they are of the first or last type. The interaction of these different fabrics with the sampling arrangement underlies the difficulty in their definition.

By ignoring this practical difficulty, the existence of orientation fabric in a rock can be analyzed hypothetically; there are two kinds of orientation recognized, the one referring to internal crystallographic structure (petrofabric) and the other related to external morphology of the particles. This leads to four classes of fabric (see Fig. 7.7). There may be no preferred orientation, either structural or morphologic, or there may be structural without morphologic or morphologic without structural; and there may be preferred orientation of both structural and morphologic properties. It is fairly simple to deduce that the fourth kind is characteristic of some metamorphic rocks (schists, gneisses, metaquartzites, etc.). Similarly, the class of structural without morphologic is likely to be largely confined to metamorphic rocks (marbles, hornfels, metaquartzites). The class morphologic without structural is probably typical of sediments, although there will be some cases in which both kinds show preferred orientation in sedimentary rocks (Ingerson and Ramisch, 1942; Rowland, 1946). The first class, no orientation, is almost impossible to achieve; the nearest example appears to be "decussate texture" (Harker, 1939, pp. 34–35), and Harker, for some obscure reason, is most careful to state that this does not imply random causes (quotation at chapter head and Harker, 1939, p. 35). Since in many cases morphology reflects internal structure, preferred orientation of one implies preferred orientation of the other; thus, many igneous rocks possess "sedi-

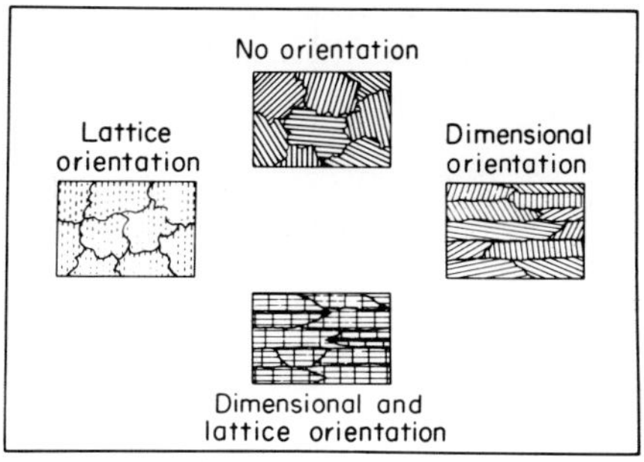

Figure 7.7 Fabric structure in aggregates. (*Drawn by L. Jacobsen.*)

mentary" structures (layered intrusions, crystal sorting, etc.); no rock formed by a simple process can avoid some degree of preferred orientation.[1] When no preferred orientation is found, this, therefore, implies some process superimposed on the primary simple process; in sediments, this generally means that the original structure has been subsequently destroyed, e.g., disturbed bedding, slumping, ingestion by worms, etc.

In developing the fabric of detrital sediments the particles, assumed to be dimensionally anisotropic, will be transported in a moving fluid with their long axes parallel to current direction if in suspension but with long axes perpendicular to the current if in traction (see Fig. 7.8). Deposition will generally maintain this parallelism and generally induce an imbrication (Cailleux, 1938), but the degree of preferred orientation may be different in different directions through the particles (see Fig. 7.8). Of course, since nearly all particles that fall out of suspension may subsequently be affected by traction, the final orientation of detritus may be determined in all, or at least most, cases by traction. While analysis to this degree is fairly straightforward, there are a number of complicating factors in actual processes. Cailleux has described how the size of the substrate upon which the particle comes to rest plays a role, and size and shape are, of course, varying; in fact, of course, the characteristic feature of orientation, *as measured*, is that it is not an independent feature but a function of kind of particle m, size s, shape sh, orientation o, and packing p; and in practice these features always interact (see, for example, Hutta, 1956).

The structural fabrics of sediments are summarized in Fig. 7.9; beach and river deposits (after Cailleux, 1938) show pronounced preferred orientation and differing degrees of imbrication of their particles. Glacial deposits also often show preferred orientation of their constituents (Holmes, 1941; Richter, 1932; Krumbein, 1939a), all influenced essentially by dominant current directions. Particles in gravity-slide deposits generally show a parallelism, with the surface of slide being the dominant reference direction (Miner, 1934). Last, sediments with disturbed structures, sometimes called *slumped*

[1] If only for the reason that Nature, even more than human beings, abhors randomness!

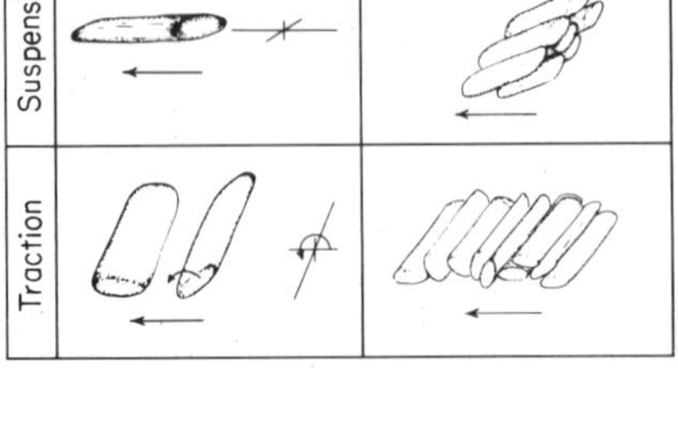

Figure 7.8 Genesis of grain orientation in sediments. (*Drawn by L. Jacobsen.*)

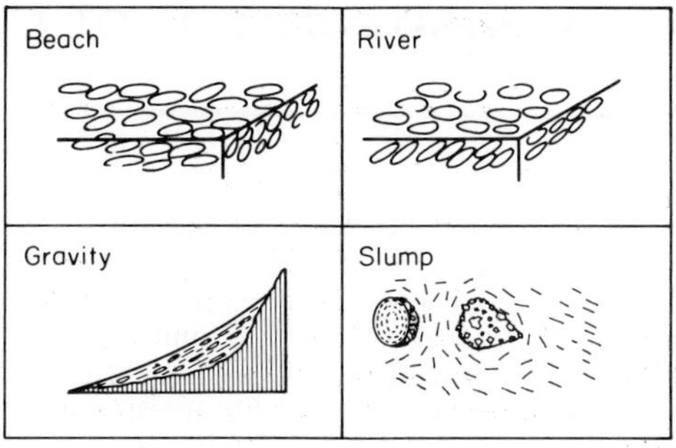

Figure 7.9 Orientation and environment. (*Drawn by L. Jacobsen.*)

sediments, show patchy orientation patterns, including circular patterns in some cases (so-called "storm rollers").[1]

From this analysis it is evident that all the preferred orientation patterns found by measurement are easily interpreted into one or another of these mechanisms. Generally, some degree of preferred orientation is to be expected, and no preferred orientation suggests that sampling arrangement has confused the existing pattern. Until some satisfactory solution to the sampling arrangement is found, only the simplest homogeneous fabric patterns will yield reliable results.

[1] It is interesting, in examining rocks in the field and in hand specimen, to estimate the number of micas per unit area showing their basal section, i.e., showing high luster, in different planes through the rock specimen. If the rock is layered, there should be a preferred direction with a pronounced maximum number of micas per unit area. On the other hand, if the texture is disturbed, there will be many lustrous mica plates on planes in each direction. In many cases this procedure can be quantified by using the field of the hand lens as unit area and sampling different fracture surfaces with three or four fields, counting all mica plates in a field. A surprising number of sediments normally classed as massive appear to have disturbed texture when this test is used.

8 Measurement of grain packing

The term *packing* will here be applied to any manner of arrangement of solid units in which each constituent unit is supported and held in place in the earth's gravitational field by tangent contact with its neighbors.

L. C. Graton and H. J. Fraser, 1935

8.1 Introduction: The Concept

The arrangement of elements (or particles) in an aggregate includes the concept of packing, which refers to spatial density of the elements in the aggregate. Such a concept embraces density of matter in the universe and packing of subatomic particles in and around the nucleus. In all cases the conceptual treatment first considers spheres of constant size and kind and thus ensures that four (composition, size, shape, and orientation) of the five fundamental properties are fixed and that only one property and no interactions arise in the subsequent treatment. Of course it is also generally recognized that such a simplification reduces the validity of the properties of the model, and complications are introduced in various ways. Packing density of ions (Fairbairn, 1943) starts with oxygen spheres and becomes more sophisticated by introducing order-disorder, polarizability, directional field strength, etc. The shapes of elements in the universe have been quite satisfactorily spherical, but with the advent of direct exploration these elements are tending to deviate more or less from true spheres.

Quotation: Reprinted from Systematic Packing of Spheres with Particular Relation to Porosity and Permeability, by L. C. Graton and H. J. Fraser, *J. Geol.*, **43:** 790 (1935), by permission of The University of Chicago Press. Copyright 1935 by the University of Chicago.

Packing in aggregates of spheres of sand size has long been the subject of theoretical and experimental treatment (Slichter, 1899; Graton and Fraser, 1935; Pettijohn, 1957; Fatt, 1957, 1958), but observations on natural aggregates are scarce although, starting with Cailleux (1938), these studies have been increasing in number (Taylor, 1950; Masson, 1951; Emery, 1954; Kahn, 1954). Such studies must await development of a suitable technique which permits some estimate of the conceptual property. Many attempts have been made to measure packing indirectly by using porosity or density (e.g., Athy, 1930; Hedberg, 1926); but as with the other properties already discussed, the complexity of interdependent relationships among the fundamental properties and the behavioral characteristics generally defy understanding in all but the simplest aggregates.

8.2 Reducing Packing Concepts to Operational Procedures

The treatment of packing as a concept and the gradual reduction to operational procedures are historically summarized in Table 8.1; most of the investigators have differentiated more than one property in attempting to reduce the concept, and, of course, any estimate of packing p is essentially a function of the five fundamental properties. In other words, contacts between grains will vary in kind (mutual proportions), size, shape, and arrangement. This is the tautologic character met with in all attempts to explain or reduce a concept by definition.

Cailleux (1938) attempted to estimate the contiguity or collective arrangement of particles in an outcrop, thus developing a field technique. The basic operation comprised the comparison of outcrop appearance with photographic standards. Such procedures of visual comparison suffer from all the disadvantages of any subjective assessment procedure (see Sec. 6.2, where visual comparison of sphericity and roundness of grains is discussed).

Most of the other analyses failed to develop into measurement procedures, as may be seen by comparing Table 8.1 with Table 8.2, in which the historical development of techniques is summarized. Taylor (1950) was the first to attempt a measurement procedure in thin section; in this technique two sources of variation in packing arrangement were considered, the shape of the contact and the number of contacts per grain. The kind of contact was classified into five classes, floating (no apparent contacts in the plane of the thin section), tangent contacts, long contacts, concave-convex contacts, and sutured contacts. No attempt was made to standardize composition, size, etc., of the grains in the rock analyzed, so that the measurement level achieved is affected by complex interactions among the fundamental properties. The objective in Taylor's work was an attempt to relate depth to pore-space reduction. Five specimens of rock were used over a depth of 5,000 ft, with results exhibited in Fig. 8.1. Trends quite obviously exist, but with five points and so many interfering sources of variation no unique interpretation is possible.

TABLE 8.1
Summary of the Evolution of the Concept of Grain Packing

Author	Date	Properties of individual grains	Properties of sets of grains	Source
Slichter	1899	Theoretical discussion of packing of spheres		Slichter, 1899
Graton and Fraser	1935			Graton and Fraser, 1935
Cailleux	1938	(*a*) Position	(*b*) Contiguity or collective arrangement	Cailleux, 1938
Taylor	1950	(*a*) Number of contacts per grain (*b*) Type of contact		Taylor, 1950
Masson	1950	Definition: Spatial density of grains in an aggregate (*a*) Number of grain-to-grain contacts (*b*) Configuration of the contacts	(*c*) Areal density of grain sections	*Bull. Am. Assoc. Petrol. Geologists,* **36**:253–277 (1952); Abstracts St. Louis meeting, 1950
Griffiths	1950–1951	Definition: The position of a particle in relation to the surrounding particles		Mimeo. syllabus: "Method of Analysis of Sediments," Mineralogy 513, The Pennsylvania State University
Krynine	1953–1954	(*a*) Contact or touch (*b*) Fit (kind of contact)	(*c*) Closeness or spread	Oral communication quoted in Rosenfeld, 1953, and in Kahn, 1954, 1956
Rosenfeld	1953	(*a*) Percent of grain-to-grain contacts (after Masson)		Rosenfeld, 1953
Emery	1953–1954	(*a*) Number of quartz-to-quartz grain contacts per traverse (× size ÷ length of traverse)		Emery, 1953; Emery and Griffiths, 1954a
Kahn	1954	Definition: Mutual spatial relationships among the grains (*a*) Grain-to-grain contacts = packing proximity (*b*) Shape of contacts (defined but not measured)	(*c*) Packing density	Kahn, 1954, 1956

TABLE 8.2
Measurement Techniques

Author	Date	Technique	Remarks	Sample unit
Cailleux	1938	Comparison of outcrop appearance with visual standards in photographs	Field technique: Contiguity or collective arrangement	Outcrop (field technique)
Taylor	1950	(*a*) Number of contacts per grain (*b*) Classification of shape of contact	Microscopic technique in thin sections: Count number of contacts per grain; classify shape of contact into 5 classes	Thin section
Masson	1951	(*a*) Number of grain-to-grain contacts	Count of grain-to-grain contacts per traverse	Thin section
Emery	1953–1954	(*a*) Number of quartz-grain to quartz-grain contacts	Count of contacts (quarts grain only) and intercept length (corrected for size and composition)	Thin section
Kahn	1954	(*a*) Grain-to-grain contacts: packing proximity (*b*) Areal density of grains: packing density	Count contacts and measure grain intercepts and traverse length (corrected for grain size)	Thin section

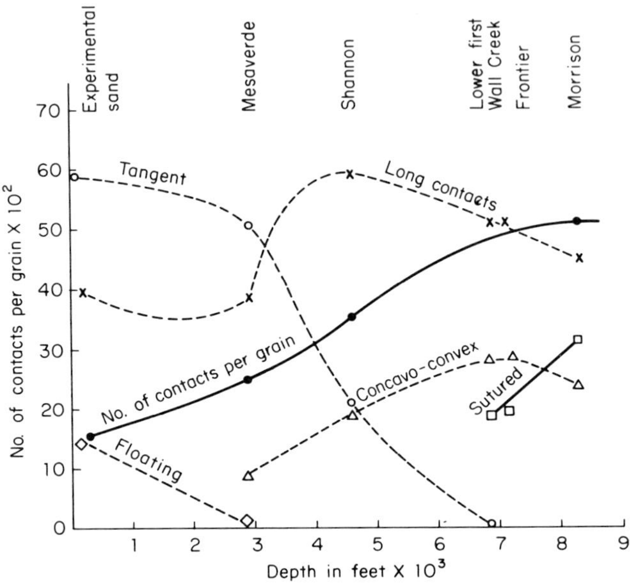

Figure 8.1 Number and type of contacts in Wyoming sands. [*Redrawn by permission from Pore-space Reduction in Sandstones by J. M. Taylor, Bull. Am. Assoc. Petrol. Geologists,* **34**:*715* (*1950*).]

The second essay on the measurement of packing of grains in sediments using thin sections was described by Masson (1951);[1] in this procedure the mineral constituents in a rock are classified into three *textural components* (Masson, 1951), framework grains, interstitial grains, and voids; and packing is defined as the number of grain-to-grain contacts in a traverse of fixed length across the thin section using only the framework grains. Thus variation in composition is reduced by confining attention to the detritus rather than including matrix or cement. Masson also selected *leaving contacts* as the basic variable, rather than using both entering and leaving contacts of grains. In discussing the nature of packing Masson recognized three aspects, the number of grain-to-grain contacts, the configuration of these contacts, and the areal density of grain sections; he also mentioned that variation in mineral composition, size, and shape affect the packing texture. In reducing the concept to an operational procedure he chose grain-to-grain contacts as the variable for measurement. The objective of this investigation was to determine a measure of packing texture which could be used to relate to other properties of sandstones, particularly density, porosity, and tortuosity (Winsauer et al., 1952).

Thirty-two sandstones from oil-producing horizons on the Gulf Coast and Midwest and one artificial sand were analyzed by using four traverses in each thin section. The number of grains counted "ranged from 200 to 700

[1] Through his courtesy the original manuscript, entitled Measurement of Grain Packing in Sandstones, describing the unpublished details of this procedure was made available to the author.

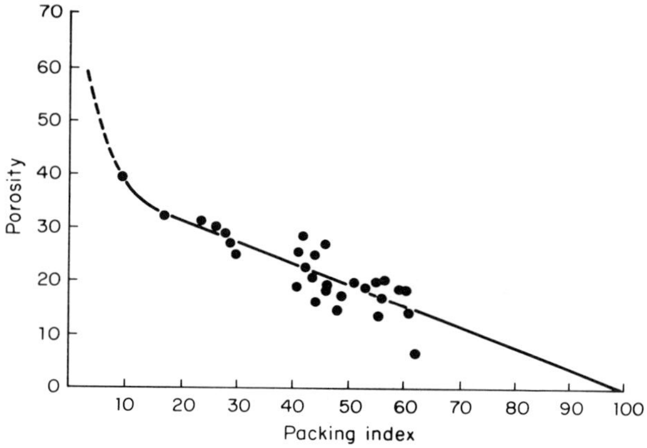

Figure 8.2 Relation of porosity to packing index. (*After P. H. Masson, Measurement of Grain Packing in Sandstones, unpublished paper given at the Annual Meeting Am. Assoc. Petrol. Geologists, St. Louis, Mo., 1951, by permission.*)

depending on grain size" (Masson, 1951). The number of grain-to-grain contacts so obtained is converted to a percentage of total contacts, and the observed values varied in the natural sandstones from 18 in the Eocene Yegua formation (Texas Gulf coast) to 62 in the Ordovician Oil Creek formation (Oklahoma); the artificially packed sand yielded a packing index of 10. Theoretically the packing index ranged from 0, in a sand where none of the framework grains are in apparent contact, to 100, in which all grains are in complete contact. The relationship between packing index and porosity on the 33 specimens analyzed is illustrated in Fig. 8.2, and the agreement between the theoretical and observed characteristics is close.

Rosenfeld, using Masson's technique, measured the packing index on 61 specimens of Oriskany orthoquartzite, and the resulting frequency distribution is shown in Fig. 8.3. The average value of the packing index is 41.4 percent with a standard deviation of 17.7 percent. It should be obvious that a wide range in packing texture (3 to 75 percent in the Oriskany) exists in any one formation.

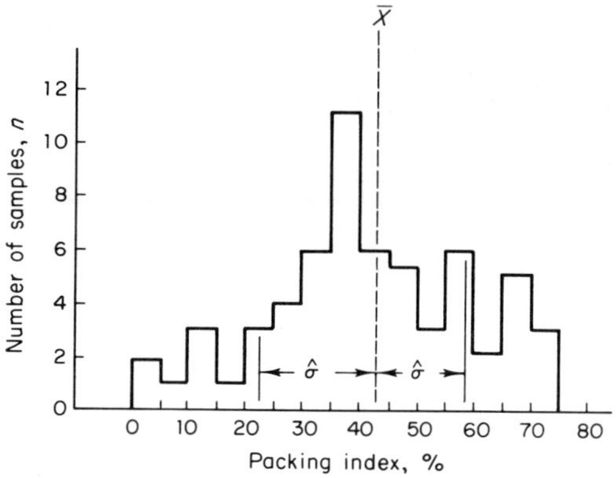

Figure 8.3 Frequency histogram of grain-to-grain contacts in 61 specimens of Oriskany quartzite. $\bar{X} = 41.4\%$; $\hat{\sigma} = 17.7\%$; $n = 61$. (*After Rosenfeld, 1953.*)

Emery (1954) in attempting to discriminate between barren Pocono sandstones and oil-producing Berea sandstones devised a packing index in which the effect of grain size was removed. Emery's packing index is defined as

$$\text{PI\%} = 100 \frac{Q - \bar{X}}{\text{LT}} \tag{8.1}$$

where PI = packing index
Q = number of quartz-to-quartz contacts
$\bar{X}$ = average length of quartz grains, mm
LT = length of traverse, mm

This packing index is derived from a comparison of expected quartz-to-quartz grain packing, which equals the length of traverse divided by the average grain size, with the observed number of quartz-to-quartz grain contacts. Emery's index is similar to Masson's except for the use of quartz-to-quartz grain (leaving) contacts and the adjustment for grain size. It is seriously affected by variations in amount of matrix, cement, and voids. Emery found that the Berea differed from the Pocono sandstones both in average packing index and in spread of packing index (see Table 8.3), and it was the most important property in differentiating the barren from the producing sandstones (Emery, 1954; Emery and Griffiths, 1954a, b). Here again a considerable variability in packing index is evident, with the packing tighter in the Berea than in the Pocono sandstones despite a wide difference in composition (particularly the proportion of matrix). The Pocono formation is much less uniform in packing index and is, on average, less tightly packed than the Berea.

The most detailed and thorough analysis of packing, both as a concept and reduced to an operation, was performed by Kahn (1954, 1956; Miller and Kahn, 1962, pp. 228–244), who defined packing as "the mutual spatial relationship among the grains" (Kahn, 1956, p. 390).

TABLE 8.3
Packing Index in Pocono and Berea Sandstones

Sandstone	$\bar{X}$	$\hat{\sigma}$	n
Pocono	38.0	31.8	16
Berea	56.4	16.9	16

Source: Emery, 1954.

This concept was then subdivided into two aspects, the first concerned with *aggregate* properties of the grains and the second with *unit* properties of the grains. The aggregate property of grain packing was reduced to a procedure for measurement of *packing density*, and the unit property was defined by a measurement procedure called *packing proximity*. A second unit property, the shape of the grain contacts (or *fit*, as described by Krynine), was not used in Kahn's analysis; it is equivalent to Taylor's classification of the shape of the contacts.

The practical definitions of Kahn's packing properties are

$$P_d = \left(m \sum_{i=1}^{n} \frac{g_i}{t}\right) 100 \qquad 0 < \sum_{i=1}^{n} g_i \leq \frac{t}{m} \tag{8.2}$$

where P_d = packing density
m = correction for magnification
$\sum_{i=1}^{n} g_i$ = sum of all the grain intercept values in a given traverse
t = total length of the traverse

$$P_p = \frac{q}{n} 100 \qquad 0 \leq q \leq n \tag{8.3}$$

where P_p = packing proximity
q = number of grain-to-grain contacts
n = total number of contacts = total number of grains

Measurements are performed in thin section, traversing the section by means of a mechanical stage and using distal margins or leaving contacts, as recommended by Masson (1951). Kahn also included a measure of grain size which he called *intercept size* g_i, very similar to the procedure of Emery (1954). Using these procedures, Kahn then analyzed a representative set of detrital sedimentary rocks, three quartzites, three arkoses, and three low-rank graywackes, yielding the results in Table 8.4.

In subsequent analysis of these results Kahn showed that all three measures were interdependent and that variation among specimens and among formations was significantly larger than among traverses but that differences among rock types was not greater than variation among formations within rock types. Since packing is largely a function of environment of deposition and diagenesis, whereas Krynine's rock-type classification, used as a basis for sampling, is a function of composition, i.e., largely source area, it is to be expected that the range of variation within rock types will be greater than between rock types.

Subsequently Kahn applied his techniques to an analysis of packing in ore-bearing and barren Salt Wash sandstones and showed that the packing textures differed in these two subpopulations; the grains tended to cluster in

TABLE 8.4
Packing Density P_d, Packing Proximity P_p, and Intercept Size I_s for Some Representative Detrital Sediments

Formation	P_d	P_p	I_s
Tuscarora	87.5	63.3	0.176
Oriskany	60.9	27.9	0.097
Cuche	55.7	36.9	0.089
Average quartzite	68.0	42.7	0.121
Pocono	72.6	39.9	0.139
Oswego	82.5	57.0	0.161
Bradford	70.8	33.6	0.062
Average low-rank graywacke	75.3	43.5	0.120
Portland	76.6	46.9	0.165
Roslyn	90.5	73.5	0.178
Stockton	78.5	58.5	0.128
Average arkose	81.9	59.6	0.157

Source: Reprinted in part from The Analysis and Distribution of the Properties of Packing in Sand-size Sediments, 2. The Distribution of the Packing Measurements and an Example of Packing Analysis, *J. Geol.*, **64**:583 (1956), by permission of The University of Chicago Press. Copyright 1956 by the University of Chicago.

the ore-bearing types and were more uniformly distributed in the barren rock, although the packing density was similar in both (Kahn, 1956).

In a subsequent investigation Kahn (1959, p. 383) showed that packing, as measured by his procedures, is an anisotropic property of a rock and, in the specimen analyzed, is related to regional deformation. He proposed the hypothesis that "the presence or absence of fractures in sediments is purely a function of the anisotropicity of these textural parameters which are capable of reflecting applied stress" (Kahn, 1959, p. 385); this is an important generalization which could form the basis for analysis of many interesting features connected with rock fractures, both local and regional.

A further study of variation in packing texture is described by Steinmetz (Griffiths et al., 1956, pt. 1, pp. 51–61).

8.3 Recommended Procedure

It seems clear that packing is an important property of aggregates, both in concept and empirically in terms of variation in measured characteristics. It is also clear that to study packing it is necessary to measure effects of composition, grain size, shape, and probably orientation. The adopted procedure should then consider materials which possess the same characteristics of composition, size, and shape; such constituents as quartz grains and the rock fragments quartzite and chert are probably the best selection. Using these components, grain-to-grain contacts can be counted to obtain an esti-

TABLE 8.5
Classification of the Configuration of Contacts

Kind of contact	Symbol	Diagrammatic appearance	Class No.
Floating	*F*	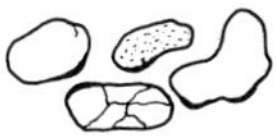	0
Tangent	*T*		1
Long	*L*		2
Complete	*C*		3
Sutured or serrated	*S*		4

mate of Kahn's packing proximity.[1] A second aspect, the configuration of the contacts, analogous to Taylor's recommendations, could also be included, again considering only these three components, and a count of the proportion of different configurations would fulfill this kind of packing index. When the constituents are limited to these three elements, a suitable classification of the shapes of contacts is presented in Table 8.5, and the analytical procedure consists in counts of the different kinds of contact configurations. This aspect could be made more precise by measurement of length of contact and specifying its arrangement, but it seems probable that counts of classes would be adequate as a first approximation, because, in most cases, there is an interdependence between the properties measured, and much of the variation is common (or redundant) in the measure of size, shape, orientation, and packing of grains. Eventually, when adequate data are available, it should be possible to reduce these interdependencies and estimate the amount of information in the measurements by multivariate analysis; such a procedure could be used to determine the smallest number of measures which can "account for" the maximum variability. Now that adequate procedures exist for measurement of all the properties, this is a feasible and necessary step in analyzing petrographic variation and translating it into petrologic dialectic (Griffiths, 1962b; Griffiths and Smith, 1964).

The configurations of Table 8.5 would also be useful in measuring the degree of consolidation of a sedimentary rock from loose modern sands, where values of $F = 0$ to $T = 1$ would be characteristic, to metaquartzites, in which the values would be 100 percent $S = 5$. It is interesting to note that in theory the contacts between grains may be independent of amount of matrix and cement, but in practice it is advisable to adjust packing measures for these materials if packing textures in different rocks are to be compared.

Measurement procedure for packing characteristics comprises a count of frequency in different classes, such as grain-to-grain contacts and the five different types of configuration. In the former case a binomial frequency distribution would be expected where counts exceed, say, 5 percent and the randomness of sampling can be checked against a binomial model.[2] In the case of configurations the proportion in any one class would approximate a binomial but using all six classes would require a multinomial frequency distribution as a model (see Chap. 14). In this sense measures of packing most closely resemble estimations of composition (see Chap. 9).

[1] Equivalent to Masson's packing index.

[2] As the number counted grows large, this frequency distribution will converge on the normal frequency distribution. When counts average less than 5 or exceed 95 percent, the Poisson model is the expected model.

9 Measurement of mineral composition: modal analysis

As time goes on qualitative methods are replaced by quantitative methods.

F. Y. Loewinson-Lessing, 1954

9.1 Introduction

Suitably quantified observations yield data varying from direct measurement, such as the measurement of length in grain-size analysis, to counting frequencies of various items, as in the measurement of packing already described (Chap. 8). Measurement of mineral composition is of the latter kind; it comprises estimates of relative frequencies of different mineral constituents.

Petrographic analysis began with a purely qualitative description, but before long such descriptions included crude estimates of proportions embraced in such terms as rare, common, or abundant. It has taken a very long time, however, for these descriptive terms to yield to more precise estimates of quantity. It is essential to realize that petrologic dialectic necessitates much more than is contained in qualitative petrographic description. Indeed, it is not too much to state that only the most obvious large-scale differences can be adequately diagnosed by qualitative analysis. It is perhaps to be expected that identification of constituents and their classifi-

Quotation: F. Y. Loewinson-Lessing, "A Historical Survey of Petrology," p. 32, Oliver & Boyd, Ltd., 1954.

cation into mineral species come first,[1] but it is certainly essential that the description of a rock specimen does not terminate with identification.

It is generally stated that because identification comes first, any error (mistake) in identification will upset the estimates of proportion, but such a concept is based on too narrow an interpretation of what constitutes an error; it is manifestly impossible to identify all the constituents in a batholith, and exhaustive identification of the constituents in a single thin section is rarely achieved. The main aim of identification is consistency, a prerequisite which may be reduced to strictly operational terms and which is subject to rigorous test for the level of achievement. In most cases petrographic analysis requires some compromise between precise identification of any and every constituent in a specimen and estimates of proportion of these constituents. This appears to be true of all the analytical techniques which have been evolved and requires a careful planning of the experimental program with due emphasis on each phase of the operation if the final data are to carry the maximum amount of information and possess a suitable degree of precision. Fortunately, quantitative data usually permit much more exacting test procedures than qualitative observation; and as long as tests of adequacy are performed, the value of such information can ultimately be measured against the final achievement of the objective.

The objective in the measurement of composition of aggregates is to estimate variation in proportions of the several constituents in a rock body; it immediately becomes obvious that this objective requires consideration of sampling procedure because only part of the rock unit (population) is analyzed.

It can readily be seen that the exhaustiveness or otherwise of the analysis of any subunit (sample), such as a specimen or thin section, must be balanced against the number of subunits analyzed. Choice between exhaustive analysis of a few thin sections or less exhaustive analysis of many thin sections will depend upon the magnitude of variability, within, as against among, thin sections. The larger number of subunits should be apportioned to the largest source of variability. In this way proper emphasis is placed at appropriate level in the analysis.

In the ensuing account it is assumed that the investigator is adequately trained in identification procedure and has ready access to reference texts (e.g., Larsen and Berman, 1934; Winchell and Winchell, 1939 and 1951; Heinrich, 1956; Milner, 1962; Moorhouse, 1959) listing the properties of minerals which may be used to diagnose the various constituents in mineral aggregates.

9.2 Historical Review of the Measurement of Composition

The evolution of quantitative techniques for the estimation of mineral composition is briefly examined by Loewinson-Lessing (1954, p. 30) and

[1] Although this is not always necessary; for example, sieving sands has supplied information of value to sedimentary petrology, and sieves do not discriminate between different mineral species.

Chayes (1956a), and a somewhat more detailed account is given by Holmes (1930, pp. 312ff.). Delesse (1848) first proposed to estimate areas of different minerals in thin section, but this technique proved too tedious for routine use. Delesse also suggested a linear technique (Holmes, 1930) now called the *micrometric* or *Rosiwal method*, developed in detail by Rosiwal (1898). Subsequently the micrometric technique was improved by the construction of special stages to ease the tedium of analysis (Shand, 1916; Wentworth, 1923; Hunt, 1924; Dollar, 1937; Hurlbut, 1939).

Glagolev (1933, 1934) proposed the use of point counting, and this is certainly the most efficient procedure in use at present. Special stages have been devised to simplify this procedure, some requiring manual manipulation and others relying on automatic translation.

These techniques are aimed at estimating the volume of constituents in specimens of rock, and the geometric basis of the procedures is reviewed by Chayes (1956a, chaps. 1 and 2), who also shows (1956a) that use of linear intercepts or sets of points to estimate areas of the constituents in a thin section leads to unbiased estimates of the respective parameters and that estimators of area are, under the specified conditions, unbiased estimators of the volume of the constituents in the rock. It has been demonstrated, however, that the geometric requirements are not always fulfilled in actual analytical procedures (Griffiths, 1960d; Wood, 1960), and the theoretical basis of estimation described by Chayes does not necessarily lead to *unbiased estimators* in practice.

When we return to the defining equation, we find that any analysis performed on a thin section of rock yields a statistical estimator p_i of a parameter P, and P is a function of at least five independent interacting properties, that is,

$$P = f(m,s,sh,o,p) \tag{9.1}$$

Thus, once more, the observation p_i obtained from estimation by point counting is affected by a change in the relative proportion of the constituents, a change in their sizes, their shapes, their orientation, and their packing. Various aspects of these sources of variation are discussed by Chayes (1956a, chaps. 8 and 9), and a correction factor (IC number) is proposed to adjust the estimators. The IC number is a function of at least size and packing of the grains and is likely to be anything but a constant. Hence once more there seems little recourse other than to estimation of all factors simultaneously and isolation of the relevant properties by suitably designed statistical analysis (Griffiths, 1960d, 1961).[1]

9.3 Defining the Classes of Elements

Estimation of the volume of constituents in an aggregate by point counting requires, first, the establishment of a set of classes, approximately

[1] See also discussion of microstructure in Chap. 7.

equivalent to the mineral species, and, second, counting a sample of points and assigning them to the appropriate classes. Establishing the classes is equivalent to defining the tabulation convention of Chayes (1956a, chap. 6). The requirements here are that the classes form a mutually exclusive and exhaustive set so that points can be assigned without ambiguity to classes and all points can be so assigned. These logical requirements take precedence over strict identification of mineral species because during point counting it is generally not feasible to use all the criteria generally available for mineral identification. The set of classes chosen represents a compromise between the mineral elements we should like to recognize and those it is possible to recognize without ambiguity on the basis of rather simple optical characteristics used in point counting.

Ideally it is possible to identify each and every mineral element, but to achieve this requires complex multiple procedures such as disaggregating the rock and separating the disaggregated assemblage into mutually exclusive classes. This step is not completely achieved by any combination of techniques in practice, although a close approximation is obtained by crushing, sieving, and using magnetic and heavy-liquid separations. After separation each mutually exclusive class must be identified by suitable criteria, which may include further physical and chemical tests and optical and x-ray analysis. To complete such an exhaustive analysis means that a very small sample of the original specimen must be used; and because sample-to-sample variation may be large, the precise identification of all constituents in a single small sample does not suffice to yield an estimation of the proportion of a class in the rock specimen and falls still further short of a suitable estimate of the volume of the classes in a rock body. Hence the estimate finally obtained is always a function of the technique used to obtain it. Since it has been shown that point counting yields meaningful data which compare favorably with those obtained from other analytical techniques (Fairbairn et al., 1951), it appears justified to use this approach for the estimation of the proportion of elements (classes) in a rock.

When it is realized that the essential requirement is a mutually exclusive and exhaustive set of classes, it becomes necessary to define the classes on the basis of these requirements, and it soon becomes evident that the attainment of this goal is beset with difficulties. Mineral identification is, in a classical sense, a mental and physical (psychological) habit developed through long and arduous training by a few dedicated specialists. It is very difficult, if not impossible, to reduce this habit to a simple routine. Nevertheless, if estimates of proportions are required, this is the goal at which to aim. In most rocks quartz and feldspar are among the most important constituents; this has been recognized for a very long time, but only recently has the immense variety shown by both these minerals been appreciated. It is obvious that confining the analytical procedure to a single instrument, and not using that to its full capacity, will enforce a much simpler and more embracing classification than can be attained by the use of detailed x-ray and chemical analysis on single mineral species. Here again, however, it seems likely that exact specification will fail to achieve all the information necessary to the solution of petrologic problems and, eventually, estimates of proportions of the constituents, however subdivided, will be recognized as a necessity.

In addition to quartz and feldspar there are minerals much more difficult to identify, and the list of classes eventually adopted must embrace them all. Constituents in sediments comprise essentially grains of quartz, feldspar, and preexisting rock fragments, micas, matrix, and cement; and an approximate estimate of what to expect is outlined in Table 9.1. It is possible to simplify the counting procedure by using a few broad classes, suitably defined, such as the grains, matrix, and cement in sediments (Krynine, 1948) or crystals and glass in igneous rocks. Subsequently each class can be subdivided and counts made of subclasses; such a procedure is essential if it is required to estimate proportions.

As a first step all the constituents in sediments can be embraced by a fourfold classification which fulfills the essential requirements of being mutually exclusive and exhaustive. The criterion forming the basis of this classification is the morphology or habit of the constituents as seen in thin section—essentially the nature of the internal and external bounding surfaces. If the bounding surface, which is generally a curved line in a thin section, is continuous and essentially everywhere convex and the interior area encompassed by this boundary is homogeneous, then the entity is internally homogeneous, and its external bounding margin is continuous (see Fig. 9.1). There are therefore two morphological criteria: First, a grain may be internally homogeneous or internally heterogeneous; second, its boundary may be continuous or discontinuous. These two criteria can be combined to diagnose four mutually exclusive and exhaustive classes (Fig. 9.1).

Thus a constituent may be an entity which is internally homogeneous, such as a grain of quartz; or a constituent may be an entity which comprises several individuals, such as a fragment of a preexisting rock. Again, matrix is internally and externally complex, whereas some carbonate cement is internally homogeneous (a single crystal) but is externally complex (luster-mottled; see Milner, 1940, p. 369; see reference for Milner, 1962). We may therefore propose the classification in Fig. 9.1 as the first basis of subdivision.

In terms of symbolic logic there are four classes, the first possessing the property A and B, the second "not A" ($\bar{A}$) and B, the third A and "not b" ($\bar{B}$), and the fourth "not A" ($\bar{A}$) and "not B" ($\bar{B}$).

Grains of single minerals such as quartz, feldspar, large micas, and the accessory minerals possess internal and external continuity; fragments of preexisting rock consist of externally continuous elements but are internally discontinuous, usually but not always containing crystals or grains of more than one mineral composition; fragments of granite, schist, shale, slate, etc., are of this kind. In many rocks these continuous entities lie in a paste

TABLE 9.1
Mineral Composition and Classification of Sedimentary Rocks: Tentative Estimate of Proportions

Rock type	Quartzite, %	Graywacke, %		Arkose, %
		Low-rank	High-rank	
Essential Minerals				
Rock Fragments	10	20–30	25–35	25–35
Quartzite	100	95	10	Very rare
Chert				Very rare
Shale to phyllite				Very rare
Schist		Rare	90	Common
Volcanic rocks		Very rare		Rare
Gneiss			Rare	95
Granite				
Feldspar	<5	<5	5–30	5–50
Potash feldspar	100	100	Very rare	90
Albite oligoclase	Very rare		100 dominant	10±
Oligoclase labradorite				Very rare
Quartz	70+	30–70	30–70	30–70
		Av. 50	Av. 40	Av. 40
Varieties	Mixed	Metamorphic	Metamorphic and volcanic	plutonic
Varietal Minerals				
Micas	Rare–absent	Common 10+	Common 20+	Common < 10
Muscovite	100	0–50	0–50	0–100
Chlorite		0–50+	0–50+	Rare
Biotite		0–50	0–50	0–100
Glauconite	Rare–common	Common	Rare–common	Rare

Accessory Minerals				
Accessory minerals	<1	1–5	1–15	1–15+
Zircon, rutile, tourmaline	100	10–50+	10–20	10–50
Garnet and apatite	Very rare	0–50	0–30	0–20
Staurolite and chloritoid		0–10	0–20	0–10
Epidote and clinozoisite		0–10	0–20	0–5
Kyanite and sillimanite		0–5	0–15	0–15
Amphibole		0–10	0–70	0–20
Pyroxene		0–5	0–80	0–15
Matrix				
Matrix	None–rare	Abundant	Abundant	Abundant
Micas		Occurs	Occurs	Rare
Chlorite		Occurs	Occurs	Rare
Biotite		Occurs	Occurs	Rare
Clays		Common	Common[a]	Common
Kaolinite	Rare	Rare–50		50+
Illite		50+	Common	Rare
Montmorillonite		Rare	Common	Common
Secondary and Authigenic				
Cements	Abundant	Common	Rare–common	Rare
Silica	50+	0–20	0–10+	Rare
Carbonate	50+	0–20	0–10	0–10
Titaniferous group:				
Anatase	Common	Common	?	Rare
Brookite	Common	Rare	?	?

[a] In high-rank graywackes most of those described have been metamorphosed, hence, do not contain clay minerals.

Note: There are two properties of frequency confused in the above table, namely, amount and occurrence; e.g., a mineral may occur in all low-rank graywackes with a percentage frequency of 0.1%. It is, therefore, characteristic, but rare.

The table represents the author's personal opinion and is subject to considerable modification when more is known of the mineralogy of sedimentary rocks.

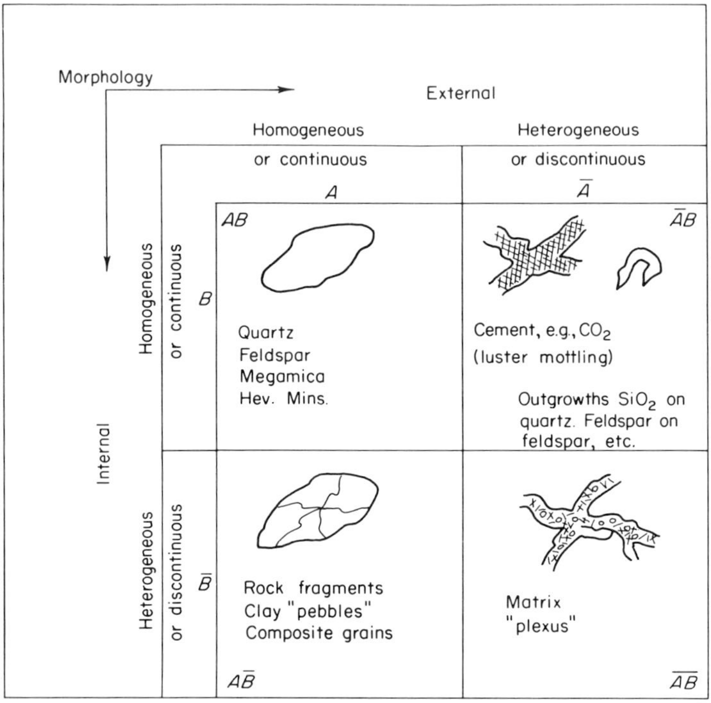

Figure 9.1 Classification of elements for the "measurement" of composition. This classification is independent of genesis but can be used to imply a genetic process.

called matrix, and this matrix is externally and internally discontinuous, thus falling into the class labeled $\overline{AB}$. Under some circumstances the matrix material is disturbed after initial deposition and then undergoes a morphological change, wherein it remains internally discontinuous but its bounding surface becomes externally continuous. Slumped sediments frequently have such elements, and the material is then classed as $A\bar{B}$. Such reconstituted matrix is generally difficult to tell from rock fragments of shale and clay and so is usefully classed with these materials.

Other elements of importance in the constitution of sediments are called cements, and these have a wide variety of habits. For example, silica cement occurs as an outgrowth around a quartz grain. Very rarely it persists throughout the rock in somewhat the same arrangement as matrix, and then it should be classed as $\overline{AB}$; more commonly it occurs as isolated outgrowths around individual quartz grains, when it may be considered as internally homogeneous, i.e., an individual in internal composition, but externally it is discontinuous. It falls, therefore, in class $\bar{A}B$. Any outgrowth type of material, such as feldspar on feldspar or tourmaline on tourmaline, is similarly classified.

The external discontinuity emphasizes that its diagnosis relies on the fact that there is a discontinuity between the parent grain and the outgrowth; where this discontinuity is lacking, it is impossible to diagnose this kind of "cement," e.g., silica cement in some specimens of Oriskany quartzite (Wood, 1960).

Carbonate cement, on the other hand, has a quite different habit; it frequently occurs as small patches in pores between the elements of classes AB and $A\bar{B}$, i.e., quartz, feldspar, and rock fragments. In such cases it may be considered as externally continuous and internally discontinuous (polycrystalline) and belongs in class $A\bar{B}$. Clastic limestone fragments and fossil fragments also show this kind of habit. On the other hand, when carbonate cement is poikilitic, it is internally continuous and externally discontinuous (class $\bar{A}B$). Quite frequently it may be both internally and externally discontinuous, as in some specimens of the Cow Run sand (Griffiths, 1958b, 1960d, fig. 4b).

The advantage of the proposed classification is that it leads to grouping together those compositional elements which "look alike" in habit and separating those elements which differ in habit, and this classification does not require a knowledge of the origin of the material. On the other hand, elements having different habits and the same chemical composition are separated, a feature generally interpreted as implying a difference in origin, e.g., quartz and silica cement or clastic and cementitious carbonate.

These classes may now be further subdivided by using operationally sound criteria which permit establishing classes which are truly mutually exclusive. While there exists at present no standard classification, as our knowledge of the mineral elements of sediments increases it will no doubt be possible to establish a general convention.[1] If it is required to compare different rocks, a standard convention becomes a necessity, and the classes and subclasses of elements must be defined on the basis of the same criteria if the comparison is to be meaningful. The classification eventually adopted will require at least the following classes of elements.

9.3.1 Elements of Class** AB **Externally** (A) **and Internally** (B) **Continuous. *Quartz.* Those elements[2] which are colorless, of low relative relief (relative to the conventional mounting medium, Canada balsam), generally clear but often containing dusty inclusions regularly or irregularly arranged and a continuous and generally distinct outline (Fig. 9.2) are assigned to the class quartz. Under crossed nicols, elements assigned to this class exhibit low polarization colors of the first order. Actually the

[1] The procedure adopted for this purpose would most usefully follow the practices of diagnosis advocated for the medical profession (Ledley and Lusted, 1959), wherein the identification of a disease is closely analogous to that of a mineral under conditions of point counting thin sections.

[2] During point counting the elements are strictly the points of the sampling grid, and each point must be assigned to some class (or subclass) and to only one class.

quartz of sedimentary rocks is generally elliptical in thin section but varies from circular (rare) to triangular and almost rodlike. Notwithstanding this variation in grain shape, the mean shape of small sets of such grains may vary little (see Chap. 6).

This class also contains some feldspar, particularly orthoclase and untwinned plagioclase. To differentiate these items it is advisable to stain the thin section with sodium cobaltinitrate (Chayes, 1952). In many sediments total feldspar is less than 5 percent of the rock, and only rarely will it be necessary to differentiate the kinds in detail, as they appear to be approximately constant in kind and relative proportion as far as is known at present. If it is necessary to obtain reliable estimates of the kinds and relative proportion of such uncommon elements, perhaps other techniques such as x-ray analysis should be used.

Another source of ambiguity arises in some sediments when the elements in this class have strain extinction; quartz of various origins tends to develop undulose patterns and also twinning, usually attributed to stress during metamorphism. In some cases the extinction pattern becomes so clear that the element can no longer be considered internally homogeneous, and under such circumstances it should be assigned to class $\bar{A}B$. It should be decided before the point count on what criteria such elements should be excluded from class AB and included in $\bar{A}B$, and those criteria should be maintained consistently during the experiment. Ambiguity will arise if the criteria are conceptual rather than operationally sound, and this is the most serious fault in diagnosis which will affect the estimates of proportion.

Feldspar. Elements which resemble those of the previous class but which have lamellar twinning or cross-hatched twinning under crossed nicols should be assigned to feldspar (Fig. 9.2). They are frequently more rectangular in outline than quartz, but this is not an invariable criterion. Again such constituents are often dusty with decomposition products, but this criterion is also subject to variation and is therefore liable to lead to ambiguity. Staining partially ameliorates this difficulty.

At first sight it is tempting to subdivide this subclass on the basis of the kinds of twinning, but it is very difficult to be consistent on the basis of such

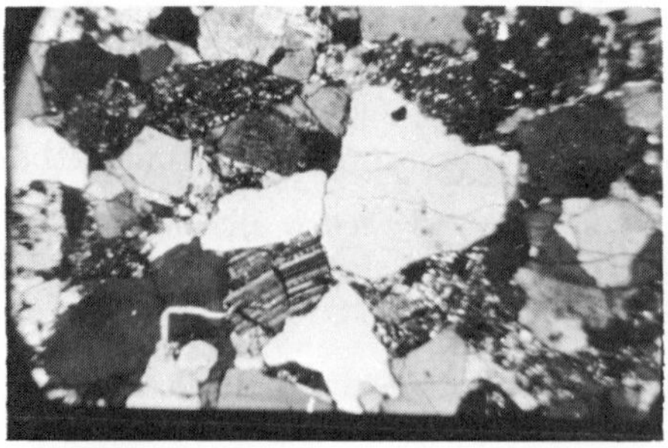

Figure 9.2 Photomicrograph of Eocene arkose, Washington, showing quartz, feldspar, and rock fragments; crossed nicols; magnification 100×; Penn. State Cat. No. 9025. (*Specimen supplied by C. T. Bressler.*)

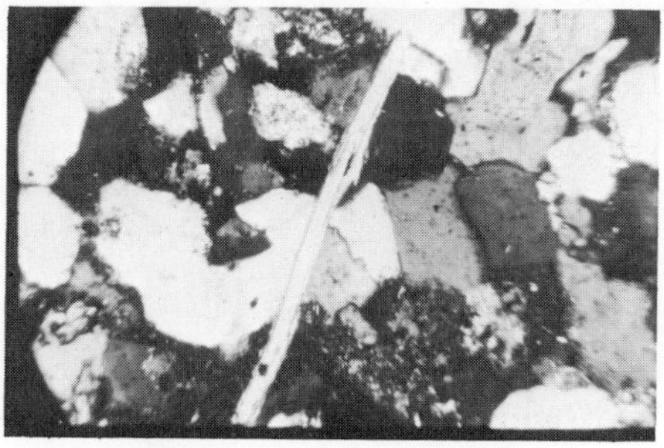

Figure 9.3 Photomicrograph of Devonian (Chipmunk) low-rank graywacke, showing quartz, silica cement, aphanitic rock fragments, and megamica; crossed nicols; magnification 200×; Penn. State Cat. No. 7816. (*Specimen supplied by F. L. Stanonis.*)

criteria because pericline twinned plagioclase frequently resembles cross-hatched microcline, and not all microcline is cross-hatched and not all sections of cross-hatched microcline show crosshatching. Perthitic intergrowths also show wide variation in appearance, so it is advisable to use a general class feldspar for the initial count and subsequently, in a separate count devoted to this subclass, to attempt to estimate the proportions of the different kinds of feldspar.

Micas. In thin section this mineral usually varies in color from clear colorless through various shades of brown to red-brown and through various shades of green to bluish green. The habit may vary from elongated rectangular grains (most often; see Fig. 9.3) to large rounded areas or even polygonal outlines (rare). Its relative relief shows greater contrast than the previous two classes, and the colorless or lightly colored varieties generally possess bright (second- to third-order) polarization colors. The extinction in rectangular grains is generally straight, but the grains are often twisted and bent. Again it is tempting, and to some extent practicable, to subdivide this subclass on the basis of color, but it should not be implied that these color classes coincide with the varieties muscovite, biotite, and chlorite without more stringent criteria than can generally be used while point counting.

Other Elements Assigned to Class AB. In some cases other elements may occur which are internally uniform and externally continuous, such as the accessory minerals zircon, apatite, garnet, and particularly opaque minerals, but in general these are rare in most sediments and during point counting could most conveniently be assigned to "others of class *AB*." Such a class will rarely exceed 2 percent in most cases, and if it does, it may then be worthwhile to differentiate a new subclass for the persistent variety.

9.3.2 Elements of Class $A\bar{B}$***, Externally Continuous*** (A)***, Internally Discontinuous*** ($\bar{B}$). The bulk of the constituents in this class are fragments of preexisting rocks; and as they are quite important in interpreting the genesis of the sediments containing them, they will almost always need subdivision. Fragments of granite, schist, phyllite, slate, shale, clay, quartz-

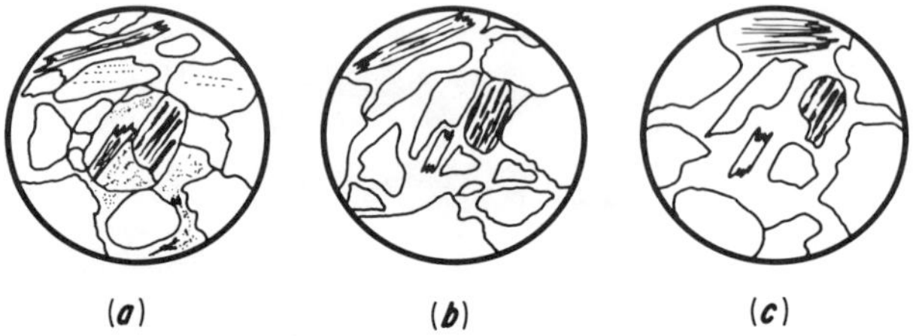

Figure 9.4 Granite fragment in arkosic sandstone. (*a*) Undoubted composite grain; (*b*) questionable composite grain; (*c*) individual grains.

ite, sandstone, chert, volcanic rocks, limestone, and dolomite occur quite frequently in many sediments. It is most advisable that in choosing a set of subclasses, rigorously operational criteria be applied, but it is also extremely difficult to maintain this classification consistently during point counting. As a general recommendation it is practicable to adopt a very simple classification to perform the initial point count and then, if necessary, to reexamine the section and count the different varieties in this class in a count specially devoted to rock fragments (Griffiths et al., 1956). A subdivision into phanerite and aphanite is quite convenient for the initial foray, and layered (or foliated) and nonlayered may also be used either alone or in conjunction with phanerite and aphanite.

Granitic fragments of various kinds (Fig. 9.4) occur in both arkoses[1] and high-rank graywackes,[1] and when these fragments are composed of quartz and feldspar, they are, of course, easy to identify. In some of these sediments, however, rock fragments composed only of fragments of quartz are not so obviously fragments of preexisting granites, and no operationally exclusive criteria exist at present for differentiating these "quartzites" from others derived from different source materials. This ambiguity, of course, is aggravated in sediments where obvious fragments of granite are not common. It should perhaps be emphasized that assigning an element (point) to "quartzite" contains this ambiguity in all cases.

Another difficulty arises when it is realized that rock fragments of granite in a sediment are often of the same grain size (and shape, approximately) as the quartz grains (of class AB) in thin section. This is emphasized when the term gneiss is used, because a fragment of "gneiss" in the usual sandstone would be very different from the rock named gneiss in the field. Thus foliated granite fragments and nonfoliated would lead to more consistent usage. In fact, of course, schist would perhaps be an appropriate term for the foliated granite fragment under such conditions.

Frequently during point counting an ambiguity tends to arise because there is a transition (Fig. 9.4) from rock fragments of granite (class $A\bar{B}$) to fragments of the minerals quartz and feldspar (class AB). Strict observance of the criteria differentiating class AB from class $A\bar{B}$ should help to avoid inconsistency in usage.

[1] Rock terms used in the general sense defined by Krynine (1948).

Fine-grained foliates would usefully include phyllite, slate, and shale; and fine-grained nonfoliates would include both clay and chert fragments.

Quartzite refers to those entities composed almost entirely of crystals of quartz and belonging to class $A\bar{B}$. A transition exists between quartzite and chert as the grain size of the internal crystals decreases to submicroscopic level. Again some criterion such as size of crystal may be established to resolve this ambiguity.

Volcanic rock fragments are generally aphanitic and vary from microcrystalline through cryptocrystalline to glass; when there are phenocrysts (Fig. 9.5), these rocks are relatively simple to identify. When phenocrysts are absent, microtexture may be used; and when neither phenocrysts nor microtexture is available, these fragments are generally classified as chert (Griffiths et al., 1956).

Fragments of limestone and dolomite also occur in some sediments, and fossil fragments are also not infrequent (Griffiths, 1958b); in all cases there are possibilities of confusion unless the criteria used to classify into the broad fourfold classification are exercised rigorously, and even then absence of specific habits may introduce inconsistency in the counting.

9.3.3 Elements of Class $\bar{A}B$, Externally Discontinuous, ($\bar{A}$), Internally Continuous (B). This class generally contains the cementitious elements of a sedimentary rock, and generally the experienced petrographer can recognize the materials fairly readily, but the term cement is apt to be misused when the "criterion" used to identify the material is based on genetic implications rather than observable morphologic distinctions. Thus silica cement differs very markedly from carbonate cement in its morphology, and classing the two together is liable to lead to confusion in subsequent interpretive petrology.

Silica cement refers to the clear quartz which is generally in optical continuity (or approximate optical continuity) with the contiguous quartz grain (Fig. 9.6). To differentiate cementitious silica from detrital silica, it is necessary to be able to observe a distinct boundary between the quartz grain and the cement. In the classical case of the Penrith sandstone (Harker, 1923, p. 211; Hatch, Rastall, and Black, 1938, p. 105) a pellicle of iron oxide

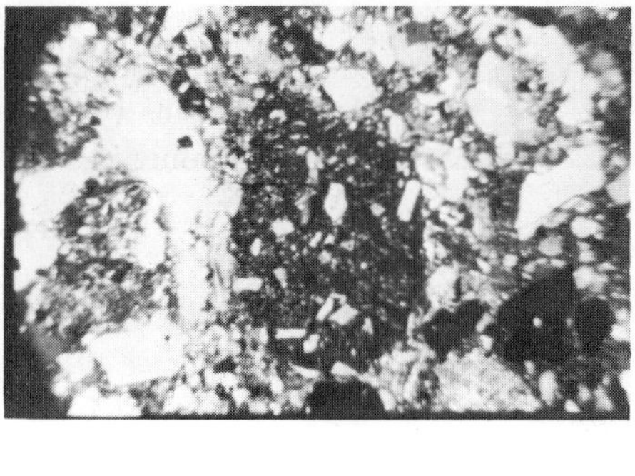

Figure 9.5 Photomicrograph of Ordovician high-rank graywacke from Wales, showing volcanic rock fragment with phenocrysts of quartz and feldspar; crossed nicols; magnification 100×; Penn. State Cat. No. 9707. *(Specimen supplied by D. A. Bassett, Museum of Wales.)*

Figure 9.6 Photomicrograph of Silurian (Tuscarora) quartzite, Mt. Union, Pennsylvania, showing silica (cement) overgrowths on quartz; crossed nicols; magnification 150×; Penn. State Cat. No. 7987. (*Specimen supplied by G. V. Wood.*)

serves to demarcate the boundary quite clearly, and in most cases a line of dust inclusions suffices to distinguish the original grain from the cement. In some cases, however, the outline is strongly crystalline and euhedral, while there is no way of distinguishing the internal boundary. This lack of distinction is frequent in some specimens of the Oriskany quartzite (Wood, 1960).

Another criterion which implies precipitation from solution is illustrated by the morphology of the contacts between neighboring grains or crystals; thus when the contacts between these neighboring grains meet at contact angles of 120°, this implies crystallization from solution (Smith, 1948, 1952) and so indicates the presence of silica cement in cases in which no (internal) dust boundary is evident.

Silica cement is always associated with quartz in this habit, but other kinds of siliceous cements are not infrequent (Hatch, Rastall, and Black, 1938). In such cases silica cement may overlap the classes $A\bar{B}$ and $\bar{A}B$ and perhaps $\overline{AB}$. The specific kind should be clearly defined on the basis of morphology and the diagnostic class indicated.

Outgrowths on feldspar are also frequent (Boswell, 1933, pp. 87ff.), and tourmaline outgrowths are well known (Krynine, 1946). Similarly, outgrowths on zircon have been described (Butterfield, 1936). All these fall in the class $\bar{A}B$.

Carbonate cement, as contrasted with clastic fragments of carbonate (including fragments of preexisting limestone, dolomite, and fossil fragments), may show equally variable habits. The habit which typifies the class $\bar{A}B$ is the variety which occurs as a relatively large patch which is internally optically continuous over the entire area but is externally discontinuous because it includes poikilitically grains of other minerals; this habit has been called luster mottling (Milner, 1940, p. 369; see reference for Milner, 1962).

9.3.4 Elements of Class $\overline{AB}$, Externally ($\bar{A}$) and Internally ($\bar{B}$) Discontinuous. Carbonate cement occurs more commonly as small areas infilling the pores in a sedimentary rock (Fig. 9.7). It is then internally discontinuous ($\bar{A}$ is polycrystalline) and externally discontinuous ($\bar{B}$ is irregular in outline). This is the variety which by increase in relative pro-

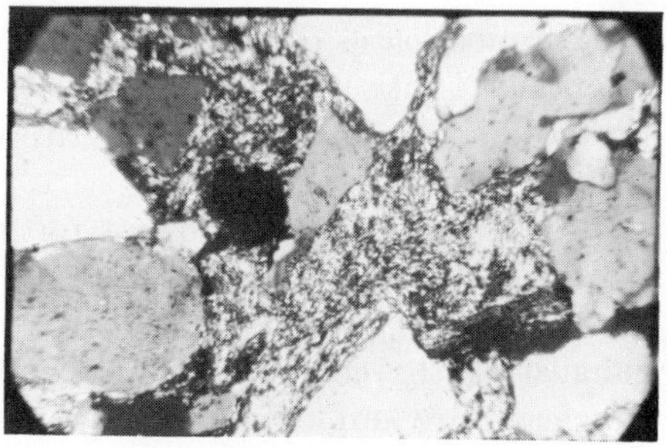

Figure 9.7 Photomicrograph of Mississippian (Pocono) low-rank graywacke, Altoona, Pennsylvania, showing typical morphology of micaceous matrix with quartz and quartzite grains; crossed nicols; magnification 150×; Penn. State Cat. No. 6290. (*Specimen supplied by H. W. Shadle.*)

portion to over 30 percent by area of section suggests that the rock is a "sandy limestone."

Another element which must be classed as internally discontinuous ($\bar{A}$) and externally discontinuous ($\bar{B}$) is the material called matrix (Krynine, 1948, p. 138). This material is complex in composition and may vary in grain size but is generally finer-grained than the elements of the class AB. It is difficult to give exact criteria for its identification because of its complex composition, comprising at least clay minerals, unobservables such as cementitious material (siliceous or carbonate), small particles of quartz, and other minerals (Fig. 9.8); Brammall's definition of plexus illustrates the potential complexity of matrix (Brammall and Leech, 1943).

This material is typical of those rocks called graywackes (low-rank and high-rank) and arkoses and is generally rare or absent from quartzites. Quite frequently matrix is strictly that material in a thin section not easily identified by optical means. Under such circumstances its habit and arrangement in the thin section are perhaps the best guide to its classification.

Matrix is differentiated from fragments of clay of class $A\bar{B}$ on the basis of morphology, and it should be remembered that the existence of this morphology does not necessarily imply far-traveled or local derivation; in other words, additional criteria are needed to discriminate between fragments of preexisting clay considered to be of detrital origin and fragments of matrix which have developed a continuous external habit due to postdepositional disturbance of the fabric of a sediment (Griffiths, 1960d, fig. 2).

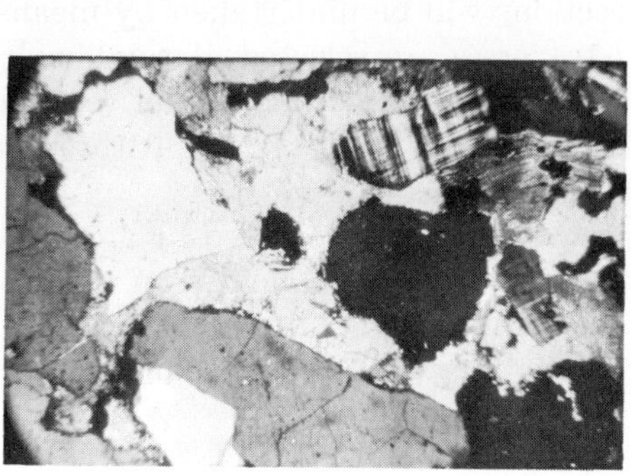

Figure 9.8 Photomicrograph of Pennsylvanian Fountain formation from Red Rocks Park, Colorado, showing morphology of crystalline carbonate cement with feldspars and quartz; crossed nicols; magnification 100×; Penn. State Cat. No. 8506. (*Specimen supplied by J. F. Hubert.*)

In most cases other items, lying outside these subclasses, can be included as "others" with a class name as qualifier. Thus they will be suitably defined by recognizable morphological criteria and can be specifically identified subsequently by more exact tests.

It is advisable not to use too many subclasses during point counting. This is particularly true if there is any ambiguity in assigning the elements to the subclasses, because the larger the number of alternatives, the more exhausting the count procedure and the less consistent the results. Eight subclasses are ample for any particular count, and if there is a need for greater detail, a recount with special attention to subdivisions within a single subclass will result in more reliable estimates than if the analysis is attempted in one count. In fact, for preliminary work the four main classes may be ample to give a basic estimate of the kinds and proportions of the commoner elements.

9.4 Testing the Consistency of the Technique

When a set of classes is established for a point-count analysis, it is advisable to start the analysis with a preliminary experiment essentially aimed at testing the consistency of the classification. For a single operator this requires replicated counting on different days to decide whether the operator is maintaining a consistent basis for decision.

A more comprehensive experiment may be performed with several operators, and, of course, this is really a necessity if the conclusions are to be extrapolated over all operators.

It is now recognized that chemical analysis requires standards and standardization of technique; such experimental analysis is capable of more precise duplication of results than a procedure like point counting, which relies on the skill of the operator and is quickly affected by operator fatigue. The examination of consistency of counting thin sections of granite has been described by Chayes and Fairbairn (1951); of sediments, by Griffiths and Rosenfeld (1954). Some general recommendations on experimental design are possible in this context.

First, we must test the consistency of certain definitions of subclasses over n different thin sections and p operators. The analysis of the thin sections will be undertaken by means of t traverses per thin section, and the t traverses are selected at random by each operator. In order to maintain the infinite-population model (see Chaps. 18 et seq.) it will be necessary to sample each thin section with replacement. The design is illustrated in Table 9.2.

Consistency is tested by a comparison of expected mean squares (strictly the variance components) of operators by thin sections versus traverses in the form of a variance ratio F

$$F = \frac{\sigma_\epsilon^2 + t\sigma_{mp}^2}{\sigma_\epsilon^2} \qquad \text{d.f.} = (p-1)(m-1) \quad \text{by} \quad mp(t-1)$$

TABLE 9.2
Experimental Design for Testing Consistency of Classification

Source of variation	d.f.	Expected mean square
Among operators	$p - 1$	$\sigma_\epsilon^2 + t\sigma_{mp}^2 + tm\sigma_p^2$
Among thin sections	$m - 1$	$\sigma_\epsilon^2 + t\sigma_{mp}^2 + tp\sigma_m^2$
Operators × thin sections	$(p - 1)(m - 1)$	$\sigma_\epsilon^2 + t\sigma_{mp}^2$
Traverses (within operators and thin sections)	$mp(t - 1)$	σ_ϵ^2
Total sources	$mpt - 1$	σ_{tot}^2

This test can be made as sensitive as required by increasing the number of traverses. Furthermore, the traverse variance can be tested by collecting the data in small sets, say five traverses each comprising 30 sets of five points each. It is essential to realize that if differences among the thin-section variances are not significant, it may be due *either* to similarity among thin-section means *or* to a large inconsistency among operators from thin section to thin section.

Again, the reproducibility among operators can be tested by making each operator follow the same traverses. It is important to use a flexible design and a fairly large number of units in the error term to establish consistency of the subclass units. Quality-control charts which graphically exhibit the analysis of variance are a great help in this respect (see Griffiths and Rosenfeld, 1954, fig., p. 80; Sec. 15.6, Fig. 15.5).

It is good practice to set up a small number of thin sections as standards to be analyzed by new operators, thus ensuring that the operator is in control before starting analysis of unknowns. These thin sections, a few from several different rock types, will become very useful standards if extensive research of this type is contemplated.

9.5 Procedure and Technique

The procedural details of point counting are reasonably simple; the required apparatus is a point-count stage and a counter, and a variety of these are readily available.

Stages can be manually manipulated or automatically controlled; a standard stage designed for manual point counting of thin sections can be purchased from the American Optical Company,[1] or a standard mechanical stage can be modified; the Rosenfeld modification of the Bausch and Lomb stage is quite suitable (Rosenfeld, 1954).

[1] American Optical Instrument Division, Buffalo, N.Y., cat. No. 496.

TABLE 9.3
Data Sheet for Mineralogy Point Count

Slide # ________ Microscope # ________ Type ________
Rock ________ Stage ________
Objective ________ Eyepiece ________
Start position: ________ V ________ H No. of traverses: ________
Slide position: ________ Length of traverse: ________
Operator: ________ Spacing of traverse: ________
Comments: ________

Key	1	2	3	4	5	6	7	8	9	10	T
Min											
1											
2											
3											
4											
5											
6											
7											
8											
9											
10											
Sum											
Mean											

Blood-cell counters[1] may be used to record the counts of frequency. They are sold in various sizes, most convenient being one with eight separate recording units and a total. Another type, the Denominator counter,[2] can also be purchased in various sizes and can be made up into banks of four, six, or eight cells. An electrically operated automatic counter, available from Swift and Son, Ltd., consists of a point-count stage coupled to a counting cell.[3]

Records of the counts are best kept on a data sheet designed for this purpose; one used for general-purpose counting is illustrated in Table 9.3.

Magnification is important in permitting rapid counting with minimum

[1] Clay Adams Co., Inc., New York, N.Y.
[2] Fisher Scientific Co., Pittsburgh, Pa.
[3] J. Swift and Son, Ltd., London, England.

errors of diagnosis, and for most sediments convenient magnification varies between 100 and 250×. The spacing of grid points, of course, is not affected by the magnification used, but close spacing is convenient in analyzing sedimentary rocks. On the stages in general use the spacing of vertical traverses is about 0.5 mm, and the horizontal-traverse interval is about 0.3 mm. In some cases closer spacing has been used by means of a square grid in the eyepiece, and this is a useful accessory for detailed sampling experiments by point counting.

Before beginning an analysis it is always advisable to examine the thin section to see that it is adequate in area and properly prepared with few holes. The grid of points is usually considered to be a rectangular arrangement, but this is not essential; traverses can be staggered, and sometimes it is necessary to traverse in different directions (see, for example, Kahn, 1959). After examination of the thin section it is possible to decide on the classes and subclasses which will be required, but before setting up the final system to be followed, a few preliminary counts to test the selected arrangement are recommended, and these data should then be discarded. For adequately precise estimates it is necessary to ensure that the operator has reached a stable level in his decision procedure, but these preliminaries may be reduced materially when operators are sufficiently experienced.

The results of point-count analysis are frequencies (counts) of elements in classes and subclasses. For stable estimates of thin-section means Chayes (1956a) has recommended 1,000 points per section, and for sediments rather more may be required. As a first step it may be desirable to decide on the sample size and arrangement, and for this purpose a few, say four, widely different thin sections may be chosen from the set to be analyzed; 10 traverses of 100 points each may be assembled on these slides with the points taken in small sets if required, and the stability of the estimates may then be tested by the dispersion index (Fisher, 1948; Student, 1919; Griffiths, 1960a, d; Chap. 14). If the estimates appear stable, the analysis may be continued; but if they are unstable, for example, χ^2 significant at the $P < 0.05$ level, it may be necessary to combine traverses. In effect this procedure is aimed at ensuring that the samples finally used are random samples of the population under analysis. It is necessary first to decide the kind of frequency distribution likely to be generated by the analytical procedure applied to thin sections of sedimentary rocks. For granites this has been discussed by Chayes (1949a), and for sediments it is described by Griffiths (1960a, d).

For subclasses which occur with proportion about 5 to 95 percent, the binomial frequency distribution is often a suitable model; and for constituents comprising less than 5 and more than 95 percent of the thin section, the Poisson may be the appropriate model. In both these cases the sampling requirements are rigidly prescribed (Griffiths, 1960a; Chap. 14).

Quite often the constituents in a thin section do not lead to fulfillment of the sampling requirements, and alternative models may be used such as the hypergeometric (Chayes, 1949a) and the negative binomial (Griffiths, 1960a, b). In layered rocks such distributions as the Lexian and Poissonian may be expected (see Sec. 14.6; Kendall, 1948; Weatherburn, 1957).

Some examples of the manipulations necessary to the fitting of a sample set of observed frequencies to expected distributions are described in Chap. 14 and in Griffiths (1960a, b).

The final form of the analytical data will be expressed as means and standard deviations, and the frequency distributions may be illustrated as histograms. Some examples are given in the following pages and in Chap. 14.

9.6 Examples of the Measurement of Mineral Composition of Sediments

Examples of the frequency distributions arising in analysis of thin sections by point counting under various conditions may be used to illustrate the kind of results to be expected.

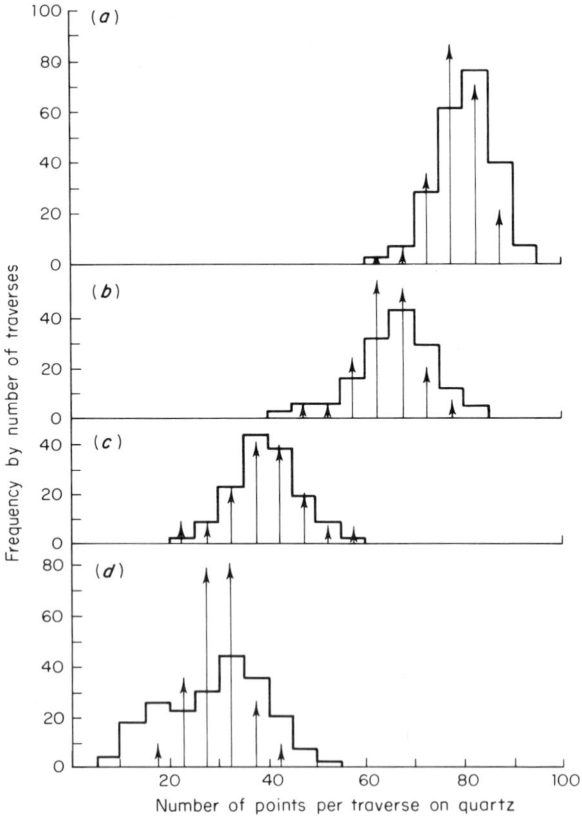

Figure 9.9 Frequency distribution of quartz proportion in sediments.

(*a*) Oriskany quartzite.
$\bar{X} = 80.4$ $\bar{X}\% = 78.8$
$\hat{\sigma} = 5.9$ $n = 10.2$
$N = 225$ $m = 3$

(*b*) Salt Wash arkosic quartzite.
$\bar{X} = 65.4$ $\bar{X}\% = 65.4$
$\hat{\sigma} = 7.9$ $n = 100$
$N = 150$ $m = 25$

(*c*) Cow Run low-rank graywacke.
$\bar{X} = 39.7$ $\bar{X}\% = 39.9$
$\hat{\sigma} = 7.0$ $n = 100$
$N = 156$ $m = 39$

(*d*) Roslyn arkose.
$\bar{X} = 29.4$ $\bar{X}\% = 32.7$
$\hat{\sigma} = 10.4$ $n = 90$
$N = 225$ $m = 3$

During a class experiment five operators performed an analysis of three thin sections from the same specimen of Oriskany quartzite; they used 15 traverses of 102 points each on each of the three sections. The frequency distribution for quartz is illustrated in Fig. 9.9*a*, and the expected frequencies from a binomial with the same mean are indicated in the figure by arrows. Discrepancies are obvious and presumably indicate that the sampling was nonrandom or that the binomial is not a suitable model.

Similarly, the quartz frequency distribution resulting from a single operator counting six traverses of 100 points on each of 25 sections of Salt Wash sandstone (Morrison, Colorado plateau; see Griffiths et al., 1956) is illustrated in Fig. 9.9*b*. Again the observed frequencies depart from those expected from a binomial with the same mean (arrows).

In the Cow Run sand, on the other hand, the frequencies of the class quartz, obtained from a point count of 39 thin sections by four traverses of 100 points each, compare closely with those of a normal distribution having the same mean and standard deviation (Fig. 9.9*c*). In this case there is no significant departure from a hypothesis stating that these 156 traverses are random samples from a normal distribution with this mean and standard deviation.

Finally, in a class experiment used to illustrate the practice of point counting, a group of five operators counted 15 traverses of 90 points each per thin section on three thin sections from the Roslyn arkose (Eocene, Washington state; see Bressler, 1951). The departures of the observed from the expected frequencies of a binomial with the same mean are striking.

In all these graphs the abscissa is subdivided on the basis of the number of points in a traverse of the sampling grid which were assigned to the class designated quartz. The ordinate is expressed as frequency by number of the traverses which contained the number of points on quartz shown on the abscissa. The area and traverse lengths vary in these four cases, so that the sampling statistics are not comparable until reduced to the same units.

The mean $\bar{X}$ is equal to np, where n is the number of points per traverse, and, in order to compare the means, each is converted to percentage by dividing $\bar{X}$ by n and multiplying by 100.

The standard deviation is also not strictly comparable without a suitable conversion to the same units. A number of other examples are given in Chap. 14 and in the literature cited.

A representative collection of point-count analyses from a number of sedimentary rocks is illustrated in Tables 9.4 to 9.6, together with relevant details of the sampling procedure used to obtain the figures. A conversion to common units is then possible by means of the relationship of the mean and standard deviation and a knowledge of n, the number of points per traverse.

TABLE 9.4
Modal Analysis of Some Detrital Sediments: Arkosic Sediments

					Salt Wash					
	Roslyn[a]		Fountain[b]	Lyons[b]	Barren[c]		Ore-bearing[c]		Waynesburg[d]	
Element	$\bar{X}$	$\hat{\sigma}$	$\bar{X}$	$\bar{X}$	$\bar{X}$	$\hat{\sigma}$	$\bar{X}$	$\hat{\sigma}$	$\bar{X}$	$\hat{\sigma}$
Quartz	30.7	6.6	27.4	58.2	60.9	9.9	56.5	9.6	49.8	8.4
Feldspar	23.7[e]	8.7	29.3	11.9					6.0	4.4
Rock fragments	18.2	12.1	20.2	9.0	13.1	9.0	10.6	5.1	13.4[f]	6.7
									7.8[g]	4.9
Micas	7.3	6.2	4.4	0.5						
Matrix	13.7	16.9	14.3	11.0	14.5	12.2	6.8	5.8	11.6	6.6
Silica			0.1	2.7	11.9	8.1	11.9	6.1	0.7	1.2
Carbonate	5.2[h]	6.7	2.6	6.0	7.6	8.1	8.5	6.9	8.9	8.3
Others	1.2	1.1	1.6	0.6						
Total	100.0		99.9	99.9	108.0		94.3		98.2	
Specimens	40		48	32	25		25		75	
Traverses per specimen	2		6	6	6		6		10	
Points per traverse	100		100	100	100		100		50	

[a] Bressler, 1951. [b] Hubert, 1958. Weighted means, no standard deviation recorded. [c] Griffiths et al., 1956. [d] Siegrist, 1961. [e] Plagioclase, 12.2%; potash feldspar, 9.5%; others undifferentiated. [f] Quartzose rock fragments. [g] Micaceous rock fragments. [h] Carbonate and sulfate.

TABLE 9.5
Modal Analysis of Some Detrital Sediments: Quartzites

	(1)	(2) Oriskany		(3)		(4) Homewood		(5)		(6) Tuscarora		(7)	
Element	$\bar{X}$	$\bar{X}$	$\hat{\sigma}$	$\bar{X}$	$\hat{\sigma}$	$\bar{X}$	$\hat{\sigma}$	$\bar{X}$	$\hat{\sigma}$	$\bar{X}$	$\hat{\sigma}$	$\bar{X}$	$\hat{\sigma}$
Quartz, Feldspar	77.7	94.1	2.0	95.6	2.7	70.1	9.6	80.2	5.4	77.9	3.3	82.1	4.2
Rock fragments[a], Micas	1.4	0.4	0.8	0.7	0.9	15.9	10.0	5.4	3.3	0.9	0.4	1.3	0.8
Matrix[b]		0.9	0.9	1.2	1.4	9.5	8.3	6.9	4.4	5.7	1.0	4.6	2.8
Silica	9.5	4.5	1.9	1.9	1.3	3.7	1.8	7.1	2.6	14.5	3.4	12.6	4.2
Carbonate	3.8												
Others	7.1[c]												
Total	99.5	99.9		99.4		99.2		99.6		99.0		100.6	
Specimens	80	20		20		20		20		20		20	
Traverses per specimen	20	80		80		80		80		80		80	
Points per traverse	25	100		100		100		100		100		100	

[a] Rock fragments refer to quartzite rock fragments only in columns 2 to 7.
[b] Matrix includes micaceous rock fragments. [c] Clastic carbonate.
Source: Column 1, Rosenfeld, 1953; columns 2 to 7, Wood, 1960.

TABLE 9.6
Modal Analysis of Some Detrital Sediments: Graywackes (Low-rank)

	Cow Run[a]		Maxton[b]		Chipmunk[c]		Berea[d]		Pocono[d]	
Element	$\bar{X}$	$\hat{\sigma}$	$\bar{X}$	$\hat{\sigma}$	$\bar{X}$	$\hat{\sigma}$	$\bar{X}$	$\hat{\sigma}$	$\bar{X}$	$\hat{\sigma}$
Quartz	39.7	7.0	51.1	7.7	53.5	6.5	63.4	8.6	61.7	10.9
Feldspar	1.0	0.9	0.4	0.7	6.4	3.3	0.4		0.5	
Rock fragments	40.1	8.9	23.7	7.3	14.1	10.6	2.1		7.7	
Micas	0.7	0.9					1.3		1.6	
Matrix	9.8	6.3	11.1	6.2	19.3	5.4	16.2		25.9	
Silica	8.4	3.7	13.5	7.4	5.3	3.2	<0.1		<0.1	
Carbonate	4.0	7.0	0.3	0.8			11.6		3.1	
Others							5.1		0.8	
Total	103.7		100.1		98.6		100.1		101.4	
Specimens	39		33		71		16		16	
Traverses per specimen	6		4		5		5		5	
Points per traverse	200		200		100		100		100	

[a] Griffiths, 1958b. [b] Griffiths, unpublished data. [c] Stanonis, 1956, Pachman, 1961. [d] Emery, 1954.

Note: The totals may vary relatively widely from 100 percent because of a cumulative grouping error.

The standard deviations indicate the degree of variability to be expected in the figures, and they are usually quite large. These figures indicate that the proportions of the various subclasses vary over a wide range in one and the same sediment, and in order to determine differences in quantity of any constituent it will be necessary to arrange a suitable sampling pattern with considerable care.

10 Measurement of mineral composition: accessory minerals

An ancient sedimentary deposit possesses a mineralogical facies that is peculiar to it, exactly analogous to the facies which we find displayed by every present-day sediment in the same maritime region, and often different from the facies of the beds immediately below and above it, in such a manner that each is truly characteristic.

J. Thoulet, 1913

10.1 Introduction: The Concept

The cult of the heavy mineral forms a special area of investigation in studying the composition of rocks; the chemical composition of a rock, its bulk composition, is equivalent to modal analysis, whereas the accessory minerals play a role analogous to that of the trace elements. Similarly, the procedures for investigation of the accessory minerals, like that of the trace elements, require special preparation and treatment of the samples before the "elements" can be isolated; in both chemical and mineralogical studies the principal procedure reduces to a count of the mutual proportions of the elements.

Accessory minerals are those minerals which are relatively rare, generally constituting less than 1 percent of igneous, sedimentary, and metamorphic rocks. Sometimes these rare constituents occur as concentrates, e.g., in pegmatites, mafic rocks formed by early crystal settling, alluvial concentrates, and generally in ore deposits, so that accessory minerals are often of considerable economic interest. The distribution of the commoner accessory minerals in different kinds of rocks is illustrated in Table 10.1; for the present purpose, the metamorphic rocks can be considered as embracing both altered igneous and sedimentary rocks.

Quotation: From Boswell's translation (1933, p. 29) of Thoulet, 1913, p. 9.

TABLE 10.1
Relationship between Accessory Minerals of Igneous and Sedimentary Rocks

	Igneous, %	Minerals	Sedimentary, %	
Essential	0–65	Orthoclase	0–60	Essential
	0–50	Plagioclase	0–30	
	0–30	Quartz	0–95	
Varietal	0–50	Amphibole	<1	Accessory
		Pyroxene		
	0–20	Micas	0–5	Varietal
	0–20	Olivine		
Accessory	<1	Zircon	<1	Accessory
		Tourmaline		
		Apatite		
		Garnet		
		Sphene, etc.		
Secondary (variable)		Chlorite		
		Epidote		

About 300 of the more than 3,000 minerals which occur in the earth's crust are relatively common as accessory minerals in sediments. Most sediments contain about 10 to 15 species of accessory minerals; but some, e.g., black sands or volcanic sands, are almost monomineralic, whereas others, e.g., glacial deposits, contain 50 or more different species. The accessory minerals may, on the basis of genesis, be subdivided into allogenic, or detrital, and authigenic, or minerals formed *in situ* in the sediment; therefore, studies of the accessory minerals have been used to differentiate source areas (problems in provenance), to deduce conditions of deposition (define different facies or environments), and to elucidate diagenetic changes in sediments (study their paragenesis). As would be expected, the degree of success in achieving these different goals has varied widely. Some workers believe that accessory minerals are sufficient to provide almost all the information, whereas other workers have been very disappointed with the results. In general, in near-source sediments and in the younger, particularly Tertiary, sedimentary rocks the accessory minerals have proved very useful. In sediments far from their source, in which the accessories may have suffered several cycles of sedimentation, and also in the progressively older sediments, in which diagenesis has reduced the variety of accessories, the amount of information supplied may be very meager. Again, generalizing broadly, the arkoses[1] and high-rank graywackes[1] are frequently relatively richer in amounts and varieties of ac-

[1] Petrographic terms used in the general sense defined by Krynine (1948).

cessory minerals than the low-rank graywackes,[1] and the quartzites[1] are usually poor in variety and amount. This is, of course, another way of stating that near-source sediments, such as the average arkose and high-rank graywacke, have much more varied accessories than those sediments, e.g., the average low-rank graywacke and the quartzites, likely to be formed after long travel, in which the constituents have suffered from a high degree of selective sorting.

It must be realized that variation in composition contains a limited amount of information at best, and more chance of confusion of this information exists in those sediments where processes of sedimentation and diagenesis have had greater opportunity to modify and disguise the original proportions and arrangement of the minerals.

10.2 Historical Review of Accessory-mineral Investigation

The evolution of accessory-mineral investigation is described in some detail by Boswell (1933, chap. 11, pp. 15ff.), who traces its beginnings to Réaumur in 1718. The early work, when accessory minerals were regarded as curiosities, was chiefly concerned with unconsolidated rocks such as beach and river sands (e.g., Damour, 1856); the records of precious stones by these early writers suggest that the interest even then was not solely academic. Until about 1870, according to Boswell (1933), the analytical techniques were largely chemical but included magnetic separation and recognition by crystal form. After 1870 special procedures for separation in heavy liquids became the mode, and the standard procedure for preparation of "heavy" residues was established; at the same time, about 1870 to 1890, the polarizing microscope came into use for petrographic analysis, so that this period became the acme of descriptive petrography. Thoulet's work on the Sahara sands is an outstanding example of heavy-mineral investigation (Thoulet, 1881).

The intensity of interest increased during the decade from 1890 to 1900, and technique became more refined and description more detailed. It was during this period that Retgers (1895) first documented the possibility of determining source area, conditions of deposition, etc., of the sediments and Artini (1898) introduced quantification with a scale of 1 to 10, representing abundant to rare, respectively. At this time, too, the carbonate rocks received attention in the classic work of Cayeux (see, for example, 1897).

From 1900 to 1920 accessory-mineral investigation flourished and became a discipline, achieving status by being taught at universities (Boswell, 1933, p. 25). Accessory minerals were found in older and older rocks, and the basic principle that each sedimentary rock possessed its own unique assemblage became established, so that stratigraphic correlation by heavy minerals proved feasible (Boswell, 1916a); Boswell's colleague Illing (1916), recognizing the

[1] Petrographic terms used in the general sense defined by Krynine (1948).

potential of this tool, applied it to correlation in the Tertiary of Trinidad,[1] thus introducing accessory-mineral analysis to the petroleum industry.

From the earlier work it had become evident that to assign the detrital heavy minerals of a sedimentary rock to a specific source area, a knowledge of source-area heavy minerals is necessary, and Rastall and Wilcockson (1917) were among the first to attempt to describe the accessory heavy minerals in igneous rocks; subsequently, the use of varietal variation, earlier suggested by Mackie's work on quartz (1896), reached the ultimate in perfection in Brammall's studies (1928) of the Dartmoor granite. A knowledge of the heavy minerals of this and neighboring granitic source areas enabled Groves (1931) to evaluate their separate contributions to the younger sediments of southern England.

Thus by the early 1930s the fundamental features of accessory-mineral investigation were fully routinized, and the basic principles involved in their applications were well established (see Boswell, 1933, pp. 47–49); source areas, conditions of transportation and deposition, and (from the relative stability of different species) diagenesis could be deduced. During the two decades from 1920 to 1940 accessory heavy minerals were adopted as a standard tool in petroleum geology, and many extensive investigations,[2] particularly in Tertiary sediments, illustrate their usefulness (Déverin, 1923; Reed and Bailey, 1927; Hedberg, 1928; Edson, 1929; Edelman, 1931; Evans, Hayman, and Majeed, 1933; Doeglas, 1940).

By the 1940s it was recognized that while accessory-mineral studies contributed to the reconstruction of the broad regional picture, and while correlation of broad stratigraphic zones over large areas was quite feasible, local variation tended to be haphazard, and the effectiveness of the tool in detailed studies was questionable. Selective sorting had early been recognized as a factor affecting the variation in proportions of accessory heavy minerals (Boswell, 1933, pp. 5, 75ff.), and Rubey (1933) was among the first to emphasize this role. Attempts were made to offset the effects of size sorting by confining attention to one or two size grades (Russell, 1936), but these approaches were largely ineffective; the exact nature and magnitude of the problem were eventually presented by Rittenhouse (1943b, 1944), and this work indicates the stringent limitations which exist in using variations in proportion of accessory-mineral species. After the early 1940s accessory-mineral studies continued to be used but with much reduced enthusiasm and only for certain very specific problems (e.g., Hutton, 1959; Allen, 1948).

It seems evident that to be able to achieve more exact and detailed conclusions a more refined approach must be used, one which is capable of extracting "the signal from the noise" in studying smaller and smaller fluctuations in proportions of the accessory heavy minerals.

[1] Throughout the book this refers to the island of the former British West Indies, now part of the country of Trinidad and Tobago.

[2] The bulk of this work is available only in the files of various oil companies,

10.3 Defining the Classes of Elements

Accessory minerals are usually classified into homogeneous groups by a set of diagnostic criteria, and a very elaborate set of "symptoms" is included in many valuable reference texts (e.g., Milner, 1962; Winchell and Winchell, 1951; Krumbein and Pettijohn, 1938). Tables such as those in Winchell and Winchell (1951, pt. 3) or Russell (1942) are also very helpful. It must be admitted, however, that the identification of accessory heavy minerals is a special skill developed essentially by long practice, and, as in the case of modal analysis, this procedure requires a psychophysical conditioning to develop a mental habit. Generally the identification procedure is much more mysteriously difficult for the beginner than almost any other routine; it is more difficult to identify accessory minerals as loose grains than minerals in thin section, because added to the variation shown by minerals in thin

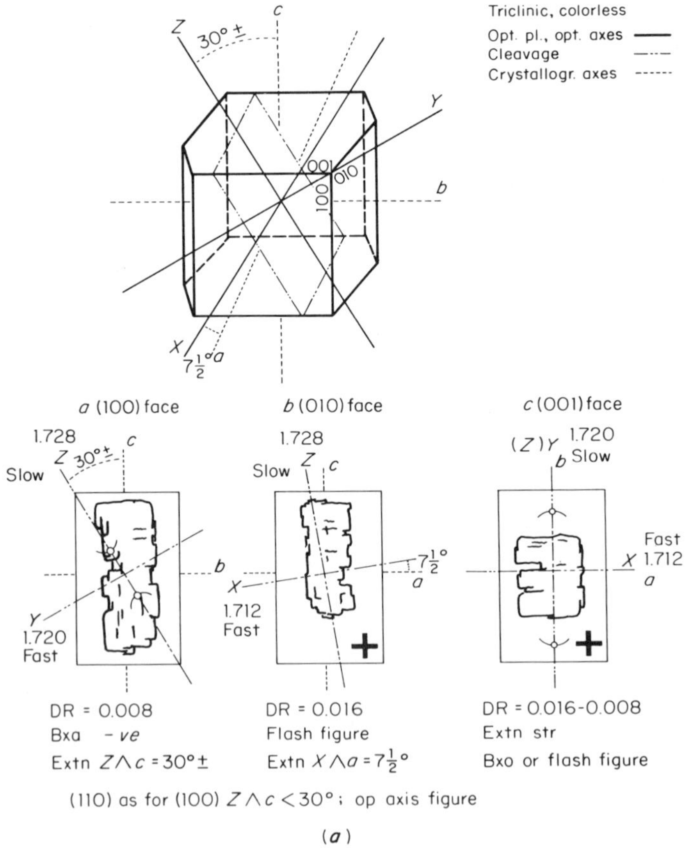

Figure 10.1a Habit and optical properties of kyanite.

section is the variation in thickness and habit of loose grains. As a result, the most important single diagnostic characteristic is the habit of the mineral;[1] an experienced heavy mineralogist has a mental picture of the appearance of each mineral and an intuitive probability distribution of the likelihood that some species of mineral will occur in a specific characteristic habit. In thin section, random sections occur across the mineral grains; but in mounting individual heavy mineral grains they tend to take up their stable position and generally expose their maximum projection area. For example, in thin section a basal pinacoid of muscovite is rare, but in loose grain mounts it is the face usually exposed. Again, in thin section, the confusion of muscovite and topaz is most unlikely, but in loose grain mounts both appear as colorless round plates of similar relief (refractive index = 1.64±) exposing the basal

[1] This assumes that the habit is a natural attribute and is unaffected by the technique of preparation of the material.

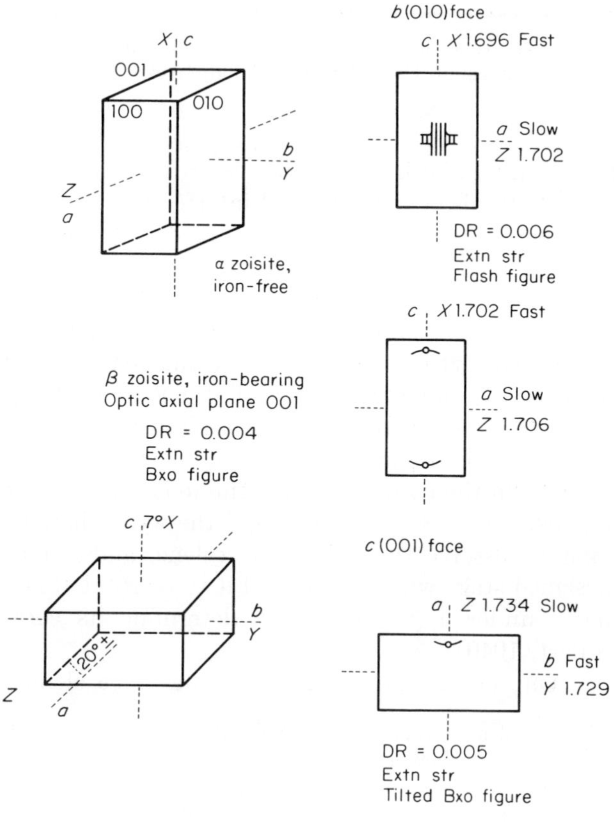

(b)

Figure 10.1b Habit and optical properties of zoisite-clinozoisite.

TABLE 10.2
Heavy Liquids for Accessory-mineral Separation

	Bromoform (tribromoethane)	Acetylene tetrabromide (sym-tetrabromoethane) (1,1,2,2-tetrabromo-ethane)	Methylene iodide (diiodomethane)
Formula	$CHBr_3$	$C_2H_2Br_4$	CH_2I_2
Molecular weight	252.77	345.70	267.87
Boiling point, °C	149.5	151.0	180
Freezing point, °C	6–7	0.1	5–6
Density	$2.890^{20°C}$	$2.9630^{20°C}$	3.325
Refractive index	$1.63795^{19°C}$	1.63795	$1.7559^{10.5°C}$
Comment	Colorless liquid or hexagonal crystals	Colorless to yellow liquid; decomposes at 239–242°C	Colorless liquid; leaflets at 0°C

Source: Information from "Handbook of Chemistry and Physics," 28th ed., Chemical Rubber Publishing Co., Cleveland, Ohio, 1944.

pinacoid, and both tend to be "isotropic" in this section: since, in conoscopic light, both also exhibit biaxial interference figures of low to moderate optical axial angle ($2V$), they are easily confused, and the critical difference is that topaz yields a positive sign and muscovite a negative one.

Diagnosis of accessory heavy minerals often requires that the crystallographic orientation of the exposed face be specifically determined in order to achieve correct identification, and this demands quite time-consuming study; many heavy mineralogists find it necessary to supplement their reference background by carefully drawn pictures of typical minerals (Fig. 10.1);[1] even with these several aids it is often necessary to isolate the individual grain from the mount and determine its precise refractive index in immersion liquids. Alan Stuart[2] instructed the writer how to drill the cover slip of the mount, dissolve the balsam in xylene, and mount the grain on a specially designed stage which permitted it to be rotated in liquids of known refractive index under a microscope to determine its correct species (Griffiths and Stuart, 1940).

One of the advantages of using trace elements as contrasted with ac-

[1] Brammall accumulated at the Royal School of Mines, London, a set of hand-painted colored charts showing each possible section of each of the common accessory minerals, illustrating the pleochroism and interference figures of each section; a set of typical heavies originally prepared by Brammall is available through Gregory and Bottley, Ltd., Chelsea, London, and is an extremely useful guide and teaching aid.

[2] Swansea University College.

cessory minerals is that identification of the elements in the population is automatically performed by machine, and estimation of proportion of the different elements may, under suitable conditions, also be automated. Precision or reproducibility of diagnosis and of estimates of proportion is therefore orders of magnitude superior, and larger numbers of specimens can be analyzed for less investment in time and money. Furthermore, trace-element analysis can be "learned" much more easily than accessory-mineral identification; unless, therefore, the study of accessory minerals can be demonstrated to supply unique information of considerable interest and value, heavy-mineral studies may never again claim the popularity they achieved in the decade around the 1930s.

10.4 Preparation, Procedure, and Technique

The segregation and concentration of the accessory heavy minerals rely essentially on their having higher density (hence the name heavy mineral) than the bulk minerals quartz, feldspar, and rock fragments. It is first necessary to disaggregate the sample of rock without, if possible, affecting the habit of the heavy minerals, and then the set of grains is added to a heavy liquid, such as bromoform or acetylene tetrabromide, and the lighter minerals float and the heavies sink.

Properties of the heavy liquids in common use are summarized in Table 10.2, and a wide variety of apparatus for separation is described in standard texts (Milner, 1962, pp. 101ff.; Krumbein and Pettijohn, 1938, pp. 335–344). For the analysis of a few samples a simple arrangement consisting of two separatory funnels and a beaker to collect the excess liquid is adequate (Fig. 10.2). When large numbers of samples are analyzed and quantitative results are required, it is advisable to organize the work more carefully, and centrifuge separation will be found more satisfactory, both to obtain a more complete and rapid separation and because the analysis is generally simpler to control.

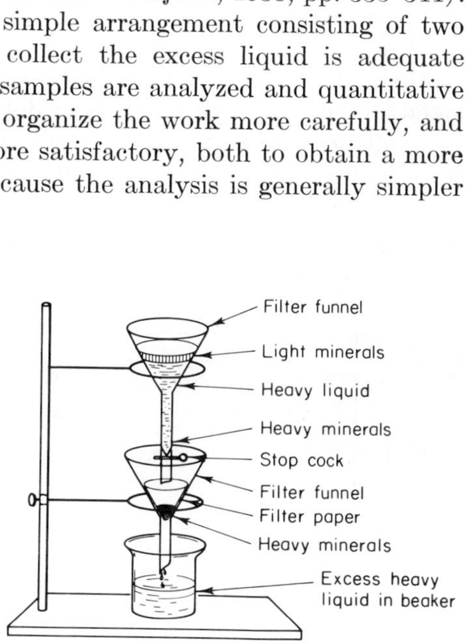

Figure 10.2 Apparatus for heavy-mineral separation by flotation in heavy liquids.

The main difficulty in centrifuge separation is recovery of the heavy residue; a technique first demonstrated by R. Mackay at the Royal School of Mines, London (personal communication, 1938), used a simple centrifuge tube (Fig. 10.3*a*) and relied upon freezing the bromoform in an ice-and-salt mixture after separation of lights from heavies. Gentle warming of the centrifuge tube in the hand frees the frozen plug of bromoform containing lights and heavies. The plug is then broken, the section with the heavies being dropped on a filter paper in a filter funnel, where the bromoform melts and passes through the filter. The section with the lights, if the light minerals are not of particular interest, is then accumulated in a suitable container and later filtered in bulk. For precise quantitative studies this technique is often inadequate because some of the light residue clings to the tube and contaminates the heavies when the plug is withdrawn. An improvement suggested by Krynine (personal communication, 1947) consists in using a centrifuge tube with a constriction at the lower end (Fig. 10.3*b*), which permits freezing only the constricted portion; and then the lights and bromoform can be cleanly washed out of the tube before extracting the heavies. An alternative procedure first mentioned by Sullivan (1927) and independently evolved by Jeffries (1941) uses a specially designed tube within the centrifuge tube (Fig. 10.3*c*). The residue is introduced into the inner tube, and, after centrifuging to separate the lights and heavies, the inner tube, containing the lighter minerals, is removed by covering the opening with the thumb. The heavies are then isolated in the outer tube and can be removed by filtering as usual.

The main practical difficulty involved in centrifuging bromoform, or any other heavy liquid, arises because the weight of the liquid often leads to tube breakage during centrifugation, but by cushioning the whole glass tube in its metal container in bromoform this difficulty can be reduced.

Since the main objective involved in these procedures is separation and concentration of the accessory minerals, many alternatives exist for improving the separation including the use of hand magnets, electromagnets, and, more recently, the Franz isodynamic separator, either alone or in combination with

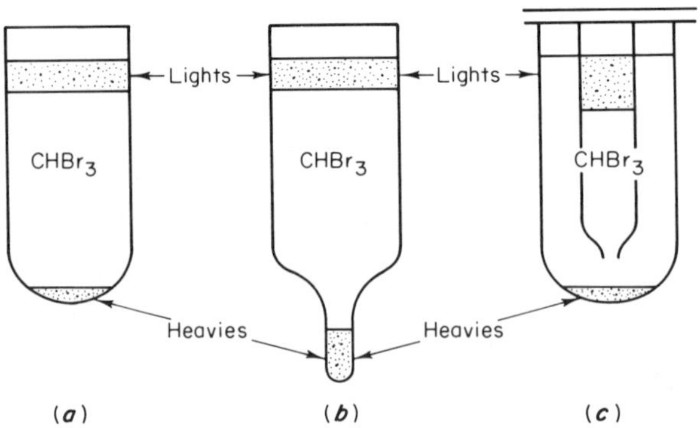

Figure 10.3 centrifuge tubes for separating heavy minerals in heavy liquids. (*a*) Freeze whole tube (*after R. A. Mackay, personal communication, 1938*); (*b*) freeze constricted section only (*after Krynine, personal communication, 1947*); (*c*) tube within a tube (*after Jeffries, 1941*).

heavy-liquid separation. The various alternative techniques are described in full in Milner (1962). It should perhaps be emphasized that these techniques are largely aimed at concentration of the accessories and, where feasible, the separation of each mineral species into fractions. Sieving of the residue to remove the coarser-sized fractions and sedimentation for the finer-sized fractions are often used, because the accessory minerals are generally concentrated in the size range between medium-grained sand and coarse silt, effectively from 500 to 30 microns (1 to 5 phi units). When the aim is to assess proportions of the accessory-mineral species by counting the final residues, the segregation into subsets inevitably introduces subsampling errors (see next section), and it is doubtful whether the procedure is of much advantage. If, on the other hand, diagnosis is the main problem, isolation of the mineral species is a great advantage, and the isolates can then be submitted to x-ray or chemical analysis. In this sense the objective, as usual, is the final arbiter, and clear recognition of the exact objective is essential to the choice of appropriate technique.

After separation, the heavy residue can be cleaned; acid removes iron staining, organic solvents remove petroleum residues, and a number of solvents overcome the various obscuring films (see Milner, 1962). Finally, the residue is mounted in a heat-setting resin[1] for permanent mounts or in a liquid of known refractive index for a temporary mount where unusual diagnostic difficulties are suspected. The final mounting of the accessory minerals is quite an important step, particularly if counts are to be made of the proportions of the different mineral species; the first difficulty arises in subsampling the heavy residue if there is more available than can be conveniently included in a single slide. Various sample splitters are available to control the subsample variation arising from this procedure (Otto, 1933; Wentworth, Wilgus, and Koch, 1934; Kellagher and Flanagan, 1956). A second feature, important in estimating proportions, is to ensure a homogeneous distribution of the heavy residue over the area of the slide under the cover slip, and Allen (1948) uses slides with specially prepared grids to simplify this mounting procedure.[2]

Certainly there is no substitute for well-made preparations to facilitate both diagnosis and quantitative estimates of variation in proportion; the quality of the data accruing from these studies is, of course, directly dependent upon successful preparation of the mounts containing the heavy residue.

10.5 Estimation of Proportions of Accessory Minerals

The earlier investigations into accessory heavy minerals in rocks were purely descriptive, but as the subject developed, the simple recording of the

[1] Such as Canada balsam (refractive index 1.54), piperine (refractive index 1.68), or hyrax (refractive index 1.70).

[2] These slides are supplied by Cutrock Engineering Co., Ltd., London, England.

presence or absence of a mineral species gave way to records of abundance (Boswell, 1933, p. 69). At first these were vague subjective assessments of relative frequency, but in 1898 Artini introduced the use of numerals to represent relative frequencies, using 10 for rare and 1 for abundant. At this stage the scale of information was modified from nominal, in which the grains are classified into groups, to ordinal, in which the frequency estimates are ordered (see Sec. 12.5). Salmojarghi (1904) reversed Artini's scale, with increasing numbers representing increase in frequency; actually, of course, both scales are equivalent in terms of information content. Subsequently Fleet (1926) proposed that frequencies should be determined by counting the number of grains in each group (mineral species). In 1931 Dryden discussed the "accuracy" (actually precision or reproducibility) of counting and suggested the use of the probable error as an estimate of "accuracy" of the count. Rittenhouse (1940), by means of log-log graph paper, transformed Dryden's curves to straight lines so that estimates of precision could be more easily achieved.

With the adoption of counting, the measurement level was again improved, and ordinal gave way to interval scales; however, the zero count equals "not detected" and so is somewhat arbitrary. Furthermore, the distance between percentages is not fixed, so that the scale level possible by counting is not likely to achieve interval level under ordinary circumstances.

When counting is adopted to estimate proportions, it is necessary to standardize procedure; the slide containing the heavy-mineral grains is mounted in a mechanical stage under the microscope and traversed by means of the stage. Counts are performed either as linear traverses or in microscope fields; the line-counting procedure is generally preferred because it is simpler to move the slide than the eye[1] and because, when using linear traverses, each grain comes to the center of the microscope field and is, therefore, in the most advantageous position for identification. In most heavy residues the opaque minerals exceed the nonopaques in abundance; and since identification of opaques requires an optical system adapted to reflection rather than transmission, it is usual to dispense with the opaques as rapidly as possible. The Dutch School[2] of petrographers has developed the procedure of classifying 100 grains into opaque and nonopaque classes and then ignoring the opaques while classifying 100 or more grains of nonopaque minerals. This has a number of advantages; e.g., suppose there are 90 percent of opaque grains in a sample of heavy residue and 100 grains of nonopaques are counted; then this is equivalent, in terms of precision of the count, to counting 1,000 grains of the original residue. Since the opaques are rarely used for interpretation, this appears to be a favorable procedure for achieving greater precision of counting. This kind of procedure has been advocated to reduce the diffi-

[1] The effects of eye fatigue arise much sooner with field counting than with line counting.
[2] This term embraces the distinguished group of heavy mineralogists led by Edelman and including Doeglas, Crommelin, van Andel, and many others from Wageningen.

culties inherent in the statistical analysis of percentage data (Krumbein and Tukey, 1956; Vistelius and Sarmanov, 1961).

Now suppose that the frequency distribution of a species of heavy mineral, say garnet, is binomial and there are P grains of garnet in N grains of nonopaques on the slide; then each linear traverse performed yields an estimate X_i of P, and if n (less than N) grains are counted, then $\bar{X}$, the average of the X_i, is a suitable estimator of P if the traverses are random samples of the residue in the slide. If all N grains are counted, the number of garnets is P, the exact number in the slide; more commonly a sample of n is counted, and we require that as n approaches N, $\bar{X}$ approaches P (see Secs. 2.3 and 12.6). Estimates of the precision of the count are generally represented as some function of the standard deviation, which, in a binomial distribution, is $\sigma = \sqrt{NPQ}$ (see Sec. 14.2). Then the estimate of the number[1] of grains of garnet is represented as $\bar{X} \pm \sqrt{\bar{X}(1 - \bar{X})}$. The probable error recommended by Dryden is a function of the standard deviation as follows:

$$\text{Probable error (number of grains)} = 0.6745\sqrt{\bar{X}(1 - \bar{X})} \qquad (10.1)$$

It may be argued that as n grows large, the normal distribution becomes a suitable approximation for the binomial (see Sec. 14.2), and the mean ± 1 standard deviation includes some 67.45 percent of the area of a normal distribution; i.e., the probability of an observation's falling in the interval $\mu \pm \sigma$ is 67.45 percent (Sec. 13.7). Then the probable error represents an interval $\mu \pm 0.6745\sigma$, or 50 percent of the area of the normal curve; therefore there is an even chance of an observation's falling within or without this interval. It is more conventional to use $\mu \pm k\sigma$, where $k = 2$ or 3 to include 95.4 and 99.7 percent of the area, respectively. If the usual recommendations, which assume a binomial distribution, are followed, then the "error of the count" $2\sqrt{\bar{X}(1 - \bar{X})}$ may be estimated quite readily from the calculated $\bar{X}$; a representative set of values is given in Table 10.3. It may be seen that the error is symmetrical around a maximum at 50. Furthermore, there is a 95 percent probability that a new estimate will fall in the interval 50 ± 10, so that for a count of 100 grains the estimated range of 40 to 60 grains is to be expected.

Under these conditions a more realistic approach is to use subsamples and the interval $\bar{X} \pm k\hat{\sigma}_{\bar{X}}$, where $k = 2$ or 3 and $\hat{\sigma}_{\bar{X}} = \hat{\sigma}/\sqrt{n}$; by judicious choice of n the "error" in the subsample mean can then be controlled (see Sec. 2.6). For example, suppose four sets of 25 grains are counted and the number of garnets is 50, that is, $P = 50$; then the interval $\bar{X} + 2\hat{\sigma}_{\bar{X}} = 50 \pm 2\hat{\sigma}/\sqrt{n}$, or $50 \pm \frac{2}{5}\hat{\sigma}$. It is advantageous therefore to count in small sets.

[1] This can be transformed to percentage units by dividing by n and multiplying by 100, that is, % garnet $= (NP/N)100$, or, in terms of the estimators, $(100\bar{X})/n$, and since $\bar{X} = np$, this becomes $100p$; similarly, $\sqrt{\bar{X}(1 - \bar{X})} = \sqrt{npq}$, and in percentage scale this becomes $\sqrt{pq/n}$.

TABLE 10.3
Estimates of "Errors of Counting"

Mean $\bar{X}$	p $\frac{\bar{X}}{n}$	q $\frac{1-\bar{X}}{n}$	npq $\bar{X}\left(\frac{1-\bar{X}}{n}\right)$	$\sqrt{npq}$ $\sqrt{\bar{X}\left(\frac{1-\bar{X}}{n}\right)}$	$2\sqrt{npq}$ $2\sqrt{\bar{X}\left(\frac{1-\bar{X}}{n}\right)}$
5	0.05	0.95	4.75	2.18	4.36
10	0.90	0.90	9.00	3.00	6.00
20	0.20	0.80	16.00	4.00	8.00
30	0.30	0.70	21.00	4.58	9.16
40	0.40	0.60	24.00	4.90	9.80
50	0.50	0.50	25.00	5.00	10.00
60	0.60	0.40	24.00	4.90	9.80
70	0.70	0.30	21.00	4.58	9.16
80	0.80	0.20	16.00	4.00	8.00
90	0.90	0.10	9.00	3.00	6.00
95	0.95	0.05	4.75	2.18	4.36

With a binomial distribution, the above argument is quite rigorous and includes random samples which yield unbiased, consistent estimators (see Sec. 12.6); but the assumption of random samples from a binomial population is equivalent to assuming a constant-probability model, and this, in turn, requires certain features in the sampled population (see Sec. 14.3). Physically this means that the occurrence of garnet is constant from trial to trial. It is therefore most necessary that the slide be adequately prepared so that the assumption of a constant-probability model is a reasonable one (see Sec. 14.3). It will be found in practice that it is difficult to prepare a mount which fulfills this requirement, and Allen (1948) used glass slides with marked grids to achieve better control of the counting and mounting procedure. Without adequate control the counts may yield biased estimators, and the estimated proportion may not represent the true proportion of the mineral species in the slide.

Large variations in proportion of the heavy minerals in different samples from the same bed are quite common, so much so that Evans, Hayman, and Majeed (1933) proposed transforming the percentage determined by counting to a new scale in which these fluctuations are subdued. Since on this new scale the zero is arbitrary and the intervals irregular, this implies a step back toward the ordinal scale; if this becomes necessary, counting is largely pointless and the counting error becomes meaningless.[1]

[1] In fact, it is doubtful whether the arithmetic procedures of picking a mean and standard deviation are justified in this case.

An attempt to isolate and compare the sources of variation in proportion of heavy minerals is described by Krumbein and Rasmussen (1941), who find that the total laboratory error is approximately equal to the field sampling error and that "errors" in weight percent of heavies are larger than those in number frequency; they recommend the use of the centrifuge for better control in separation of heavy residue (see also Rittenhouse and Bertholf, 1942). In a further breakdown of laboratory sources of variation (error) Krumbein and Rasmussen (1941) find that "errors of microsplitting, mounting and counting" are about three times as large as errors from "sample splitting and sieving" and errors from bromoform separation.

It seems evident from these analyses that only the larger changes in proportion, which remain consistent over a region, may be used to postulate features of geological interest. Furthermore, if variation in proportions is of real interest, subsampling must be controlled; and to achieve this, the simplest preparation procedure is the most advantageous. Since, after disaggregation, the effects of preparation of a sample on variation in frequency of a heavy-mineral species can be represented as equivalent to a transformation from Poisson to binomial frequency distributions (see Sec. 14.7), it may be as well, to avoid any further problems, to use the disaggregated sample without modification. Then the Poisson distribution would be a suitable model for testing the adequacy of the mount (Griffiths, 1960a), and many sources of subsampling error would be avoided.

If the estimates obtained by counting are truly representative of the frequency of the mineral species in the samples under analysis, the variation from sample to sample may be studied. Again, in order to establish that this variation is truly representative of the variation of frequency of the mineral in a rock body, it is necessary to find the frequency distribution of the mineral species in "nature" ("real world"). Some possible frequency distributions are discussed in Secs. 14.5 to 14.10 and in Griffiths (1960a), and, so far at least, the negative binomial appears to be the most likely real-world model (see also Allen, 1945b).

10.6 Applications of Accessory Heavy-mineral Analysis

The detrital accessories in sediments are quite generally used to specify source areas from which the detritus is derived; in addition, the accessories are used to postulate conditions of sedimentation including erosional, transportational, and depositional conditions. The state of preservation of detrital accessories and the presence of authigenic accessories are used to reconstruct diagenetic processes subsequent to deposition. Again, the heavy minerals are used for stratigraphic correlation by establishing that a layer of rocks has a mineral assemblage which persists and which differs from the adjacent overlying and underlying layers of rocks.

Accessory minerals have also been used for problems in alluvial prospecting for gold, diamonds, and platinum (Raeburn and Milner, 1927); for investigations of glass and refractory sands (Boswell, 1916b); and to map igneous terrains from the distribution of heavy concentrates in streams where outcrop information is not available.

The pattern of application is quite typical; the first procedure is to search for a unique mineral which ties the detritus to a unique source. Derivation of chromite, occurring in the insoluble heavy residue of the Chalk at Beer Head in Devon, from the Lizard serpentine (Davies, 1919) is an outstanding example; riebeckite in the glacial deposits of Wales is derived from either Mynydd Mawr, north Wales, or Ailsa Craig in the Firth of Clyde, Scotland (Griffiths, 1940); blue tourmaline from the Dartmoor granite is used as a marker mineral in the Mesozoic and Tertiary sediments of southern England (Groves, 1931).

Unique minerals are rare, but unique varieties of common minerals frequently may be used for the same purpose; purple zircons were first recorded as indicators of the Precambrian Lewisian gneiss by Mackie (1923); zoned zircons of the Dartmoor granite (Brammall, 1928) were also used by Groves (1931), and numerous other examples are mentioned by Boswell (1927). Another similar alternative is the use of unusual morphologies or inclusions, such as saw-edged staurolite in the Triassic of England (Thomas, 1909), etched garnets (Milner, 1940, see ref. to Milner, 1962), and cored apatites (Smithson, 1931). Some morphologies are considered to reflect diagenetic processes, such as frayed amphiboles and pyroxenes (Edelman, 1931), outgrowths on zircons (Butterfield, 1936), and outgrowths and abraded outgrowths on tourmaline (Krynine, 1946).

When these unusual features proved too rare for general use, emphasis changed from unique minerals to unique assemblages of minerals, following similar developments in paleontological investigations, and this extended the applications of accessory minerals to many more sedimentary rocks. The individuality of the different stratigraphic horizons in Britain, as described by Boswell (1933, chap. 6), is one of the best examples. Stratigraphic correlation depends essentially upon the limited vertical, and extended lateral, range of typical mineral assemblages. Many stratigraphic breaks or unconformities are marked by abrupt changes in the accessory-mineral assemblages (e.g., Hedberg, 1928). The "first appearance" of a new mineral in a succession of beds is usually used as a marker horizon; e.g., in southern Trinidad the top of the Oligocene sandstones is characterized by the first appearance of glaucophane and the youngest Miocene cycle by the first appearance of actinolitic amphibole (Suter, 1951). Similarly, the nonopaque assemblage of zircon, rutile, and tourmaline is characteristic of the La Carrière formation, and, with the addition of staurolite and garnet, the assemblage typifies the lowermost Oligocene in southern Trinidad. The succession of changes in mineral assemblage may be used to zone the Tertiary of Trinidad (see Fig. 10.4), as first mentioned by Illing (1916). Beginning with the

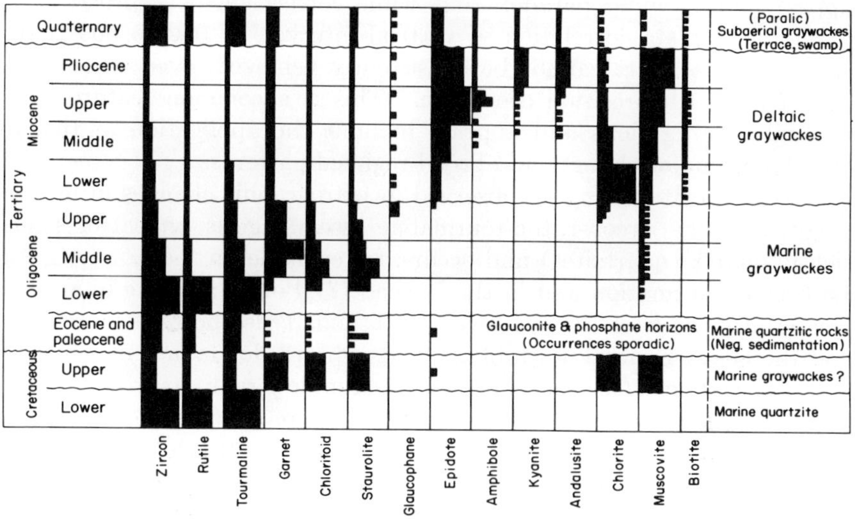

Figure 10.4 Accessory-mineral distribution in the Cretaceous Tertiary of Trinidad.

Lower Cretaceous or Jurassic and proceeding through the Eocene to Quaternary, about seven distinct zones can be established on the basis of relative frequency of the different minerals, and these zones are more or less persistent for about 700 miles into Venezuela in an east-west direction. An increasing complexity of assemblage is evident in the Eocene-Oligocene sequence, from a simple zircon-rutile-tourmaline suite through successive additions of staurolite-chloritoid, then garnet, which gradually increases to about 30 percent of the nonopaque residue, and finally to persistent but rare glaucophane (< 5 percent). This order of appearance represents the gradual "unroofing" of a metamorphic sequence with increasing intensity of metamorphism toward an igneous core, so that the sequence implies a specific source area in folded mountains with regional metamorphism such as the Sierra de la Costa, of northern Venezuela. There is then a marked change to an epidote-rich residue in the Miocene sediments, which terminates with the appearance of amphibole in the youngest Miocene beds. The bulk of the residues in the Miocene are composed of muscovite, biotite, and chlorite, with cyclical fluctuations in dominance of muscovite and chlorite, which suggest low-rank metamorphic rocks associated with altered volcanic rocks (epidote and amphibole).

These zones and the accompanying reflections of source areas can be established without any very precise quantitative analysis; they depend essentially on major changes in mineral frequency. When an attempt is made to increase precision so that local correlation can be accomplished, the fluctuations in frequency of the counts are much too haphazard ("noisy"). There appears the seeming paradox that while correlation over broad regions is feasible, complete correlation of vertical intervals of 300 to 500 ft in wells spaced 220 ft apart cannot be consistently achieved, even when counts of apparently high precision are made. This is a common feature in heavy-mineral investigations and appears to limit the application of this tool to elucidating major changes and broad regional patterns.

Similar features can be observed in petrographic changes in bulk mineralogy; thus the zircon-rutile-tourmaline assemblage is typical of sandstones rich in quartz (quartzites) and occurs, for example, in the Lower Cretaceous La Carrière formation and in the Eocene (?) Pointe-à-Pierre quartzites; the Quaternary sands of the southwest of Trinidad also become more quartzose and the heavy-mineral assemblage tends toward dominance of zircon, rutile, and tourmaline. On the other hand, the low-rank graywackes of both Oligocene and Miocene age contain much more complex heavy-mineral assemblages. If, therefore, a quartzite is compared with a low-rank graywacke, the heavy-mineral assemblages will almost always be very different in appearance, and changes in both bulk mineralogy and the accessory mineral assemblages are evidently highly correlated in a statistical sense, i.e., the changes are interdependent.

10.7 The Principle of Hydraulic Equivalence

It is, of course, clearly recognized that heavy minerals occur as "pay streaks," implying special concentrations by natural processes of selective sorting (e.g., Boswell, 1933, chap. 7); as mentioned earlier, Rubey (1933) calculated the magnitude of these effects, and Rittenhouse (1943b, 1944) evaluated in detail several aspects of hydraulic equivalence in modern rivers, such as the Rio Grande. In addition, Allen (1945b, 1947, 1949) emphasized the patchy nature of the frequencies of heavy minerals in the Ashdown sand of southern England. The principle of hydraulic equivalence is also discussed in terms of grain size in Sec. 4.10. Let us suppose that where two detrital minerals accompany each other in a sedimentary rock, they are hydraulically equivalent; then, as is well documented, the heavier minerals will generally be smaller than the lighter minerals. For example, if zircon, quartz, and muscovite are compared in terms of this principle, zircon will be the smallest, quartz will be intermediate, and muscovite largest in "size"; it is evident that the features which lead to hydraulic equivalence include at least effective density, size, and shape, and this assumes equal availability and no diagenetic removal of either mineral. If, under these conditions, there are changes in current velocity in the medium transporting and depositing the detritus, there will be changes in average size and shape and change in the ranges of sizes and shapes. Since we know, from grain-size measurements and observation on present-day environments, that such changes are common, fluctuations in *relative* proportions of the minerals will also be common. This

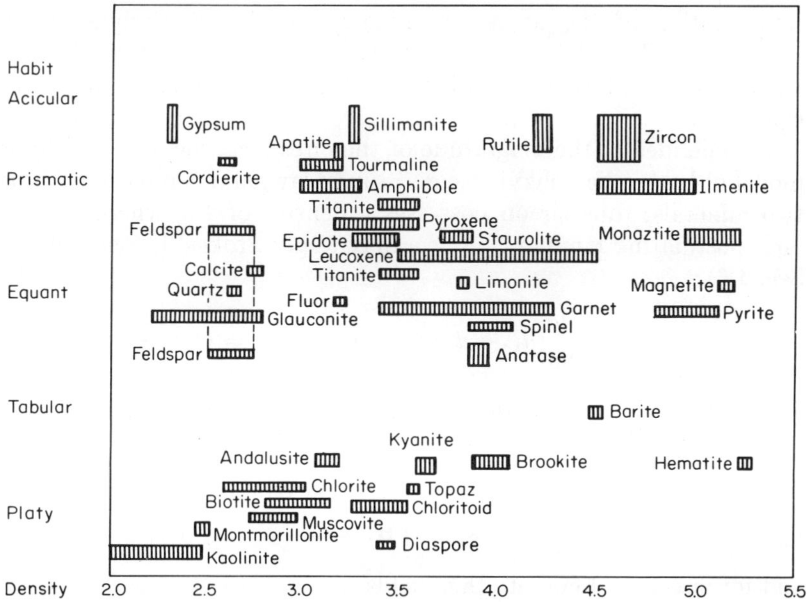

Figure 10.5 Diagrammatic representation of hydraulic equivalence of the commoner accessory heavy minerals.

means, in effect, that if a source area contains, say, 2 percent of staurolite,[1] the heavy residue in a sandstone derived from it may contain no staurolite or 30 percent or more, depending on conditions of sedimentation including concentration by weathering in the source area, erosion, transportation, and deposition. Since these conditions fluctuate rapidly, such variations may take place in one and the same sandstone within an outcrop. It becomes evident that if we are to interpret moderate fluctuations in relative proportions of accessory minerals in sediments, it will be necessary to adjust for conditions of sedimentation.

One solution, which is at least intuitively appealing, is to use minerals which are likely to behave in a similar manner and study their fluctuations in proportion. Thus, if a pair of minerals have the same density and habit, they should react similarly under the same hydraulic conditions; and if they differ in either density or habit or both, they should behave differently. The commoner minerals in detrital sediments are classified on the basis of density and habit in Fig. 10.5; and since pyroxene and amphibole appear reasonably close together and both differ from garnet, we expect that the fluctuations in proportions of the former pair will be similar, while both should differ from garnet. A test of this hypothesis, using Russell's data from his

[1] It seems worth mentioning that we have no firm information on the composition of source areas, so that staurolite may vary from 0 to 30 percent or more in the source rock as far as we know.

investigation of the accessory minerals in the Mississippi River sand, is described in Sec. 21.14; and the closer association of amphibole and pyroxene when contrasted with amphibole and garnet and pyroxene and garnet is confirmed.

Some idea of the magnitude of the effects can be obtained by choosing a model of hydraulic behavior, e.g., Stokes' law, and comparing the behavior of two minerals; thus zircon possesses a density of 4.6, whereas epidote is 3.2, and, postulating laminar flow, we may apply Stokes' law as follows (see also Sec. 4.6):

$$V = \frac{2g}{9\eta}(d_1 - d_w)r^2 \qquad \text{or} \qquad V = C(d_1 - d_w)r^2$$

Then, since velocity is equal for an associated pair of minerals, and taking the density of water d_w as constant at 1 gm/cm³, we have

$$V = C(d_1 - d_w)r_1^2 = C(d_2 - d_w)r_2^2$$

or

$$(d_1 - 1.0)r_1^2 = (d_2 - 1.0)r_2^2$$

and for zircon and epidote this yields

$$3.6r_1^2 = 2.2r_2^2 \qquad \text{or} \qquad 1.28r_1 = r_2$$

It is difficult to realize that this apparently small difference is so sensitive to fluctuations in current velocity, but, in extensive testing of the zircon-epidote ratio in the Miocene sediments of southern Trinidad, where both minerals are common and persistent constituents of the heavy residue, apart from a systematic increase of epidote from the older to the younger Miocene, the fluctuations were large and appeared to be quite haphazard, i.e., were out of control.

The use of varietal variation as advocated by Brammall (1928) would appear to offer a suitable basis for comparison; two varieties of the same mineral should behave more sympathetically than different species, but here again extensive testing of different varieties of zircons and tourmalines in the Tertiary of Trinidad failed to remove quite large fluctuations in relative proportion of the different varieties. In both species the proportions of the different varieties (using color, habit, and inclusions to identify varieties) were very closely tied in with variation in grain size.

Selective sorting on the basis of size, shape, and effective density is therefore an extremely efficient process; and, in the average detrital sediment, the process has proceeded sufficiently long and/or at sufficient intensity to overwhelm all but the major changes in mineral composition.[1] Recent attempts to overcome these sources of variation by using variation in size and, particularly, in shape of accessory heavy minerals are described by McIntyre (1959, 1961). For well-defined specific questions, such as the

[1] That is, to mask all but the major geological events.

comparison of different environments—and therefore different degrees of selective sorting—this approach offers considerable promise.

Ultimately it must be realized that variation in proportion of the different mineral species is controlled by a very large number of interdependent factors, and the problem is a multivariate one; to a first approximation the relationship may be written

$$\text{Var}\ (X_i) = f(Sa, Wth, Er, Tr, Dep, Di)$$

so that variation in proportion X_i of a mineral is a function of variation in source area Sa, weathering in the source area Wth, erosion Er, transportation Tr, deposition Dep, and diagenesis Di. It is doubtful whether this six-factor (a six-way classification; see Chap. 20) experiment can be solved solely by study of changes in proportion of accessory heavy minerals.

11 Measurement of derived properties

In respect of weighing, a process of addition and a relation of equality can be found which are similar . . . to the arithmetical process of addition and the arithmetical relation of equality; but a process of addition and a relation of equality which are thus similar to the arithmetical process and relation cannot be found in respect of determinations of density.

N. R. Campbell, 1957

11.1 Introduction

Measurement of the five fundamental properties has now been considered, but aggregates, such as rocks, have many other properties which are used to characterize them. The properties are called *derived properties* because a knowledge of the fundamental properties is necessary for their understanding (Campbell, 1957, pp. 346ff.). The derived properties may be looked upon as *qualities* of the material; or, in many cases, they are behavioral characteristics which change as the environmental conditions are changed. In general, the derived properties are dependent upon the fundamental properties, and an important part of the analysis of aggregates is the evaluation of the nature of this relationship of dependence. The first step in such an analysis is to attempt to formalize a functional relationship between derived and fundamental properties and subsequently to establish the empirical values involved in the functional relationship. In practice this order is often reversed, and the functional relationship emerges only after the empirical relationship has been established and used for some time.

Quotation: N. R. Campbell, "Foundations of Science," p. 281, Dover Publications, Inc., 1957.

While there are only five fundamental properties, there appears to be no limit to the number of derived properties which may be used to describe an aggregate and predict its behavior under changing circumstances. In fact, however, variation in the character of an aggregate, as reflected in variation in its derived properties, tends to contain increasing redundancy as more and more dependent properties are added, and eventually the new information contained in additional derived properties becomes vanishingly small. Measurement of derived rather than fundamental properties is often more convenient and, at least in the early stages of analysis, appears to be the only practicable route to follow; ultimately, however, when it is essential to understand, modify, and control the behavior of aggregates, there is no alternative to a study of variation in the fundamental properties. In the present work density, porosity, and permeability will be briefly considered, along with some other dependent properties which are important in characterizing sedimentary rocks as well as other aggregates.

11.2 Measurement of Bulk Density

The density of an aggregate is an interesting and important derived property and one which may be used both in the field on large rock bodies (geophysical measures by gravimeter) and in the laboratory on small rock specimens. Density of an object, in the sense used here, is its weight per unit volume,[1] and two kinds are generally differentiated, bulk density and grain density. The former is a characteristic of the aggregate in its "natural" condition, while the latter is the weighted average density of the constituents of the aggregate. The difference between the two properties is largely due to the presence or absence of voids in an aggregate; and the three properties, bulk density BD, grain density GD, and porosity P, or volume of voids, are related as follows:

$$P = \frac{\mathrm{GD} - \mathrm{BD}}{\mathrm{GD}} \tag{11.1}$$

or

$$\frac{\mathrm{BD}}{\mathrm{GD}} = 1 - P \tag{11.2}$$

Grain density can be measured by pycnometer after "complete" disaggregation of the aggregate into its constituent elements. The weight of the disaggregated material can be determined on a balance, usually using a controlled atmosphere, and then its volume is determined by displacement in a vessel (pycnometer) of known volume. The procedure is quite difficult if a high degree of reproducibility is to be attained. Automatic devices exist

[1] Specific gravity is the ratio of the mass of a body to the mass of an equal volume of water at 4°C, whereas density is the mass per unit volume of a body (in grams per cubic centimeter).

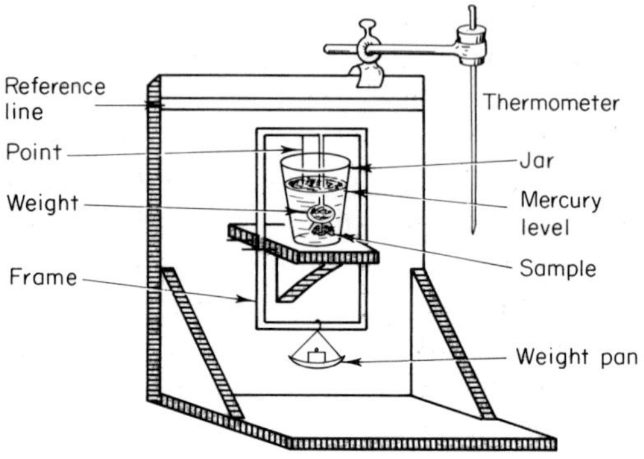

Figure 11.1 Mercury balance for the determination of bulk density of rock specimens. (*G. K. Jones, 1947, private report, United British Oilfields of Trinidad, now Shell Trinidad Ltd.*)

for this purpose, e.g., the Beckman air pycnometer, but are difficult to standardize to the level of reproducibility necessary for studying the variation in grain density shown by rock specimens.

Quite frequently the disaggregated constituents are assumed to be all quartz, with grain density of 2.65; this is rarely adequate for most studies, however, as sedimentary rocks vary in density from 1.5 to 2.8, and variation in the amount of quartz may be quite large (see Chap. 9).

Bulk density is measured in the same way as grain density; the mercury balance has been found to be a simple and controllable technique for this purpose. This procedure has been described by Westman (1926) for the determination of the bulk density of brick, and a modified version was proposed by Jones (1947).

The apparatus shown in Fig. 11.1 has been found effective over extensive trials (Griffiths, 1948; Griffiths, Groff, and Kahn, 1953; Griffiths et al., 1954a, b, 1956; Rosenfeld, 1953; Cochran, 1960) and can be used to determine the bulk density of rocks over the range 1.6 to 3.0 with reproducibility of 0.01 unit if a suitable experimental design is adopted (see below).

The procedure comprises weighing the rock specimen on an analytical balance and then determining its bulk volume by displacement in mercury, using the mercury balance. The mercury balance consists of a jar to hold the liquid and a known mass (a stainless-steel ball is convenient) attached to a frame which permits the addition of weights. First the weight necessary to sink the known mass in the liquid is determined, and then the rock specimen is introduced under the ball, and the weight necessary to sink both known mass and rock specimen is determined.[1] In both cases the mass is depressed to the same level in the mercury, as measured by the point in Fig. 11.1.

[1] H. Meyer, Department of Geochemistry and Mineralogy, The Pennsylvania State University, has added the refinement of using fine sand to bring the known mass and subsequently the mass plus specimen to the same level in the mercury bath. The container and sand are subsequently weighed on an analytical balance, thus materially improving the precision of weighing.

Some skill is necessary to ensure that the frame attached to the known mass remains horizontal and parallel to the reference line throughout the experiment, and it is frequently necessary to remove entrapped air bubbles by tilting the frame before weighing.

The data comprise the weight w_d of the specimen in milligrams, the temperature $t°$ of the mercury; the specific gravity d_t of the mercury at that temperature (read from a table such as that in the "Handbook of Chemistry and Physics"); the weight c to sink the known mass, and the weight w_m to sink the mass and specimen. The bulk density BD is then found as follows:

$$\mathrm{BD} = \frac{d_t w_d}{w_d + w_m - c} \qquad \mathrm{g/cm^3} \tag{11.3}$$

The bulk volume BV may also be found as

$$\mathrm{BV} = \frac{w_d}{\mathrm{BD}} \qquad \mathrm{cm^3} \tag{11.4}$$

It is advisable to weigh the specimen before and after immersion in mercury to determine whether mercury has penetrated into the specimen (increase in weight) or whether the specimen has lost some of its mass (decrease in weight). The data sheet for recording and calculating the results (Table 11.1) can be designed for five replicates on the same specimen, as shown, but generally three are adequate unless a very high degree of precision is required.

In many experiments in which large numbers of rock specimens are analyzed, the choice of an efficient experimental design is an important part of the analysis; this is particularly true when, to reduce the time involved in many determinations, a number of operators are used to perform the experiment (Griffiths and Rosenfeld, 1954). A variety of designs exist and are described in later chapters (Chaps. 18 et seq.) together with the appropriate statistical analysis. While simple replication in balanced designs is adequate for general use, as the size of the experiment grows, unbalanced designs with incomplete replication are very much more efficient if the analysis can be programmed for a computer (Griffiths, Groff, and Kahn, 1953; Griffiths et al., 1954a, b; Cochran, 1960). The statistical analysis of a large experiment using unbalanced incomplete blocks can be very tedious and expensive without the use of automatic electronic data processing (see particularly Cochran, 1960).

11.3 Bulk-density Variation in Sediments

In theory, bulk density may vary from unity to infinity, but in practice the range in most rocks is from 1.3 to 3.0; in sediments 1.5 to 2.9 is about the range, and for random samples from a "homogeneous" set of specimens

TABLE 11.1
Mercury-balance Data Sheet for Determination of Bulk Density

Sample # ____________

Preliminary treatment: ______________________________

____

____

____

	1	2	3	4	5
Date					
Run by					
t					
d_t					
w_d					
w_m					
c					
w_d					
BD					
BV					

BD = bulk density

BV = bulk volume

d_t = sp. gr. Hg at temp. t°C

w_d = wt. dry sample in air (before and after immersion in Hg)

w_m = wt. to sink sample in Hg

c = wt. to sink known mass in Hg

$$BD = \frac{d_t w_d}{w_d + w_m - c} \quad g/cm^3$$

$$BV = \frac{w_d}{BD} \quad cm^3$$

1.	2.
3.	4.
5.	Remarks

the frequency distributions tend to normality, so that analysis of variance is a suitable statistical procedure. An example of an analysis of this kind is illustrated in Table 18.7, and the range of variation represented by a rock body is illustrated in Fig. 11.2 for 343 specimens from three wells through the Salt Wash member of the Morrison Jurassic of the Colorado Plateau (Griffiths et al., 1956). A more specific example is given in Table 11.2 based on analysis of 161 specimens from the Salt Wash member as represented in well *C*, Bull Canyon, Colorado Plateau (see Griffiths, Groff, and Kahn, 1953, pt. 1; Cochran, 1960, p. 45).

It is interesting to note that the experimental error $\hat{\sigma}_\epsilon$ is generally of the order of 0.01 bulk density unit, whereas the variation between core specimens $\hat{\sigma}_m$ is from one to ten times larger in magnitude. It is essential to use efficient experimental design with adequate replicates to achieve this level of error variation (Griffiths, Groff, and Kahn, 1953; Cochran, 1960). The differences between mean bulk density among zones are, of course, very real and reflect the differences between mudstones (zones 1, 3*a*, 5) and sandstones (zones

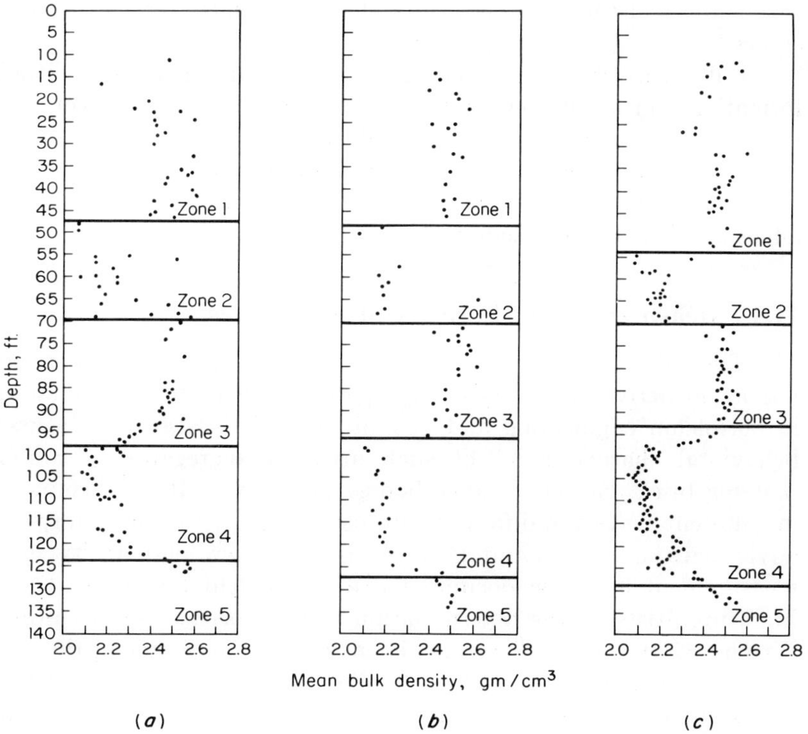

Figure 11.2 Bulk density of samples from wells *A*, *B*, and *C*, Bull Canyon, Colorado. (*After Griffiths et al., 1956.*)

TABLE 11.2
Bulk Density of 161 Core Specimens from Bull Canyon Well *C*, Montrose County, Colorado

Zone	$\bar{X}$	$\hat{\sigma}_m$	$\hat{\sigma}_\epsilon$	n
1	2.427	0.074	0.007	30
2	2.222	0.116	0.010	25
3*a*	2.477	0.014	0.012	16
3*b*	2.490	0.036	0.009	12
4	2.195	0.084	0.017	72
5	2.491	0.037	0.047	6

Source: Modified from Cochran, 1960.

2 and 4). Zone 3*b* consists of sandstones cemented by varying amounts of carbonate, and the bulk density of these materials exceeds that of the mudstones.

Bulk density is quite useful in tracing the changes in lithology of a formation and is sufficiently sensitive so that faults and unconformities can be detected by relatively sudden breaks in an otherwise smooth trend. There is a clear trend between density and depth as evaluated by Hedberg (1926) and Athy (1930), which is also useful in computing overburden and the effect of compaction in sediments.

11.4 Measurement of Porosity of Aggregates

The value of voids, usually expressed as the porosity or pore space, is an important derived property of aggregates and is useful in estimating reserves in petroleum engineering and in hydrology; it is also used in assessing the behavioral characteristics of such artificial aggregates as refractory and building brick, various ceramic bodies, and slags. It is, of course, measured in different ways for different purposes, and all the procedures have some advantages and disadvantages. A useful review of porosity and porosity measurement has been documented by Rosenfeld (1950), and an extensive literature exists on the determination and use of this derived property (e.g., Pirson, 1950). It is important, in making a choice among the wide variety of procedures, to decide the exact objective involved in collecting the data; for example, porosities are sometimes differentiated as the *absolute porosity*, or the total volume of voids, and the *effective porosity*, or the volume of voids effective in the process used to measure it. In petroleum engineering the effective porosity is generally more important than the absolute; whereas, when the density of the material is the most important feature, the absolute porosity becomes the more important aspect of void space. The effective porosity is of greatest interest in reservoir engineering and will be the aspect of porosity dealt with in the succeeding account. It is clear, of course, that absolute porosity can be found from the grain and bulk densities as defined in Eqs. (11.1) and (11.2).

Even when attention is restricted to effective porosity, a wide range of procedures exists for its measurement. It has been demonstrated that effective porosity will vary with the fluid used to determine it (Rosenfeld, 1950, pt. 2; Licastro and Griffiths, 1953). It is therefore necessary to decide what the measurement is to be used for before selecting appropriate experimental conditions. A popular procedure described by Yuster (1937) is suitable for determining effective porosity; the fluid used may be varied, and the effective porosity will depend on the presence or absence of interaction between rock and fluid. Quite frequently the fluid used is a gas, and then the interaction between fluid and rock is reduced to negligible importance, so that techniques using this approach lead to highly reproducible measurements (a high precision). Nevertheless, if the laboratory determination of porosity is to reflect "natural conditions," i.e., accurate and meaningful conditions, it is essential that the environment used in the laboratory closely simulate such conditions.

Effective porosity entails finding the weight of the specimen in air, saturating the specimen, and weighing the saturated specimen both in air and immersed in the saturating fluid. The most difficult steps in the technique are ensuring reproducible saturation and handling the saturated specimen so that there is no loss of fluid. Saturation is usually achieved by evacuating a number of specimens simultaneously in a vacuum desiccator; when evacuation is judged complete, the fluid is introduced, and the specimen is allowed to imbibe until saturated. The rate of imbibition will depend on permeability, and elaborate precautions are often taken to ensure "complete" saturation. Usually, all that can be achieved is a reproducible result, and this can be controlled by careful experimental procedure and by replication. Once the specimens are saturated, it is advisable to store them in the saturating fluid and remove them only for subsequent weighing.

After saturation is achieved, the specimen is wiped with a dry cloth to remove excess fluid and is weighed in a wire sling on an analytical balance. It is then weighed again immersed in a beaker of the saturating fluid. Let the weight of the dry sample in air be w_d and the weight of the saturated sample be w_s and w_i in the saturating fluid; the weight of the wire sling (immersed) is w_w, and d_t is the specific gravity of the fluid at the temperature t°C) of the experiment; then the effective porosity P_{eff} is

$$P_{\text{eff}} = \frac{w_s - w_d}{w_s - (w_i - w_w)} \qquad (11.5)$$

TABLE 11.3
Data Sheet for Determination of Effective Porosity

Sample # ____________

Preliminary treatment: ____

Fluid used: ____________

	1	2	3	4	5
Date					
Run by					
w_d					
w_s					
w_i					
w_w					
$w_i - w_w$					
t					
d_t					
PV					
BV					
BD					
P_{eff}					

PV = pore volume
BV = bulk volume
BD = bulk density
P_{eff} = effective porosity
w_d = wt. dry sample in air
w_s = wt. saturated sample in air
w_i = wt. sample
w_w = wt. wire sling immersed
d_t = sp. gr. of fluid at temp. t°C

1.	2.
3.	4.
5.	Remarks

$$\text{BV} = \frac{w_s - (w_i - w_w)}{d_t} \quad \text{cm}^3$$

$$\text{BD} = \frac{w_d}{\text{BV}} \quad \text{gm/cm}^3$$

$$P_{eff} = \frac{w_s - w_d}{w_s - (w_i - w_w)} \quad \%$$

$$\text{PV} = \frac{w_s - w_d}{d_t} \quad \text{cm}^3$$

The pore volume PV is

$$PV = \frac{w_s - w_d}{d_t} \tag{11.6}$$

It is, of course, possible to find the *effective* bulk volume BV and *effective* bulk density BD from these figures. A suitable data sheet for recording the measurements and performing the calculations is illustrated in Table 11.3.

11.5 Porosity Variation in Sediments

The range of porosity to be expected in aggregates of spheres packed in different ways has been computed theoretically as 26 to 47 percent. The most open, cubic packing, yields the 47 percent value; and the closest, rhombohedral packing, yields 26 percent (Fraser, 1935; Graton and Fraser, 1935). In sediments porosity may vary from zero to about 50 percent, although in general the range 2 to 35 is rarely exceeded in either direction. The porosities of oil sands vary generally between 10 and 30 percent, with figures between 10 and 25 being the most common.

From Eqs. (11.1) and (11.2) it can be seen that porosity is inversely related to bulk density, and this relationship is quite generally linear (Davis, 1954); in fact, this inverse relationship is usually almost 1:1, so that either of these variables can be used in place of the other with only a change in sign.

TABLE 11.4
Effect of Fundamental Properties on Variation in Porosity

Rock property	Source of information[a] Theoretical analysis	Artificial mixtures	Natural sediments
Coarse grain size	0	0 + −	0 + −
Good size sorting	+	+	+
Slight skewness toward finer sizes	?	+	+
High "sphericity" and "roundness" (confused)	?	+ −	0 + −
Intermediate "roundness"	?	+	?
Open packing	+	?	?
Low chemical cement	+	?	+
Low clay content	+	?	+

[a] 0, +, − indicate relative change in porosity for each rock property *assuming* the others are constant.
Source: After Rosenfeld, 1950.

Attempts to demonstrate a relationship between porosity and the fundamental properties have had very contradictory results; many of these attempted explanations have been summarized by Rosenfeld (1950) (see Table 11.4). The difficulties which lead to apparent contradictions arise from the interdependences between the measured properties of sediments, which, in turn, reflect the interactions between aggregates and environmental conditions in both laboratory and field observations. Apparently the only recourse is to use empirical determinations on suites of rock specimens (e.g., Griffiths, 1958b, 1961).

Porosity variation, like bulk-density variation, converges on normal frequency distributions as n grows large if the specimens are random samples from "homogeneous" populations; standard parametric statistical analysis can therefore be used in analyzing variation in the porosity of aggregates (see, for example, Rosenfeld and Griffiths, 1950; Emery, 1954; Stanonis, 1956; Pachman, 1961; Griffiths, 1958b, 1960b).

11.6 Measurement of Permeability of Aggregates

Permeability is a measure of the rate of flow of fluid through a unit cross section of an aggregate; it is therefore very similar to porosity and may be considered a dynamic measurement of porosity under fluid flow conditions. It therefore contains all the difficulties of measuring porosity plus sources of variation that arise from the difficulty of measuring performance in a dynamic system. As a guide to the choice of procedure, it is again of first importance to decide the objective for which the measurements of permeability will be used; and again, a very wide range of techniques is available for measurement. Interactions between fluid and aggregate are well known and led initially to standard techniques using flow of gasses (Muskat, 1937). In many cases this kind of measurement proved inadequate for reflecting the behavior of an oil-reservoir rock under conditions in the field, so that liquids of various kinds were used. Permeability of the Miocene low-rank graywacke reservoirs in Trinidad were of the order of 3,000 millidarcys to air but were reduced to 200 to 300 millidarcys to liquids such as distilled water. Nevertheless, the use of air permeabilities is standard practice in routine core analysis of oil-reservoir rocks, although various modifications are now in quite general use, including the use of one- and two-phase fluids (Pirson, 1950). Another feature which tends to complicate the measurement of permeability is its directional character; it was recognized at an early date that fluid flow along bedding planes was easier than flow across them (Muskat, 1937), but it is only relatively recently that it has been recognized that flow rate varies in different directions within the bedding planes (Johnson and Breston, 1951). It is therefore very difficult to choose a suitable technique for the measurement of permeability until the criteria for deciding what is meant by suitable are very precisely specified. One suitably simple procedure is to measure the permeability to air on a drilled cylindrical sample of known dimensions. The specimen is mounted in a neoprene-rubber bung drilled with a hole just large enough to hold the specimen. An effective seal is obtained by mounting the cylindrical core in the rubber bung in a pressured vise, which seals the rubber to the core. Air is then passed through the specimen and the volume, in milliliters per unit time, is measured by displacement in a collecting vessel. The permeability K is calculated according to the relationship

$$K = \frac{QuL}{(P_i - P_o)A} \tag{11.7}$$

where K = permeability, darcys
Q = flow rate, cm^3/sec
u = viscosity of air, centipoises
L = length of specimen, cm
A = cross-sectional area, cm^2
P_i = inlet pressure, atm
P_o = outlet pressure, atm

The flow rate should be low enough to avoid turbulence.

There are, of course, many modifications of this procedure described in the literature (Muskat, 1937; Pirson, 1950; etc.), and very elaborate experimental arrangements are sometimes used in order that the measured permeability shall reflect natural conditions; restored-state technique (Bruce and Welge, 1947) and directional permeability (Johnson and Breston, 1951; Hutta, 1956) are two examples.

11.7 Permeability Variation in Sediments

Theoretically, permeability may vary from zero to infinity, but in practice it ranges from zero to several thousands of millidarcys. Oil-reservoir rocks vary from 1 to about 3,000 millidarcys, with common values in the tens and hundreds. The prolific oil-producing reservoir known as the Bradford Third sand, Upper Devonian of Pennsylvania, averages 1 millidarcy, but this is an unusual sand.

The frequency distribution of permeability for random samples from a homogeneous aggregate measured in millidarcys is usually very highly skewed toward high values of permeability; i.e., low values are very much more common than high values. Generally, when logarithms of the permeability are used as the independent variable, the frequency distribution tends to become approximately normal, so that it is usual for log permeability to be used in analysis rather than permeability in millidarcys (e.g., Stanonis, 1956; Pachman, 1961; Griffiths, 1958b).

Ryder (1948) was the first to show that the relationship between porosity and log permeability is approximately linear in the Bradford sand. Such a relationship may be deduced from the postulated functional relationship, e.g., the Kozeny equation, between porosity and permeability (Pirson, 1950, p. 47)

$$K = Cd^2 \frac{\phi^3}{(1 - \phi)^2} \tag{11.8}$$

Then

$$\log K = \log C + 2 \log d + 3 \log \phi - 2 \log (1 - \phi)$$

This approximately linear relationship between log permeability and porosity is characteristic of many oil-reservoir sands (e.g., Headlee and Joseph, 1945; Stanonis, 1956; Griffiths, 1958b; Pachman, 1961), and frequently the degree of common variation is about 60 to 70 percent when estimated by the coefficient of determination r^2 (see Chap. 21). This implies that about 60 to 70 percent of the factors that influence variation in porosity are common to variation in log permeability and only 30 to 40 percent are independent.

Thus it can be seen that derived properties show marked interdependence, and this is to be expected if they are functions of the same five fundamental properties. Common factors may be postulated and have been empirically demonstrated in relating these derived properties to the fundamental properties (Griffiths, 1958b, 1960b; Stanonis, 1956; Pachman, 1961). It is also obvious that adding to the number of derived properties increases the degree of redundancy in the information, so there is a cutoff point in terms of how many additional derived properties to measure per unit of new information accrued. This does not imply that once the bulk density, porosity, and permeability have been measured there is no point in going further, but it does suggest some circumspection in adding property after property when the variations reflected in some are over, say, 90 percent in common with others. Additional information can be obtained only if the new derived property shows some degree of independence when compared with those properties already measured. Thus, for example, if bulk density is linearly related to porosity in 20 specimens of an oil sand and the common variation is, say, 95 percent in terms of the coefficient of determination r^2 between the two properties, there is little point in measuring both porosity and bulk density on additional specimens from this rock; it would be better to measure one of these properties on more specimens than both on fewer specimens.

11.8 Other Derived Properties of Aggregates

There is, as mentioned earlier, no limit to the number of interesting derived properties which can be used to characterize the behavior of an aggregate; but unless there is a very clear reason why these properties should be determined, the analysis would be more efficiently accomplished by emphasis on more specimens and fewer properties. In any case the analysis of additional properties may usefully follow similar lines: the concept should be clearly defined and reduced to an operational definition, a procedure should be devised which fulfills the operational requirements, and ultimately the variation in the property should be related to the concept where possible. For data analysis it is advisable to determine the range of variation and the frequency distribution of the variate in random samples from representative homogeneous populations; then the most suitable data-analysis procedure can be selected. All the experimental procedures will require well-defined experimental designs and estimates of reproducibility, and, ultimately, the collection of data should be subjected to reduction by multivariate procedures such as multiple regression and correlation for the study of dependencies; as a measure of redundancy, component analysis is a useful procedure (Griffiths, 1960b, 1961; Griffiths and Smith, 1964).

There are many derived properties which are considered only in a qualitative manner, particularly in handling sedimentary rocks in the field. For example, the most characteristic property of sediments is *bedding* or *stratification,* and this is often left undefined or is qualified as "expressed by rock units of general tabular or lenticular form that have some lithological or structural unity" (Pettijohn, 1957), or again "stratification . . . is due to many different causes—differences in texture, cohesion or cementation, color, mineralogical or lithological composition and internal structure" (Shrock, 1948, p. 10). Probably the best conceptual definition of a sedimentation unit is given by Otto (1938)[1] as "that thickness of sediment which was deposited under essentially constant physical conditions"; he further explained that within a sedimentational unit "chance deviations of all sedimentary characteristics about a mean value may be present; these deviations, however, should themselves constitute a unimodal distribution." Thus Otto's definition is equivalent to von Mises' "collective" (1957); indeed, such a reduction is necessary if random sampling of a homogeneous unit is to be unequivocally specified. Bedding or stratification may be conceptually defined as due to some change in one of the five fundamental properties, as illustrated in Fig. 11.3; but this conceptual picture must be modified, since we know that when the concepts are reduced to operational procedures, the properties do not vary independently but show varying degrees of interdependence. Each unit layer will show some degree of variation around a mean value for all five properties, and thus we may define the basic unit in terms of minimum variance.

In other words, for a layering to be present the variation between mean values per unit must exceed variation within units, and this definition of stratification is equivalent to the statistical as well as the geological impli-

[1] Apfel (1938) defined a similar unit and called it a phase.

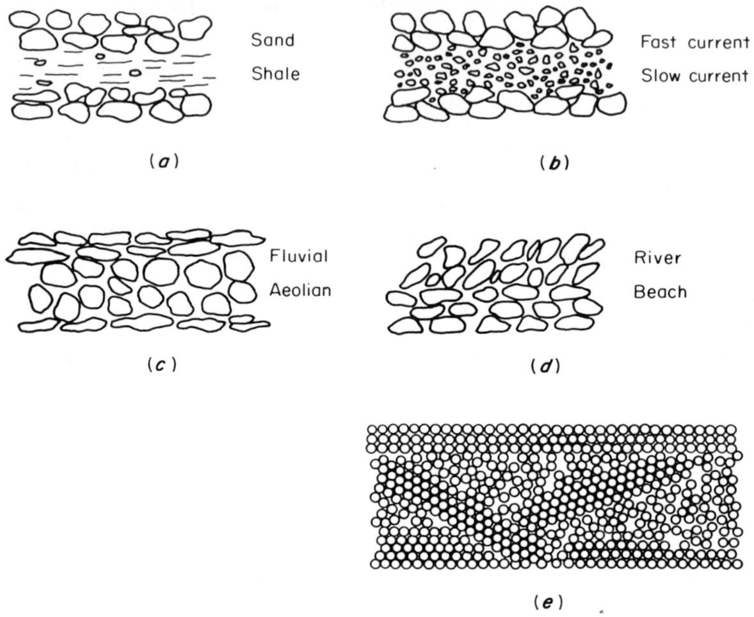

Figure 11.3 Fundamental basis of bedding or stratification. (*a*) Change of composition; (*b*) change of size; (*c*) change of shape; (*d*) change of orientation; (*e*) change of packing. [*Reprinted from Measurement of the Properties of Sediments by J. C. Griffiths, J. Geol.,* **69:***495 (1961), by permission of The University of Chicago Press. Copyright 1961 by the University of Chicago.*]

cations of stratification. Furthermore, this also supplies an operational basis for determining the presence and magnitude of stratification; e.g., in stratified deposits, sampling across the units, or channel sampling, will yield different results than sampling within the units, or stratified sampling (Griffiths, 1959b), and so it is possible by sampling in small enough units to establish a quantitative measure of stratification. If, on the other hand, the population sampled is massive, i.e., isotropic in the property under study, sampling in different arrangements will yield similar results, and no stratification will be detected. In fact, of course, by sampling a population in small enough units, say two or three contiguous elements, it will be possible to establish the internal structure as the patterned variability of the population. Analysis of variance, particularly when extended to multivariate analysis of variance, would be the ideal tool for discriminating the structure of patterned variability in any aggregate. The model which emerges from this kind of analysis forms the basis for the sampling recommendations of Chap. 2 (see Fig. 2.3) and is again exemplified in the sampling problems illustrated in Fig. 7.2; tests of the model have already established the effectiveness of this procedure

(Griffiths, 1959b; Steinmetz, 1956, 1962; Wood and Griffiths, 1963). It is worth emphasizing that these preliminary tests have included both measurement and counting techniques and are invariant to changes of scale from thin sections to outcrops; they may ultimately be applied to regional mapping and form the unambiguous basis for definition of strata at all levels of subdivision. It is also obvious that not all the changes which yield similar megascopic and field appearance will necessarily turn out, upon rigorous analysis, to represent changes similar in magnitude or in the same properties, so that refinements will be possible in structural analysis. For example, to take an oversimplified and extreme case, it is not necessarily correct to align two sets of beds in which the layering in one is due to a change in composition and that in the other is due to a change in packing. This implies that dips measured along "beds" may have different significance and not be equivalent, a feature already recognized, for example, in the contrasted significance of bedding and false bedding.

12 Data processing and reduction

It is these rules that make the measureable properties so similar to numbers, that it is possible and useful to represent them by numerals the primary purpose of which is to represent numbers. It is because of this that it is possible to find one, and only one numeral, which will fitly represent each property; and it is because of them, that these numerals, when they are found, tell us something useful about the properties.

N. R. Campbell, 1952

12.1 Introduction

The organized plan of experimental investigation summed up as the scientific method is outlined in Chap. 1. The succeeding account emphasizes the steps in this plan, starting with the basic philosophy of method, the formulation of questions, and the reduction of conceptual properties to operational procedures and techniques which produce measurements; these measurements constitute the raw data (measures or observations) upon which the succeeding analysis is performed. Techniques are chosen which lead, as far as possible, to numerical data, and simpler techniques are advocated for efficient compilation of relatively large amounts of such data. Large amounts of data, or, equivalently, large collections of numbers, are quite incomprehensible without adequate treatment. As a result the succeeding chapters are largely concerned with processing these "large" collections of data so that they can be summarized and condensed and the contained information distilled into a readily digestible form. The procedures are, however, all part of the "organized plan," and it is necessary to understand that the path of development is uniformly continuous from the beginning to the conclusion of any experimental investigation. Thus the data are collected according to

Quotation: N. R. Campbell, "What Is Science," p. 118, Dover Publications, Inc., 1952.

prescribed techniques, which in turn are selected, at least in part, so that they lead to data suitable for the subsequent statistical manipulations; this implies that both procedures and data processing are fitted together as an integral part in the consistent whole. The next step, therefore, concerns the preliminary preparation prior to treatment of the data, a feature quite frequently neglected as rather tedious and bothersome.

Since the collection of an extensive body of data is expensive in terms of cost and/or time, it is advisable to select the treatment which makes the most efficient use of the collected information, and this will comprise the first step; subsequently, a measure of the amount of information contained in the data will be described as a guide to the succeeding procedures. Statistical analysis cannot extract more from the data than they contain, and it is most advisable to "fit" the analytical tool to the purpose of the investigation and the nature of the material upon which the tools are to be used. In this sense an elementary discussion of some aspects of measurement theory will follow, and here again the choice of which procedure to adopt will be seen to depend on the objective of the inquiry and the success of the initial planning and techniques used to collect the data.

12.2 Collecting and Recording Data

The initial steps in the treatment of data are concerned with the manner in which they are recorded, and these procedures have undergone a drastic revolution in the twentieth century. It used to be customary to record data haphazardly, analyze them, and publish the conclusions; frequently, even when the data were initially recorded in published works, the subsequent reprints and new editions omitted this vital aspect of the investigation in favor of more speculation, and as published reports of scientific investigation multiplied, editorial pressure frequently led to the suppression, or at least omission, of data. Fisher (1953), quoting an example from Lyell's early work on establishing the stratigraphic succession in the Tertiary, states:

> Not only did Lyell immortalize these statistical estimates in the names still used for the great divisions of the Tertiary Series, but in an Appendix in his third volume he fills no less than 56 pages with details of the classification of each particular form, and of the calculations based on the numbers counted. There can be no doubt that, at the time, the whole process, and its results, gave to Lyell the greatest intellectual satisfaction.
>
> Now the point of this little history, its point for statisticians, is that the statistical argument by which this revolution in geological science was affected was almost immediately forgotten. In later editions of the *Principles* this great Appendix, on which so much labor had been expended, has disappeared; it survived indeed only two years. It had served its purpose; the ladder by which the height had been scaled could now be kicked down.[1]

[1] R. A. Fisher, The Expansion of Statistics, *J. Roy. Statist. Soc.*, **A116**(1):1–6 (1954).

			X_1	X_2	X_3	X_4	X_5	X_6	X_7	
Sample	Op.	Fld.	a	b	b/a_μ	Incl.	$a\phi$	$b\phi$		

Figure 12.1 Data card 1; measurement of grain size, shape, and orientation in thin section. a, long axis; b, short axis; b/a, axial ratio (axes in microns); incl., inclination of long axis to fixed direction (in degrees); ϕ, Krumbein's phi scale.

Perhaps if more emphasis had been placed on the collection and treatment of data in such geological research, geological theory would today be more successful in its prediction of petroleum and mining prospects. As it is, the experimenter rarely arranges to preserve the records of his observations and measurements, and much costly information has thereby been lost. From the historical development of scientific investigation it seems clear that theories and hypotheses, however attractive and interesting, are somewhat illusory and are soon superseded by other more comprehensive hypotheses. In this evolutionary sequence the basis of any hypothesis is usually the preceding collective observations, and the test of a hypothesis is whether it can both comprehend existing information and lead to prediction of additional corroborative observations. It seems necessary therefore to pay the painfully and painstakingly collected data their due.

Collections of data require some organized means for storage and record, and any sizable collection of data soon tends to become difficult to use unless its arrangement is systematized. Initially, tabular sheets are useful, and these are filed in various ways, forming a voluminous paper record. Sooner or later the investigator wishes to collate and arrange these data, and continued copying not only is tedious but introduces copying errors; few things are more exhausting than editing large collections of tabulated data. As a first step the investigator will generally design forms upon which the data can be recorded as collected; such formalized arrangements are of great help

	Sample:			Operator:							
Grain	a_μ	b_μ	c_μ	ψ	$a\phi$	$b\phi$	$c\phi$	S_r	R_k		

Figure 12.2 Data card 2; measurement of size and shape of individual grains and pebbles. a, b, c, long, intermediate, and short axes, respectively; μ, microns; ϕ, Krumbein's phi scale; ψ, $\sqrt[3]{bc/a^2}$ (*after Krumbein, 1941b*); S_r, visual sphericity (*Rittenhouse, 1943a*); R_k, visual roundness (*Krumbein, 1941b*).

in field and laboratory. Forms which have been found useful in measuring the size and shape of grains in thin section and as loose mounts, or of pebbles in the field, are illustrated in Figs. 12.1 and 12.2, respectively. Sooner or later the investigator will require a better-organized approach which makes the subsequent analysis somewhat simpler; one such step is represented by the use of punched cards.

12.3 Hand Punched-card Systems

In describing cores from the Salt Wash member of the Morrison formation on the Colorado Plateau a Keysort[1] hand-punched card was used for each 3 in. of well core (Griffiths et al., 1953, 1954a, b). The characters selected for study were reduced to a code, and the cards were manufactured with this code printed along the margins. In such a project the design of the code is

[1] Keysort cards are manufactured by the McBee Keysort Company.

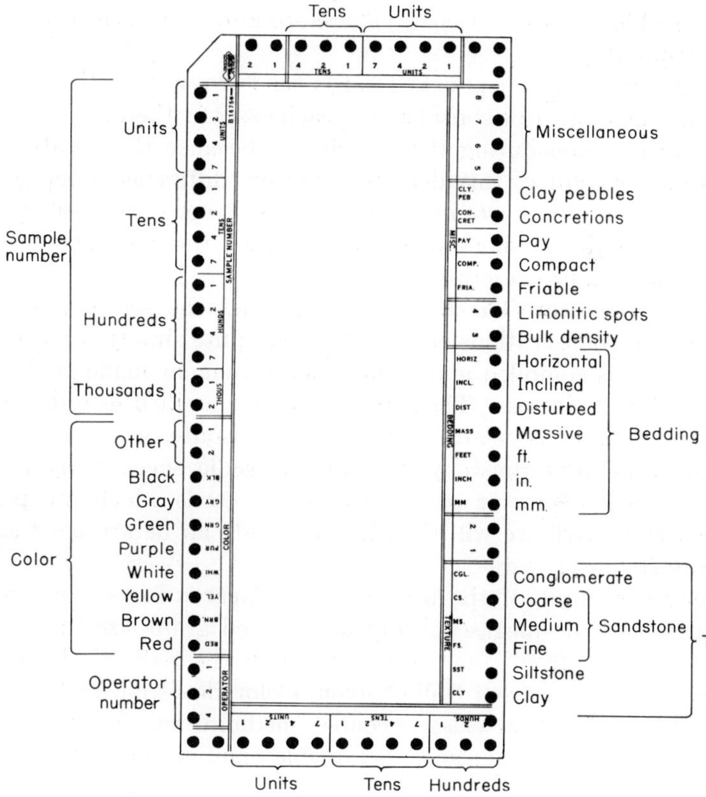

Figure 12.3 Keysort punched-card code for classifying sedimentary rock samples.

most important; each property studied is allocated a *field*, or set of holes (positions), along the margin; for example, suppose it is desired, as is generally the case, to assign each sample a number; then a field is allocated for this purpose. The conventional procedure is to assign each digit a field of four holes, as in Fig. 12.3. The four holes are assigned numbers 1, 2, 4, 7, from which can be obtained, by punching the appropriate hole or combination of holes, any number from 0 (no hole punched) to 9 (hole 7 and 2 punched). This then represents the units column, and a similar field is assigned to the tens and another for the hundreds. This set of three fields, comprising 12 holes in all, will then permit the use of sample numbers up to and including 999.[1] The next step is to assign a field to some property such as color (see Fig. 12.3); in this case the field comprises eight holes, each with a specific color designation and two extras for unusual colors not foreseen on the basis of a rapid preliminary examination of the cores. A field of six holes is assigned to grain size (texture in Fig. 12.3), and another of seven holes to bedding character. Several other features of specific interest in describing these cores were included in special fields, e.g., pay sand, friable or compact, presence of clay pebbles, or concretions. There are also a number of fields remaining for subsequent use.

Recording data on punched cards tends to systematize description because each field must be considered in each case; furthermore, it is not possible to change notation during the description because the notation is fixed on the card. Of course, this does not remove differences among and within operators if more than one investigator is used; nor does it suffice to maintain consistency within a single operator, although the systematization certainly helps to standardize description.

When each core has been described and the set of punched cards is available, the observations can be classified quite simply; first, the frequency in each class is compiled and a summary statement made of the proportions in the different classes. Then the cards are classified in terms of more than one attribute, and associations among the attributes evaluated. This sorting is most easily performed by inserting a needle successively in consecutive holes for each character of interest; those cards which are perforated for some specific attribute will then fall out and can be arranged as a two-way or n-way frequency table.

As an example of the use of punched cards for studies in lithology a comparison of the megascopically determined grain size in 3-in. (by $1\frac{5}{8}$ in. diameter) cores from some wells drilled in the Salt Wash member of the Morrison formation from Bull Canyon, Colorado, is illustrated in Table 12.1; here there are five size classes: coarse, medium, and fine-grained sandstone, siltstone, and mudstone; and all possible combinations of these five classes

[1] It was anticipated that in this investigation a few thousands would be needed in this field; hence a field of two holes marked 1, 2 in Fig. 12.3 yields up to 3,999 possible sample numbers for the cards illustrated.

TABLE 12.1
Frequency Distribution of Grain Sizes Based on Megascopic Examination of 3-in. Cores from Wells in 155*A*, *B*, and *C*, Bull Canyon, Colorado

		Well, %		
Class	Texture	*A*	*B*	*C*
1	Coarse sand (Cs)			
2	Medium sand (Ms)	20.68	16.63	6.8
3	Fine sand (Fs)	5.60	6.26	1.5
4	Siltstone (St)		1.51	1.3
5	Mudstone (Mst)	32.85	22.03	33.9
6	Cs + Ms	8.76	9.07	1.8
7	Ms + Fs	5.60	14.47	6.4
8	Ms + St			0.2
9	Ms + Mst		0.22	
10	Fs + St	11.92	5.40	16.5
11	Fs + Mst	3.89	4.32	2.2
12	St + Mst	7.54	15.12	5.7
13	Cs + Ms + Fs		0.43	0.2
14	Cs + Ms + Mst	1.46	0.65	0.2
15	Ms + Fs + St	0.73		18.8
16	Ms + Fs + Mst			0.2
17	Ms + St + Mst		1.30	
18	Fs + St + Mst	0.97	1.94	4.2
19	Cs + Ms + Fs + St			0.2
20	Cs + Ms + Fs + Mst		0.22	
21	Cs + Ms + St + Mst		0.22	
22	Ms + Fs + St + Mst		0.22	
	Total	100.00	100.01	100.[illegible]1

yield $2^5 = 32$ possibilities. One of these, the null case, i.e., samples showing grain size different from any of these classes, is not considered, so that any sample, i.e., 3-in. well core, could fall in any of the remaining 31 classes. The cores examined fall in only 21 of these, and the frequencies of the 1,331 cores from three wells are expressed as percentages in the table (for details see Griffiths et al., 1953, p. 17). The frequency distributions of these classes of grain size indicate that there is no simple and consistent gradient in size from well *A* (in uranium ore) through well *B* to well *C* (100 ft from *A*). The commonest classes are clearly indicated, so that the majority of cores are seen to fall into the group consisting of medium-grained sandstone, mudstone, siltstone and fine sandstone or the group consisting of siltstone and mudstone.

A bivariate table of attributes is displayed in Table 12.2; here the relationship between color and grain size is of interest, and it can readily be seen from the arrangement of the frequencies across the table and the complementary empty cells that red tends to be associated with mudstones, where white is associated with medium- to coarse-grained sandstones. The chi-square criterion may be used to establish that this association is significant ($\chi^2 = 771.8$ with 20 d.f.; therefore $P < 0.001$), and the intensity of the association may be indicated by the coefficient of contingency, $C = 0.79$ (see Kendall, vol. 1, 1948, p. 319, and Sec. 17.5 for explanation of the use of chi-square and contingency coefficients). Elaborate analyses of multiple hierarchical dichotomies can be performed in this way (Yule and Kendall, 1948, chaps. 1–5), and the testing for association and the estimation of its intensity are materially simplified by a systematic procedure of this kind (see also Sec. 17.5).

Punched cards also represent a suitable permanent record, and they are easily stored; furthermore, the recall time for information stored in this way is small and the cost of the entire system and its maintenance is low, while the procedure is quite easily learned. A catalog for the entire collection of samples in the Department of Geochemistry and Mineralogy at The Pennsylvania State University is exemplified by the punched card shown in Fig. 12.4. Descriptions of the principles, procedure, and design of punched cards, as

TABLE 12.2
Association between Color and Grain Size in Some 3-in. Well Cores from the Salt Wash Member of the Morrison Formation, Bull Canyon, Montrose County, Colorado

A = all other classes

	Texture						
Color	Cs	Ms	MsA	Fs	FsA	St and Ms	Total
Red						38	38
Red and green					16	116	132
Green			5	21	4	25	55
Red, white, and green	10	22	1	1	6		40
White and green	1	116	2	30	2		151
White	38	6	2	1			47
Total	49	144	10	53	28	179	463

Source: Data modified from Griffiths et al., 1953, table 7.

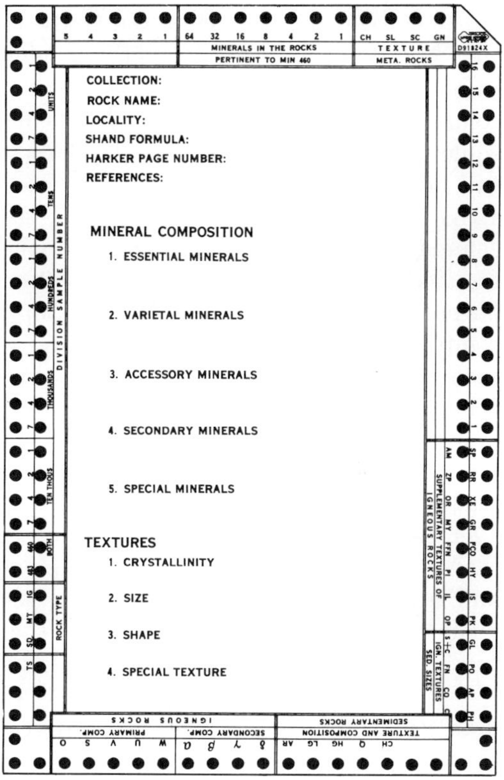

Figure 12.4 Catalog card used for specimens in the collection of the Department of Geochemistry and Mineralogy, The Pennsylvania State University.

well as examples of their use, are contained in a number of books and articles (Bush, 1945; Casey et al., 1958; Casey, Bailey, and Cox, 1946; Donnay, 1938; Hurlbut, 1948; Kirkpatrick, 1948; Parker, 1952, 1957).

12.4 Automatic Data Processing

One of the limitations of the hand-punched card system is that the holes are confined to the margins of the cards and are therefore limited in number by the size of the card; double holes[1] multiply the available positions by a factor of 3, and by efficient coding a very large amount of information can be processed in this way. Attribute data which are not recorded on a continuous scale are particularly amenable to this kind of treatment. If the data items are numerals, the limited number of holes on a punched card tends to restrict its usefulness. Furthermore, when the data are quantitative, there is usually the requirement that they be reduced by the standard statistical techniques, and hand-punched cards are not particularly adapted to these procedures. At this stage the investigation graduates to the automatic

[1] Either hole may be punched individually or both may be punched together.

punched-card system. Here the Hollerith card is usually standard, and the code notation and use are somewhat more narrowly prescribed; these cards (see Fig. 12.5) have 80 columns of 10 positions numbered 0 to 9 in each column. The first entry, the sample number, requires a field of columns equal to the number of digits in the largest sample number; i.e., four columns will take sample numbers up to and including, 9,999. For sample number 4,763 the fourth position in column 1, the seventh in column 2, the sixth in column 3, and the third in column 4 will be punched. Subsequent columns are available for data; at first sight 80 columns appear to be ample for any purpose, but suppose, by way of example, that three sets of measurements comprising figures of the type 3.45 are included; then the original data will take 9 columns, which means that a total, with the sample number, of 13 columns will be required; in data processing, to obtain the sum of squares each item will require at least 6 columns and, for the paired products, at least 6 more for each of those, leading to a total of 49 columns in use. It can be readily appreciated that 80 columns are not by any means too many for the average research problem. Of course, once the raw data are punched on cards, the computer can reproduce its own extra sets for further processing, and, apart from the duplication of cards, there is no serious problem for the analyst. An example of the use of such cards and the subsequent data-processing procedure is described by Krumbein and Sloss (1958).

At this stage cards represent only one of a number of alternatives for data storage and processing; alternative procedures include coding the data for paper tape or magnetic tape; and the storage of, and access to, very large quantities of data is certainly more efficiently performed by means of tapes than cards. There are, however, advantages to both procedures, and the choice frequently depends on the kind of computer facilities available. Some computer installations take coded data in the form of cards and tape; others take only one or the other but not both. Again, machines exist for con-

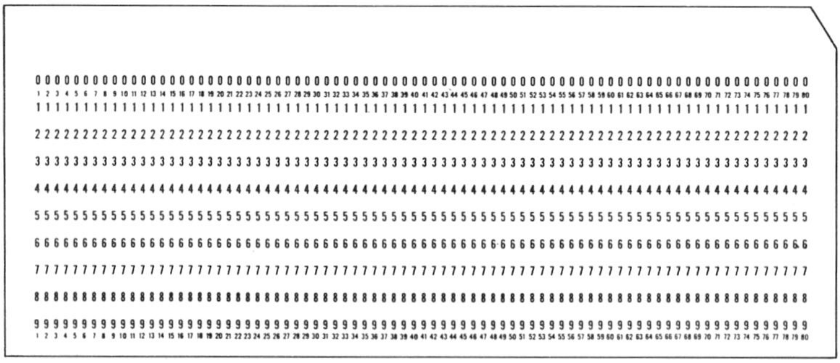

Figure 12.5 Hollerith punched card for automatic data processing by means of electronic computer.

verting data coded on cards to tapes and vice versa; in fact, a very wide choice of procedure now exists for data handling and storage (see, for example, Dixon and Massey, 1951, chap. 15).

It is perhaps worthwhile to emphasize that if the most efficient and simplest procedure is to be used, it is necessary to foresee the complete program from the specified objective through the collection, processing, and analysis of the observations and including the form of the answers likely to be finally achieved. The more powerful the data-processing installation, the more necessary it is to know how to process the data and transform it into a notation suitable for the machine on which it is to be analyzed. It should be reasonably obvious that the modern research project, generally including a team of specialists in different branches, has to be extremely well organized if it is to function efficiently and effectively; the day when specialization in one narrow field was adequate has passed. It is now necessary to be familiar with the entire plan of research if successful solutions to the complex problems facing the researcher are to be attained in a reasonable time and at reasonable cost.

It has often been contended that research in the earth sciences cannot be quantified because of the complexity of the problems and the large number of variables to be observed; it appears that the modern research program is devised essentially just for this kind of problem, and with adequate modern facilities such a complaint is no longer justified.

12.5 Aspects of Measurement in the Geosciences

Attempts to quantify observations generally involve the use of numbers to represent the observations; quantification may be generally defined as the assigning of symbols or numerals to observations according to certain rules. Under these circumstances a number of alternatives exist in the assignment dependent upon the rules adopted, and this area has received increasing attention as its importance is realized (Campbell, 1957; Stevens, 1946, 1959; Scates, 1937). In broad terms the entire process of data reduction, starting with classification and proceeding to finer and finer subdivision of the observations, represents stages in the quantification of observations (Hempel, 1952).

Fom the definition of quantification as a process of mapping observations onto symbols according to certain rules it becomes apparent that the rules of assignment will determine the information value of the assigned symbols. If the rule is to assign the observations randomly to numbers, then the resulting symbols carry no implications other than as labels, and little information can be gained by the process. On the other hand, if the rules of assignment lead to a "mapping" of observations onto numbers such that the resulting numbers have a structure isomorphic with arithmetic, the data-reduction problem is simplified, and very powerful procedures exist for analyzing the information. This continuum of procedures for quantification is usefully subdivided into four successive steps conventionally labeled as nominal, ordinal, interval, and ratio scales (Table 12.3).

In the first case, nominal scaling, the observations are mapped onto symbols, and each symbol represents a class; the requirement is that the rules of assignment lead to a set of mutually exclusive and exhaustive classes so that every observation can be assigned to some one class. The underlying requirement is equivalence; i.e., the criteria by which each observation is assigned to a class should confer equality upon the observations in each class. For example, if each member of a team of researchers is issued a letter and each letter is assigned to one and only one member of the team, the scaling is nominal. Each class has only one member and the classes are equivalent; no ordering is implied. Many examples of classification in the geosciences achieve this level of quantification, such as assigning specimens of igneous rocks to the classes granite, syenite, diorite, etc. Similarly, the classification of fossils into species is an example of nominal scaling (see also Table 9.1).

The equivalence requirement for nominal scaling may be summarized as reflexive ($a = a$ for all a), symmetrical (if $a = b$, then $b = a$), and transitive (if $a = b$ and $b = c$, then $a = c$). It is of considerable interest to note that criteria used to fulfill these requirements may be simple, as in classifying hand specimens of rocks by megascopic examination, or complex, as when a clay mineral is given a name by means of exhaustive optical, x-ray, and chemical analyses. The procedures which lead to nominal scaling are procedures for diagnosis of the classes and are the routine identification procedures used throughout the geosciences. An example is described in Sec. 9.1 in setting up the classes of elements for modal analysis by point counting. In general, the identification, or diagnostic procedure, is followed by counting the frequency in each class. This counting procedure has proved quite refractory in many cases and is a late development in much of the area of interest to the geosciences.

If the scaling procedure of mapping observations onto symbols achieves, in addition to equivalence, the properties of being irreflexive and asymmetric, the classes may be ordered. These requirements may be summarized as irreflexive: $a \neq a$ for all a; asymmetric: if $a > b$, then $b \ngtr a$; and transitive: if $a > b$ and $b > c$, then $a > c$. The symbol $>$ may mean greater than, coarser than, harder than, preferred to, and so on. Fulfillment of these requirements means that the mapping has resulted in an ordinal scale.

Field observation frequently leads to observations which are symbolically represented on an ordinal scale; when rocks are classified as phanerite, aphanite, and glass or as gravel, sand, and clay, ordinal representation is implied. It is noteworthy that sandstone, limestone, and shale represent nominal scaling, and either no ordering or, at most, partial ordering is indicated.

Typical examples of ordinal scaling are Mohs' scale of hardness and the assignment of numerals to relative frequencies in accessory-mineral studies.

TABLE 12.3
Scales of Measurement

Scale[a]	Basic empirical operations[a]	Mathematical group structure	Permissible statistics	Examples of scales
Nominal	Determination of equality	Permutation group, $x' = f(X)$; $f(X)$ means any one-to-one substitution	Number of cases (counts); mode; contingency, correlation independence	Counts of frequency of minerals, rock types, and other attributes; surface texture
Ordinal	Determination of $>$ or $<$	Isotonic group $x' = f(X)$; $f(X)$ implies any monotonic increasing function	Median; rank; percentiles; correlation	Mohs' hardness; abundance of heavy minerals; ranked-data; stratigraphic scale
Interval	Determination of equality of intervals or differences	General linear group $x' = BX + a$	Mean; standard deviation. Rank order; correlation; product moment correlation Phi or log scale	Temperature scale; sphericity, roundness; oxygen isotope scale; "absolute" time scale
Ratio	Determination of equality of ratios	Similarity group $x' = ax$	Coefficient of variation. Geometric mean	Millimeters, grams, milliliters, Kelvin temperature

[a] These scales represent a continuum, and the operations are cumulative.
Source: Adapted from Stevens, 1946.

The use of relative relief in place of measurement of refractive index of minerals during thin-section analysis is an example of replacing a higher scaling by ordinal measures. Ordinality is also achieved by visual comparison of the properties of rocks and fossils against standard charts, in which the standards are ordered on a higher level. Visual sphericity and roundness are examples of such procedures, and comparison of sedimentary rocks with grain size or color standards also leads to ordinal measures.

On an ordinal scale the intervals between the symbols are not fixed; thus diamond, of hardness 10, is not twice as hard as apatite, of hardness 5, etc. As a result the analysis of such data is restricted to procedures in which ranking or ordering is all that is required. In particular, "order statistics" such as the quartile and percentiles are the appropriate summary statistics to use with measures that achieve ordinality. It should perhaps be emphasized that if the data are ordinal in value, use of parametric statistics is not justified, because parametric procedures require information on a higher-level metric; the converse is also important: if the metric achieved is higher than ordinal, the use of ordinal statistical procedures will often lead to a loss of information.

If the data achieve ordinality and the intervals on the scale are defined, the scale achieves interval rank; the zero point and scale unit are arbitrary in interval scaling. Transformation of one scale to another can be performed either by multiplying by a constant or by adding a constant or both. A typical example of interval scaling is contained in the temperature scales Fahrenheit and centigrade, and the transformation from one to another is exemplified by multiplying and adding a constant

$$°F = \tfrac{9}{5}°C + 32$$

One of the characteristic features of this scale is that differences between successive units are constant.

It appears likely that when sphericity and roundness are measured by Wadell's procedures, the measures so obtained are interval in scale, whereas if visual comparison charts are used, the scale achieves ordinality (see Chap. 6). The advantage of increased information level leads to the use of parametric statistical analysis.

Trace-element concentrations determined by spectrographic procedure and isotopic abundance wherein the zero is arbitrary[1] (Ahrens, 1954) and the interval fixed probably achieve interval scale.

Finally, if the scale has all the qualities of an interval scale together with a fixed zero point, the information level is considered to be represented by a ratio scale. The only transformation applicable without change of information level is multiplication by a constant. Measurements of grain size, bulk density, volume, etc.—indeed, most direct measuring techniques—achieve ratio-scale level.

[1] Equivalent to "not detected."

It is of interest to note that transformation to a logarithmic scale, while qualifying as a one-to-one mapping of a ratio scale to its logarithmic equivalent, does perhaps imply some loss of information, because the transformation is not simply a multiplicative factor and the zero is somewhat arbitrary. With data on a ratio scale, both the geometric mean and the coefficient of variation may be used as representative summary statistics, but when the ratio scale is transformed to a logarithmic scale, the arithmetic mean of the logarithms equals the geometric mean of the original data. The coefficient of variation, which expresses the standard deviation as a proportion of the mean and is usually stated in percentage terms, is represented as

$$CV\% = \frac{\hat{\sigma}}{\bar{X}} 100$$

and if the original observations possess mean values in the range of unity, the CV on the logarithmic scale approaches infinity (e.g., size versus sorting in Fig. 15.3, and Fairbairn, 1953). Some care must be exercised in manipulating transformed data to ensure that there is no distortion of information.

Campbell (1957) has subdivided observations which achieve ratio scale level into *fundamental* and *derived* measurements, wherein the latter require knowledge of fundamental measurements for their determination. Thus, if the thicknesses of two beds of rock are determined, the thickness of both is found by adding the two thicknesses and the total truly represents the total thickness of both beds. If two volumes of liquid of density 3.00 are added, we obtain twice the volume but the density remains 3.00; that is, $3 + 3 = 3$. Thus direct measures of fundamental properties achieve ratio level, and the observations mapped onto numbers lead to a structure isomorphous with arithmetic; e.g., the property of additivity is preserved. On the other hand, measurements of derived properties, while they achieve ratio scale, may still possess restrictions in their manipulation. The use of such derived shape factors as sphericity and Zingg values thus appears to be less informative than the uncombined "raw" measurements of the length of axes. Even the phi transformation may result in some loss of information, and adoption of a transformation of this kind must be carefully considered by comparison of its advantages and disadvantages.

Measurement theory should be familiar in some degree to all experimenters, and introduction to the subject can be obtained by consulting such references as Stevens (1946), Campbell (1957), Siegel (1956), Churchman and Ratoosh (1959), and Griffiths (1960c).

12.6 Choice of Statistics

Experiments are performed to evaluate information on some specific objective; in general, a relatively small set of observations is collected and is considered to represent a much larger body of observations which could have been collected. In other words, the *sample* of information collected is generally used to reflect characteristics of a larger *population*, which the sample is presumed to represent. Statistical analysis is generally used to achieve this purpose; the arithmetic mean, for example, can be found from any set of data, but it can be used as a basis of logical induction from sample to population only if the arithmetic mean is truly representative. Given two frequency functions, both symmetrical, one unimodal and bell-shaped and the other U-shaped, the arithmetic mean may be calculated for both; when this mean is physically manipulated, however, it is somewhat more reassuring to know that the curve is bell-shaped rather than U-shaped. In other words, it is of considerable advantage to be able to determine what the frequency function of the variate is likely to be. Again, not all statistics are suitable for induction or for sophisticated statistical analysis, and so it is advantageous to obtain statistics of a prescribed character to ensure that subsequent analytical manipulations are reasonably justified. Characteristics of some common frequency functions are discussed in Chaps. 13 and 14, but the specification of the most desirable statistics may give some point to the necessity for finding the frequency function. In addition, the neophyte is quite justifiably confused by the plethora of "statistics" available (Krumbein, 1939b; Krumbein and Pettijohn, 1938; Inman, 1952; Folk and Ward, 1957), and some guidance in choosing the most effective procedure may assist in rejecting those which are least likely to be of value. In many cases the nature of the observations, particularly the scale level achieved, prescribes the analytical procedure, and the choice is then simplified. Frequently, however, no prior knowledge of the numbers obtained is available, and it is then most necessary to be aware of the pitfalls which exist in adopting some apparently innocous practice which, in fact, may lead to contradictory results. Fortunately, the specification of best statistics has been given considerable attention, and procedural rules exist which permit the comparison of one course with another. There is little doubt that knowledge of the frequency function is of the greatest value, and assumption of normality, either directly or after transformation of the variate, is one of the commonest and most useful steps. Indeed, whether an investigator agrees with this procedure or not, nearly all bases of evaluation are defined in terms of comparison with this distribution.

The term *best statistics* may be defined by means of the requirement of unbiasedness, consistency, efficiency, and sufficiency. If statistics are required which may be used for inductive argument from sample to population, they should be estimators of their corresponding population parameters. If a sample statistic converges stochastically (Bennett and Franklin, 1954, p. 44) on the relevant population parameter, it is generally both consistent and unbiased. A statistical estimator may stochastically converge on some fixed value which differs from the true population value, in which case the statistic is consistent but biased.

Consider the population variance σ^2 of a normal distribution; if a series of small samples is taken and their sample variances s^2 computed according to the formula

$$s^2 = \frac{\sum_{i=1}^{n} (X_i - \bar{X})^2}{n} \tag{12.1}$$

where $\sum (X_i - \bar{X})^2$ represents the sum of the squared deviations from the sample mean $\bar{X}$ divided by the number n of observations in the sample, then this statistical estimator is consistent, i.e., it converges statistically on some fixed value. However, this value is not the population variance σ^2 defined as

$$\sigma^2 = \frac{\sum_{i=1}^{N} (X_i - \mu)^2}{N} \tag{12.2}$$

In other words, s^2 is consistent but in *small* samples is a biased estimator of σ^2. Now in this case the bias can be found (Sec. 16.9), and it is customary to define a *best estimator* of the population variance as $\hat{\sigma}^2$, where

$$\hat{\sigma}^2 = \frac{\sum_{i=1}^{n} (X_i - \bar{X})^2}{n - 1} \tag{12.3}$$

In small samples the best estimator $\hat{\sigma}^2$ is an unbiased, consistent estimator of the population variance. The bias is evidently $n/(n - 1)$, where s^2 underestimates σ^2, and this is corrected by multiplying s^2 by $n/(n - 1)$ to yield $\hat{\sigma}^2$.

As the samples get larger, $n - 1$ approaches n, and so, in the limit, s^2 is also unbiased and consistent, which explains the emphasis on small samples as the basis of the estimation procedure described above.

There may be several unbiased, consistent estimators, and it is then necessary to establish a criterion for selecting the "best" of these; each will possess a sampling variance around the parameter it is intended to estimate, and the best will be that one which, for a given number of observations, possesses the least sampling variance. This is an *efficient* estimator. Thus the sample range and the sample variance may be used as estimators of the population variance in a normal distribution; since the former needs much less arithmetic to compute than the latter, why use the sample variance? It is found that as the sample size increases, the sampling variance of the range rapidly stabilizes while that of the sample variance improves; thus the sample variance is an efficient estimator compared with the sample range. The ease of calculation can here be balanced against the loss of information, with due allowance for sample size. Comparison of the sample mean and sample median as estimators of the population mean in a normal distribution illustrates the concept of efficiency (Dixon and Massey, 1957, p. 71); for equivalent precision the median requires 1,000 observations compared with 640 for the sample mean.

Finally, there may be a number of statistics which are unbiased, consistent, and efficient, and of these the most desirable are those which are sufficient (Fisher, 1948, p. 14). Thus in a normal distribution the sample mean is a sufficient estimator, in that no other statistic can throw additional light on the parameter of location, the population mean. Similarly, the best estimator $\hat{\sigma}^2$ is a sufficient statistic in that no other estimator will add to information on the dispersion of the distribution. There are, unfortunately, few sufficient estimators, but the arithmetic mean fulfills the requirements of sufficiency in the normal and the Poisson frequency distributions. Most order statistics, such as the median and quartile statistics, are inefficient when compared with appropriate estimators in a normal distribution (Dixon and Massey, 1957, pp. 69ff.; McCammon, 1962a). No comparison, of course, is possible unless the form of the frequency function is specified, so that in many cases it is not possible to decide whether some advocated statistic is a suitable estimator or not. In general, it is advantageous to select those statistics which fulfill the above criteria of "best" or at least fulfill as many of these requirements as possible; perhaps, at the very least, the properties of unbiasedness and consistency should be required.

As a general recommendation, following upon these considerations, it is evident that knowledge of the frequency function is certainly to be recommended and that without such information conclusions can only be considered tentative. Most of the techniques recommended in this book, when appropriately used, produce well-known frequency functions; then the entire set of criteria described above can be applied, and statistical analysis of the most sophisticated kind can be used with considerable confidence. The less that is known about the frequency function of the variate, the more difficulty arises in attempting to achieve reliable conclusions and solutions to problems of geological interest: in this sense "nonparametric" or "order" statistics should be looked upon as stopgaps, and conclusions based on such analysis should be regarded with considerable circumspection.

13 Frequency distributions: the normal distribution

A variate is said to be normally distributed when it takes all values from "$-\infty$ to $+\infty$" with frequencies given by a definite mathematical law, namely that: the logarithmn of the frequency at any distance from the center of the distribution is less than the logarithmn of the frequency at the center by a quantity proportional to d^2.

R. A. Fisher, 1948

13.1 Introduction

Frequency distributions play a central role in the statistical treatment of data, and their usefulness may be considered under a variety of subheadings. They are often used as graphic, pictorial summaries of sets of data. In such cases the accompanying summary statistics are perhaps easier to understand as characteristic of the "shape" of the frequency distribution. When an experimental process or treatment is known to generate a certain type of frequency distribution, the comparison of the observed data with those compiled under experimental sampling constitutes a test of randomness of the sample of observations.

When these aspects related to the physical situation have been exhausted, the same frequency distributions are, in addition, of special value in formulating the guiding rules of statistical analysis; and the known form of the distributions permits the prediction of sampling distributions of certain statistics such as the mean and variance of a normal distribution. This in turn leads to significance testing and problems of estimating parameters from sample statistics.

Quotation: R. A. Fisher, "Statistical Methods for Research Workers," 10th ed., p. 43, Oliver & Boyd Ltd., 1948.

Some sets of observations are not suitable for certain specific kinds of analysis without transformation to a new scale, and in such cases a knowledge of the form of the frequency distribution is useful in selecting the appropriate transformation (see Chap. 15).

The succeeding account will briefly survey some of the commoner frequency distributions which arise in performing experiments and which have properties known to be useful in describing sampling situations that result either from physical experiments or from statistical analysis. It should be remembered that one frequency distribution may play many roles, and it is generally necessary to distinguish which role the analyst has in mind if clarity and understanding are to be maintained.

13.2 The Normal Distribution

Perhaps the most popular and generally useful frequency distribution is the normal, or Gaussian, frequency distribution, or *error curve*. One of the earliest statistical tools, it was discovered by Moivre about 1753 (see Walker, 1931, pp. 4ff.), and it plays an important role in the development of *error theory* (Beers, 1957). This frequency distribution can often be used as a model of physical situations, and it is, of course, the central frequency distribution in parametric statistical analysis. It is also generally used as a comparison standard for nonparametric tests to indicate their power and efficiency.

Let us consider the family of curves

$$y = \frac{1}{k^{t^2/2}} = k^{-t^2/2} \qquad (13.1)$$

where k is a constant and y and t are variables. Suppose $k = 1$; then the curve is a straight line parallel to the t axis (see Fig. 13.1). For $k < 1$ the curve is U-shaped and extends above the value $y = 1$. For $k > 1$, it takes on the bell-shaped form illustrated in Fig. 13.1.

It is customary to select the constant k to yield a curve with total area unity, and thus $k = 1.0872$; the form of this constant is $(1/\sqrt{2\pi})e$, where e is the base of natural logarithms and π is also a constant.

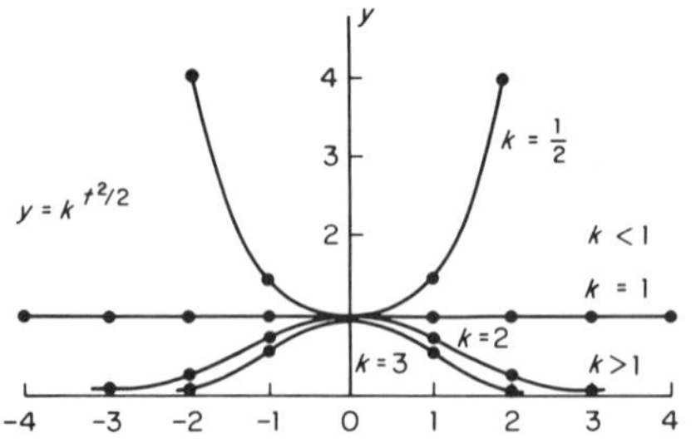

Figure 13.1 Graphic representation of the frequency function $y = k^{-t^2/2}$.

The equation for the frequency curve becomes

$$y = \frac{1}{\sqrt{2\pi}} e^{-t^2/2} \tag{13.2}$$

which is known as the normal curve in standard measure. The conventional expression for the normal frequency distribution is obtained from this by substituting for t as follows:

$$t = \frac{X - \mu}{\sigma} \tag{13.3}$$

whence the frequency function becomes

$$y = \frac{1}{\sigma\sqrt{2\pi}} \exp\left[-\frac{1}{2}\left(\frac{X - \mu}{\sigma}\right)^2\right] \tag{13.4}$$

The parameters of a normal distribution are the mean μ and the standard deviation σ. Different normal distributions have different values for μ and σ. With one particular normal curve given, μ and σ are fixed, and the variation of y as X changes can then be studied. The area under this frequency curve is reduced to unity, so that each increment of area represents a proportion, and the areas between different ordinates are of considerable importance. Unfortunately, the expression on the right-hand side cannot be integrated in its present form to determine the increments of area, and the areas lying between specific ordinates have been found by numerical analysis by using approximations.

Tables of the area between specific ordinates of the normal distribution are readily available, and in order to obtain maximum usefulness the curve is generally standardized. Suppose by way of example that a curve is known to be normal with $\mu = 10$ and $\sigma = 5$. Then by substituting $t = (X - \mu)/\sigma = (X - 10)/5$, the frequency function

$$y = \frac{1}{5\sqrt{2\pi}} \exp\left[-\frac{1}{2}\left(\frac{X - 10}{5}\right)^2\right] \tag{13.5}$$

reduces to

$$y = \frac{1}{\sqrt{2\pi}} \exp -\frac{t^2}{2} \tag{13.6}$$

which has $\mu = 0$ and $\sigma = 1$. A similar substitution is used for all normal curves by using appropriate values of μ and σ; whence the tables are expressed in terms of x/σ when $x = X - \mu$. Thus one table contains the information on areas for all normal curves (see, for example, Pearson and Hartley, 1954, table 1, pp. 104–109; Arkin and Colton, 1950, table 10, pp. 114–115), and the variates are expressed in standardized form.

13.3 Characteristics of the Normal Distribution

The normal frequency curve is symmetrical around the mean (see Fig. 13.1); the area under the curve is equal to the total frequency, i.e.,

$$y = \frac{N}{\sqrt{2\pi}} \exp\left[-\frac{1}{2}\left(\frac{X-\mu}{\sigma}\right)^2\right] \tag{13.7}$$

When in standardized form, the area under the curve is unity.

The forms or shapes of frequency curves have been variously described; one of the most useful representations is given by the moment measures and thence to Pearson's beta statistics (Elderton, 1938, p. 42; and examples in Secs. 5.4 et seq.), which may be used for a wide variety of frequency curves. In the normal curve the derived parameters based on the successive moments are:

Parameter		Corresponding moment
Area	$= 1$	Zeroth-order
Mean	$= \mu$	First-order
Standard deviation	$= \sigma$	Second-order
Asymmetry[a]	$= \sqrt{\beta_1} = 0$	Third-order
Peakedness[a]	$= \beta_2 = 3$	Fourth-order

[a] There are a number of alternative measures of asymmetry and peakedness; and, of these, Fisher's g statistics based on the cumulants k_i have certain useful properties (Fisher, 1948, p. 52).

The area under the curve is the moment of zero order.

13.4 Conditions Leading to the Generation of a Normal Distribution

Use of a frequency curve as a physical model of some specific experimental sampling situation requires a knowledge of the conditions which must be fulfilled in the development of a normal distribution.

It is simplest, perhaps, first to set up the requirements for a symmetrical (binomial) distribution as follows:[1]

1 In the absence of certain specific "causes" of variation, or in the event of perfect balancing of these effects, the data may assume a fixed central value.
2 Deviations from the central value result from certain causes of variation, the effect of any cause being either to add a fixed quantity or to subtract the same quantity.

[1] These conditions are examined in detail by Smith and Duncan (1945, pp. 37ff.).

3 The probability that a cause of variation will produce a positive effect equals the probability that it will produce a negative effect:

$$P(+ve) = P(-ve) = \tfrac{1}{2}$$

4 The net effects of all contributing causes of variation are of equal magnitude in either direction.
5 The contributory causes are independent in their action; the $P(+ve)$ or $P(-ve)$ effect from any causal factor is independent of previous contributions.
6 The total deviation of any element from its central value is the algebraic sum of positive and negative contributions of the individual causal factors.

Then if, in addition to these, the following conditions apply, a normal curve will be generated:

7 The number of contributory causes is very large.
8 The positive and negative contributions of each cause are very small.

Actually, since a normal frequency curve arises under quite general circumstances, many of the listed conditions may be relaxed. The most important conditions are 6, 7, 8, and 4 approximately. If conditions 6 to 8 hold, then conditions 2 and 3 may be largely relaxed. If the variate under study fulfills these requirements, then its population frequency distribution may be considered to be normal, and the normal model may be used as the basis for analysis of sampling situations arising during experimental investigation (see, for example, Secs. 5.5 et seq.).

13.5 Fitting the Normal Distribution to a Set of Observations

One of the first requirements in applying the normal model is to test whether a given set of observations could represent a sample set from some normally distributed parent population. The procedure begins with a calculation of the sample statistics; the mean $\bar{X}$; the standard deviation $\hat{\sigma}$; the asymmetry, either $\sqrt{b_1}$ or g_1; and the peakedness, or kurtosis, either b_2 or g_2.

If the observations represent random samples from some normal population, then the mean $\bar{X}$ is an estimator of the population mean μ, and the standard deviation $\hat{\sigma}$ may be used as an estimator of the population value σ. The procedure of reducing this curve to standard form then comprises reducing the mean to zero by subtracting it from each observation and scaling the deviation $X - \mu$ by dividing by the standard deviation $\hat{\sigma}$. This is a device frequently used to produce a normal deviate with mean zero and standard deviation unity, and the result is often referred to as the Z score (Dixon and Massey, 1951, p. 24).

In practice the observations are graduated into classes (the grade scale), and the areas between class limits are found; for this purpose the class limits are standardized as in the first three columns of Table 13.1 and Fig. 13.2.

The standardized normal deviates in column 3 may then be used to find the area lying between the respective class limits and minus or plus infinity, as the case may be, from tables of the area of a normal distribution (Arkin and Colton, 1950, pp. 114–115; Pearson and Hartley, 1954, pp. 104–110). In the present example the former tables were used, and the areas are listed in column 4; it should be noted that because the normal curve is symmetrical, only half the area need be tabulated. The areas of column 5 are then found by difference; the first value is found by subtracting 0.49966 from the total area (0.5 unit) under half the curve ($\mu = 0$ to $-\infty$); the second by subtracting $0.49966 - 0.4948 = 0.0049$; and so on until one reaches the central class, which contains the mean. The area subtended by these class limits comprises the area from μ to the class limit on the negative side plus the area from μ to the positive class limit, which equals $0.2794 + 0.0517 = 0.3311$. The remaining areas are again found by difference.

The total area represented by the sum of the figures in column 5 should equal unity. The area under the "expected" curve is then found from column 5 by multiplying by the total area (number of observations) under the observed curve; thus column 6 is derived from column 5 by multiplying by $N = 450$. The observed frequencies are entered in column 7 for comparison. The expected frequencies are shown as the continuous curve in Fig. 13.2, and the observed sample of observations is drawn as a histogram. The excesses and deficiencies of observed compared with expected are also shown.

TABLE 13.1
Fitting a Normal Distribution to a Set of Observations of Short *c* Axes of Quartz Grains from Homewood Quartzite[a]

(1) Class limits X	(2) $x = X - \mu$	(3) x/σ	(4) Area $+\infty - x/\sigma$	(5) Difference	(6) Expected $\times N = \times 450$	(7) Observed
$-\infty$ to -0.5	-1.9278	-3.453	0.49966	0.00034	0.14	
-0.50 to 0.0	-1.4278	-2.577	0.4948	0.0049	2.21	6
0.00 to 0.50	-0.9278	-1.662	0.4515	0.0433	19.49	12
0.50 to 1.00	-0.4278	-0.766	0.2794	0.1721	77.45	59
1.00 to 1.50	$+0.0722$	$+0.129$	0.0517	0.3311	149.00	191
1.50 to 2.00	$+0.5722$	$+1.025$	0.3485	0.2968	133.56	117
2.00 to 2.50	$+1.0722$	$+1.920$	0.4726	0.1241	55.85	55
2.50 to 3.00	$+1.5722$	$+2.816$	0.4976	0.0250	11.25	7
3.00 to 3.50	$+2.0722$	$+3.712$	0.4999	0.0023	1.04	2
3.50 to $+\infty$				0.0001	0.05	1
Total				1.0000	450.04	450

[a] See also Fig. 13.2.

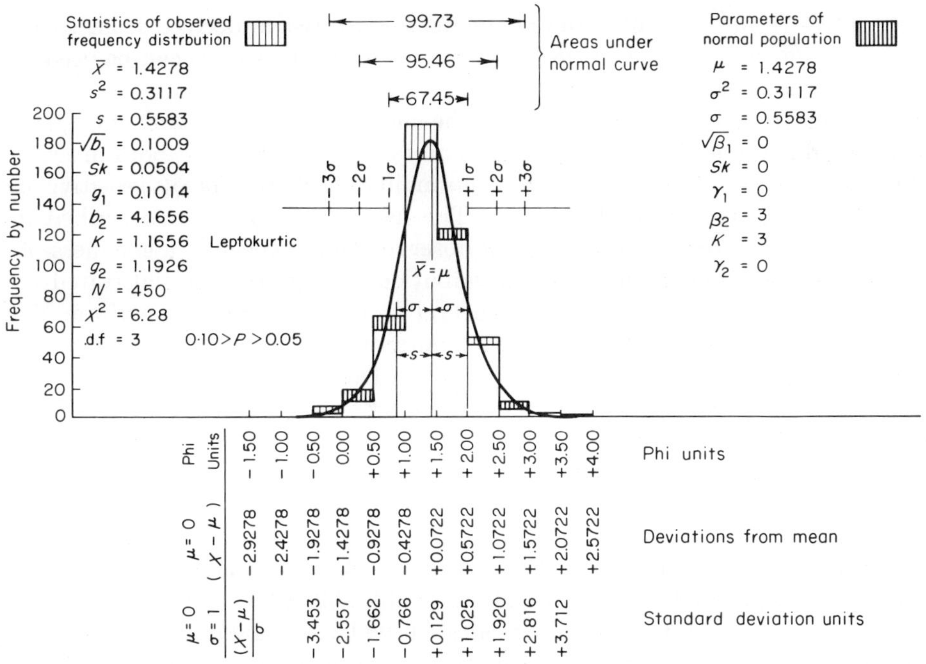

Figure 13.2 Size-frequency distribution of quartz grains in Homewood quartzite; c_ϕ axis.

If the theoretical, or expected, now fits the observed, the continuous curve may be considered to represent the population distribution with mean μ and standard deviation σ.

13.6 Testing a Set of Observations to Decide whether They Are Approximately Graduated by a Normal Curve

Qualitative comparison of two frequency functions can be performed by means of histograms, as in Fig. 13.2, but a somewhat simpler pictorial comparison can be obtained by using probability paper and changing the frequency distributions. Probability paper is designed in such a way that a normal curve plots as a straight line, and so the expected normal curve can be inserted, together with its two parameters, μ and σ, as in Fig. 13.3. The observed cumulative frequency distributions can now be superimposed and the departures of observed from expected can easily be noted. In Fig. 13.3 there is a slight excess around the central tendency, deficiency in the shoulders, and excess at the coarse tail (say 0 phi and a deficiency at the fine end, beyond 2.5 phi).

More rigorous test procedures to decide whether these departures are significant can be performed by using the moments or the chi-square criterion (see also Sec. 17.4). These two tests yield independent information on departures from normality; the moments, and statistics derived from them, describe the shape of the frequency curve, whereas chi square can be used to compare the specific departures class by class and can be used with a wide variety of frequency distributions. The moment measures and chi-square tests should be looked upon as complementary tests in this case; both measure different aspects of departures from normality, each test being sensitive in different situations from the other. The use of chi square is described in Sec. 17.4, and to test the departure of observed from expected, a goodness-of-fit test, in the case of the c axis distribution of Fig. 13.2, is illustrated on page 345 and in Table 17.2. In general it is convenient to perform the chi-square test on the tabular sheet containing the computation used to determine the expected normal distribution (Table 17.1).

Moment measures will be used in the present instance; with a normal distribution, the moment values, translated into measures of asymmetry and peakedness, are known; and for a normal distribution with $\mu = 1.4278$ and $\sigma = 0.5583$, they are listed along with the corresponding sample statistics in Table 13.2. The moment measures are found as described earlier, and Fisher's g statistics can be calculated from the moments around the assumed mean n_i, as described by Griffiths (1962a, tables 32, 33).

The observed statistical values can then be tested against their appropriate standard errors. In practice this test is simplified by tabulations of the expected values of the sampling distributions of the respective parameters;

TABLE 13.2
Theoretical Normal and Observed Values of the Parameters and Statistics Respectively of the Set of Data Represented by 450 c Axes of Quartz Grains from the Homewood Quartzite (Phi Scale)

Theoretical = expected normal		Observed sample statistics
Mean	$\mu = 1.4278$	$\bar{X} = 1.4278$
Variate	$\sigma^2 = 0.3117$	$\hat{\sigma}^2 = 0.3117$
Standard deviation	$\sigma = 0.5583$	$\hat{\sigma} = 0.5583$
Asymmetry	$\sqrt{\beta_1} = 0$	$\sqrt{b_1} = 0.1009$
	$\gamma_1 = 0$	$g_1 = 0.1014$
Skewness	$Sk = 0$	$Sk = 0.0504$
Peakedness	$\beta_2 = 3$	$b_2 = 4.1656$
	$\gamma_2 = 0$	$g_2 = 1.1926$
Kurtosis	$K = 0$	$K = 1.1656$

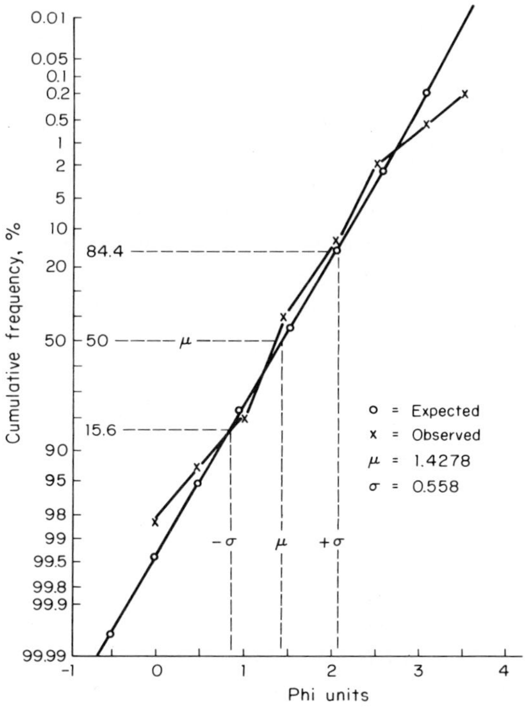

Figure 13.3 Cumulative frequency distribution of short *c* axes of quartz grains from Homewood quartzite plotted on phi-probability graph paper.

thus the sampling distributions of b_i around their respective β_i values are listed by Pearson and Hartley (1954), and the sampling variation of g_i around its respective γ_i is given by Griffiths (1962a).[1] In the present case with $n = 450$ the 5 percent value of $\sqrt{b_1}$ is 0.188, and the 1 percent value of $\sqrt{b_1}$ is 0.269 (see Pearson and Hartley, 1954, table 34, p. 183); since the observed value of $\sqrt{b_1}$ is 0.1009 and is less than the 5 percent tabulated value, the asymmetry as measured by the $\sqrt{b_1}$ statistic is not so large as would be expected 5 times in every 100 on random sampling a (symmetrical) distribution with samples of size 450. Hence this skewness (asymmetry) value is not indicative of a departure from normality, and we may assume that the sample estimate $\sqrt{b_1}$ is an estimator of $\sqrt{\beta_1} = 0$. Similarly, if this set of 450 observations is a random sample from a normal population in which $\gamma_1 = 0$, then a value of $g_1 = 0.22$ or larger would be expected about 5 times in every 100 (Griffiths, 1962a, p 578). There is again no evidence of departure from a symmetrical distribution contained in the observed g_1.

We may now test the peakedness, using the approach that if this set of 450 observations is a random sample from a normal population, then the expected variation of b_2 around the population value of β_2 (or g_2 around γ_2) is recorded in the appropriate tables.

[1] A superior table is given in Bennett and Franklin (1954, table 4.5, p. 95).

For the β_2 value, with $n = 450$, the 1 percent level of b_2 is 3.63 and the observed value of b_2 is 4.1656, which far exceeds it; similarly, the expected sampling variation of g_2 is 0.45 (approximately) around a γ_2 value of zero, whereas the observed value is 1.193, again excessively large. Hence this sample departs from normality in being much too peaked. If we now return to Fig. 13.2, where expected and observed values are graphically portrayed, it can be seen that there is indeed an excess of observed over expected in the central class and in the tails, but the curve is reasonably symmetrical about its mean.

It may be of interest to note that the chi-square criterion (see Table 17.1) yields a statement that such a departure in terms of class-by-class comparison would be expected less often than 1 time in every 1,000 such samples, and the departure reflects the effect of leptokurtosis in the observed frequency distribution.

In practice one would have to decide the reason for this departure; presumably it is a reflection of nonrandomness in the sampling. It is not clear whether this nonrandomness is to be attributed to "natural" selective sorting or to predilection of the investigators in selecting grains for measurement. This dilemma should be resolved before attempting to assign conclusions of geological importance to the departure from normality.

In applying such tests of the moment measures an approximate test for n moderately large is immediately available for the g statistics. Thus the expected values for γ_1 and γ_2 in a normal distribution are zero; the variance of the sampling values of g_i around their respective γ_i is solely a function of n, as follows (Fisher, 1948, p. 75):

$$\sigma^2_{g_1} = \frac{6n(n-1)}{(n-2)(n+1)(n+3)} \approx \frac{6}{n}$$

$$\sigma_{g_2}^{\ 2} = \frac{24n(n-1)^2}{(n-3)(n-2)(n+3)(n+5)} \approx \frac{24}{n}$$

where n is large, say greater than 100.

As an example, the values of the variances in the present case are $\hat{\sigma}_{g1}^{\ 2} = 6/n = \frac{6}{450} = 0.0133$, and $\hat{\sigma}_{g2}^{\ 2} = 24/n = \frac{24}{450} = 0.0533$; then $\hat{\sigma}_{g1} = 0.115$ and $\hat{\sigma}_{g2} = 0.231$, and using the relationship $t = (X - \mu)/\sigma$, where $X - \mu$ is the observed value of g and σ is its standard error, yields

$$g_1 = \frac{0.1014}{0.115} = 0.882$$

and

$$g_2 = \frac{1.1926}{0.231} = 5.163$$

Then the value of g_1 is 0.9 standard deviation from its mean whereas g_2 is 5.2

standard deviations from its mean. It is reasonably obvious that g_1 does not significantly exceed (say twice) its standard error, whereas g_2 is much larger than twice its standard error. One of the advantages of the g statistics is that in large samples they are normally distributed around their respective parametric γ values, and the areas of the normal curve can be used as approximate estimates of their significance (Bennett and Franklin, 1954, p. 95).

13.7 Some Special Properties of the Normal Distribution

The normal distribution has a number of useful properties which are of very considerable value to an experimenter, so much so that it is worthwhile, in many cases, to choose a technique which generates this kind of frequency function. In addition to offering the usefulness of a rigorously prescribed procedure, the normal distribution, once established, may be used in the following manner.

Means of random samples drawn from a normal distribution with mean μ and standard deviation σ are normally distributed with mean μ and standard deviation $\sigma/\sqrt{n}$, where n is the sample size. This property may be generalized and the strict requirement of normality relaxed to the following extent.

If X has a distribution with mean μ and standard deviation σ, for which the moment-generating function exists, then the variable $t = (\bar{X} - \mu)\sqrt{n}/\sigma$ has a distribution which approximates the standard normal distribution as n becomes infinite (Hoel, 1947, p. 69). Another form of this statement leads to the Chebyshev inequality and also represents a considerable relaxation of the normality requirement; thus, if an arbitrary population distribution has a mean μ and finite variance σ^2, the distribution of the sample mean approaches the normal distribution with mean μ and variance σ^2/n as the sample size increases (Anderson and Bancroft, 1952, p. 71). This theorem[1] may be expressed in the form of a probability statement

$$P(|X - \mu| > \delta\sigma) \leq \frac{1}{\delta^2}$$

i.e., the probability that the absolute value of a deviation $X - \mu$ exceeds a constant δ times the standard deviation σ is less than or equal to the fraction $1/\delta^2$; by arranging δ small, the probability of a deviation can be controlled. Thus with a normal distribution with mean μ and variance σ^2,

$$P(|X - \mu| > 2\sigma) = 0.05$$

whereas if we relax normality and require solely finite variance, the statement becomes

$$P(|X - \mu| > 2\sigma) = \frac{1}{2^2} = 0.25$$

[1] The central-limit theorem, or law of large numbers.

TABLE 13.3
Areas of Normal Distributions between Certain Limits

Range	Area, %
$\mu \pm \sigma$	67.45
$\mu \pm 2\sigma$	95.45
$\mu \pm 3\sigma$	99.73
$\mu \pm 4\sigma$	99.994

The loss in efficiency is considerable, but the less one knows about one's observations, the greater the price one pays in interpreting them.

In practice this theorem, leading to the Chebyshev inequality, implies that although a variate may possess a nonnormal frequency distribution, unless it is a very peculiar type in which the second moment is not finite, random samples from this frequency function will yield means which tend to be normally distributed as the sample size increases; a sample size of, say, 50 items will, in many cases, ensure that means of such samples will exhibit a distribution which is closely normal (Hoel, 1947, p. 69).

The area of a normal distribution lying between specific ordinates, expressed as a function of the standard deviation on either side of the mean, is given in Table 13.3. If we take sample means $\bar{X}$ in place of the variate X, we can use μ and $\sigma/\sqrt{n}$, thus narrowing the interval $\mu + k(\sigma/\sqrt{n})$; by increasing n we may make as precise a statement about the interval in which $\bar{X}$ lies as we wish—or, at least, as we can afford.

Consider the distribution of medians illustrated in Fig. 5.4; the overall mean $\bar{X}$, which is an estimator of μ, is 3.34 phi, and the standard deviation $\hat{\sigma}$ is 2.44 phi. Then by substituting $\bar{X}$ for μ and $\hat{\sigma}$ for σ, we may infer that the majority of medians will fall in the intervals:

k	$\mu \pm k\sigma$	Percentage of medians
1	0.90–5.78	67.45
2	−1.54–8.22	95.45
3	−3.98–10.66	99.994

If we take samples and calculate the mean of each set of four samples, we can express the range as $\mu \pm k\sigma/\sqrt{n}$; that is, we can reduce the interval by one-half, and by increasing the number of observations entering the sample mean, we can effectively reduce this interval within any limits we please. Such a procedure rapidly becomes expensive, because a mean based on sample size $n = 16$ reduces the interval to only one-fourth its initial value.

As another example suppose we know that the frequency function of a variate X_i, where X_i is the Krumbein intercept sphericity, is approximately normally distributed with mean $\mu = 0.75$ and standard deviation $\sigma = 0.08$ (see Fig. 6.13 and subsequent discussion), and we measure sphericity of a small sample of 16 quartz grains, yielding $\bar{X} = 0.68$; then, if this sample is random, we can predict that 3 in 1,000 such samples will on the average lie outside the limits $\mu \pm 3\sigma/\sqrt{n}$, or 0.75 ± 0.06, or 0.69 to 0.81. From previous experiments, there seems little reason to question the normality of the frequency function, and so the assumption of randomness is violated. Indeed, we expect that natural processes of selective sorting on the basis of shape are most unlikely to be random, and so the process leading to this sample with mean sphericity 0.68 has led to a concentration of unusually low sphericity. The mean sphericity of arkoses is 0.733 based on a sample size of 900 ($\sqrt{n} = 30$); then, by using the large sample estimators as population parameters we expect 3 in 1,000 to lie outside the sphericity range 0.75 ± 0.008 or outside 0.742 to 0.758. Hence the arkoses have lower sphericity than we expect from random samples of this size taken from the specified population.

13.8 The Concept of Confidence Limits

We may now use the characteristics of the normal distribution to make probability statements concerning the variation likely to be observed in continued sampling. Perhaps it should be emphasized that no statement can be made about a single experiment but only about experiments which can be (and in general are) repeated a number of times. We can make use of the property of normal distributions that means of random samples from a normal distribution with mean μ and standard deviation σ will themselves be distributed in a normal distribution with mean μ and standard deviation $\sigma/\sqrt{n}$. From the table of areas of the normal distribution we may state that 95 percent of the area of the distribution will be contained in the interval $\mu \pm 1.96\sigma$, and therefore $\bar{X} - \mu$ will fall in the interval $\mu \pm 1.96\sigma/\sqrt{n}$ 95 percent of the time, on an average. Writing this as an inequality gives

$$-\frac{1.96\sigma}{\sqrt{n}} < \mu - \bar{X} < +\frac{1.96\sigma}{\sqrt{n}}$$

which, upon adding $\bar{X}$ to each side, becomes

$$\bar{X} - \frac{1.96\sigma}{\sqrt{n}} < \mu < \bar{X} + \frac{1.96\sigma}{\sqrt{n}}$$

Suppose, by way of example, that we have a variate X_i representing the long axis of quartz grains, where X_i is the logarithm of the diameter or is expressed in phi units; then X is approximately normally distributed, and let us suppose that, on the basis of a large sample, we have a firm estimate of $\sigma = 0.5$ phi. A sample of $n = 25$ is taken and yields a mean value 3.0 phi. The 95 percent confidence limits are

$$\bar{X} - 1.96 \frac{0.5}{5} < \mu < \bar{X} + 1.96 \frac{0.5}{5}$$

or

$$3.0 - 0.1 \times 1.96 < \mu < 3.0 + 0.1 \times 1.96$$

$$2.804 < \mu < 3.196$$

and we expect that random samples of size $n = 25$ will yield means which will fall in this interval 95 percent of the time. An alternative form of this statement of confidence intervals is somewhat more revealing of the nature of inferential statements. If random samples of size 25 are taken and each time the confidence limits are constructed as above, then on an average, 95 in every 100 such intervals will contain the true mean μ and 5 will not (Fig. 13.4). Any particular interval will or will not contain μ, and the statement is not concerned with a single interval; but if this statement is made about each interval, the claim will be incorrect 5 times in every 100 on an average.

13.9 Applications of Confidence Limits

An interesting example of the application of confidence limits is illustrated in Fig. 13.5. In an extensive experiment aimed at determining the most efficient sampling pattern to use in following changes in the size and shape properties of quartz grains in a quartzite, some 1,500 quartz grains were examined from five specimens from each of three blocks at three outcrops (Griffiths, 1959a); the long axis measured by the technique of Hulbe (1955) is approximately normally distributed, or at least the frequency function is the kind which obeys the central-limit theorem, and so we expect that random samples of size n will fall within certain confidence limits. Since there are five specimens from two outcrops and four from a third, we have fourteen samples of size $n = 100$. The grand mean 0.111 phi is an estimate of μ,

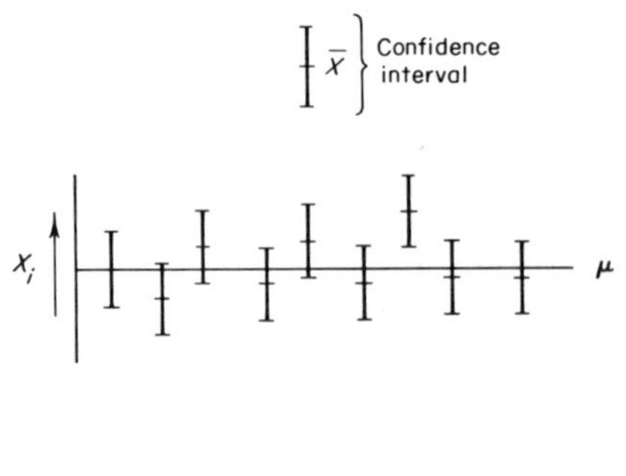

Figure 13.4 Confidence limits and their relationship to the "true mean" μ.

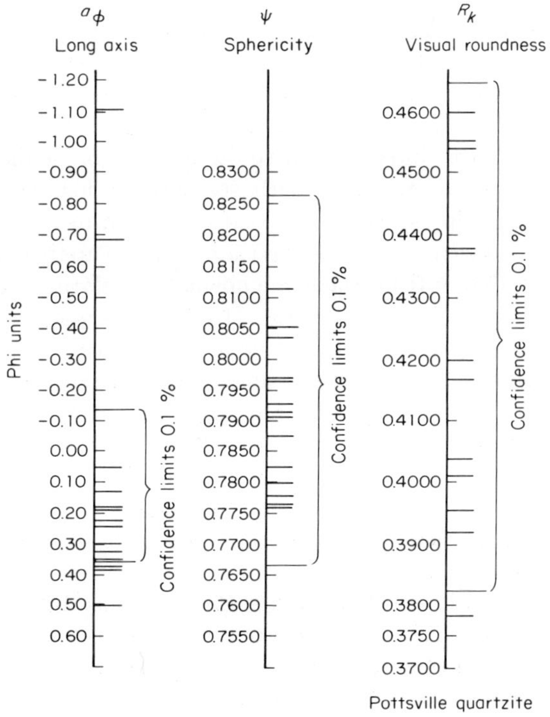

Figure 13.5 One-tenth percent confidence limits for sample means of quartz grains from Homewood quartzite.

and the confidence interval is $\bar{X} \pm \hat{\sigma}/\sqrt{n} = 0.24$. When the sample means are plotted on this scale (Fig. 13.5), 5 out of 14 fall outside the specified interval. Thus, either the assumption of normality or the randomness requirements or both have been violated. More extensive analysis on these outcrops suggests that the specimens are nonrandom, not that the frequency function is nonnormal. In the case of the Krumbein intercept sphericity on the same quartz grains there is no evidence against the hypothesis that these are random samples from a normal distribution. Finally, although we know that the frequency function of visual roundness is unlikely to be normal, the roundness of these same quartz grains, expressed in terms of confidence limits around the mean, yields 1 out of 14 outside the limits. It seems likely that quite serious departures from normality will be needed to upset these probability statements.

Other examples, in which the basic requirement of knowing the population value of σ may be relaxed, are described in Sec. 16.3.

13.10 Comparing Two Sample Means

Statements about a single value of a variate and the mean of a sample from a normal distribution in which μ and σ are known have been discussed, and the next procedure of interest requires the comparison of two sample means. Again we are assuming that the frequency function is normal and that μ and σ are known; and while these requirements are restrictive, they are, on occasion at least, approximately fulfilled and are likely to occur more frequently as more investigators add to our knowledge of the frequency functions likely to arise in studies of sedimentary rocks. They may be relaxed quite often with some small loss in efficiency by using a different approach (see Student's t distribution in Sec. 16.4).

Let us now suppose that we have analyzed two samples which we suspect are drawn from the same normal population and we wish to test the difference between the two means. In fact, we may consider a real situation in which the long axis of white quartzite pebbles (expressed in phi units) represents the random variate X_i and the two means arise because the outcrop face was sampled by two different sampling plans. The frequency function of the variate X_i is known to be approximately normal; and the appropriate statistics are $\bar{X}_1 = -4.887$, $\hat{\sigma}_1 = 0.8516$, $n_1 = 336$ for stratified sampling and $\bar{X}_2 = -5.682$, $\hat{\sigma}_2 = 0.8070$, and $n_2 = 300$ for spot sampling. We now wish to know whether these represent two random samples from the same normal population.

First we shall express the means as deviations from the population mean; i.e., the difference between the means is $(\bar{X}_1 - \mu) - (\bar{X}_2 - \mu)$; and it can be seen that μ disappears from the expression, leaving $\bar{X}_1 - \bar{X}_2$. If these two are not different, we expect that $\bar{X}_1 - \bar{X}_2 = 0$, and so we wish to express confidence limits which are centered around zero and which will indicate how large a difference is to be expected in random sampling from a normal population with $\mu = 0$. It is necessary to obtain an appropriate estimate of the standard deviation σ from the two estimates we have and then convert it to the standard error.

The variance of the sum or difference of two random variates is equal to the sum of the variances of the two variates together with the covariance, as follows:

$$\sigma_{\bar{X}_1 \pm \bar{X}_2}^2 = \sigma_{\bar{X}_1}^2 + \sigma_{\bar{X}_2}^2 \pm 2\rho\sigma_{\bar{X}_1\bar{X}_2}$$

If the random selections for each sample are independent, then $\rho = 0$, and the expression becomes

$$\sigma_{\bar{X}_1 \pm \bar{X}_2}^2 = \sigma_{\bar{X}_1}^2 + \sigma_{\bar{X}_2}^2$$

$$\sigma_{\bar{X}_1}^2 = \frac{\sigma_{X_1}^2}{n_1} \qquad \text{and} \qquad \sigma_{\bar{X}_2}^2 = \frac{\sigma_{X_2}^2}{n_2}$$

$$\sigma_{\bar{X}_1 \pm \bar{X}_2}^2 = \frac{\sigma_{X_1}^2}{n_1} + \frac{\sigma_{X_2}^2}{n_2}$$

$$\sigma_{\bar{X}_1 - \bar{X}_2} = \left(\frac{\sigma_{X_1}}{n_1} + \frac{\sigma_{X_2}}{n_2}\right)^{1/2}$$

In practice, then, we may write that a test of the difference between two means from the same normal population can be performed by calculating the test criterion

$$t = \frac{\bar{X}_1 - \bar{X}_2}{(\sigma_{X_1}{}^2/n_1 + \sigma_{X_2}{}^2/n_2)^{1/2}}$$

In our example

$$t = \frac{-4.887 - (-5.682)}{(0.7253/336 + 0.6510/300)^{1/2}}$$

$$= \frac{0.795}{(0.002159 + 0.001822)^{1/2}}$$

$$= \frac{0.795}{0.0631} = 12.6$$

Then if these are indeed two random samples from the same normal population, such a difference will be 12.6 times its standard error and we evaluate the meaning as shown in Fig. 13.6.

It is quite evident that under our suppositions such a value is likely to be rare; indeed, the area of the curve between $\mu + 12.6\sigma_{\bar{X}}$ and $\mu - 12.6\sigma_{\bar{X}}$ is 99.99999999 and beyond $\mu \pm 12.6\sigma_{\bar{X}}$ is 0.0000000001. That is, we expect to fall outside these limits $\pm 12.6\sigma_{\bar{X}}$ less than once in every 10^{10} trials. Either such a rare event is a very unusual accident, or else we must reject some or all of our suppositions that these are two random samples from the same normal population. Since we know that they are from the same population, and normality is an adequate approximation and even wide departures from it will have only minor effects, we are led again to question the randomness of the sampling. A more extended discussion is included in the original article (Griffiths, 1959a).

Before we can comment on the nature of significance testing, we must first set up a null hypothesis H_0 and then deduce from it certain likely circumstances which will arise upon performing a prescribed experiment. Then the experiment is performed and the set of observations compared with those predicted from the null hypothesis; if the experimental findings are unlikely

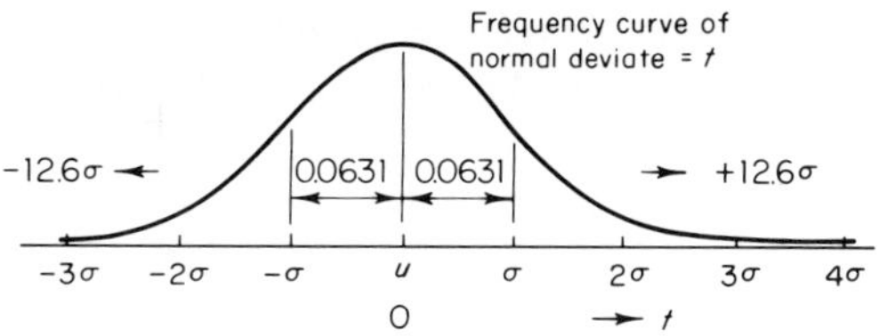

Figure 13.6 Diagram representing significance of $t = 12.6$.

on the basis of the hypothesis, we must decide whether to believe that an unusual event has occurred or to reject the hypothesis. This decision cannot be made without risk, and so we always formulate the result as a probability statement. Second, the experimental result may reject the hypothesis, but if it does not, the converse is that there is no evidence against the hypothesis. The null hypothesis is always capable of rejection but never of more than tentative acceptance. An experiment is successful when it permits a very rigorous test of the null hypothesis and is capable of rejecting a variety of circumstances at one time.

The establishment of efficient experiments which will really test any null hypothesis requires considerable skill and represents quite a challenge (see Chap. 22). When there exist ambiguities in the hypothesis, there will always be ambiguities in the conclusions based on tests of the hypothesis. One of the most important requirements in the use of statistics is to be able to choose the most appropriate analytical tool to test any specific hypothesis; broad, general questions lead to vague, ambiguous answers and are very poor substitutes for exactness and rigor.

13.11 The Log-normal Distribution

If an observation X depends on the sum of the effects of a large number of small causes acting at random, and if each effect is independent of X, the distribution is likely to be normal. If it is not normal, we assume that this is because X is not measured in the right way. We find that $f(X)$ is normally distributed, and we therefore transform our observations of X by converting them into observations of $f(X)$ (Gaddum, 1945).

The log-normal frequency distribution has been proposed as appropriate for a number of variates in the geosciences such as grain size (Udden, 1898; Krumbein, 1938, 1954b), abundance of trace elements (Ahrens, 1954c, chap. 4), and the pennyweight value of small samples of gold ore (Sichel, 1949; Krige, 1951; Ross, 1950; Slichter, 1960); there are also many examples in other fields (Gaddum, 1945; Kapteyn, 1916; Aitchison and Brown, 1957). In each case it is possible to cite critics of the proposal. Indeed, it is impossible to prove beyond a doubt that some variate is distributed in the form of any definite frequency function. For example, when Student's proposal of the Poisson model was challenged (Whitaker, 1914), he subsequently explained that while an exact fit is unlikely in practice, the Poisson model is, nevertheless, a useful approximate model. The only logical justification for adoption of some model frequency function is that an investigator can advance toward his goal better with than without it. In this sense the log-normal frequency function appears to be a useful representation of the variation of particle sizes in sediments. Since this is one of the most valuable applications and the one which is most familiar, it will be well to examine in some detail the implications of adopting the model.

Justification for adopting a model frequency function may be approached along three main avenues. The first is to postulate theoretically the physical conditions characterizing the (natural) process which leads to the distribution of the variate and then to show that these are analogous to the requirements which generate some specific frequency function. An example, using the Poisson model as a test of satisfactory preparation of a mineral mount in accessory-mineral investigations, is described in Sec. 14.7. An alternative route is to examine the frequency distribution of the variate in a relatively large sample and then show by appropriate tests, such as moment analysis and/or a chi-square goodness-of-fit test, that the empirical frequency function is a good fit to some known distribution function. The difficulty in this empirical approach is that the sample should be a random sample, and if the selection is not random, the frequency function will display bias and will not fit the expected frequency function.

Finally the deductive approach can use both of these avenues: a suitable theoretical model is predicated on the basis of what is known of the variation in a variate, and then predictions of sampling distributions of statistical characteristics are made and the predictions tested by sampling. It is, of course, nearly always possible to adjust the sampling arrangements so that a frequency distribution of a variate meets any required specification, and so the final justification is largely dependent on whether it can be agreed that the model and the argument leading to it and from it are indeed an adequate representation of the observations. Arguments about the normal model, extending from Poincaré's classical criticism (1892, quoted in Gaddum, 1945, p. 463), which is cited by every skeptic of the existence of normality, all the way down to modern times (e.g., Bagnold, 1941; Doeglas and Smithuyzen, 1941; Doeglas, 1946; Irani, 1959), reflect the fact that proof is indeed difficult and perhaps impossible to attain in a finite number of steps. Again, then, let us follow Student and insist that what we seek is an adequate approximation which will enable us to advance our purpose and that when an apparently superior alternative is proposed, it will be given due consideration in the same manner and as far as possible without prejudice.

13.12 Some Theoretical Considerations Suggesting the Log-normal Model for the Particle-size Frequency Function

Measurement of grain size in sedimentary rocks is usually a means to an end, and the end in view is generally an attempt to predict, from the size frequency distribution, the conditions of formation of the sediment. Thus grain size is presumed to be related to current velocity, and extensive experiments have been made to elucidate the relationships between size and velocity (e.g., Gilbert, 1914; Nevin, 1946). Descriptions of the relationships between current velocity and particle size are usually idealized as Stokes' law and the impact law, the former describing the relationship in the finer grain sizes under laminar flow and the latter describing the relationship in the region of coarser particles and under turbulent flow (Nevin, 1946, pl. 1). Both "laws" are linearized by the logarithmic transformation, as shown in Table 13.4.

In the regions where these laws hold there is a direct one-to-one relationship between the logarithm of the velocity and the logarithm of the radius; and if the radius is log-normal, so is the velocity, and vice versa.

The justification for assuming that the log-normal model is adequate depends on how realistic the log-normal law is in describing the variation expected in current velocities under ordinary conditions of fluid flow. Since the two laws cover limited portions of the size-velocity relationship and there exists a transition (saltation) region between them, the model is only an approximation; and its limitations depend, in the first place, on whether the transition region is small enough to be ignored for most practical purposes, i.e., whether the change from laminar flow to turbulent flow is abrupt under most conditions of natural fluid flow (see Nevin, 1946, pl. 1). If this is considered to be a reasonable assumption, we may examine the further implications of the log-normal model in terms of velocity and we may transform the velocity from its logarithm back to the value on the original scale and see what the resulting curve portends. An example of what this means in terms of grain size is given in Griffiths et al. (1956, pt. 2, figs. 2, 3, pp. 15, 16). The resulting curve will be of the form shown in Fig. 13.7.

This curve implies that there are no velocities of zero or less; it is obvious that under natural conditions some currents cease to flow for longer or shorter periods, but this would not be considered as "natural" fluid flow. Very high velocities are also rare, and the higher they are, the rarer they become, which appears to be a reasonable representation. The larger number of velocities lies below the mean, and the curve is strongly skew toward high velocities. This representation is certainly a reasonable approximation to conditions in most rivers of water, air, ice, or even conditions of gravitational sliding, so that it does appear to be a fair approximation to fluid flow under most natural

TABLE 13.4
Relationship between Current Velocity and Size, Measured as Particle Radius

Law	Relationship	Logarithmic relationship
Stokes'	$V = Cr^2$	$\log V = \log C + 2 \log r$
Impact	$V^2 = Cr$	$2 \log V = \log C + \log r$

Note: V = velocity, cm/sec; C = constant; r = radius, cm.

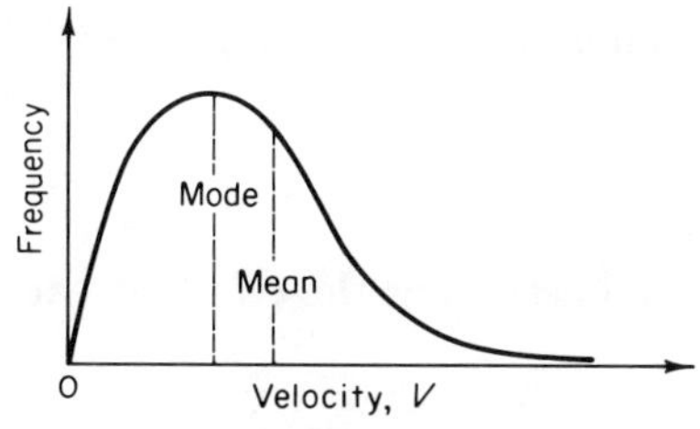

Figure 13.7 Hypothetical frequency distribution of velocity in fluid flow under natural conditions.

conditions. Actual observations can be used to check the form of the curve in specific cases (Krumbein, 1938; Searcy, 1960).

Hence we are led to believe that the log-normal model of velocities, and therefore the log-normal model of particle size, is a reasonable approximation on the basis of hypothetical physical conditions. If we now attempt to examine some empirical size frequency distributions to determine whether they agree with these theoretical expectations, we must be careful to fulfill the model requirements. Therefore it is necessary to examine what these requirements are; fortunately, they are exceedingly well documented and have been used to cover many situations which appear analogous to that of particle size (e.g., Aitchison and Brown, 1957; Kapteyn, 1916; Gaddum, 1945).

One of the essential requirements is that at the jth step the change in the variate be a random proportion of a function $\phi(X_{j-1})$ of the value already attained; thus, following Kapteyn (1916) and Aitchison and Brown,[1]

$$X_j - X_{j-1} = \epsilon_j \phi(X_{j-1}) \tag{13.8}$$

where the set $\{\epsilon_j\}$ is mutually independent and is also independent of the set $\{X_j\}$. The change at any step is therefore not independent of the value attained except for the cases $\phi(X) = 1$ and $\phi(X) = 0$. We may specify the special case where $\phi(X) = X$, that is, the change in the variate is a random proportion of the momentary value of the variate. The law of proportionate effect describes this relationship as follows.

A variate subject to a process of change is said to be described by the law of proportionate effect if the change in the variate at any step of the process is a random proportion of the previous value of the variate, and we may write

$$X_j - X_{j-1} = \epsilon_j X_{j-1} \tag{13.9}$$

or

$$\frac{X_j - X_{j-1}}{X_{j-1}} = \epsilon_j \tag{13.10}$$

[1] The following derivation is taken by permission from J. Aitchison and J. A. C. Brown, "The Lognormal Distribution," pp. 22–23, Cambridge University Press. 1957,

Summing over all n values of the variate,

$$\sum_{j=1}^{n} \frac{X_j - X_{j-1}}{X_{j-1}} = \sum_{j=1}^{n} \epsilon_j \tag{13.11}$$

and assuming the effect at each step to be small,

$$\sum_{j=1}^{n} \frac{X_j - X_{j-1}}{X_{j-1}} \approx \int_{X_0}^{X_n} \frac{dX}{X} = [\log X]_{X_0}^{X_n}$$

$$= \log X_n - \log X_0 \tag{13.12}$$

Thus, returning to Eq. (13.10), we have

$$\log X_n = \log X_0 + \epsilon_1 + \epsilon_2 + \cdots + \epsilon_n \tag{13.13}$$

By the addition form of the central-limit theorem, $\log X_n$ is asymptotically log normally distributed in two-parameter form.

It is, of course, possible to generalize the conditions and use distributions with more than two parameters, but it seems adequate for the present to adopt this model for the velocity and equivalent particle-size frequency distributions. It should be emphasized that for this model to be reasonably representative the variate used to measure particle size should reflect variation solely in detrital constituents,[1] and it will almost certainly be necessary to assume that neither composition nor fabric of the aggregate affects the (measured) variation in the variate; in other words, this model will be applicable only to one-component systems in which the variation is randomly distributed. This effectively removes indirect size-measurement techniques and probably thin-section technique unless the material subjected to analysis is approximately monomineralic and displays no systematic arrangement of the constituents in terms of the size measure. In fact, we are led quite clearly to the measurement of size (and/or shape) by direct means.

If we now proceed to examine some of the attempts to determine particle-size distributions, we find that some are indeed log normally distributed but that there are many more which fail to achieve log normality. A large number of the failures may be attributed to incorrect choice of size measure because the material is polycomponent, as indeed are most sedimentary rocks. But even where the sediments approach monomineralic compositions, failure to achieve the normal model will still occur due to nonrandomness in the sampling (Irani, 1959). Examples of the former are plentiful in cases where graywackes and arkoses have been subjected to indirect size measurement (Doeglas, 1946; Bressler, 1951); even when the composition factor is reduced in importance, as in many beach and dune sands,[2] unless adequate precautions are taken to ensure random sampling with constant probability, the frequency

[1] Actually law of wear can also be expressed in log-normal form (Epstein, 1947).
[2] And artificial aggregates (Ames, Irani, and Callis, 1959; Irani, 1959, 1961).

distributions will again show significant departures from normality (Bagnold, 1941; Otto, 1939). Where adequate sampling arrangement is adopted and the sediment or aggregate is monodisperse, very many materials yield size frequency distributions which are log-normal (Hatch, 1933; Hatch and Choate, 1929).

It can be very simply demonstrated that even when reasonable precautions are taken to keep composition constant, if the sampling departs from randomness, the frequency distributions will depart from log normality; whereas if the same population is sampled randomly and the measurement procedure is performed in the same way, there is no obvious departure from log normality even under stringent testing (Griffiths, 1959a, b).

Some examples of size frequency distributions which appear to fulfill the model requirements have already been described (Secs. 5.5 et seq.), and the suitability of the log-normal model has been effectively substantiated. If we accept this model as a reasonable approximation, we may then use the known properties of normal distributions to predict the expected outcomes of experimental analysis and effectively use the testing procedure as a check on the randomness of the sampling. In fact, as pointed out on page 234, given the model frequency function we can arrange a theoretically random sampling process and determine from the bias revealed in the measurements the pattern of variability in the population, i.e., determine the structure of the material by sampling.

In random samples from a normal frequency function the mean is independent of the variance, and so we can postulate that under this hypothesis there should be no relationship between mean and variance, and particularly between statistical estimators of these parameters. Hence we are led to study relationships between mean and variance (e.g., size and size sorting; see Chap. 15) to characterize the processes of sedimentation in terms of their characteristic biases. It is noteworthy that in three regions of this relationship (Fig. 15.1) even indirect measurement suffices to produce normal distributions; and in each case the sediments approach one-component systems, the clays, the quartzites (including beach and dune sands), and some gravels containing a preponderance of one kind of element (particle). Again, when thin-section measurement is examined in the same way, the mean is found to be almost always independent of the variance (Griffiths, 1958b; Griffiths et al., 1956; and Sec. 5.7), suggesting that under these conditions of measurement we are approximating random samplings from a normal parent population.

Finally, when direct measurement is adopted, the relationships between mean and variance can be used to illuminate the structural arrangement of size within the population by means of the quality-control-chart representation (Fig. 2.5).

It seems, then, that adoption of the log-normal function as a model frequency distribution in the case of grain size, when adequate precautions are taken in performing the experiment (see Griffiths, 1962a), does materially assist us toward our objective of understanding the processes of sedimentation. The essential conditions required in achieving the log-normal model are:

1 The size measure chosen is physically simple to interpret, e.g., an operationally defined intercept of a single homogeneous mineral constituent. For this requirement to be fulfilled the size measure chosen must be directly interpretable in terms of the variate (concept) it is supposed to reflect, e.g., which is equivalent to size, magnitude, which in turn is equivalent to velocity.

2 Probability sampling is attained; i.e., simple random sampling requires constant probability from trial to trial. For stratified populations this requires one of two alternatives, either "channel" samples, which fulfill simple random sampling requirements by sampling the entire population each time, or, for stratified random sampling, random samples from within single homogeneous sedimentation units.

It is of some interest to note that if the logarithm of size is indeed normal, any linear compound of random normal variates is also normally distributed; e.g., suppose that the volume V of irregular particles is expressed as a function of the axes a, b, c of this particle

$$V = abc$$

Then

$$\log V = \log a + \log b + \log c$$

and if $\log a$, $\log b$, $\log c$ are normally distributed, so is $\log V$. This is the basis of relating log volume to axes measured in phi units, and variation in log volume is very closely related to variation in the axes (Griffiths and Smith, 1964).

Similarly, the investigations of Sichel (1949), Ross (1950), and Krige (1951) have established the log-normal model as a suitable population frequency distribution of pennyweight value of gold samples.

So far the claim that log normality may be the best model for the abundance of a trace element in the earth's crust has not been adequately established, largely because of inadequate sample sizes and lack of attention to the problem of attaining a random sample (Ahrens, 1954a, b, 1957; Chayes, 1954b; Miller and Goldberg, 1955; Miller and Kahn, 1962, pp. 33, 34). At present, the model frequency distribution which most closely expresses the abundance of trace elements is quite likely to be Poisson rather than log-normal (Secs. 14.4 et seq.), and to determine which model is "best" will require a thorough investigation of a few representative trace elements with very careful control of the sampling program to ensure random samples from "homogeneous populations." Until this desirable status has been achieved, the interpretation of frequencies of trace elements will be subject to an unknown degree of inexactitude[1] (Vistelius, 1960).

[1] There is some difficulty too in deciding the measurement level attained by current procedures for determining the concentration of trace elements; the scale may lie somewhere between ordinal and interval, and exact treatment of such data is difficult until these questions have been clarified (Landy, 1961).

13.13 An Alternative Model

Let us suppose that log normality is unsuitable as a model representing the frequency distribution of particle size; we may use as the basis of an alternative approach the fact that the distributions of means of samples tend not to converge as required by the central-limit theorem. It is quite possible to explain this lack of convergence on the basis of the fact that sets of grains which form "natural" subpopulations (e.g., layers and laminae or, more exactly, sedimentation units; Otto, 1938; Apfel, 1938) are not random but systematically biased samples from the parent population, which is nevertheless log-normal. On the other hand, it is useful to examine an alternative model which may also embrace this situation.

Let it be assumed that the variate representing log particle size is composed of a ratio of two random normal variates defined as (Curtiss, 1941; Hoel, 1947; Aitken, 1957, p. 70)

$$x = \frac{X - \mu_x}{\sigma_x} \qquad \text{and} \qquad y = \frac{Y - \mu_y}{\sigma_y}$$

Then if x and y are randomly selected, $Z = x/y$ is a variate possessing a Cauchy distribution

$$f(z) = \frac{1}{\pi(1 + z^2)} \qquad -\infty < z < +\infty$$

For this distribution the mean of a sample of n observations is no better than a single observation for estimating the value of the population mean μ; the median is the maximum-likelihood estimator of the central tendency, and no moments of higher order than the first exist; in particular, the variance is infinite, and the central-limit theorem does not apply. A representative Cauchy frequency curve is illustrated in Fig. 13.8, and its superficial resemblance to the normal distribution is obvious.

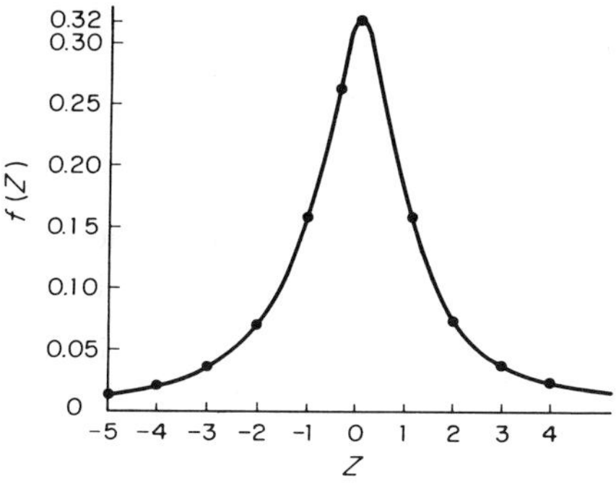

Figure 13.8 Example of a frequency distribution following the Cauchy function

$$f(z) = \frac{1}{\pi(1 + z^2)}$$

Variates which follow the Cauchy frequency function are not nearly so common as those which approximate the normal, binomial, Poisson, etc.; but they do exist, and whenever a ratio of two variates is contemplated as the basic variate, it is well to realize that if the original variates are normal, the new one may be best represented by a Cauchy distribution.

One of the main disadvantages of adopting a Cauchy model is the difficulty that arises in the problem of estimation because of the lack of convergence and the inapplicability of the central-limit theorem.

14 Discrete frequency distributions: the binomial, Poisson, etc.

. . . situations of the following type will be considered:

1. There exists a conceptually repeatable experiment, a single trial of which results in either an S or an F. [Where S = success and F = failure.]
2. On any single trial of the experiment the probability of obtaining an S is equal to p. That is, $\Pr(S) = p$ and $\Pr(F) = 1 - p = q$.
3. n independent trials of the experiment will be performed, and the number of S's occurring will be recorded.

These three conditions define the *binomial model*.

P. J. McCarthy, 1957

14.1 Introduction

The normal distribution is the model most frequently used for measurement data, principally because it is most suitable for representing observations related to a continuous scale. In other types of experiments the principal operation is a breakdown of objects into discrete classes, followed by a count of frequencies of the objects allocated to each class. Hence the scale is discontinuous, and a model which embraces this aspect is more appropriate.

A number of models are available which may be related to different experimental situations, such as the binomial, Poisson, negative binomial, and contagious distributions; these will be considered as suitable alternatives for counts of observations subdivided into mutually exclusive classes. These models represent the sampling distributions of observations estimating mineral composition by point count in thin section or of counts of proportions of accessory heavy minerals and so on (Griffiths, 1960a, d).

Quotation: From "Introduction to Statistical Reasoning," p. 177, by P. J. McCarthy.

14.2 The Binomial Distribution

The binomial expansion is perhaps the best known of these distributions and is generally written as

$$(p + q)^n = p^n + np^{n-1}q^1 + \frac{n(n-1)}{2!} p^{n-2}q^2 + \cdots + \frac{n!}{(n-r)!r!} p^r q^{n-r} \tag{14.1}$$

This expression, suitably interpreted, represents a convenient procedure for counting the incidence of events of a certain kind and may be developed from purely algebraic considerations; however, as a probability distribution, it is perhaps more illuminating to express the experimental situation as representing the description of the frequency distribution of tosses of a fair coin. In this case an event such as the occurrence of a head equals a success S. A head and a tail (not a head) are the only two possible outcomes of the experiment; and as the probability of either event is equally likely, it is customary to write the binomial relevant to this situation as $(p + q)^n = (\frac{1}{2} + \frac{1}{2})^n$. The probability of occurrence of the event is thus $p = 0.5$, and the probability of its not happening is $1 - p = q = 0.5$.

Let us now toss two coins where the relevant binomial is $(p + q)^n = (0.5 + 0.5)^2$, where n is the number of trials. The possible outcomes are given in Table 14.1, and the outcomes in terms of the frequency of occurrence of heads are: there is one way of obtaining two heads, two ways of obtaining a head and a tail (irrespective of order), and one way of obtaining no heads. It is possible to count the outcomes, using four coins, as shown in Table 14.2, but it rapidly becomes a tedious procedure for large numbers of coins (or large numbers of throws). The expansion of the relevant binomial helps to ex-

TABLE 14.1
Frequency of Number of Heads in Tossing Two Fair Coins

Coin				
1	2	Probability	No. of heads	Frequency of occurrence
H	H	$\frac{1}{2}\cdot\frac{1}{2} = \frac{1}{4}$	2	1
H	T	$\frac{1}{2}\cdot\frac{1}{2} = \frac{1}{4}$	1	2
T	H	$\frac{1}{2}\cdot\frac{1}{2} = \frac{1}{4}$	1	
T	T	$\frac{1}{2}\cdot\frac{1}{2} = \frac{1}{4}$	0	1

TABLE 14.2
Frequency of Occurrence of Heads in Tossing Four Coins

Coin				Probabilities, head	No. of heads	Frequency of occurrence
1	2	3	4			
H	*H*	*H*	*H*	$(\frac{1}{2})^4(\frac{1}{2})^0 = \frac{1}{16}$	4	1
H	*H*	*H*	*T*	$(\frac{1}{2})^3(\frac{1}{2})^1 = \frac{1}{16}$	3	
H	*H*	*T*	*H*	$(\frac{1}{2})^3(\frac{1}{2})^1 = \frac{1}{16}$	3	
H	*T*	*H*	*H*	$(\frac{1}{2})^3(\frac{1}{2})^1 = \frac{1}{16}$	3	4
T	*H*	*H*	*H*	$(\frac{1}{2})^3(\frac{1}{2})^1 = \frac{1}{16}$	3	
H	*H*	*T*	*T*	$(\frac{1}{2})^2(\frac{1}{2})^2 = \frac{1}{16}$	2	
H	*T*	*H*	*T*	$(\frac{1}{2})^2(\frac{1}{2})^2 = \frac{1}{16}$	2	
H	*T*	*T*	*H*	$(\frac{1}{2})^2(\frac{1}{2})^2 = \frac{1}{16}$	2	6
T	*H*	*T*	*H*	$(\frac{1}{2})^2(\frac{1}{2})^2 = \frac{1}{16}$	2	
T	*T*	*H*	*H*	$(\frac{1}{2})^2(\frac{1}{2})^2 = \frac{1}{16}$	2	
T	*H*	*H*	*T*	$(\frac{1}{2})^2(\frac{1}{2})^2 = \frac{1}{16}$	2	
H	*T*	*T*	*T*	$(\frac{1}{2})^1(\frac{1}{2})^3 = \frac{1}{16}$	1	
T	*H*	*T*	*T*	$(\frac{1}{2})^1(\frac{1}{2})^3 = \frac{1}{16}$	1	
T	*T*	*H*	*T*	$(\frac{1}{2})^1(\frac{1}{2})^3 = \frac{1}{16}$	1	4
T	*T*	*T*	*H*	$(\frac{1}{2})^1(\frac{1}{2})^3 = \frac{1}{16}$	1	
T	*T*	*T*	*T*	$(\frac{1}{2})^0(\frac{1}{2})^4 = \frac{1}{16}$	0	1

tend the counting reasonably simply; thus

$$n = 1 \qquad p + q$$

$$n = 2 \qquad p^2 + 2pq + q^2$$

$$n = 3 \qquad p^3 + 3p^2q + 3pq^2 + q^3$$

$$n = 4 \qquad p^4 + 4p^3q + 6p^2q^2 + 4pq^3 + q^4$$

and when the coefficients of the terms are arranged in the form of Pascal's triangle, the rule for counting is fairly obvious:

```
                1       1
            1       2       1
        1       3       3       1
    1       4       6       4       1
1       5      10      10       5       1
1   6      15      20      15       6       1
```

This procedure is, however, confined to the case where the probability of an event's happening, i.e., a head on a tossing of a coin, or, more generally, a success, is equal to the probability of its not happening. Thus the probability of occurrences of a 5 on a fair die is $\frac{1}{6}$, and the probability of its not occurring is $\frac{5}{6}$, leading to an asymmetric binomial form, and the counting becomes slightly more complex. Fortunately, expansion of the relevant binomial is also representative of this kind of experiment, so that the general rules for finding the coefficients of the terms in the appropriate binomial lead to the frequency of occurrence of the event. The general form of this expansion is

$$C_r^n p^r q^{n-r} = \frac{n!}{(n-r)!r!} p^r q^{n-r} \tag{14.2}$$

where C_r^n represents the combinations of n things taken r at a time and is the coefficient representing the frequency of occurrence of the event r times; p^r expresses the probability of the event's occurring, and q^{n-r} the probability of the event's not occurring.

Suppose that the objective of the experiment is to estimate the number of zircons in an accessory heavy-mineral residue or the number of points on

TABLE 14.3
Evaluation of Binomial $N(p+q)^n = 1{,}000(0.4+0.6)^5$

(1) No. of events r	(2) C_r^n	(3) p^r	(4) q^{n-r}	(5) $C_r^n p^r q^{n-r}$	(6) 1,000× col. 5
0	$\frac{5!}{5!0!} = 1$	0.4^0	0.6^5	$1 \times 1 \times 0.07776 = 0.07776$	77.76
1	$\frac{5!}{1!4!} = 5$	0.4^1	0.6^4	$5 \times 0.4 \times 0.1296 = 0.2592$	259.20
2	$\frac{5!}{2!3!} = 10$	0.4^2	0.6^3	$10 \times 0.16 \times 0.216 = 0.3456$	345.60
3	$\frac{5!}{3!2!} = 10$	0.4^3	0.6^2	$10 \times 0.064 \times 0.36 = 0.2304$	230.40
4	$\frac{5!}{4!1!} = 5$	0.4^4	0.6^1	$5 \times 0.025 \times 0.6 = 0.0768$	76.80
5	$\frac{5!}{5!0!} = 1$	0.4^5	0.6^0	$1 \times 0.01024 \times 1 = 0.01024$	10.24
Total				1.00000	1,000.00

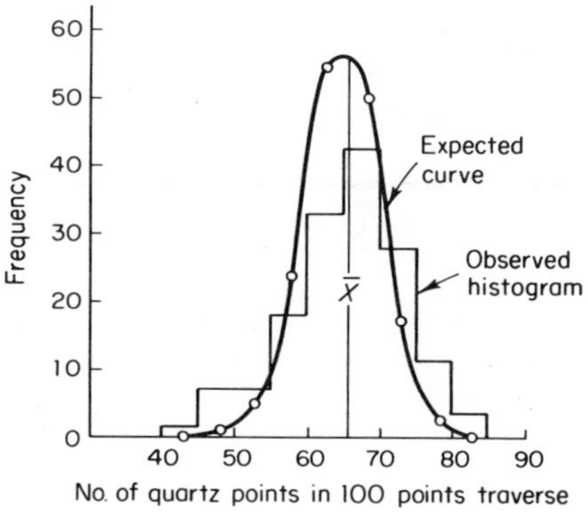

Figure 14.1 Frequency distribution of quartz proportion in thin sections of barren Salt Wash sandstones from the Colorado plateau.

$$(p + q)^n = (0.65 + 0.35)^{100}$$
$$\bar{X} = 65.4$$

quartz in the traverses of a thin section by point counting; it is known that the constituent (event, success, zircon grain, or point on quartz) is binomially distributed with $p = 0.4$, and a traverse comprises five grains or five points, respectively; we wish to know the probability distribution of the outcome of the event (zircon or quartz point, respectively). In all it is intended to record 1,000 (N) traverses of five (n) points (or grains), and it is required to determine the theoretical or expected frequency distribution of the event (quartz point or zircon grain) per traverse throughout the 1,000 traverses. The relevant binomial is

$$N(p + q)^n = 1{,}000(0.4 + 0.6)^5$$

The expansion may be evaluated as shown in Table 14.3.

Therefore we expect, under the conditions of the experiment, that 77.76 traverses of five points will be nonquartz, 259.20 traverses will contain one point on quartz (or one zircon), and so on. In practice, a whole number of traverses will contain 0 quartz points, 1 quartz point, etc., but the observed number should approach the expected as N grows large. Again, in practice our problem is generally very different; we may well suspect that the constituent is binomially distributed, and we can fix $n = 5$ and $N = 1{,}000$, but p is generally unknown and must be estimated from the data.

As a more realistic example the observations in Table 14.4 were obtained by point counting the quartz grains in 25 thin sections of barren Salt Wash sediments (Griffiths et al., 1956, fig. 1, p. 13). The sampling process comprised six traverses of 100 points each for each thin section, leading to $n = 100$, $N = 150$. The moment measures and statistical estimates are calculated in Table 14.4, and the frequency distribution is displayed in Fig. 14.1.

The relevant parameters based on the first four moments which characterize a binomial distribution are listed in Table 14.5.

The observed values for these statistics computed by moments and those calculated from the formula are listed in Table 14.6, and it can be seen that the variance of the observations is about 2.8 times larger than the expected

TABLE 14.4
Frequency of Quartz in Thin Sections of Barren Salt Wash Sandstones from the Colorado Plateau[a]

Class limits	Frequency f	d	fd	fd^2	fd^3	fd^4	$f(d-1)^4$
40–45	1	−4	−4	16	−64	256	625
45–50	7	−3	−21	63	−189	567	1,792
50–55	7	−2	−14	28	−56	112	567
55–60	18	−1	−18	18	−18	18	288
60–65	33	0	0	0	0	0	33
65–70	42	1	42	42	42	42	0
70–75	28	2	56	112	224	448	28
75–80	11	3	33	99	297	891	176
80–85	3	4	12	48	192	768	243
Total	150		86	426	428	3,102	3,752

$$f(d-1)^4 = fd^4 - 4fd^3 + 6fd^2 - 4fd + f$$

$$3{,}752 = 3{,}102 - 1{,}712 + 2{,}556 - 344 + 150 = 3{,}752$$

$$n_1 = 0.5733 \qquad n_2 = 2.8400 \qquad n_3 = 2.8533 \qquad n_4 = 20.6800$$

$$\bar{X} = \bar{X}' + cn_1 = 62.5 + 5 \times 0.5733 = 65.365 = np$$

$$m_2 = c^2(n_2 - n_1^2) = 25(2.84 - 0.3287) = 25 \times 2.5113 = 62.7825 = npq = \hat{\sigma}^2$$

$$\hat{\sigma} = npq = 7.9235$$

$$m_3 = c^3(n_3 - 3n_2n_1 + 2n_1^3) = 125(2.8533 - 4.8845 + 0.3768)$$

$$= 125(-1.6544) = -206.8000$$

$$\sqrt{b_1} = \frac{m_3}{m_2^{3/2}} = -\frac{206.8}{497.457} = -0.4157$$

$$m_4 = c^4(m_4 - 4n_3n_1 + 6n_1^2n_2 - 3n_1^4)$$

$$= 625(20.68 - 6.5432 + 5.6010 - 0.3240)$$

$$= 625 \times 19.4138 = 12{,}133.6250$$

$$b_2 = \frac{m_4}{m_2^2} = \frac{12{,}133.625}{3{,}941.6423} = 3.0783$$

[a] See Fig. 14.1.

TABLE 14.5
Summary Statistics and Parameters for the Binomial Frequency Distribution

Mean	$= \mu = np$	(14.3)
Variance	$= \sigma^2 = npq$	(14.4)
Skewness	$= \beta_1 = \dfrac{(q-p)^2}{npq}$	(14.5)
Kurtosis	$= \beta_2 = 3 + \dfrac{1-6pq}{npq}$	(14.6)

binomial; similarly, the asymmetry of the observations is some 40 times larger than expected, whereas the kurtosis is very similar. On the basis of comparison of statistics this frequency distribution departs widely from a binomial with the same mean.

From the expected values of the parameters of a binomial in Table 14.5 it can be seen that an increase in n for fixed p will lead to reduced skewness (smaller β_1), and the kurtosis β_2 will approach 3; in the limit this suggests that even the asymmetric binomial will approach the normal curve as n (or N) increases. The example of quartz frequency for the Salt Wash sandstones illustrates that even with asymmetry $p = 0.654$ and $N = 150$ the expected skewness is very near zero and the expected kurtosis near the value 3. In general, then, as n (and/or N) grows large, the binomial will approximate to the normal distribution, and in fitting the frequencies of observed to expected the table of areas of the normal curve can be used as previously described (Sec. 13.5).

TABLE 14.6
Observed and Expected Summary Statistics for Frequency Distribution of Quartz in Barren Salt Wash Sandstone

Statistic		Expected	Observed
Mean	$= \mu = np$	65.37	65.37
Variance	$= \sigma^2 = npq$	22.6284	62.7825
Skewness	$= \beta_1 = \dfrac{(q-p)^2}{npq}$	0.0042	0.1728
Kurtosis	$= \beta_2 = 3 + \dfrac{1-6pq}{npq}$	2.9842	3.0783

14.3 Fitting the Binomial Frequency Distribution to a Set of Observed Frequencies

When a number of trials of an experiment is performed and the number of successes is recorded in classes of zero successes, one success, two successes, etc., the frequency distribution may approximate a binomial. The conditions under which such a set of trials yields a binomial are:

1 A number of trials are performed in which the outcomes may be classified as successes or failures; i.e., in counting the number of zircons in a heavy residue the outcome is classifiable into "zircon" (success) or "not zircon" (failure).
2 The trials are independent; i.e., occurrence of a zircon as the second grain is independent of the occurrence of a zircon as the first, and so on.
3 The probability of zircon's occurring is constant from trial to trial.

In investigating the frequency of occurrence of zircon in an accessory-mineral residue the population comprises the grain mount or slide; and if the mount is properly prepared, zircon will be randomly distributed throughout the other grains. A trial consists of a traverse of n grains in which each grain is classified as "zircon" or "not zircon," and as zircon is randomly distributed, its probability will be constant from trial to trial. A number N of traverses will be performed. In this case each traverse will yield an estimate of the probability (frequency) of zircon in the slide, and the frequency distribution of the observed p_i will be binomially distributed around a central value $\bar{p}$, which is an estimate of the true proportion P of zircon in the slide. Under these conditions, fitting the relevant binomial becomes a test of a satisfactory slide or, equivalently, a test of the postulate that zircon is randomly distributed throughout the slide.[1] If the binomial does not fit, the estimates p_i will yield a biased estimate of P, and the sampling pattern must be adjusted or a new mount prepared.

On the other hand, when this procedure is applied to point counting a thin section (Griffiths, 1960d), the arrangement of the constituent, e.g., quartz grain, is unlikely to be random, and so the fitting to a binomial again serves to characterize the distribution of the constituent in the thin section. It is also necessary in this case to adjust the sampling pattern to obtain stable statistical estimates p_i of the proportion P. In either case the information obtained from fitting the frequency distribution is of considerable importance in establishing the petrographic data in the form necessary for petrologic interpretation (Griffiths, 1960d).

To fit a binomial frequency distribution to a set of observed frequencies it is first necessary to determine the probability P; in practice the observed p_i are obtained and their mean $\bar{p}$ found from $\bar{X}$ calculated from the observed frequencies. The relevant binomial $(p + q)^n$ is then determined, and the

[1] For example, see Fig. 14.2, in which the model is the Poisson distribution but the testing is similar to that described in the text; the dispersion index for the binomial model is defined as $\chi^2 = Sx_i^2/\bar{X}(1 - p)$.

frequencies in the various classes may be obtained from tables of the binomial frequency distribution (National Bureau of Standards, 1950; Romig, 1947). As an example we may take the frequency of quartz grains in the Salt Wash sandstones recorded in Table 14.4. From $\bar{X} = 65.365$ and the known traverse length of $n = 100$ we have $\bar{X} = n\bar{p}$, where $\bar{p} = 0.654$; the relevant binomial is $(0.654 + 0.346)^{100}$, and the required frequencies are found in Romig (1947, p. 167). Thus we require the number of times 40 or fewer quartz points occur in a 100-point traverse, and the transformation is

$$B(X,n,p) = 1 - B(n - x - 1,n,q) \tag{14.7}$$

In our case $n = 100$, $p = 0.65$, and $q = 0.35$, yielding

$$B(X,100,0.65) = 1 - B(100 - x - 1,100,0.35)$$

The required values for X may now be substituted and the cumulative values, up to each value, read from Romig's tables (1947, p. 167, col. 3, under $p = 0.35$). The work can be arranged as in Table 14.7. Column 1 is the X value, column 2 the $n - x - 1$ value. Column 3 contains the cumulative frequencies up to x or less (actually $n - x - 1$ or less) from Romig's tables. The fractional frequencies in the classes found by difference are in column 4. To obtain the expected frequencies of column 5, the frequencies in column 4 are multiplied by the number of trials N (150 in the example). The observed frequencies are inserted in column 7 and the corresponding class intervals in

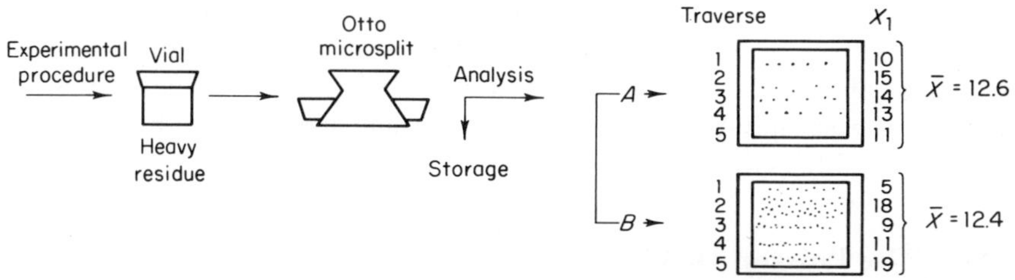

Figure 14.2 The Poisson model as a test of technique. Sampling requirements: random in traverses and grains. Model requirements: probability (=frequency) of mineral $= p =$ constant from traverse to traverse. Hypothesis (H_0): constituent is uniformly random throughout slide. Frequency distribution: Poisson (n very large; p very small; np finite). Criterion: dispersion index $\chi^2 = Sx_i^2/\bar{X}$; d.f. $= N - 1 = 4$. χ^2 under H_0 follows χ_ν^2 distribution.

Case A	Case B
$\chi^2 = 1.37$	$\chi^2 = 11.55$
$P < 90 > 80$	$P < 0.05 > 0.02$
Hypothesis accepted: uniform distribution. Mean is stable estimator.	Hypothesis rejected: nonuniform distribution. Mean is biased estimator.

TABLE 14.7
Calculating Expected Frequencies for Binomial Distribution with $(p + q)^n = (0.65 + 0.35)^{100}$

(1) X or less	(2) $n - X - 1$	(3) $B(n - X - 1, 100, 0.35)$	(4) $B(X,100,0.65)$	(5) Expected frequency 150 × col. 4	(6) Class	(7) Observed
40	59	0.999 995	0.000 005	0.0008	0–40	
45	54	0.999 961	0.000 034	0.0051	40–45	1
50	49	0.998 545	0.001 416	0.21	45–50	7
55	44	0.975 393	0.023 152	3.47	50–55	7
60	39	0.827 581	0.147 812	22.17	55–60	18
65	34	0.462 430	0.365 151	54.77	60–65	33
70	29	0.123 598	0.338 832	50.82	65–70	42
75	24	0.012 127	0.111 471	16.72	70–75	28
80	19	0.000 343	0.011 784	1.77	75–80	11
85	14	0.000 002	0.000 341	0.05	80–85	3
90	0		0.000 002	0.0003	85–100	
Total			1.000 000	149.9862		150

column 6. The observed and expected frequencies are illustrated in Fig. 14.1. The discrepancies are quite large;[1] they are all connected with the asymmetry of the observed frequencies when compared with the expected. Apparently the frequencies of quartz points are not binomially distributed in this set of traverses and/or thin sections. It seems likely that once again nonrandomness in the sampling is the main feature accounting for the discrepancies, and this will be subsequently discussed.

14.4 The Poisson Distribution

Let us now suppose that we maintain effectively the same model as before but decrease the probability of success p toward a limit of zero; in order to maintain a finite value for $np = \bar{X}$, n must increase as p decreases. Under such conditions the binomial is approximated by the Poisson exponential frequency distribution. Thus by writing the binomial as

$$\frac{n(n-1)\cdots n-r+1}{r!}\,p^r q^{n-r} \tag{14.8}$$

and increasing n and decreasing p so that $np = m$ is moderate and r is negli-

[1] When tested by chi square (see Sec. 17.4) the discrepancies in frequency are highly significant ($P < 0.001$).

gible compared with n, then q approaches unity, and the expression (14.8) becomes

$$\frac{n^r p^r}{r!}(1 - p)^n = \frac{(1 - m/n)^n m^r}{r!} \tag{14.9}$$

In the limit this yields (Tippett, 1952, p. 60; Student, 1907, p. 352)

$$C_r{}^n p^r q^{n-r} = \frac{(1 - m/n)^n m^r}{r!}$$

$$= \frac{e^{-m} m^r}{r!} \tag{14.10}$$

This frequency distribution has been used as an approximation to the binomial when p is very small and may be used in its own right as a description of the frequency of occurrences of certain events such as counting blood cells with a hemacytometer (Student, 1907), yield of α tracks during radioactive decomposition (Rutherford and Geiger, 1910; Smith, 1959), accidents to personnel under certain conditions, and equally rare events such as rare minerals in loose sands or rare constituents in thin-section point counting for modal analysis of sediments (Griffiths, 1960d). Student first proposed the use of this frequency distribution, as a model for blood counts using a hemacytometer, in a classical article (1907, p. 351); he subsequently discussed, in considerable detail, the reasons for departure from the Poisson distribution under many experimental conditions (Student, 1919, p. 211; Pearson and Wishart, 1942, p. 65).

If we consider counting situations which may arise in sedimentary petrography, the requirements for the generation of a Poisson distribution, as given by Student, may be modified as follows:

1 The probability of occurrence of a specific mineral constituent during counting is constant for an individual.
2 The probability of occurrences of an individual mineral grain of this species is constant for each trial, e.g., traverse or field.
3 The fact that an individual of the species has occurred in a trial does not affect the chance that other individuals may occur in the same trial.

These conditions match almost exactly those described for the binomial (page 286), with binomial condition 3 corresponding with Poisson conditions 1 and 2 and the condition of independence being 2 in the binomial and 3 in the Poisson.

The Poisson frequency distribution may be expanded into

$$\frac{e^{-m} m^r}{r!} = e^{-m}\left(1 + m + \frac{m^2}{2!} + \frac{m^3}{3!} + \cdots + \frac{m^r}{r!}\right) \tag{14.11}$$

Hence we may again look upon the probability as equivalent to the area under the specified frequency distribution.

The characteristics of the Poisson frequency distribution are:

$$\text{Mean} = \mu = np = m \tag{14.12}$$

$$\text{Variance} = \sigma^2 = np = m \tag{14.13}$$

$$\text{Skewness} = \beta_1 = \frac{1}{np} = \frac{1}{m} \tag{14.14}$$

$$\text{Kurtosis} = \beta_2 = 3 + \frac{1}{np} = 3 + \frac{1}{m} \tag{14.15}$$

Under the conditions that p becomes very small, q approaches unity, and $np = m$ remains finite, the parameters for the binomial approach those of the Poisson; and again if p is fixed and n made very large, $1/m \approx 0$, $\beta_1 \approx 0$, and $\beta_2 \approx 3$; that is, the Poisson approaches the normal distribution.

14.5 Fitting the Poisson Frequency Distribution to a Set of Observed Frequencies

Here again it is necessary to compute the moment measures and put $np = m$ to find the relevant Poisson. The frequencies are the coefficients in the expansion of

$$e^{-m}\left(1 + m + \frac{m^2}{2!} + \frac{m^3}{3!} + \cdots\right)$$

and they may be multiplied by the number of trials to obtain equal area under

TABLE 14.8
Computing the Frequencies for a Poisson Distribution with Mean $m = 1.0676$ **and** $N = 340$ $e^{-m} = \text{antilog}\,(-1.0676 \times 0.43429) = 0.34383$

(1) No. of micas r per traverse	(2) $r!$	(3) m^r	(4) $\frac{m^r}{r!}$	(5) $\frac{e^{-m}m^r}{r!}$	(6) $Ne^{-m}\frac{m^r}{r!}$
0	1	1.0	1.0	0.34383	116.90
1	1	1.0676	1.06760	0.36707	124.80
2	2	1.1398	0.56990	0.19595	66.62
3	6	1.2169	0.20282	0.06974	23.71
4	24	1.2992	0.05413	0.01861	6.33
5	120	1.3870	0.01156	0.00397	1.35
6+	720	1.4808	0.00206	0.00083	0.28
Total				1.00000	339.99

TABLE 14.9
Comparison of Computed and Observed Values of Summary Statistics of Frequency Distribution of Large "Micas" from Pocono Graywacke

Statistic	Observed	Expected
Mean $= m = \bar{X}$	1.068	$m = 1.068$
Variance $= \sigma^2$	1.2514	$\sigma^2 = 1.068$
Skewness $= \beta_1$	1.9665	$\frac{1}{m} = 0.9363$
Kurtosis $= \beta_2$	6.2105	$3 + \frac{1}{m} = 3.9363$

the computed Poisson and observed frequency distribution. The expected frequencies may be computed by logarithms (Snedecor, 1956, p. 484) or obtained from tables of Poisson's exponential binomial limit (Molina, 1942).

The frequency distribution of large "micas" (megamicas) in five thin sections from Pocono low-rank graywacke sandstone from Route 322 north of Port Matilda, central Pennsylvania, was analyzed by four operators, who used 17 traverses of 90 points each across each thin section, yielding $N = 340$. The observed frequencies are given in Table 14.10, and they yield the summary statistics listed in Table 14.9, which differ widely from the expected values calculated for a Poisson with the same mean.

The frequencies obtained for a Poisson with the mean value 1.0676 may be found as follows: first find e^{-m} as antilog $(-m \log e)$ = antilog $(-m0.43429)$. In the present case this becomes

$$\text{antilog}\,(-1.0676 \times 0.43429) = \text{antilog}\,-0.46365$$

or

$$\text{antilog}\,\bar{1}.53635 = 0.34383 = e^{-m}$$

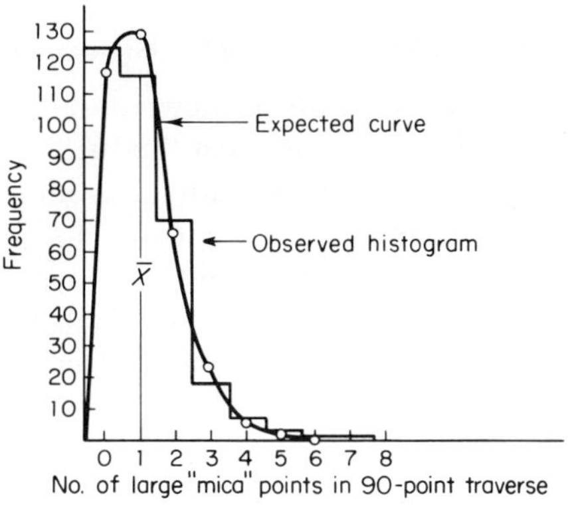

Figure 14.3 Frequency distribution of large "micas" in thin sections of Pocono graywacke from central Pennsylvania.

$$e^{-m}\frac{m^r}{r!} = e^{-1\cdot 0676}\frac{(1.0676)^r}{r!}$$

$$\bar{X} = 1.06 \qquad n = 340$$

TABLE 14.10
Comparison of Observed and Expected Frequency Using the Chi-square Test

No. of micas per traverse	Frequency: Observed O	Frequency: Expected E	$O - E$	$(O - E)^2$	$\frac{(O - E)^2}{E}$
0	125	116.90	+8.10	65.61	0.56
1	115	124.80	−9.80	96.04	0.77
2	70	66.62	+3.38	11.4244	0.17
3	19	23.71	−4.71	22.1841	0.94
4	11	7.96	+3.04	9.2416	1.16
Total	340	339.99	0.01		$3.60 = \chi^2$

Note: d.f. $= c - r = 5 - 2 = 3$; $50 > P > 30$.

The requisite expansion of

$$Ne^{-m}\left(1 + m + \frac{m^2}{2!} + \frac{m^3}{3!} + \frac{m^4}{4!} + \frac{m^5}{5!} + \frac{m^6}{6!}\right)$$

where $m = 1.0676$, is then found in Table 14.8, and the expected and observed histograms are given in Fig. 14.3.

The fractional Poisson frequencies listed in column 5 of Table 14.8 may be read directly from Molina's tables (1942, p. 3) with mean 1.1, and the differences between the computed frequencies of column 6, Table 14.8, and those of Molina's tables are not significant when tested by chi square ($P > 0.995$).

Similarly, when the observed frequencies are compared with the expected Poisson, as in Table 14.10, the discrepancies are not significant ($\chi^2 = 3.60$, d.f. $= 3$; $50 > P > 30$). The large "micas" in these thin sections of Pocono graywacke are randomly distributed with Poisson frequency.

14.6 Extensions of the Binomial and Poisson Models

It is generally somewhat difficult to maintain the probability constant from trial to trial in most "natural" sampling situations, and "simple" random sampling is very difficult to achieve in practice. The effects of deviations from the model requirements in the case of the Poisson distribution have been very thoroughly dealt with by Student (1919, para. 19, p. 211; Pearson and Wishart, 1942, p. 65), who emphasized that in the case of the Poisson or of the normal distribution, the frequency model is an approximation to the "real" situation, and its utility does not necessarily demand an exact fit (Pearson and Wishart, 1942, p. 69).

Suppose, by way of example, that the population it is desired to sample is composed of a number of subpopulations within each of which p is constant from trial to trial but p varies from subpopulation to subpopulation. A simple model may be appropriate for any individual subpopulation, but the pooled sampling distribution represents a heterogeneous set of distributions; this is called a *Lexian series* (Weatherburn, 1957, p. 115; Kendall, 1948, vol. 1, p. 122).

Let p_i be the value of the ith sample ($i = 1, 2, \ldots, N$). It is required to find the standard error of the number of successes per sample when the records of all the samples are pooled. Let p be the mean value of the probability, so that

$$Np = \sum_{i=1}^{N} p_i \tag{14.16}$$

Then the expected value (Weatherburn, 1957, p. 24) of the number of successes in the whole series of samples is equal to the sum of the expected values for the individual samples

$$\sum np_i = nNp$$

Hence the expected value of the number of successes per sample is np.

The variance ϵ^2 of the number of successes per sample is found as follows (Weatherburn, 1957, p. 115). The mean-square deviation from np in the ith sample is

$$np_iq_i + (np_i - np)^2$$

or, in words, is equal to the variance of the ith sample (np_iq_i) plus the square of n times the deviation of the sample mean from the general mean; summing over all N samples yields

$$\begin{aligned} N\epsilon^2 &= n \sum p_iq_i + n^2 \sum (p_i - p)^2 \\ &= n \sum p_iq_i + n^2N\sigma_p^2 \end{aligned} \tag{14.17}$$

where σ_p^2 is the variance of the quantities p_i.

The remainder of the expression may be written[1]

$$\begin{aligned} p_iq_i &= \sum p_i(1 - p_i) = \sum p_i - \sum p_i^2 \\ &= Np - N(\sigma_p^2 + p^2) \\ &= Np - Np^2 - N\sigma_p^2 \end{aligned}$$

[1] *Note:*

$$\begin{aligned} N\sigma_p^2 &= \sum p_i^2 - \frac{(\sum p_i)^2}{N} = \sum p_i^2 - \frac{N^2p^2}{N} \\ &= \sum p_i^2 - Np^2 \end{aligned}$$

Substituting these values in Eq. (14.17) yields

$$N\epsilon^2 = Nnpq - nN\sigma_p^2 + n^2N\sigma_p^2$$

$$\epsilon^2 = npq + n(n - 1)\sigma_p^2 \qquad (14.18)$$

This exhibits the fact that the variance of the number of successes in a Lexian series is larger by a factor of $n(n - 1)\sigma_p^2$ than the variance in a binomial distribution.

Similarly, it can be shown that if the sampling is represented by a Poisson series in which the probability of a success changes with selection of each individual in a sample, the expected value of the number of successes is

$$\sum p_i = np \qquad (14.19)$$

while the expected variance if each drawing is independent is (Weatherburn, 1957, p. 115)

$$\epsilon^2 = npq - n\sigma_p^2 \qquad (14.20)$$

The importance of these models in petrographic analysis arises in cases where the population is stratified; then if each sample is taken from within a layer and each layer fulfills the requirements for a binomial or Poisson probability model but the p_i are not constant from layer to layer, the resulting variate will approximate a Lexian series. If, on the other hand, the sample is taken, say, perpendicular to these layers so that each individual comes from a single layer, the sampling model will approximate a Poisson series. In the first case the variance will be larger than that expected for a binomial; in the second, smaller. In sedimentary rocks the situation giving rise to excessive variance appears to suggest the Lexian model as more appropriate than the Poisson series.

14.7 The Negative Binomial Distribution

Student describes the effect of departures from the Poisson frequency distribution in somewhat different terms (Pearson and Wishart, 1942, pp. 65ff.) and exhibits the changes in the parameters for the frequency distributions generated under different sampling conditions, such as the positive binomial, Poisson, and negative binomial. He shows that relationships between the moment measures can be used to characterize these three sampling models. Starting with the binomial, he designates the moments about zero as

$$n_1 = np$$

$$n_2 = npq + n^2p^2$$

whereas the variance $m_2 = n_2 - n_1^2 = npq + n^2p^2 - n^2p^2 = npq$. Then using the ratio m_2/n_1, he shows that for the positive binomial, $m_2/n_1 < 1$; for the Poisson, $m_2/n_1 = 1$; and for the negative binomial, $m_2/n_1 > 1$.

This test is related to the subnormal and supernormal dispersion of Lexis (Yule and Kendall, 1948, p. 368) and has been used to characterize the variations in frequency of accessory heavy minerals by Allen (1945b, p. 76). Fisher introduced the dispersion index, which plays a similar role (1948, pp. 48, 688). The dispersion index is a useful test of random sampling for a binomial and Poisson frequency distribution in the case of small samples suspected to have arisen from such distributions. Suppose, for example, that an accessory heavy-mineral count is performed and a success is defined as the occurrence of a zircon in a traverse of n grains; since the distribution of zircon in the slide is uniformly random if the slide is properly prepared, we may assume a Poisson distribution if n is very large and p very small (say $p < 5$ in 100 grains), and p will be constant from trial to trial. The results of a number N of traverses yield 10, 15, 14, 13, 11 ($N = 5$). If the dispersion index χ^2 is defined as (Fisher, 1948, p. 58)

$$\chi^2 = \sum_{i=1}^{N} \frac{(X_1 - \bar{X})^2}{\bar{X}}$$

then

$$\chi^2 = \frac{17.2}{12.6} = 1.37 \tag{14.21}$$

Under the hypothesis of Poisson frequency and random sampling χ^2 follows the chi-square distribution with degrees of freedom one less than the number of traverses; that is, d.f. = 4. Then from a table of chi square (Fisher and Yates, 1948, p. 33; Arkin and Colton, 1950, p. 121) $\chi^2 = 1.37$ with d.f. = 4 yields the probability $90 > P > 80$, and there is no evidence for rejection of the hypothesis; i.e., these five traverses are random samples from a Poisson distribution, and a stable estimate of the amount of zircon in the slide is $\bar{X} = np = 12.6$ (see Fig. 14.2).

Suppose, however, that the five traverses yield 5, 18, 9, 11, 19 zircons, respectively; then

$$\chi^2 = \frac{143.2}{12.4} = 11.55$$

and with d.f. = 4 this leads to $0.05 > P > 0.02$, and the hypothesis is rejected; i.e., the probability is not constant from trial to trial, and the mean is not necessarily a stable estimate of the amount of zircon in the slide. In particular, the variance is much larger, leading to a large dispersion for the same mean. It is necessary either to make a new slide or else to adjust the sampling pattern until the estimate is stable (see Fig. 14.2).

If we now alter the initial conditions and suggest that while p remains constant from trial to trial, the trial size is adjusted to the value $n = 50$ and two slides are analyzed, we obtain the same results as before. In this case the binomial dispersion index is appropriate (Fisher, 1948, p. 68), as follows:

$$\chi^2 = \frac{\sum_{i=1}^{N} (X_i - \bar{X})^2}{\bar{X}(1 - p)} \tag{14.22}$$

$\bar{X} = 12.6$ is equated to np; and with $n = 50$, $p = 0.25$, and $1 - p = 0.75$,

$$\chi^2 = \frac{17.2}{12.6 \times 0.75} = \frac{17.2}{9.45} = 1.82$$

and with d.f. $= 4$, $80 > P > 70$, whence the hypothesis is accepted as in the Poisson case. Again with $np = \bar{X} = 12.4$, then $p \approx 0.25$ and $1 - p \approx 0.75$ and $\chi^2 = 143.2/9.3 = 15.40$; again $0.01 > P > 0.001$, and the hypothesis is rejected. The conclusions are essentially the same as in the Poisson case.

Actually in making mounts of heavy residues of sandstones, if the usual preparation procedure is followed and the accessory minerals are concentrated so that with traverse length of, say, 100 grains the probability of the mineral constituent being evaluated is between, say, 5 and 95, then the binomial model is appropriate, and the dispersion index is a suitable test of technique. If, on the other hand, the accessory minerals are not concentrated, a count of the original disaggregated sedimentary detritus being used, the sample unit becomes a traverse length n composed of many hundreds of grains, and the probability of occurrence of one accessory is very small, but np is finite because of the large n. Under these conditions the frequency distribution of the mineral constituent is likely to be Poisson, and the appropriate dispersion index is again a suitable test of technique. Here the experimental procedure of sieving, bromoform separation, and so on is equivalent to increasing the probability of occurrence and results in a transformation of Poisson to binomial frequency distribution in the mount.

When an attempt was made to apply these tests, to establish the frequency distribution of an accessory mineral in a series of specimens of Oswego sandstone, the hypothesis of constant probability had to be abandoned, and it seemed likely that the probability of occurrence of zircon was not constant from specimen to specimen. Indeed, it seems likely that while zircon is generally absent from much of the sandstone, it sometimes occurs in relative abundance. This is the expected mode of occurrence where there has been concentration of the constituent by "natural panning" (Boswell, 1933, p. 48). Under such conditions the model frequency distribution will be best represented as a contagious type of distribution or negative binomial, as originally suggested by Burr (in Krumbein, 1954b, pp. 58, 59). We may therefore consider next the procedure for fitting the negative binomial to a set of observations.

It is of some interest at this stage to consider the results of relaxing the conditions initially stipulated for the generation of a variate which follows the Poisson frequency distribution, and for this purpose the discussion by Student (1919, p. 69) succinctly presents the possible alternatives:

1 If the restriction of probabilities of each individual's falling into a trial is relaxed to the extent that, even if the probabilities are unequal, they all remain small, then the frequency distribution will still remain Poisson.
2 If, however, the probabilities of some individuals are large, the frequency distribution will approach the positive binomial.
3 If the different trials have different probabilities of containing individuals (as is usual), a negative binomial may be generated (except insofar as condition 1 may interfere).
4 When the condition of independence is relaxed, then
   a If the presence of an individual in a trial increases the chances that others will occur therein, a negative binomial will result.
   b If the presence of an individual in a trial decreases the chances that another will fall therein, a positive binomial will result.

All these instances are of potential interest to a sedimentary petrographer; but in particular, as pointed out by Student (1919), the concentrating of events, such as the occurrence of a number of zircons in one place and their extreme rarity in another, leads to a negative binomial. On the other hand, if the element is of similar magnitude to the sampling unit (trial, such as in point counting very coarse-grained rocks), then condition 4*a* applies, and its resulting distribution will be best approached by a positive binomial.

Since condition 3 is quite frequent in practice, the negative binomial will be a common frequency distribution (Student, 1919).

14.8 Fitting the Negative Binomial Frequency Distribution to a Set of Observed Frequencies

The expression which represents the positive binomial is $(p + q)^n$; the analogous expression for the negative binomial is (Bliss and Fisher, 1953, p. 177)

$$(q - p)^{-k} \tag{14.23}$$

and the general term of this expansion is

$$P_X = \frac{(k + X - 1)!}{X!(k - 1)!} \frac{R^X}{q^k} \tag{14.24}$$

where P_X represents the probability that a trial (traverse) will contain $X = 0, 1, 2, \ldots$ individuals, i.e., events, or successes. This should be compared with the analogous expression for the positive binomial

$$P_X = \frac{n!}{X!(n - X)!} p^X q^{n-X}$$

In the expression $(q - p)^{-k}$, p is, as usual, the probability of occurrence of the event and q is the probability of nonoccurrence; and $p = m/k$ and

$q = 1 + p$; $R = p/q = m/(k - m)$. The two parameters m and k must be estimated from the observations; the arithmetic average $\bar{X} = S(fX)/n$ is an efficient estimator of m (Bliss and Fisher, 1953, p. 179), and there are a number of alternative procedures for finding k', the statistical estimator of the parameter k. Examples of the estimation procedures and their limitations are described by Bliss and Fisher (1953, p. 179), and an example using accessory-mineral frequencies is given by Griffiths (1960a).

Other examples of the negative binomial distribution arise in estimating mineral composition by point counting thin sections, and it has been suggested that where the observations so determined fit the negative binomial, the pattern of arrangement of the constituent in the sediment reflects the effect of two or more processes (Griffiths, 1960d, 1963a). In particular, when the constituents are cements, if their frequency distribution tends to the negative binomial, it may reflect secondary reprecipitation of the cementitious material (Griffiths, 1958b, 1960d).

14.9 The Hypergeometric Frequency Distribution

In most of the previous model frequency distributions the condition of independence is generally maintained, but an important model, which forms the basis of the Pearsonian frequency system, is represented by a model frequency distribution in which this consideration is relaxed. The illustration generally taken is a deck (or several decks) of playing cards, and cards (samples) are drawn without replacement. This model is essentially concerned with a finite population sampled without replacement. Suppose, for example, that the probability of drawing a spade is required; then, since there are 13 possible elements of a total of 52 which can be classified as successes, the probability $p = 13/52$. On the second draw the population consists of 51 elements and there may be only 12 spades; hence the probability is $p = 12/51$. On the other hand, if the first draw results in any card but a spade, the probability of a spade on the second draw is 13/51. By extending the argument, it can be seen that the probabilities are not constant from trial to trial but depend on the outcome of each preceding draw.

The general term of this expansion is (Smith and Duncan, 1945, p. 54)

$$P(N_1) = \frac{(p_1S)!(p_2S)!(S - N)!N!}{(p_1S - N_1)!(p_2S - N + N_1)!S!N_1!(N - N_1)!} \tag{14.25}$$

where the probability P of the combination of elements N_1 of a certain kind is p_1S. There are N packs of cards (finite populations) and each contains S cards (elements), p_1S of which represent the occurrence of the event (success) and p_2S the occurrence of the complementary event (failure).[1]

The characteristic parameters of this frequency function contain the

[1] An alternative derivation is succinctly stated by Aitken (1957, p. 56).

values of N and S and are quite complicated (Smith and Duncan, 1945, p. 56). They tend, however, to the equivalent binomial as S increases and becomes very large compared with N (or as the sample of elements N is small relative to the population S). The frequency distribution resembles the corresponding binomial, and it is possible to find the class frequencies from tables (e.g., Romig, 1947, p. xix). The importance of this frequency function is, however, as a model for sampling a finite population without replacement, and Chayes has suggested that selected traverses in point-counting a thin section represent a likely example (Chayes, 1949a).

Another interesting feature of this distribution is its possible use in representing the experimental situations which lead to nonnormal frequency functions even with large populations; this aspect is discussed by Smith and Duncan (1945, pp. 60ff.) based on Pearson's (1895) description. It is now clear that if the number of contributory causes is large and their magnitudes are equal, the frequency function asymptotically approaches the normal form; even if they are not equal but are all small, either the Poisson or normal will, in general, be generated, and asymptotically the Poisson approaches the normal. Thus, as the discrete frequencies are made continuous with increasing sample size, they lose their discreteness and become continuous, and apparently most distributions tend in the same direction with increasing sample size. However, if the number of causes of variation remains small and their contributions are not constant but vary, the discreteness will again be blurred, but the continuous frequency distribution will not tend to the normal but remain skew and/or peaked. In Pearson's words: "It may possibly be that continuity in biological or physical frequency curves may arise from a limited number of contributory causes with a power of fractionizing the results" (Pearson, 1895, p. 358).

14.10 The Multinomial Distribution

While variation in the proportion of any single constituent in a population may be considered in terms of the binomial (or Poisson approximation), quite frequently the population is classified into several constituent classes during counting, and the investigator may wish to consider a model in which events may be classified into several classes in which frequencies vary simultaneously. A suitable model may then be the multinomial distribution.

Suppose that an observation (event) may be classified into one of k categories $A_1, A_2, \ldots, A_k$ and that the probability of falling into a class such as A_i is P_i $(i = 1, 2, \ldots, k)$; $X_1, X_2, \ldots, X_k$ are random variables representing the observed frequencies of events falling into the k categories, respectively. Then for any particular sample set when there are n_1 of the X_1's, n_2 of the X_2's, ..., and n_k of the X_k's, the probabilities are

$$p_1^{n_1}, p_2^{n_2}, \ldots, p_k^{n_k}$$

Since any arrangement of the n_1X_1's and n_2X_2's, etc., is equally likely, the number of possible arrangements must be counted which yields the permutation of n things, n_1 of which are of one kind, n_2 of which are of another kind, and so on, or

$$\frac{n!}{n_1!n_2!\cdots n_k!}$$

The general term of the multinomial is therefore

$$\frac{n!}{n_1!n_2!\cdots n_k!}\,p_1^{n1}p_2^{n2}\cdots p_k^{nk}$$

where
$$\sum_{i=1}^{k} n_i = n,$$

and this is analogous to the general term of the binomial when $i = 2$. The multinomial distribution is written as

$$(p_1 + p_2 + \cdots p_k)^n$$

TABLE 14.11
Multinomial Distribution with $n = 4$, $p_1 = 0.5$, $p_2 = 0.3$, $p_3 = 0.2$; n Points per Set and 300 Sets Counted

	p_1	p_2	p_3	$P(p_1,p_2,p_3)$	$300P(p_1,p_2,p_3)$ = exp frequency
	4	0	0	0.0625	18.75
	3	1	0	0.1500	45.00
	3	0	1	0.1000	30.00
	2	2	0	0.1350	40.50
	2	1	1	0.1800	54.00
	2	0	2	0.0600	16.20
	1	2	1	0.1080	32.40
	1	1	2	0.0720	21.60
	1	0	3	0.0160	4.80
	0	4	0	0.0081	2.43
	0	3	1	0.0216	6.48
	0	2	2	0.0216	6.48
	0	1	3	0.0096	2.88
	0	0	4	0.0016	0.48
Total				1.0000	300.00

and may be expanded in a fashion strictly analogous to the binomial (Hoel, 1947, p. 54; Johnson and Tetley, 1950, pp. 67ff.).

The relevant parameters of the multinomial are

$$\bar{X}_i = E(X_i) = np_i$$

$$\sigma_i = np_i(1 - p_i) = np_iq_i$$

and since the $\sum p_i = 1$, the different values are not independent, so that the covariance becomes

$$\text{Cov }(X_iX_j) = -np_ip_j \qquad i \neq j$$

By way of example, suppose that a detrital sediment is known to contain 50 percent quartz, 30 percent rock fragments, and 20 percent other constituents; counts are performed in sets of four, and 300 such counts are made. What are the expected frequencies for comparison with those observed? The computation is illustrated in Table 14.11, representing the expansion of the multinomial

$$(p_1 + p_2 + p_3)^n = (0.5 + 0.3 + 0.2)^4$$

$$= \frac{4!}{r_1!r_2!r_3!} 0.5^{r_1}0.3^{r_2}0.2^{r_3} \qquad r_1 + r_2 + r_3 = 4, r_i = 0, 1, \ldots, 4$$

It is obvious that computations of this kind become very tedious as the number of categories increases; approximations are frequently helpful, such as use of tables of factorials (Barlow's tables).

14.11 Conclusions

A number of possible model frequency distributions have now been presented, together with descriptions of possible experimental situations giving rise to them. Examples of all these are known, or can be conceived as existing, in the realm of geological investigation, and this does not by any means exhaust the possibilities (e.g., Aitken, 1957; Otto, 1939; and Tanner, 1958).

Two items need emphasis. First, it is almost always possible to find a frequency function which will fit an observed set of data, for example, by using the Pearsonian or Gram Charlier series of frequency curves (Kendall, 1948; Smith and Duncan, 1945). It is, however, not necessarily any advantage if the representation is not fruitful in indicating the likely physical model. The principle of parsimony suggests that there is some advantage to reducing a mass of observational data to a few appropriately informative estimators of parameters, where the parameters are of clear physical significance; otherwise the fitting of a frequency function is not of very great value except as a convenient data-reduction device. Second, it is sometimes possible to synthesize a model which may approximate the experimental situation; and although the set of data may depart from the model, the reason for departure may be attributed to nonrandomness in the sampling arrangement. It is, of course, possible to adjust the sampling in some cases so as to obtain a set of data which agrees with the proposed model, but here again some ambiguity arises. Whether one begins with a model and defines a random sample, or whether one draws a sample and, by assuming it is random, searches for the appropriate model frequency distribution, this dilemma arises; and perhaps the only practical solution is the requirement that both approaches should converge on the same final approximation.

15 Relationship between mean and dispersion

If it were the normal custom, when scientific observations show uncontrolled variations large compared with the observations themselves, to convert them to logarithms before estimating their mean or variance, the usual result would be an increase in the accuracy and scope of the conclusions drawn from them.

J. H. Gaddum, 1945

15.1 Introduction

When a frequency distribution is decided upon as a model to represent some physical (real-world) process, the parameters necessary for specifying the model are known, and data are collected to estimate, from the sample statistics, the values of the parameters. In most cases four parameters based on the first four moment measures (Sec. 5.4) represent the most general approach. It is then necessary to examine the estimators to see whether they are independent or related in some way. Suppose the normal model is chosen; then the two parameters μ and σ^2 are required to specify the distribution, and likely values for these are obtained from their corresponding sample estimators, the mean $\bar{X}$ and the variance $\hat{\sigma}^2$. In random samples from a normal population the mean is independent of the variance, and so a test of independence represents a test of randomness of sampling.

Relationships between the statistical estimators may arise from various causes; e.g., in a binomial distribution the mean np is related to the variance npq whether estimators or parameters are used, although if the estimators are not derived from random samples, the relationship may be different from that expected. Again, whereas the expected behavior of the parameters

Quotation: Lognormal Distributions, *Nature*, **156:** 466 (1945).

may be postulated on the basis of the model, the measurement procedure may induce a bias which leads to a different relationship. It is, then, of considerable interest to examine the relationships between the estimators as a check on the model requirements. This becomes necessary, not solely from the desire to detect experimental difficulties (assignable causes), but also from the need to select a suitable procedure so that, in the subsequent statistical analysis, the requirements of the analytical procedure are fulfilled;[1] transformations are devised in order that the transformed variables may possess the characteristics required for the application of hypothesis testing and to ensure that the probability statements accompanying the conclusions based on the tests are reasonably correct.

In general, it is not possible to select, a priori, a measurement procedure which leads to data on a "natural" scale, and the measurement units selected are largely a matter of convention and convenience. Transformations may then be looked upon as devices to obtain a better control of measurement data. Quite frequently, for example, "nature" appears to operate on a geometric scale, and the measurements performed on an arithmetic scale behave in an intractable manner; when the measurements are transformed to a logarithmic scale, the variables may then follow the normal law or some other well-known frequency function (Kapteyn, 1916). It seems unlikely that the scales we select to classify data are, *ab initio*, the "best" scales, so that transformation is a very general procedure (Gaddum, 1945).

Studies of the relationships among estimators are, then, of interest both in attaining theoretical model requirements and in understanding the physical meaning of variation in a measured variate.

15.2 Model Relationships of Mean and Dispersion

As mentioned above, once the model is known, the relationships between the mean μ and variance σ^2 or the mean μ and standard deviation σ are also generally known. The mean μ is independent of the variance σ^2 in random samples from a normal frequency distribution. With a normal distribution, if the sample means $\bar{X}$ are related to their variance $\hat{\sigma}^2$, the samples are not random. For parametric statistical analysis it is then necessary to repeat and modify the sampling until randomness is obtained, to change the procedure so that bias is removed, or to use a transformation (see next section) to ensure that the mean is independent of the variance.

When discrete scales are used and the binomial is a suitable model, the mean μ is related to the variance σ^2, where $\mu = np$ and $\sigma^2 = npq$. From this relationship it can be shown that the variance increases as p increases from 0 to 50 units and then decreases symmetrically from 50 to 100 units (see Table 10.3); the variance is largest at 50. When the scale is changed to a

[1] Particularly, for example, to ensure that variances are homogeneous, homoscedastic, and additive.

proportion by dividing by n or to a percentage by dividing by n and multiplying by 100, the relative variance is smallest at 50 and increases symmetrically in either direction away from 50 percent. If a set of sample data shows these relationships and it is intended to use, say, analysis of variance, then transformation to a scale in which the mean is independent of the variance is necessary (see Bartlett, 1947; and Table 15.1).

In a Poisson frequency distribution the mean $\mu = np$, and the variance $\sigma^2 = np$, so that the variance increases with the mean; here again any parametric analysis performed on data conforming to the Poisson frequency function requires a transformation to a scale in which the mean is independent of the variance (see Bartlett, 1947; and Table 15.1).

In most frequency distributions when asymmetry or skewness is present, the mean is related in some way to the variance; similarly, if the peakedness or kurtosis is significant, then the "variance of the variance" affects subsequent comparisons among means and variances. Quite often when relationships occur among the means and variances of sample data, they are characterized as heteroscedastic; and in data in which variance increases with increase in mean or decreases with increase in mean, the variance is heterogeneous. This can be quite serious in upsetting statistical analysis of bivariate data (e.g., Griffiths, 1959a; Griffiths and Smith, 1964).

Various other relationships between mean and variance may appear, as discussed in Secs. 13.13 and 14.6, and require some stabilizing transformation. In general, when it can be achieved, the experimental procedure may be adapted to generate the frequency-distribution model by suitable modification of sampling procedure. When a set of data is collected and the model frequency function is not known, it is prudent to examine the various statistical estimators to evaluate any lack of independence; and if a relationship appears, it is advisable to take steps to modify the experimental procedure or to apply a transformation so that the data comply, as closely as possible, with the model requirements incorporated in the statistical tests which are applied to analyze the data. An excellent discussion of the assumptions underlying the analysis-of-variance procedure is given by Eisenhart (1947), and the effect of departures on the results of the testing is described by Cochran (1947); a suitable set of transformations which may be used to fulfill the assumptions when the data as collected are in unsuitable form to apply analysis of variance is derived and discussed by Bartlett (1947).

15.3 Transformations Aimed at Attaining Independence of Mean $\bar{X}$ and Variance $\hat{\sigma}^2$

When the model frequency distribution is known and the samples are random samples, it is possible to postulate the relationships among the mean μ and the variance σ^2. For those frequency distributions in which the relationship of the mean μ to the variance σ^2 takes a known form, a suitable

TABLE 15.1
Transformation of Variates Arising from Measurement of Petrographic Properties by Different Techniques

(1) Property	(2) Frequency distribution[a]	(3) Relationship of mean and variance	(4) Transformation[b]
Compositional or count data	(5–95%) binomial	$\mu = np;\ \sigma^2 = npq$	$\sin^{-1}\sqrt{X}$
Thin section	(<5 or >95%) Poisson	$\mu = np;\ \sigma^2 = np$	$\sqrt{X}$ or $\sqrt{X + \frac{1}{2}}$
Accessory minerals	Negative binomial		
Any classified variate	Variable		Information H: $H = -\sum_i p_i \log p_i$
Grain size	Unimodal (?); varies in skewness and kurtosis	Varies	Empirical determination when relationship of $\bar{X}$ to $\hat{\sigma}^2$ known
Indirect			
Thin section	Positively skew	μ proportional to σ^2	$\log X$ or $X = 2^{-\phi}$
Direct in loose grains	Positively skew	μ proportional to σ^2	$\log X$ or $X = 2^{-\phi}$
Grain shape			As above or empirical function
Wadell sphericity and roundness	Binomial or Poisson	As above	
Visual sphericity and roundness	Binomial or Poisson (?) or information-transformation	As above	As above or empirical function
Linear compound of axes in phi	Normal	Independent	None
Orientation	Rectangular to	$\bar{X}$ and $\hat{\sigma}^2$ in small random samples tend to independence	None
	Normal	Independent	None
Packing	Binomial or Poisson	As above	As above
Density	Binomial (as n grows large, approx. normal)	Independent	None
Porosity	Binomial (%) as n grows large, approx. normal	Independent	None
Permeability	Positively skew	$\bar{X}$ proportional to $\hat{\sigma}^2$	$\log X$ or $X = 2^{-\phi}$

[a] All these frequency distributions are embraced by the central-limit law, so that means of random samples tend to normality as n grows large.
[b] From Bartlett, 1947.

transformation has been theoretically derived. A list of such transformations for the commoner frequency models is included in Table 15.1, and a similar list has been tabulated by Krumbein and Miller (1954), Krumbein (1955), and Miller and Kahn (1962, table 7.19, p. 176). Tables to facilitate these transformations are listed in Fisher and Yates (1948) and Snedecor (1956). Included in Table 15.1 are the model frequency distributions expected from random sampling geological populations by using the measurement procedures described in the earlier chapters (see also Krumbein, 1955). For example, if a layered population such as a detrital sedimentary rock is sampled by an appropriate stratified random sampling design, using sedimentation units as strata, and the individual grains of quartz are mounted to achieve their stable position, then the logarithms of the long axes of the quartz grains, if directly measured as described in Sec. 4.10, will be normally distributed. Bias, of course, may be introduced by deviations from any of the recommended procedures, and, generally, if bias is present, the frequency function of the long axes will show departures of varying degree from the normal model; in particular, some relationship between mean $\bar{X}$ and variance $\hat{\sigma}^2$ will almost certainly be induced by the sampling bias. It is not by any means a trivial problem to ensure that the data are collected in the appropriate manner.

When data are collected and the frequency function is not known and a relationship between mean $\bar{X}$ and variance $\hat{\sigma}^2$ appears, it is necessary to devise a transformation to convert the data to a scale in which the mean $\bar{X}$ is independent of the variance $\hat{\sigma}^2$. Bartlett (1947) recommends the use of the following procedure. Suppose the variance σ^2 is some function f of the mean m, that is, $\sigma^2 = f(m)$; then for any function $g(x)$, $\sigma_g{}^2 = \int (dg/dm)^2 f(m)$, approximately.

If the variance is to be constant, say C^2, then

$$g(m) = \int \frac{C\,dm}{\sqrt{f(m)}} \tag{15.1}$$

Suppose that $f(m)$ is proportional to m^2; then an approximate transformation is $g(m) = \log m$. An example of the use of this procedure is given in Rosenfeld and Griffiths (1953, p. 565). When an empirical procedure of this kind is used, it should be realized that the transformation, while useful for the set of data on which it is based, may not apply outside this set. Thus in the example quoted above (Rosenfeld and Griffiths, 1953, p. 565) any roundness value greater than 0.7 unit yields a negative adjusted value, so that the transformation is not suitable for all roundness data.

A number of relationships between mean $\bar{X}$ and variance $\hat{\sigma}^2$ or standard deviation $\hat{\sigma}$ are graphed by Quenouille (1950, chap. 8), and suitable transformations are recommended; quite frequently (see below) these relationships are of value in understanding the nature of variation in the variate and/or in isolating the source of the bias.

15.4 Examples of Real-world Relationships between Estimators

15.4.1 Grain Size and Size Sorting. The average grain size and the spread or dispersion of grain size are of considerable interest in geological studies because they are often taken to represent reflections of local conditions, e.g., changes in current velocity (Sec. 13.12). The spread or dispersion in grain size is generally called *sorting* (a measure of uniformity in grain size) and is considered to typify certain geological processes; e.g., aeolian sands are well sorted, whereas glacial sands are poorly sorted, a reflection of the contrast in efficiency of sizing by wind and ice, respectively. If the frequency function of the variate used to measure size is normal, then the mean $\bar{X}$ and variance $\hat{\sigma}^2$ or standard deviation $\hat{\sigma}$ are adequate measures of size and size sorting, respectively; but if the frequency function is not normal, then various other estimators will affect the measure of sorting, and, in particular, the kurtosis will be important. The best-sorted sediment is one in which all grains are of the same size; then the mean equals each measure, and variance is zero. If a small range in size is present, the variance becomes positive, and kurtosis is infinite. It is therefore necessary to use caution in discussing size sorting if the frequency function is nonnormal.

Another feature of considerable interest in measurements of grain size is that different techniques yield different results (Chap. 3) and, in particular, yield different relationships among mean $\bar{X}$ and variance $\hat{\sigma}^2$ or size and size sorting, respectively.

Relationship between Size and Size Sorting when Measurement Is Indirect. When indirect measurement procedures, e.g., sieving and pipette sedimentation, are used to measure grain size, the frequency function observed has open ends (Sec. 5.2); the order statistics, median Md_ϕ and quartile deviation QD_ϕ or percentile deviation PD_ϕ, are then used as estimators of the mean μ and standard deviation σ, respectively. It is generally found that a relationship exists between the median and the estimator of dispersion (Griffiths, 1951a; Inman, 1949; Shukri and Higazy, 1944). This trend may be treated as linear over limited ranges of grain size (e.g., Griffiths, 1951a), but is nonlinear over the entire size range (Griffiths, 1951a, b). The entire range of size is not adequately represented, probably because of sampling bias (Sec. 5.6), but the hypothetical relationship over the range -6 to 14 phi (64 mm to 0.06 micron) may be constructed from published data, as in Fig. 15.1. If the entire population is considered as a whole, the median (average) grain size is independent of the phi percentile deviation (dispersion, or sorting), as would be expected from random sampling a normal population (Fig. 5.4 and Sec. 5.6). This general trend may be subdivided into two humpbacked curves in terms of characteristics of the transporting medium; the finer-grained sizes from 4 to 14 phi (62.5 to 0.06 micron) represent detritus transported in suspension, whereas the coarser-grained sizes from 2 to -6 phi (250 microns to 64 mm) represent detritus transported in traction. Overlap

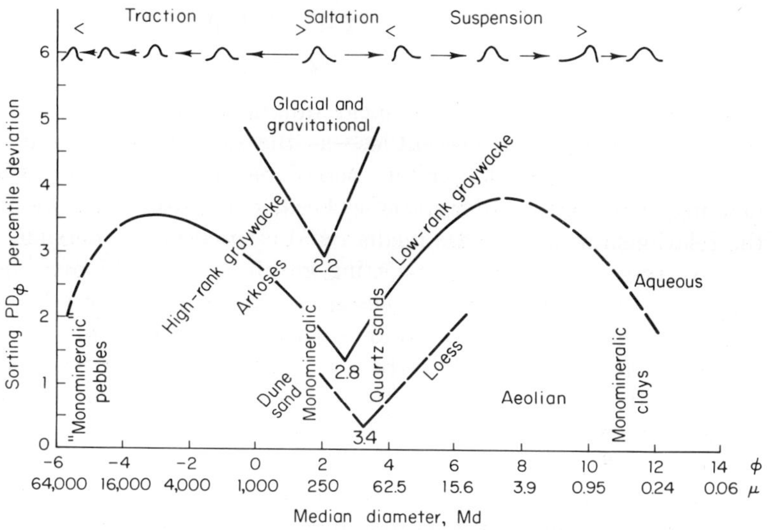

Figure 15.1 Size Md_ϕ and size sorting PD_ϕ in detrital sediments; indirect measurement procedure.

of the two trends occurs in the region 2 to 4 phi (250 to 62.5 microns), representing saltation. The entire trend may be further subdivided into a tripartite series on the basis of the size-sorting characteristics of different transporting media; the best-sorted sediments are aeolian (dune sand and loess), the loess representing suspension processes and the dune sand saltation processes. There appear to be few examples of aeolian detritus outside the size range of 1.5 to 6 phi units (35 to 15.6 microns).

The aqueous detritus may be similarly classified, and the entire range appears to be represented in detritus deposited from an aqueous medium. Finally, a very poorly defined trend is represented as attributable to glacial and gravitational deposits. The vague character of this trend is largely due to paucity of information on the size distributions of such deposits.

The best-sorted sediments are around 3.4 phi (95 microns), 2.8 phi (144 microns), and 2.2 phi (218 microns) in the three different media. As would be expected, the agencies and their transportational characteristics act as nonrandom sampling processes, thus introducing different degrees of bias, which are representative of the efficiency of the selective sorting intensity of each process.

If we next consider skewness and kurtosis estimated by appropriate order statistics and defined as (Krumbein, 1939b; Krumbein and Pettijohn, 1938)

$$Skq_\phi = \frac{Q_3 + Q_1 - 2\mathrm{Md}}{2} \qquad (15.2)$$

and
$$Kq_\phi = \frac{2(P_{90} + P_{10})}{Q_3 - Q_1} \tag{15.3}$$[1]

it can be seen that these are linear functions of the same parameters—the median, quartiles, and percentiles—as the size (median) and sorting (QD_ϕ and PD_ϕ); they cannot therefore be independent, and there is little point to plotting these higher-order statistics because the results can be predicted from the relationships among the median and estimate of dispersion.

By returning to the size-sorting graph of Fig. 15.1 and considering the frequency distributions of samples associated with the trends, it will be seen that over the size ranges around −6 phi, 2 to 4 phi, and 12 to 14 phi the frequency functions tend to be symmetrical and at least approximately normal. In all three ranges the detritus tends to be monodisperse, i.e., monomineralic, e.g., pebbles, which behave similarly toward sedimentational processes in the region of −6 phi, quartz grains, in the range of 2 to 4 phi, and single species of clay minerals in the region 12 to 14 phi units. If one begins with a typical sand composed of quartz of average grain size (median) equal to 3 phi and adds to it successive increments of clay mineral, the frequency function departs from symmetrical, and at first it increases in skewness with added increments of clay mineral. Eventually it decreases in skewness, becoming symmetrical again at the peak of the aqueous curve in the suspension

[1] This ratio has been reversed from that of Krumbein in order that increasing values of kurtosis represent increasing peakedness.

TABLE 15.2
Relationship between Size $\bar{X}_\phi$ and Size Sorting $\hat{\sigma}_\phi$ Measured as the Correlation Coefficient r; Axes of Quartz Grains from the Maxton Sand, West Virginia; Measurements Performed in Thin Sections

Statistic	$\bar{X}_a$	$\bar{X}_b$	$\hat{\sigma}_a$	$\hat{\sigma}_b$
$\bar{X}_a$		+0.993**	−0.263	−0.255
$\bar{X}_b$			−0.276	−0.280
$\hat{\sigma}_a$				+0.724**
$\hat{\sigma}_b$				

Note: For a = long axis, phi units

b = short axis, phi units

n = 33 thin sections, $r > 0.349$*

$r > 0.449$**

region; continued addition of increments of clay now reverses the sign of the asymmetry, and the skewness increases and then decreases again to a symmetrical (normal) curve around 12 phi. A similar display of change in symmetry occurs from 2 to −6 phi units as successive increments of "gravel" (rock fragments) are added to the quartz sand. When clay or gravelly components are added to sand, the change in skewness represents a concomitant change in mineral composition, size, and shape; hence variation in the so-called size frequency function is really a reflection of changes in three interdependent variables, as described in Chap. 3.

There seems little point to pursuing these studies further, as the model permits the prediction of relationships over the entire range of detrital sediments; and attempts to interpret the "meaning" of the size frequency distribution cannot overcome the confounding of variation from the three different sources. Indeed, this is not truly a size measure, and variation in the descriptive statistics merely reflects this fact; as mentioned in Chaps. 3 and 4, this size measure cannot in general be related solely to characteristics of the transporting media, because of the polydisperse nature of the detritus, and no rigorous mathematical analysis is applicable. This size measure produces data which achieve ordinality at best.

Relationship between Size and Size Sorting when Measurement is Performed in Thin Section. Where grain size is measured as the long a and/or the short b axes of quartz grains in thin section, as recommended in Sec. 6.3.1, the model is again the phi-normal frequency function. It would therefore be expected that if random sampling is achieved, the average grain size, now represented by the sample mean $\bar{X}_\phi$, and the sorting, represented by the sample standard deviation $\hat{\sigma}_\phi$, would be independent.[1] As a measure of linear association, the correlation coefficient may be used (Sec. 21.12), and an example using quartz-grain axes from thin sections of the Mississippian Maxton sand of West Virginia is the source of information in Table 15.2.

Similar results are found when other sandstones are treated in this way, e.g., the Salt Wash sandstone (Griffiths et al., 1956), the Cow Run sandstone (Griffiths, 1958b), the Chipmunk sandstone (Stanonis, 1956; Pachman, 1961), the Waynesburg and Middle Kittaning sandstones (Siegrist, 1961), and the Tuscarora, Oriskany, and Homewood quartzites (Wood, 1960). It may be said, therefore, that the mean is generally independent of the standard deviation when size is measured as the axes of quartz grains in thin section. Occasionally a low degree of linear association is found to be significant, but the coefficient of determination r^2 (see Sec. 21.2), expressed as a percentage, rarely exceeds 10 percent; or not more than 10 percent of the variation in size is common to the average and standard deviation of axes of quartz grains. When the correlation coefficient is significant in one or other of the pairs of axes and standard deviations listed in Table 15.2, it may arise because there is a low degree of linear association between the pair, it may be

[1] The range of $\hat{\sigma}$ is from 0.25 to 1.5 phi units; see Secs. 5.7 and 17.6.

an accident of sampling (which will occur 5 percent of the time), or it may be due to nonrandomness in the physical sampling process.

As an example, consider the measurements of the axes of "quartz" grains performed upon thin sections of specimens from the Pottsville and Tuscarora quartzites of eastern Pennsylvania (Landy, 1956); the results from 54 specimens in the Pottsville show no relationship between mean and variance of long axes. In 46 specimens from the Tuscarora, however, the phi medial correlation ϕ_r (Quenouille, 1952), an order statistic approximating the product moment correlation r, between the mean size $\bar{X}_\phi$ and the standard deviation $\hat{\sigma}_\phi$ of long axes yields a significant value $\phi_r = -0.409$ (for $n = 46$, significant at the 5 percent level; see Quenouille, 1952, table V). This implies that about 16 percent of the variation in the two measures is common and 84 percent is independent. If the correlation coefficient is real, then the variance increases with decrease in mean phi value or with increase in mean grain size (Fig. 15.2); in other words, there is a tendency for coarser "quartz" grains to vary more than finer ones. However, this relationship, significant but small, may be a reflection of biased sampling.

Relationship between Size and Size Sorting when Direct Measurement Is Performed on Individual Grains. Here again the model frequency function is phi-normal, and in random samples we expect the mean to be independent of the variance. These expectations are borne out in the limited data available, e.g., in quartz grains from a beach and associated dune sand (Hulbe, 1957), in quartzose pebbles from some New Jersey gravels (Steinmetz, 1956, 1962), in fragments of quartzite from Tuscarora scree, in quartzite pebbles from the Olean conglomerate, and in quartz grains from the Homewood quartzite (Griffiths, 1959a; Griffiths and Smith, 1964). A detailed study of the size and shape variation of quartz, garnet, and hornblende grains from beach, delta, and till deposits of the Pleistocene of Ontario, Canada, is described by McIntyre (1961); here again no obvious trend appears, and the linear association of mean and variance estimated by the phi medial correlation coefficient ($\phi_r = -0.288$, $n = 81$) is not significant at the 5 percent level (Quenouille, 1952). Even if this value were significant, the degree of common association between mean and variance would be less than 10 percent. Here

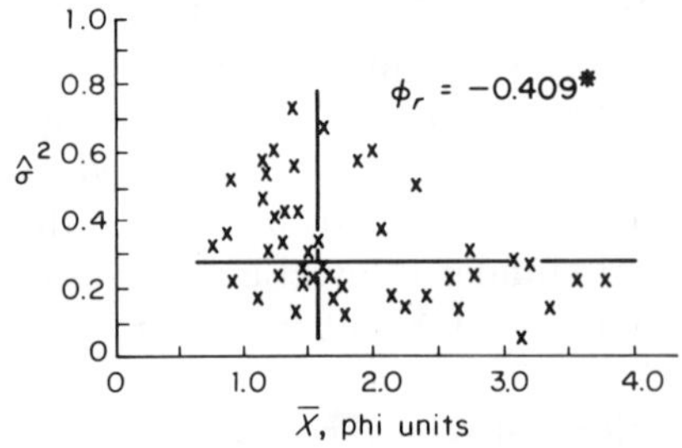

Figure 15.2 Size $\bar{X}_\phi$ versus size sorting $\hat{\sigma}_\phi$ in quartz-grain long axes from Tuscarora quartzite. (*Data from Landy, 1956.*)

again the standard deviation shows a very limited range similar to that shown by thin-section measurement; that is, $\hat{\sigma}$ varies from 0.2 to 1.0 phi unit.

15.4.2 ***Shape and Shape Sorting in Sediments.*** The relationships between mean $\bar{X}$ and variance $\hat{\sigma}^2$ of the sphericity and roundness of quartz grains in detrital sediments are described in Chap. 6; when visual estimation of these properties is used as measurement procedure, there appears to be no simple relationship in sphericity, but a relationship appears to be present for visual roundness. This is too limited a set of data upon which to generalize, but from other investigations the frequency distribution of roundness is generally strongly skewed toward smaller roundness values. Some transformation is necessary, but, so far at least, no general recommendations can be made.

When the measurements of shape are performed in thin section, the phi means of lengths and widths are independent of their respective phi standard deviations; if the measurements are expressed on a millimeter scale and shape is estimated by bivariate plots of the long and short axes, the graphs exhibit heteroscedasticity (Griffiths, 1959a; Griffiths and Smith, 1964). The alternative to logarithmic transformation is to use ratios of long to short axes (a/b) or short to long (b/a); in the latter case the range becomes 0 to 1 and in the former, 1 to ∞, both ratios containing the same information and b/a being the one in common use. The frequency distributions are generally approximately normal, although varying degrees of cutoff may sometimes appear, particularly an excess accumulation of frequency in the classes near a ratio value $b/a = 1$ (circular grain sections).

For the immediate future, the phi transformation for the axes, a linear compound such as the regression model (Fig. 6.8) to describe the relationship between the axes, and the correlation coefficient as a measure of degree of association appear suitable analytical tools. If this procedure is followed, the ordering of the axes as $a > b$ in millimeters ($b < a$ in phi units) induces cutoffs in the bivariate plots. This, in turn, affects the correlation coefficient and induces a correlation of 0.66 in the bivariate case (Smith, personal communication). However, since correlations among means of small sets of data generally exceed $r = 0.75$, some degree of association is present despite the pseudocorrelation. It may well be that the association among long and short axes of grains is effectively zero, whereas, among means of sets of grains, as either the number in the set or the number of sets is increased, the association increases (see Griffiths et al., 1956, and Sec. 21.15).

Since the measurement of axes in thin section really represents measurements of "apparent" long and short axes, the results of such studies are based on empirical data and may not bear any simple relationship to actual (real-world) variation in shape.

When shape measurements are performed directly on individual particles, the three axes, long a, intermediate b, and short c, may be transformed from the millimeter or micron scale to phi units; the frequency model for each axis is then the phi-normal distribution. Any linear compound of these measurements is also normally distributed; e.g., the sphericity (ellipsoidicity) measure ψ recommended by Krumbein (1941b) is defined as

$$\psi = \sqrt[3]{\frac{bc}{a^2}}$$

which upon transformation becomes

$$\psi' = \tfrac{1}{3}(b_\phi + c_\phi - 2a_\phi)$$

and this will be normally distributed; empirical studies confirm this expectation (Curray, 1949; Curray and Griffiths, 1955). It may be postulated then that means $\bar{X}$ and variances $\hat{\sigma}^2$ of random samples will be independent, and this is also confirmed in Fig. 6.15. However, where the samples are grouped on the basis of petrographic composition into arkoses, low-rank graywackes, and quartzites, the means and variances show definite relationships. For arkoses, which in the three examples chosen for study represent near-source, short-traveled detritus, the sample means vary little, whereas the variances tend to be highly variable. On the other hand, for the three quartzites, representing far-traveled detritus, the variances are less variable and the means much more variable. The low-rank graywackes exhibit an intermediate stage, in which there is a trend (here linear) between mean and variance. These relationships form the basis of the sampling model of Fig. 2.3 and are a reflection of the degree of selective sorting on the basis of shape, which the detritus (quartz) has undergone from source area to the locus of deposition. Again the relationships between mean and variance represent varying degrees of bias which may be used to characterize various detrital processes, i.e., environments or facies.

Roundness measurements on the same quartz grains show similar relationships between means and variances and again reflect the same processes of selective sorting. It seems likely from these data that roundness, as is expected from laboratory experiments (Sec. 6.7), is more sensitive to environmental changes than sphericity.

In future investigations of variation in shape[1] of individual particles it would be advisable to adjust the sampling arrangements to offset the various mean and variance relationships within the subpopulations if comparisons between subpopulations are to be meaningful; this is equivalent to recommending that sampling be random, i.e., that the equal-probability model can be applied to the sample data.

[1] The same sampling recommendations apply to all properties which have suffered varying degrees of selective sorting, so that the recommendations are quite general.

15.4.3 Composition and Compositional Sorting in Sediments.

The relationships among mean and variance may be looked upon as equivalent in all variates, and so the concept of size and size-sorting relationships is not restricted to grain-size studies. When the variates used to investigate variation in composition are considered, the procedure of counting suggests that the discrete frequency distributions are the appropriate models. When there are k ($k = 2$) classes, the binomial or Poisson frequency distributions are appropriate; and when there are more than two classes ($k > 2$), the multinomial frequency function is the appropriate model. The former models have already been discussed, and they have quite definite relationships between mean and variance; in this respect so does the multinomial (Sec. 14.10), wherein the mean is related to the variance in a form similar to that of the binomial (see page 301). In addition, however, a covariance relationship appears among the different classes, and this further complicates the handling of such data. Chayes has described the effect this relationship has upon comparisons of variation in frequency between two or more classes and, in particular, has emphasized the presence of pseudocorrelation (Chayes, 1949b, 1960, 1961, 1962). Suppose there are three variables which total 100 percent, i.e., sum to a constant; then, where any two increase in frequency, the third must decrease, leading to various degrees of negative association between the variates. This feature makes the study of variation in composition, beyond the simplest qualitative comparisons, very difficult.

It is also clear that no monotonic transformation will be effective. Recently the development of information theory (Shannon, 1948) has achieved the level wherein applications to statistical analysis (Kullback, 1959; Miller and Kahn, 1962) are available, so that the Shannon-Wiener entropylike function

$$H = -\sum_i p_i \log p_i \tag{15.4}$$

may well be an appropriate transformation for closed data (Griffiths, 1960). In this transformation logarithms to the base 10, to the base e, or to the base 2 may be used, the last-mentioned being the base for binary scales and forming the binary relationship underlying the logic design of the electronic computer. Kullback (1959, table 1, pp. 367ff.) includes a table of $p \ln p$ to facilitate the application of this transformation in statistical analysis of data. When the base 2 is used, the resulting transformed data are directly measured as "bits" (short for binary digits), and it has been proposed to call the equivalent transformed data "nits" when the base e is used and "logits" when the base 10 is used (MacDonald, 1952).

The phi transformation first recommended by Krumbein (1938) may be considered as

$$d = 2^{-\phi}$$

where d is the diameter in millimeters; upon transformation, using base 2, this becomes

$$\log_2 d = -\phi \qquad \text{in bits}$$

If, therefore, the information transformation to base 2 is used on percentage data representing variation in composition, the resulting transformed data are in bits, the same scale as the grain-size (and shape) data.

Application of information theory to geological problems was first proposed by Pelto (1954), and some examples are discussed in Miller and Kahn (1962); an example of the use of the information transformation on percentage data is described by Griffiths (1960).

15.5 The Coefficient of Variation

Another aspect of the relationship between mean and dispersion is included in the use of the coefficient of variation CV, which is defined as

$$CV\% = 100 \frac{\hat{\sigma}}{\bar{X}} \tag{15.5}$$

representing the ratio of the mean $\bar{X}$ to the standard deviation $\hat{\sigma}$ in percent. The coefficient of variation is insidiously attractive because it apparently permits the comparison of variation in different variates free from scale effects; i.e., it is dimensionless. However, it has the appropriate attribute only if the data upon which it is based achieve ratio level (Sec. 12.5), and its use with data which fail to achieve this information level can be misleading.

The coefficient of variation is plotted as a graph of mean versus standard deviation in Fig. 15.3; when the graph is appropriately used, only the right-

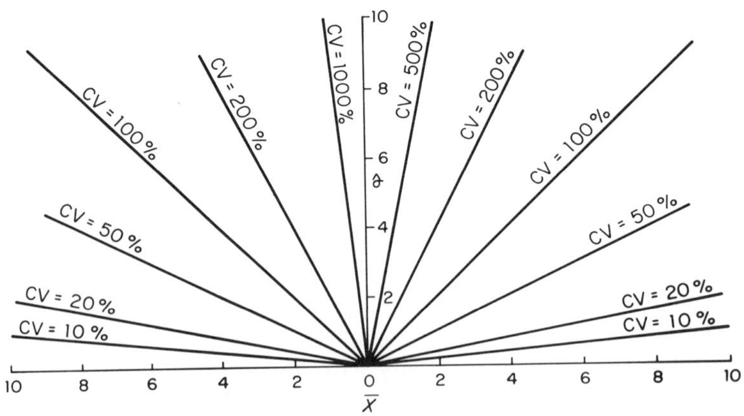

Figure 15.3 Coefficient of variation as an expression of the relationship between mean and standard deviation.

hand half applies, a variate with value $\bar{X} = 0$ having $\hat{\sigma} = 0$ because the value zero is physically meaningful on a ratio scale. However, if the variate is transformed, to achieve normality, for example, as in the case of the use of the phi transformation on grain-size data, then negative values of X_i and $\bar{X}$ are permissible, and the data may plot on the entire range of the graph of Fig. 15.3. The graph of size versus size sorting in Fig. 15.2 is an example, but the coefficient of variation should not be applied to these data. As the mean becomes smaller in absolute value for fixed standard deviation, the coefficient of variation shows an increase, until near zero the CV percent is infinite. Since it is often suggested that a CV percent exceeding, say, 30 suggests that variation is "out of control," the variability in grain size (in phi units) rapidly becomes large as sizes in the range of 1,000 microns (0 phi unit) are approached. The coefficient of variation should not be used on such data.

Similar features apply to any variate having a mean below one unit unless the variation (standard deviation) is correspondingly small. For example, when using the weight percent of heavy residue and its variation, since the weight percent is fractional, even relatively small scale variation will result in a very large CV percent. It may well be that this really reflects lack of control in the measure, but it also may be an artifact, because the scale does not represent an adequate information level for this kind of comparison. Similar features arise whenever percentage data are used and the coefficient of variation is applied as a summarizing statistic (see, for example, Fairbairn, 1953).

15.6 The Concept of Control and the Use of Control Charts

The development of mass-production engineering processes relies on a suitable measure of "control" to ensure that the products are, within specified limits, identical. Quality control, which is the process underlying this approach, was first systematized in the classical work of Shewart (1931, 1939). This concept of control may be extended to methodological applications; for example, Churchman offers this definition of the controlled experiment:

> An experiment is said to be controlled if we state all the formal conditions under which a mathematical function of a series of observations approaches a limit stochastically. The definition now becomes our criterion of meaning . . . . No question of fact can be said to have meaning unless there exists a controlled experiment for its answering.[1]

Here again the fundamental basis of the analytical procedure is a study of relationships of mean and dispersion. The concept of control depends on the fact that in a normal distribution the probability of an observation's falling in the range $\bar{X} \pm k\hat{\sigma}$ may be very precisely specified for any selected k,

[1] C. W. Churchman, "Theory of Experimental Inference," pp. 182–183, The Macmillan Company, 1948.

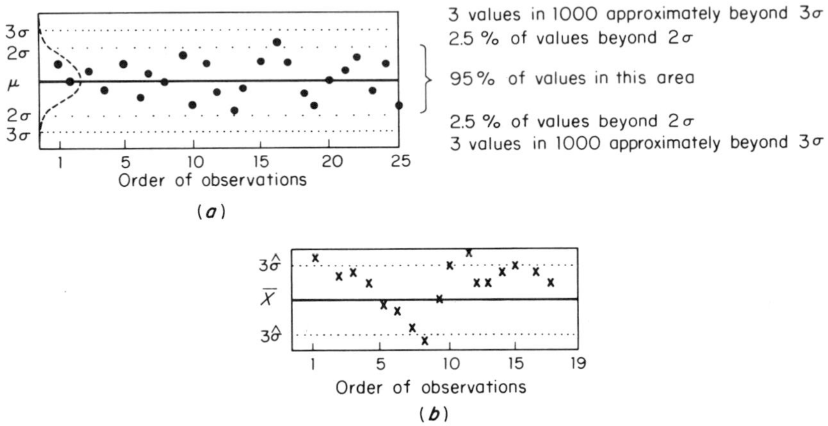

Figure 15.4 The concept of control. (*a*) Observations in control; (*b*) observations out of control.

and it is conventional to use the values $k = 2$ or 3, with associated probabilities of 95.4 and 99.7, respectively. In studies of control, however, the sample values are ordered, and, generally, small related sets are averaged and a control chart is applied to study variation in the means by using the control limits $\bar{X} \pm k\hat{\sigma}_{\bar{X}}$ (where $k = 2$ or 3). Time of appearance is the basis for ordering the observations in industrial quality control, but the procedure may be extended to embrace any basis of classification which leads to rational grouping.

Suppose a series of observations is arranged in order of their appearance, the order being determined on the basis of time or on depth in a well core or on geographical location in a linear traverse; then if the variation in the variate follows the normal law, the mean and standard deviation may be calculated, and the observations may be plotted on a control chart, as in Fig. 15.4*a*. If the observations are in control, they will fluctuate randomly around the mean value, and some 5 in 100 will fall outside the 2σ limits or 3 in 1,000 will be expected outside the 3σ limits. If a number of observations exceeding these expectations fall outside the control limits, these observations may be looked upon as not belonging to the same population.

The fact that the observations are ordered, however, permits an extension of this generalization; the observations need not fall outside the limits to indicate lack of control. For example, in an industrial process which produces a homogeneous product when the process is in control, the wear on a machine may develop gradually; then a series of observations will show a systematic trend, i.e., the characteristic of random variation will be lost. Such trends arise from a systematic bias, and it is customary to search for an "assignable cause" and remove the interfering source of variation by replacing worn parts in the machine and so on. The systematic trend may appear as in Fig. 15.4*b*, and the presence of such a trend warns the observer that his observations are not in control.

An example of the application of the concept of quality control to improvement in the use of a sample splitter is described by Otto (1937).

One of the most effective areas of application of the concept of control is that of testing the consistency of operator performance in the various measurement procedures used in collecting the data upon which analysis is performed, and a number of examples are given by Griffiths and Rosenfeld (1954); in particular, an example of the use of the quality-control graph to judge when an operator is out of control in point counting thin sections to determine the proportion of different mineral constituents is illustrated in Fig. 15.5 (Griffiths and Rosenfeld, 1954, fig. 1). The character of the fluctuations over time is quite characteristic; thus, most observers tend to fluctuate widely at first but gradually settle down to limited variation around some mean. Then, and only then, can the means for different operators be compared; and if large-scale differences exist, assignable causes can be searched for, and the criteria for assigning the points to classes of elements can be refined until the operators are performing consistently. In the example quoted, one operator (operator V, Fig. 15.5) obtained estimates of quartz

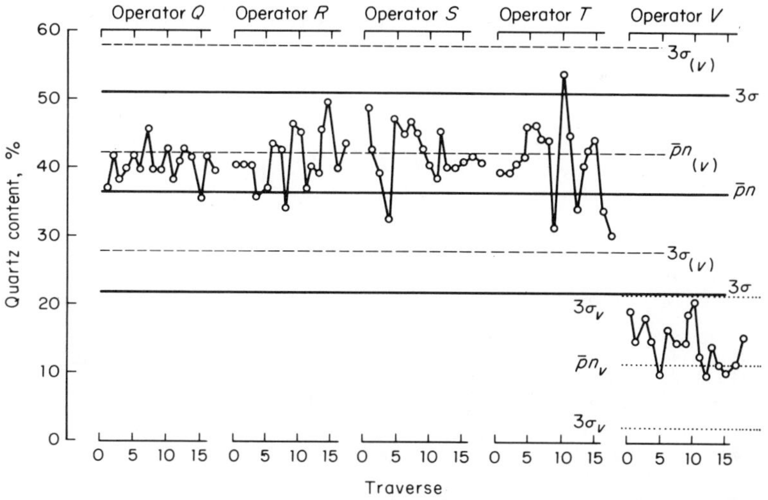

Figure 15.5 The concept of control applied to improvement in operator performance in estimating proportions of quartz in thin sections of detrital sediments. $\bar{p}n$, mean for all operators; $\bar{p}n_V$, mean for operator V; pn_V, mean for all operators without operator V. [*Reprinted from Operator Variation in Experimental Research, J. C. Griffiths and M. A. Rosenfeld, J. Geol.*, **62**:*80 (1954), by permission of The University of Chicago Press. Copyright 1954 by the University of Chicago.*]

well beyond the three standard-deviation limits of the ensemble of estimates from all operators. His results belong to a different population from those of the other operators, and these data should be discarded for subsequent analysis. It was found that in this case the aberrant operator was using a set of criteria for recognizing quartz different from that of the others, and once this had been pointed out, the results became more consistent.

However, equally important features appear in plots of the other operators; for example, operator Q is extremely consistent and shows little fluctuation. Operator S represents the more normal case, starting out with wide fluctuations and then gradually achieving consistence. Operator R is evidently deteriorating, and his fluctuations are increasing with time; obviously such a case demands action to modify this trend. Similarly, operator T shows unusually large-scale fluctuations, and steps should be taken to improve this situation. All these observations are of interest, but from the control limits based on operators Q, R, S, T, without V (symbolized as $3\sigma_{(V)}$ in Fig. 15.5), none of these variations lie outside the 3σ limits, so that here the ordering of the observations is the important feature in displaying lack of control. It is pointed out in Chaps. 19 and 20 that operator inconsistency reduces the sensitivity of the experiment and implies the presence of interaction between main effects which, if extreme, may entirely vitiate the experiment. It seems advisable, therefore, to apply the control-chart technique to ensure that operator performance is in control whenever new techniques or new operators are incorporated in a research program.

16 The commoner tests of significance: Student's t test

Again, although it is well known that the method of using the normal curve is only trustworthy when the sample is "large," no one has yet told us very clearly where the limit between "large" and "small" samples is to be drawn.

The aim of the present paper is to determine the point at which we may use the tables of the probability integral in judging of the significance of the mean of a series of experiments, and to furnish alternative tables for use when the number of experiments is too few.

Student, 1908

16.1 Introduction

The preceding statistical analyses have been mainly concerned with the reduction of sets of data to some single frequency distribution. The essential aim is to represent the set of data by an appropriate model; then characteristic descriptive statistics may be used to estimate the model parameters. In this way a large collection of data can be summarized in a few representative statistical estimates. Some tests of significance are performed to check on the representativeness or otherwise of the statistical estimators, and if the selected model is considered appropriate, the samples are tested for representativeness, i.e., randomness. The next step is to consider the application of frequency distributions to tests of significance and to determine what kinds of conclusions may be drawn from sets of data, particularly concerning their means and variances, and how to extend the tests from a single sample to the comparison of two or more sample means and variances.

16.2 Student's t Test for Specifying Characteristics of Sample Means

The fundamental procedure has already been discussed when assumptions of normality are made and it is assumed that the population mean μ and

Quotation: "Student," The Probable Error of a Mean, in E. S. Pearson and J. Wishart (eds.), "Student's Collected Papers," Biometrika Office, University College, London.

variance σ^2 are known. In general, the procedure consists of estimates of the probability of occurrence of some value of a variate X_i falling into an interval $X_i + dX_i$; the probability was defined as equivalent to the ratio of the area $X_i + dX_i$ to the total area under the frequency curve (Secs. 13.3 et seq.). The requirement of known μ and σ^2 restricts the application of such test procedures to those unusual cases where the model is completely known and/or a sufficiently large sample of data has been collected to yield firm estimates of the parameters. A more general experimental situation comprises the features that the normal model is an adequate approximation and some suitable guess or assumption may be made about μ but σ^2 must be estimated from small sets of samples of data. In such a case the use of the standardized transformation $t = (X - \mu)/\sigma$ or $t = (\bar{X} - \mu)\sqrt{N}/\sigma$ is not feasible, and it is necessary to substitute

$$t = \frac{(\bar{X} - \mu)\sqrt{N}}{\hat{\sigma}} \tag{16.1}$$

The discovery of this form of t was first described by W. S. Gossett, who published under the famous pseudonym of Student, and the development of small-sample statistical procedures is generally understood to have begun with his classical article on the distribution of t of Eq. (16.1) (Student, 1908; Pearson and Wishart, 1942). It will be convenient to define this t distribution by means of chi square as follows:

THEOREM 1: If X is normally distributed with mean μ and variance σ^2 and $t = (X - \mu)/\sigma$, then t is normally distributed with mean zero and unit variance. Then the sum of squares of n random sample values of t has a chi-square distribution with ν d.f. (Hoel, 1947, p. 136).

Under the stipulated conditions

$$\chi^2 = t_1^2 + t_2^2 + t_3^2 + \cdots + t_\nu^2 \tag{16.2}$$

with ν d.f.

Now consider a test criterion

$$t = \frac{(\bar{X} - \mu)\sqrt{N}}{\hat{\sigma}}$$

to be used in testing the hypothesis that a sample set of X_i is drawn from a normal population with a specified mean.

The sample mean $\bar{X}$ and variance $\hat{\sigma}^2$ are calculated from a random sample of size n drawn from the parent population, which is normal with mean μ and variance σ^2. Two properties of t should be noted (Anderson and Bancroft, 1952, p. 77):

1 t is the ratio between a normal deviate and the square root of an unbiased estimate of its variance.

2 t is the ratio between the square of a variate $y = \sqrt{n}(x - \mu)/\sigma$, which is normally distributed with mean $\mu = 0$ and variance $\sigma^2 = 1$, and a variate $\hat{\sigma}^2/\sigma^2$, which is distributed as χ^2/ν with ν d.f.

Then
$$t^2 = \frac{n(\bar{X} - \mu)^2/\sigma^2}{\hat{\sigma}^2/\sigma^2} \tag{16.3}$$

and
$$t = \frac{\sqrt{n}(\bar{X} - \mu)}{\hat{\sigma}} = \frac{\bar{X} - \mu}{\hat{\sigma}/\sqrt{n}}$$

or
$$t = \frac{\bar{X} - \mu}{\hat{\sigma}_{\bar{X}}} \tag{16.4}$$

This criterion has a Student's t distribution with $\nu = n - 1$ d.f.

It will be noted that σ^2 does not enter into this expression, and the only parameter required is μ. By means of this criterion, therefore, it is possible to substitute an expected value for μ and determine whether the set of samples X_i are random samples from a normal population with given μ and $\hat{\sigma}^2$ calculated appropriately from the set of observations.

Again, suppose an experiment yields two sets of data n_1 and n_2, respectively, from which are calculated $\bar{X}_1$ and s_1 and $\bar{X}_2$ and s_2; then the means may be compared and the difference tested against a suitable estimate of the variance. The mean difference may be expressed as

$$(\bar{X}_1 - \mu) - (\bar{X}_2 - \mu) = \bar{X}_1 - \bar{X}_2 \tag{16.5}$$

By choosing an appropriate unbiased estimate of the population variance $\hat{\sigma}^2$, no further parameter is required for this test.

Student's t test requires that the variate be expressed as a ratio of the measure to the square root of an *unbiased estimate* of its variance, or the variate is transformed to

$$t = \frac{\bar{X} - \mu}{\hat{\sigma}_{\bar{X}}}$$

Then, under the stated restrictions, the X_i are independent random samples from some normal population with mean μ, and t is distributed as Student's t with $\nu = n - 1$ d.f. This transformation is called *Studentizing* a statistic and is of wide application in significance tests of statistics.

The frequency distribution of t may be represented as (Tippett, 1952, p. 143)

$$\text{d.f.} = \left\{\frac{[(\nu - 1)/2]!}{[(\nu - 2)/2]!\,\sqrt{2\pi}}\right\}\left(1 - \frac{t^2}{\nu}\right)\exp\left(-\frac{\nu + 1}{2}\right) dt \tag{16.6}$$

The only two variates are the degrees of freedom ν and the transformed statistic Student's t.

This distribution is symmetrical and approaches the normal as the degrees of freedom approach infinity. In general, the curve is somewhat less peaked and has a larger spread than the equivalent normal distribution (Hoel, 1947, fig. 5, p. 144). The areas under these frequency curves have been calculated and are published in tables, the most popular form being that due to Fisher (1948, table IV, p. 174), which has been widely reproduced (Fisher and Yates, 1948, table III, p. 32; Arkin and Colton, 1950, table 12, p. 116). The values of t are recorded in the body of the table, and the columns represent probability levels, whereas the rows are designated as degrees of freedom.

16.3 Examples of Application of Student's *t* Distribution

It is necessary in all cases to obtain an unbiased estimate of the population variance $\hat{\sigma}^2$ from the samples to use Student's t test, and some care is necessary to ensure the appropriate procedure; the best unbiased estimate of the population variance σ^2 obtainable from the sample is

$$\hat{\sigma}^2 = \sum_{i=1}^{n} \frac{(X_i - \bar{X})^2}{n - 1} \tag{16.7}$$

where n is the number of X_i in the sample (see Sec. 16.9 for derivation).

Suppose we have measured the grain size, i.e., the long axes, of nine quartz grains by Hulbe's technique (1955) and obtained

$$\bar{X} = 2.00 \text{ phi} \qquad \hat{\sigma} = 0.6 \text{ phi} \qquad n = 9$$

We wish to test the hypothesis that this sample set of measurements is a random sample from a population in which $\mu = 0$. Calculate

$$t = \frac{(\bar{X} - \mu)\sqrt{n}}{\hat{\sigma}}$$

then

$$t = \frac{(2 - 0)\sqrt{9}}{0.6}$$

$$= \frac{6}{0.6} = 10$$

The mean $\bar{X}$ was calculated from the sample; hence the degrees of freedom ν are 1 less than the number of measurements: $\nu = n - 1 = 8$. Referring to the t tables (Fisher and Yates, 1948; Arkin and Colton, 1950) indicates that for $\nu = 8$, $P_{01} = 3.36$ (or $P_{001} = 5.041$), and a value of 10 is well beyond either; therefore, if the hypothesis is correct, such a set of measurements would arise much less often than once in 1,000. We may reject the hypothesis and claim that this set of variates is not likely to have risen from random sampling a normal distribution with mean $\mu = 0$.

If in another case we find $\hat{\sigma} = 6.0$, where $\bar{X} = 2$ phi and $n = 9$, then

$$t = \frac{(2 - 0)\sqrt{9}}{6} = 1.0$$

Such a value of $t = 1$ with $\nu = 8$ would arise more often than 10 times in every 100 (Arkin and Colton, 1950); more specifically, by using the Fisher and Yates table (1948) this value of t arises under the conditions stated in the hypothesis between 40 and 30 times in every 100. Such an event is quite common and constitutes no evidence against the hypothesis. In this case, then, $\mu = 0$ is a likely possibility, and the contrast between these two tests indicates the effect of variability on tests of hypotheses.

A statement of somewhat greater generality may be made by means of confidence limits; thus in a normal distribution some 95 percent of the measurements are included between the limits

$$\mu \pm 1.96\sigma$$

and this relationship may be stated as

$$\bar{X} + 1.96\sigma > \mu > \bar{X} - 1.96\sigma \tag{16.8}$$

Here the interval $\mu \pm 1.96\sigma$ includes some 95 percent of the area of a normal curve, and we may now replace this coefficient by the equivalent value of Student's t, thereby allowing the expression to be written with $\hat{\sigma}_{\bar{X}}$ in place of σ as follows:

$$\bar{X} - t\hat{\sigma}_{\bar{X}} < \mu < \bar{X} + t\hat{\sigma}_{\bar{X}} \tag{16.9}$$

Now we may substitute from the previous values $\bar{X} = 2$ phi, $\hat{\sigma}_{\bar{X}} = 0.2$, and $n = 9$; then

$$2 - 0.2t < \mu < 2 + 0.2t$$

and for 8 d.f. and the 95 percent probability level $t = 2.31$, whence

$$2 - 0.2 \times 2.31 < \mu < 2 + 0.2 \times 2.31$$

$$2 - 0.462 < \mu < 2 + 0.462$$

or

$$1.538 < \mu < 2.462$$

The mean μ may lie within this interval or outside it; if the experiment is repeated and the interval recalculated with the new estimate $\hat{\sigma}$, then, on an average, 95 percent of such intervals will contain μ. Alternatively, the probability that any one of the intervals contains μ is 95 percent (McCarthy, 1957, pp. 195–207).

Suppose now that a second random sample of nine quartz-grain long axes yields a mean of 3.00 phi units; then this series differs from the previous one at the 95 percent probability level.

In the second case with $\bar{X} = 2$, $\hat{\sigma} = 6.0$, and $n = 9$ the confidence limits become

$$2 - 2.0 \times 2.31 < \mu < 2 + 2.0 \times 2.31$$

$$-2.62 < \mu < 6.62$$

In this case $\mu = 0$ is a likely value; to obtain a mean with the same precision in terms of the spread of these confidence limits requires increase in sample size, proportional to the square of n. Suppose we increase the sample size to $n = 81$, with the same standard deviation and mean; then $\hat{\sigma}_{\bar{X}} = \frac{6}{9} = 0.667$ and, at the 95 percent probability level, the value of t for 80 d.f. is 1.99. Hence

$$\bar{X} - t\hat{\sigma}_{\bar{X}} < \mu < \bar{X} + t\hat{\sigma}_{\bar{X}}$$

$$2 - 1.99 \times 0.667 < \mu < 2 + 1.99 \times 0.667$$

$$2 - 1.327 < \mu < 2 + 1.327$$

$$0.673 < \mu < 3.327$$

We may now reject the hypothesis of $\mu = 0$ at the 5 percent probability level; with large variability it becomes quite expensive to reduce the confidence limits.

16.4 Comparison of the Means of Two Samples

The t test is also appropriate to test the hypothesis that "these two samples are random samples from the same normal distribution." Initially we may consider that the two samples of n_1 and n_2, respectively, are independently drawn from a normal population with mean μ and variance σ^2. Then the difference between the means $(\bar{X}_1 - \mu) - (\bar{X}_2 - \mu)$ is to be compared with the best unbiased estimate of the variances; μ disappears from the numerator, and the mean difference becomes $\bar{X}_1 - \bar{X}_2$. Then

$$t = \frac{\bar{X}_1 - \bar{X}_2}{\hat{\sigma}_{\bar{X}_1 - \bar{X}_2}} \tag{16.10}$$

It is now necessary to calculate an unbiased estimate of the variance of the difference between $\bar{X}_{1i}$ and $\bar{X}_{2i}$. The variance of the sum or difference of two random variates from a normal distribution may be defined as (Snedecor, 1946, p. 159)

$$\mathrm{Var}\ (\bar{X}_{1i} + \bar{X}_{2i}) = \sigma_1^2 + \sigma_2^2 + 2\rho\sigma_1\sigma_2 \tag{16.11a}$$

$$\mathrm{Var}\ (\bar{X}_{1i} - \bar{X}_{2i}) = \sigma_1^2 + \sigma_2^2 - 2\rho\sigma_1\sigma_2 \tag{16.11b}$$

Then if the two random samples are independent, the association between the $\bar{X}_{1i}$ and $\bar{X}_{2i}$, measured by ρ, is zero, and the variance of the sum or difference becomes

$$\mathrm{Var}\ (\bar{X}_{1i} + \bar{X}_{2i}) = \sigma_1^2 + \sigma_2^2 \tag{16.12}$$

Similarly, the variance of the sum or difference of means from independent random samples from the same normal distribution is

$$\mathrm{Var}\ (\bar{X}_1 \pm \bar{X}_2) = \sigma_{\bar{X}_1}{}^2 + \sigma_{\bar{X}_2}{}^2$$

$$= \frac{\sigma_{X_1}{}^2}{n_1} + \frac{\sigma_{X_2}{}^2}{n_2} \tag{16.13}$$

If the hypothesis is correct and these two random samples are drawn from the same normal population,

$$\sigma_{X_1}{}^2 = \sigma_{X_2}{}^2 = \sigma^2 \tag{16.14}$$

and $$\mathrm{Var}\ (\bar{X}_1 - \bar{X}_2) = \sigma^2\left(\frac{1}{n_1} + \frac{1}{n_2}\right)$$

or $$\sigma_{\bar{X}_1 - \bar{X}_2} = \sigma\sqrt{\frac{1}{n_1} + \frac{1}{n_2}} = \sigma\sqrt{\frac{n_1 + n_2}{n_1 n_2}} \tag{16.15}$$

Now from samples 1 and 2 we may calculate

$$\hat{\sigma}_1^2 = \frac{Sx_1^2}{n_1 - 1} \qquad \text{and} \qquad \hat{\sigma}_2^2 = \frac{Sx_2^2}{n_2 - 1}$$

or, combining the two estimates, we have

$$\hat{\sigma}^2 = \hat{\sigma}_1^2 + \hat{\sigma}_2^2 = \frac{Sx_1^2 + Sx_2^2}{n_1 + n_2 - 2}$$

But $$Sx_1^2 = n_1 s_1^2 \qquad \text{and} \qquad Sx_2^2 = n_2 s_2^2$$

Then $$\hat{\sigma}^2 = \hat{\sigma}_1^2 + \hat{\sigma}_2^2 = \frac{n_1 s_1^2 + n_2 s_2^2}{n_1 + n_2 - 2}$$

(in terms of sample variances);[1] or

$$\hat{\sigma} = \sqrt{\frac{n_1 s_1^2 + n_2 s_2^2}{n_1 + n_2 - 2}}$$

and

$$\hat{\sigma}_{\bar{X}_1 - \bar{X}_2} = \sqrt{\frac{n_1 s_1^2 + n_2 s_2^2}{n_1 + n_2 - 2}} \sqrt{\frac{n_1 + n_2}{n_1 n_2}} \tag{16.16}$$

or

$$\hat{\sigma}_{\bar{X}_1 - \bar{X}_2} = \sqrt{\frac{s_1^2}{n_2} + \frac{s_2^2}{n_1}} \sqrt{\frac{n_1 + n_2}{n_1 + n_2 - 2}} \tag{16.17}$$

We now return to the hypothesis that these two samples are random samples from the same normal population and calculate the studentized statistic t as follows:

$$t = \frac{\bar{X}_1 - \bar{X}_2}{\sqrt{s_1^2/n_2 + s_2^2/n_1}\ \sqrt{n_1 + n_2/n_1 + n_2 - 2}} \tag{16.18}$$

which follows the Student t distribution with degrees of freedom $\nu = n_1 + n_2 - 2$

If $n_1 = n_2$, this reduces to

$$t = \frac{\bar{X}_1 - \bar{X}_2}{\sqrt{s_1^2 + s_2^2/n}\ \sqrt{2n/2n - 2}}$$

$$= \frac{\bar{X}_1 - \bar{X}_2}{\sqrt{s_1^2 + s_2^2/n - 1}} \tag{16.19}$$

which follows the Student t distribution with degrees of freedom $\nu = 2(n - 1)$.

16.5 Example of the Application of Student's t to the Comparison of Two Means

In an experiment comprising the measurement of the long axes of quartzite pebbles from the Montoursville gravel near Williamsport the following statistics were obtained by different sampling plans applied to the same outcrop face (the means are expressed in minus phi units):

$$\bar{X}_1 = 4.571 \qquad \hat{\sigma}_1^2 = 0.6673 \qquad n_1 = 336$$

$$\bar{X}_2 = 5.547 \qquad \hat{\sigma}_2^2 = 0.5628 \qquad n_2 = 324$$

The null hypothesis H_0 to be tested is that "these two samples are random samples from the same normal population."

[1] Note the differences between sample variances s^2 and best unbiased estimate of the population variance from the sample $\hat{\sigma}^2$ (see Sec. 16.9).

It will be noted that the variances s^2 in Eqs. (16.18) and (16.19) are sample variances, whereas the values calculated from the Montoursville gravel are best estimates. It is necessary, therefore, to multiply these best estimates $\hat{\sigma}^2$ by $(n_1 - 1)/n_1$ and $(n_2 - 1)/n_2$, respectively, to obtain $s_1^2 = 0.665314$ and $s_2^2 = 0.561063$. Then

$$t = \frac{\bar{X}_1 - \bar{X}_2}{\sqrt{s_1^2/n_2 + s_2^2/n_1}\sqrt{n_1 + n_2/n_1 + n_2 - 2}}$$

$$= \frac{5.547 - 4.571}{\sqrt{0.6653/324 + 0.5611/336}\sqrt{660/658}}$$

$$= \frac{0.976}{\sqrt{405.3372/108.864}\sqrt{1.003039}}$$

$$= \frac{0.976}{\sqrt{0.003723}\sqrt{1.003039}}$$

$$= \frac{0.976}{0.06111} = 15.971$$

with degrees of freedom $\nu = n_1 + n_2 - 2 = 324 + 336 - 2 = 658$. Such a t value is highly significant ($P < 0.001$). In other words, if the null hypothesis is correct, such a difference would arise very much less often than 1 time in 1,000. We are led to reject the hypothesis. In this particular case since these are two samples from the same population and the true frequency distribution is likely to be phi-normal (Griffiths, 1959b), we are led to deduce that the sampling is nonrandom.

16.6 Testing a More Complex Hypothesis

Bokman, describing the petrography of the Stanley and Jackfork formations, states that

> In connection with the size-distribution study an attempt was made to determine whether the sandstones of the Stanley and Jackfork are graded. . . . Chips were collected from the tops and bottoms of beds of various thickness, and thin sections were made from them. . . . Seven beds, ranging in thickness from 9 to 52 inches, were studied. Table I tabulates the results of the measurements. . . . Secondly, whether grading is present is equivocal, although the beds appear to show a slight grading effect from bottom to top.[1]

[1] Reprinted from Lithology and Petrology of the Stanley and Jackfork Formations, *J. Geol.*, **61:** 164 (1953), by permission of The University of Chicago Press. Copyright 1953 by the University of Chicago.

If we designate as a criterion for graded bedding that the top is generally finer than the bottom, we may use these data (Table 16.1) to test the effectiveness of the criterion. The null hypothesis H_0 becomes "these sample means are two independent random samples from the same normal population."

Now the normality assumption is a reasonable approximation (Sec. 13.12), but if the beds are graded and the suggested criterion is an effective measure of grading, then the samples are neither random nor independent. There are several possible hypotheses and procedures for analyzing this set of data. If these beds are not graded, the top may be coarser than the bottom or vice versa with equal probability on average; or, in other words, the difference between top and bottom is positive or negative with equal probability. We may then count the number of positive (six) and negative (one) signs and test whether they represent a random sample from a binomial population with $p = 0.5$; or, the null hypothesis becomes "these signs represent random samples from a binomial population with $(p + q)^n = (0.5 + 0.5)^7$." The frequencies expected on the basis of this hypothesis are summarized in Table 16.2 and Fig. 16.1.

Since 8 samples out of every 128 would be expected to contain one or no negative signs by chance and, at the 95 percent level, 6.4 out of 128 would contain one or none, this sample of seven with one negative sign does not constitute evidence for rejection of the hypothesis that the signs are distributed with equal probability positive and negative in the parent population. Actually, a larger sample is necessary if this test is to be sensitive enough to detect

TABLE 16.1
Bed Thickness and Grain Size

Thickness of bed, in.	Mean size, phi units		Top − base, phi units
	Base	Top	
12	2.81	3.13	+0.32
13	3.95	4.13	+0.18
15	3.75	3.88	+0.13
19	2.68	2.91	+0.23
26	3.25	3.61	+0.36
40	3.90	4.20	+0.30
52	3.30	3.12	−0.18

Source: Reprinted from Lithology and Petrology of the Stanley and Jackfork Formations, by J. Bokman, *J. Geol.*, **61**:164 (1953), by permission of The University of Chicago Press. Copyright 1953 by the University of Chicago.

TABLE 16.2
Frequencies of Successes (negative signs) in a Binomial with $(p + q)^n = (0.5 + 0.5)^7$

Number of negative signs	Individual term in binomial expansion $\frac{N}{X!(N-X)!}p^X q^{N-X}$	Frequency multiplied by 128
0	$1(\frac{1}{2})^0(\frac{1}{2})^7 = \frac{1}{128}$	1
1	$7(\frac{1}{2})^1(\frac{1}{2})^6 = \frac{7}{128}$	7
2	$21(\frac{1}{2})^2(\frac{1}{2})^5 = \frac{21}{128}$	21
3	$35(\frac{1}{2})^3(\frac{1}{2})^4 = \frac{35}{128}$	35
4	$35(\frac{1}{2})^4(\frac{1}{2})^3 = \frac{35}{128}$	35
5	$21(\frac{1}{2})^5(\frac{1}{2})^2 = \frac{21}{128}$	21
6	$7(\frac{1}{2})^6(\frac{1}{2})^1 = \frac{7}{128}$	7
7	$1(\frac{1}{2})^7(\frac{1}{2})^0 = \frac{1}{128}$	1
Total	1.0	128

a difference at the 5 percent level. The sign test can be performed very simply by means of tables (e.g., see Dixon and Massey, 1951, table 10, p. 324; and text pages 247 to 249 for explanation).

Since in this case the magnitudes of the differences are also available, a more sensitive test can be performed by using Student's t. Thus if the null hypothesis H_0 is defined as "these differences (column 4, Table 16.3) are a

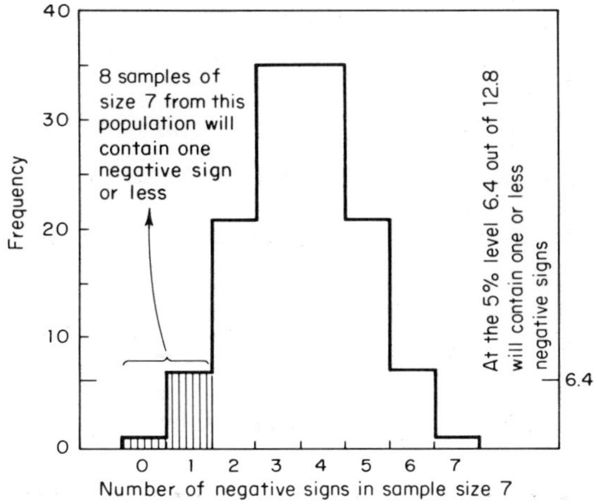

Figure 16.1 Expected binomial frequency distribution of negative signs; sample size 7 where $(p + q)^n = (0.5 + 0.5)^7$.

random sample from a normal population with mean difference zero, that is, $\mu = 0$," the test statistic becomes

$$t = \frac{(\bar{X} - \mu)\sqrt{n}}{\hat{\sigma}}$$

where $\mu = 0$, $n = 7$, and $\hat{\sigma} = \sqrt{0.03335} = 0.1826$ (see Table 16.3). Then

$$t = \frac{0.1914 \times 2.646}{0.1826} = \frac{0.5064}{0.1826} = 2.773$$

From the table of Student's t in Arkin and Colton (1950, p. 116) this value of t lies between 2.45 at $P = 0.05$ and 3.14 at $P = 0.02$. Therefore, if the null hypothesis is correct, this set of observations will occur more often than twice and less often than five times in every 100 such trials; that is, $t = 2.773$ is significant at the 5 percent level, and the hypothesis is rejected. The mean is evidently significantly different from zero. In other words, this set of differences is firm evidence for nonrandom distribution of mean difference. The top is finer than the bottom too often to be attributed to chance, and the criterion suggests that grading is present.

TABLE 16.3
Computation of Means and Variances from Data of Table 16.1

Statistic	Top	Base	Top − base
ΣX^2	90.7828	81.4060	0.4566
$\frac{(\Sigma X)^2}{N}$	89.1429	79.8357	0.2565
Sx_i^2	1.6399	1.5703	0.2001
$Sx_1^2 + Sx_2^2$	3.2102		
ΣX	24.98	23.64	1.34
$\frac{\Sigma X}{n} = \bar{X}$	3.5685	3.37714	0.1914
$\hat{\sigma}^2 = \frac{Sx_i^2}{n-1}$	0.273316	0.261716	0.03335
$s^2 = \frac{Sx_i^2}{n}$	0.2343	0.2243	0.028586

This test could also have been applied to the means, and the null hypothesis would be "these two sample sets are random samples from the same normal distribution."

Since $N_1 = N_2 = N$, we could use

$$t = \frac{\bar{X}_1 - \bar{X}_2}{\sqrt{s_1{}^2 + s_2{}^2/n - 1}}$$

with degrees of freedom $\nu = 2(n - 1) = 2(7 - 1) = 12$; or, using the data in Table 16.3,

$$t = \frac{3.5685 - 3.3771}{\sqrt{0.2343 - 0.2243/6}}$$

$$= \frac{0.1914}{\sqrt{0.4586/6}}$$

$$= \frac{0.1914}{\sqrt{0.07643}} = \frac{0.1914}{0.2765}$$

$$= 0.692$$

with degrees of freedom $\nu = 2(n - 1) = 12$; this value of t leads to a probability of $60 > P > 50$. In other words, such a difference would be expected between 50 and 60 times in every 100 trials, and so this constitutes no evidence for the rejection of the hypothesis. These two samples would be considered as random samples from the same normal population, and the difference among the means is not significant.

These two tests represent completely different questions; in the first case the question relates to whether the top differs from the base of the beds irrespective of the absolute value of the grain size in the beds. In the second the seven samples of top and of base are considered as two independent samples of the same population, and the variability within each sample is then large enough to mask the differences among mean sizes between samples. In the present context the first test is the more appropriate, and the samples of top and base of each bed are not independent—a requirement in the second null hypothesis.

16.7 Some Advantages Obtained from Experimental Design

The contrast between these two results is worthy of emphasis; thus in the second case the mean difference was compared with a combined estimate of the variance from the two samples, and the variability within samples was sufficiently high to mask any difference among the means. When, however, the samples are paired and each top is compared with its corresponding base, the variability among the two sets is much reduced; i.e., the within-sample variability of the first test is removed. In general, it may be recommended that, when possible, paired samples should be used; the advantages of this experimental design over the second one can be very considerable. It should also be noted that there is a marked increase in the degrees of freedom from 6 in the first case (paired samples) to 12 in the second. If pairing is effective, the accompanying reduction in the degrees of freedom is more than compensated for by removal of variability; but if pairing is not effective in removing the within-bed variability, the paired arrangement leads to a less sensitive t test. Another example of this procedure is described in Milner (1962, vol. 1, p. 585). The device known as *pairing* is a simple example of stratification or *blocking*, and when data are classified (by stratification), if the variability is large between strata and small within strata, the classification is highly effective in leading to more sensitive statistical tests.

Another aspect of experimental design is contained in arranging that each sample be of the same size. Equal-sized samples, even when not paired, possess distinct advantages; for example, if we return to Eq. (16.15) when $\sigma_{\bar{X}_1-\bar{X}_2} = \sigma\sqrt{(n_1+n_2)/n_1n_2}$ and then express t as

$$t = \frac{\bar{X}_1 - \bar{X}_2}{\sigma\sqrt{(n_1+n_2)/n_1n_2}}$$

or

$$t = \frac{\bar{X}_1 - \bar{X}_2}{\sigma}\sqrt{\frac{n_1n_2}{n_1+n_2}}$$

then we see that if $(\bar{X}_1 - \bar{X}_2)/\sigma$ is fixed, t will vary as

$$\sqrt{\frac{n_1n_2}{n_1+n_2}}$$

Suppose $n_1 = n_2 = 10$; then

$$\sqrt{\frac{n_1n_2}{n_1+n_2}} = \sqrt{\frac{10}{20}} = \sqrt{5} = 2.236$$

On the other hand, if $n_1 = 5$ and $n_2 = 15$, then

$$\sqrt{\frac{n_1n_2}{n_1+n_2}} = \sqrt{\frac{75}{20}} = \sqrt{3.75} = 1.936$$

In other words, the multiplying factor will be largest and hence t will be largest when $n_1 = n_2$, and this difference will be more marked in small samples. In discussing this aspect of experimental arrangement Youden states that "this merely verifies an established belief that it is no use to know one of the quantities very precisely if the other one is subject to large errors" (Youden, 1951, p. 74).

Tippett points out still another advantage: "However Welsh (1938) has shown that where $n_1 = n_2$ the t test of significance is not sensitive to differences in population variances . . ." (Tippett, 1952, p. 148).

Attention to such improvements in the arrangement of experiments can quite frequently lead to marked improvement in tests of significance from both practical and theoretical points of view.

16.8 Estimating the Sample Size to Achieve a Significant Difference with Stipulated Probability

Another advantage of Student's t is that it can be used to approximate the sample size necessary to establish a significant difference among means at a required probability level. Thus by writing

$$t^2 = \frac{(\bar{X} - \mu)^2 n}{\hat{\sigma}^2}$$

we can rearrange the expression to

$$n = \frac{t^2 \hat{\sigma}^2}{(\bar{X} - \mu)^2} \tag{16.20}$$

With a small sample of data it is possible to calculate $\bar{X}$ and $\hat{\sigma}^2$ and, from the degrees of freedom and selected probability level, to find t (and hence t^2); if the required value for $\bar{X} - \mu$ is substituted in Eq. (16.20), a value for n can be found.

The data in Table 16.4 are recorded for the porosity of two sets of samples of sandstones determined by two techniques.

The sums of squares are $SS_{x1} = 128.808$ and $SS_{x2} = 129.4311$, and the differences are $SS_d = 1.9147$. In the last case $\hat{\sigma}_{\bar{X}} = \sqrt{1.9147/n(n-1)} = \sqrt{1.9147/20} = 0.3094$.

Suppose we now test the null hypothesis H_0, as before, that these differences are drawn from a population in which the mean difference $\mu = 0$. Then

$$t = \frac{\bar{X} - \mu}{\hat{\sigma}_{\bar{X}}}$$

or

$$t = \frac{0.112}{0.3094} = 0.362$$

with degrees of freedom $\nu = n - 1 = 4$; then $80 > P > 70$, or under H_0 such a difference would arise between 70 and 80 times in every 100, and there is no evidence to deny that the differences are random samples from a population in which $\mu = 0$. It is of some interest to attempt to decide how many specimens would be needed to establish this difference as significant at a 95 percent probability level; for $\nu = 4$, $t_{05} = 2.776$,

$$n = \frac{t^2 \hat{\sigma}^2}{(\bar{X} - \mu)^2}$$

where $t^2 = 7.706 \qquad \hat{\sigma}^2 = 0.4787 \qquad \hat{\sigma} = 0.6919$

$$(\bar{X} - \mu)^2 = 0.112^2 = 0.125$$

Then $n = \dfrac{7.706 \times 0.4787}{0.125} = \dfrac{3.688}{0.125} = 29.5$

This estimate can be improved in several ways, e.g., by iteration (Milner, 1962, vol. 1, p. 589) and as described by Cochran and Cox (1957, pp. 17–29). A number of assumptions are involved in such a procedure; in particular, as more information accrues, the estimate of the variance $\hat{\sigma}^2$ improves. It is assumed constant in the estimation procedure proposed above.

This information can be found graphically by means of Fig. 16.2 and, as the estimate of N is generally approximate, this graphic procedure is probably adequate. Some examples are (1) given $s = 2.0$; $\bar{X} - \mu = 0.5$, then $n = 64$, and (2) with $n = 40$, $s = 2.0$; then at the 5 percent probability level $\bar{X} - \mu$ must exceed 0.64 to achieve significance.

TABLE 16.4
Comparison of Techniques for the Measurement of Porosity

	Porosity, %		
Specimen	Technique 1	Technique 2	Difference
1	17.1	16.92	+0.18
2	20.1	20.68	−0.58
3	19.3	18.05	+1.25
4	10.2	10.39	−0.19
5	25.9	26.00	−0.10
Total	92.60	92.04	+0.56
Mean	18.520	18.408	+0.112

Source: Data from H. C. Pyle and J. Sherborne, Core Analysis, *Trans. AIME Petrol. Tech.*, **132**:45 (1939), by permission of the Society of Petroleum Engineers of AIME.

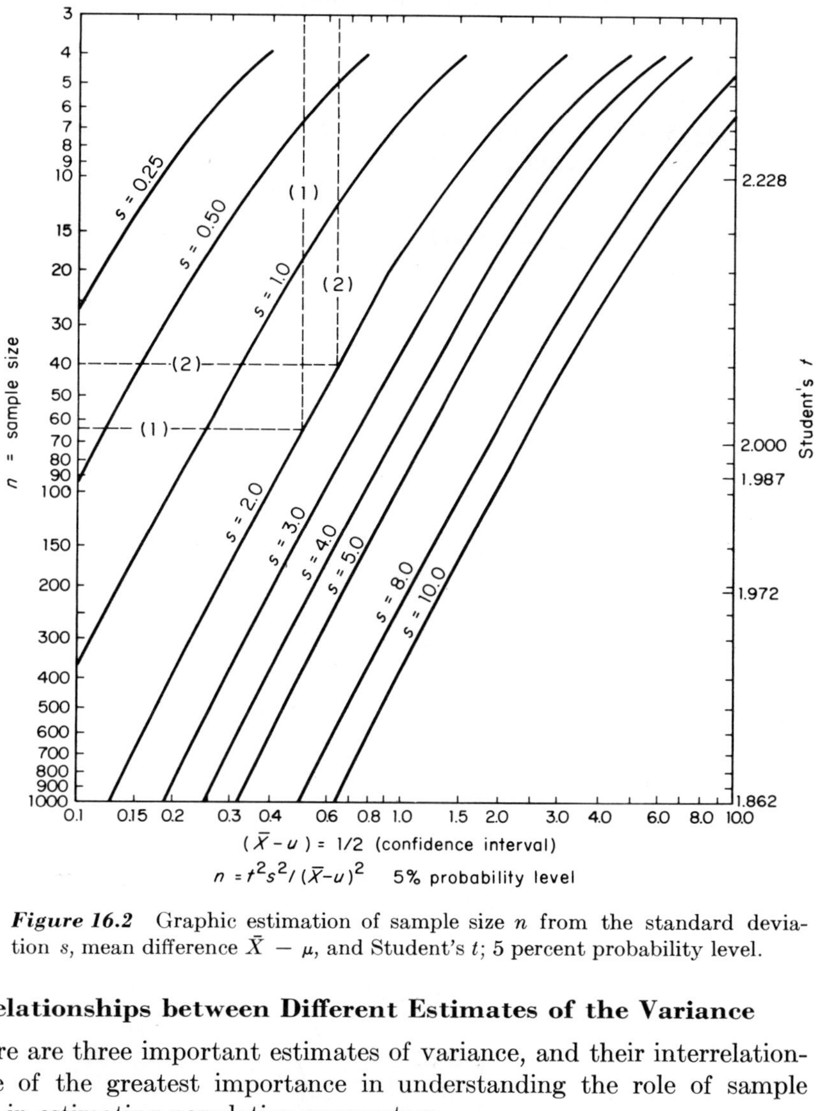

Figure 16.2 Graphic estimation of sample size n from the standard deviation s, mean difference $\bar{X} - \mu$, and Student's t; 5 percent probability level.

16.9 Relationships between Different Estimates of the Variance

There are three important estimates of variance, and their interrelationships are of the greatest importance in understanding the role of sample statistics in estimating population parameters.

Thus we may define the population variance as σ^2, where

$$\sigma^2 = \sum_{i=1}^{N} \frac{(X_i - \mu)^2}{N} \tag{16.21}$$

and the sample variance is then defined as s^2, where

$$s^2 = \sum_{i=1}^{n} \frac{(X_i - \bar{X})^2}{n} \tag{16.22}$$

Now it is necessary to define the best unbiased estimate of the population variance obtained from sample data, and we may proceed as follows (see Hoel, 1947, p. 128, for example).

The *expected* value of a statistic is its mean value in a large number of trials, that is, $E(X) = \sum X/N = \bar{X}$. It follows that we can find the expected value of the sample variance

$$E(s^2) = E\left[\frac{1}{n}\sum(X_i - \bar{X})^2\right]$$

$$= \frac{1}{n}E\sum[(X_i - \mu) - (\bar{X} - \mu)]^2$$

$$= \frac{1}{n}E\sum[(X_i - \mu)^2 - 2(X_i - \mu)(\bar{X} - \mu) + (\bar{X} - \mu)^2]$$

$$E\sum(X_i - \mu)^2 = n\sigma^2 \qquad \text{and} \qquad E(\bar{X} - \mu)^2 = \frac{\sigma^2}{n}$$

$$= \frac{1}{n}\left[n\sigma^2 - 2E\sum(X_i - \mu)(\bar{X} - \mu) + \frac{n\sigma^2}{n}\right]$$

But $$\sum(X - \mu) = \sum x - \sum\mu = n\bar{X} - n\mu = n(\bar{X} - \mu)$$

$$= \frac{1}{n}\left[n\sigma^2 - 2nE(\bar{X} - \mu)^2 + \frac{n\sigma^2}{n}\right]$$

$$= \frac{1}{n}\left(n\sigma^2 - \frac{2n\sigma^2}{n} + \frac{n\sigma^2}{n}\right)$$

$$= \frac{1}{n}(n\sigma^2 - \sigma^2)$$

$$= \frac{1}{n}(n - 1)\sigma^2 = \left(\frac{n-1}{n}\right)\sigma^2$$

Then $E(s^2) \neq \sigma^2$ but $E(s^2) = [(n - 1)/n]\sigma^2$, which is smaller by a factor $(n - 1)/n$; then

$$E\left(\frac{n}{n-1}\right)s^2 = E\left[\frac{n}{n-1}\frac{\sum(X - \bar{X})^2}{n}\right]$$

$$= E\left[\frac{\sum(X - \bar{X})^2}{n - 1}\right]$$

Therefore the best unbiased estimator of the population variance for a small set of values is

$$\hat{\sigma}^2 = \frac{\sum_{i-1}^{n} (X_i - \bar{X})^2}{n - 1}$$

It should be noted that as n increases, the correction for bias $n/(n - 1)$ approaches unity, indicating that in large samples the sample variance s^2 converges on the population variance σ^2. The sample variance is consistent but biased in small samples; the best estimate is consistent and unbiased in small samples.

17 The commoner tests of significance: chi-square and F tests

The term Goodness of Fit has caused some to fall into the fallacy of believing that the higher the value of P the more satisfactorily is the hypothesis verified. Values over .999 have sometimes been reported which, if the hypothesis were true, would only occur once in a thousand trials. . . . In these cases the hypothesis considered is as definitely disproved as if P had been .001.

R. A. Fisher, 1948

17.1 Introduction

The use of Student's t test is restricted to comparison of a mean with some expected value or to the comparison of two means, and the measures used should achieve at least interval level of information (Sec. 12.5). A significance test[1] which is less restricted in its application and which may be used with data on a nominal or ordinal level is the chi-square statistic. This statistic is of much wider application than Student's t test, but it lacks the specific character of the t test; in other words, these two statistics lead to tests of significance which should be used when, and only when, they are appropriate. As an approximation it may be said that whereas Student's t is associated with tests of means, the chi-square statistic is more generally associated with comparisons of sums of squares.

The chi-square statistic may be defined via the same theorem as the t statistic (Sec. 16.2).

If X_i is normally distributed with mean μ and variance σ^2 and $t = (X - \mu)/\sigma$, then t is normally distributed with mean zero and unit variance.

Quotation: R. A. Fisher, "Statistical Methods for Research Workers," 10th ed., pp. 80–81, Oliver & Boyd Ltd., 1948.

[1] Note that in this chapter sample variance s^2 is used frequently, and this may be adequate in large samples, i.e., where $n > 30$.

Then the sum of squares of n random sample values of t has a chi-square distribution with ν d.f. or

$$\chi^2 = t_1^2 + t_2^2 + t_3^2 + \cdots + t_\nu^2$$

with ν d.f.

Then we may write

$$\chi^2 = \frac{(X_1 - \mu)^2}{\sigma^2} + \frac{(X_2 - \mu)^2}{\sigma^2} + \cdots + \frac{(X_n - \mu)^2}{\sigma^2}$$

$$= \frac{1}{\sigma^2}\left[\sum_{i=1}^{n}(X_i - \mu)^2\right]$$

$$= \frac{n}{\sigma^2}\sum_{i=1}^{n}\frac{(X_i - \mu)^2}{n} = \frac{ns^2}{\sigma^2} \tag{17.1}$$

with ν d.f., where $\nu = n - 1$.

17.2 Some Characteristics of the Chi-square Frequency Function

The chi-square frequency function and its use as a statistical test of significance were developed by Pearson around the 1900s, and Fisher (1922) improved its applications by introducing the concept of degrees of freedom. The frequency distribution of chi square represents the distribution of the sum of squares

$$Sx_i^2 = \sum_{i=1}^{n}(X_i - \mu)^2$$

of standardized normal deviates and may be written (Rider, 1939, p. 100)

$$dP = \frac{1}{2^{n/2}[(n-2)/2]!}\exp\left(\frac{-\chi^2}{2}\right)\chi^{2(n-2)/2}\,d(\chi^2) \tag{17.2}$$

which expresses the probability that a value of chi square found from a random sample will fall in the interval $d\chi^2$.

The frequency curves of (17.2) extend from zero to infinity and are usually considered to be of Pearson type III form (e.g., Hoel, 1947, p. 134); when $\nu = 1$ the corresponding chi-square frequency function is the positive half of the normal curve (Kendall, 1948, vol. 1, p. 293). Tables of the chi-square frequency function are expressed in different forms; e.g., in Fisher and Yates' tables (1948, table IV, p. 33), the probability values (areas under the curves) are listed at the column heads, and the degrees of freedom n (ν in the present case) signify the rows. Values of chi square are included in the body of the table; this table is repeated in Arkin and Colton (1950, table 14, p. 121). Kendall (1948, p. 446) gives a graphic form equivalent to that of Fisher and Yates' table. In most cases the tables end at $\nu = 30$, and it is necessary to use an approximation to estimate the value of chi square beyond this sample size; some alternatives are discussed by Kendall (1948, vol. 1, pp. 295ff.); and, following the recommendation of Fisher (1948, p. 81), we may use the fact that $\sqrt{2\chi^2}$ is normally distributed about a mean $\sqrt{2\nu - 1}$ with unit variance. The value of chi square computed and the accompanying degrees of freedom may be inserted in the formula

$$t = \sqrt{2\chi^2} - \sqrt{2\nu - 1} \tag{17.3}$$

and the computed t may be compared with the values in a table of the standardized normal deviate (e.g., Arkin and Colton, 1950, table 10, pp. 114–115). It should be noted that the probabilities involved refer solely to one-half the standard normal curve; an example is given in Sec. 17.7.

In using the chi-square distribution it should be noted that the probability of $P > 95$ percent is equal to the probability of $P < 5$ percent. A high value of chi square, say exceeding 19.679 with $\nu = 9$, implies a probability value of 2 percent or less and is as unlikely to occur upon random sampling a population as a low value, say 2.532 with $\nu = 9$, and a probability value of 98 percent or more (Fisher, 1948, p. 80). While the high value of chi square will occur less than 2 percent of the time and is used as a basis for rejecting a hypothesis, the low value of chi square is also expected 2 percent or less of the time, and such homogeneity leads one to suspect nonrandom sampling.

17.3 Application of Chi Square to Finding the Confidence Limit for a Variance

Defining $\chi^2 = ns^2/\sigma^2$, we can find from tables of chi square (Arkin and Colton, 1950, p. 121) for a given ν a value χ_1^2 which is less than ns^2/σ^2 and a second value χ_2^2 which exceeds ns^2/σ^2, and this may be written as an inequality

$$\chi_1^2 < \frac{ns^2}{\sigma^2} < \chi_2^2$$

If, for example, we choose the 98 and 2 percent levels of chi square for some value of ν, the interval between the two chi-square values will include 96 percent of the area of this chi-square distribution. The inequality is more suitably expressed as

$$\chi_1^2 < \frac{ns^2}{\sigma^2} < \chi_2^2$$

$$\frac{1}{\chi_2^2} < \frac{\sigma^2}{ns^2} < \frac{1}{\chi_1^2}$$

$$\frac{ns^2}{\chi_2^2} < \sigma^2 < \frac{ns^2}{\chi_1^2} \tag{17.4}$$

Given a set of sample values, we can then evaluate s^2 and ν and insert appropriate values of χ_1^2 and χ_2^2 in this inequality to obtain the necessary confidence interval which will, on average in a selected proportion of the cases, contain the value of σ^2.

For example, suppose we have a sample of $n = 20$ which yields a value of $s = 0.65$; then $\nu = n - 1 = 19$; we now choose the probabilities 5 and 95 percent and look up the values of χ^2 (Arkin and Colton, 1950, p. 121) where the 95 percent value $\chi_1^2 = 10.117$ and the 5 percent value $\chi_2^2 = 30.144$; then substituting the tabulated values in the inequality yields

$$\frac{20 \times 0.65^2}{30.144} < \sigma^2 < \frac{20 \times 0.65^2}{10.117}$$

$$\frac{8.450}{30.144} < \sigma^2 < \frac{8.450}{10.117}$$

$$0.280 < \sigma^2 < 0.835$$

and this interval will contain the true value of σ^2 90 percent of the time (or 90 percent of such inequalities are true statements). The intervals are generally quite wide, indicating the instability of variance estimates from

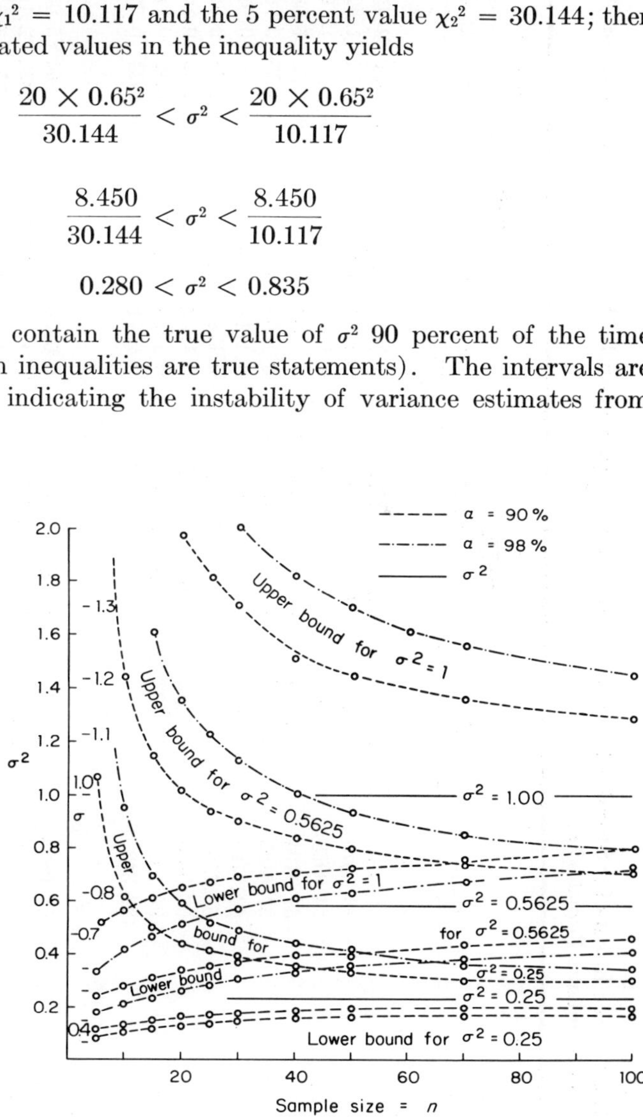

Figure 17.1 Confidence limits for standard deviations of quartz-grain lengths measured in thin section.

TABLE 17.1
Confidence Limits for Standard Deviations of Quartz Grain Length Measured in Thin Section; Measurements on Homogeneous Populations[a]

Sample size n	$\chi^2_{1-\alpha,\nu_1}$	χ^2_{α,ν_1}	$\hat{\sigma} = 0.25$ Lower	$\hat{\sigma} = 0.25$ Upper	$\hat{\sigma} = 0.50$ Lower	$\hat{\sigma} = 0.50$ Upper	$\hat{\sigma} = 0.75$ Lower	$\hat{\sigma} = 0.75$ Upper	$\hat{\sigma} = 1.00$ Lower	$\hat{\sigma} = 1.00$ Upper
$\alpha = 95\%$	$\nu = n - 1$									
5	11.0705	1.1455	0.0282	0.2728	0.1129	1.0912	0.2541	2.4553	0.527	7.035
10	18.3070	3.9403	0.0341	0.1586	0.1366	0.6345	0.3073	1.4276	0.591	3.007
15	24.9958	7.2609	0.0375	0.1291	0.1500	0.5165	0.3376	1.1620	0.633	2.283
20	31.4104	10.8508	0.0398	0.1152	0.1592	0.4608	0.3582	1.0368	0.664	1.977
25	37.6525	14.6114	0.0415	0.1069	0.1660	0.4277	0.3735	0.9624	0.687	1.805
30	43.7729	18.4926	0.0428	0.1014	0.1713	0.4056	0.3855	0.9125	0.705	1.694
40	55.7585	26.5093	0.0448	0.0943	0.1793	0.3772	0.4035	0.8488	0.717	1.509
50	67.5048	34.7642	0.0463	0.0899	0.1852	0.3596	0.4166	0.8090	0.741	1.438
70	90.5312	51.7393	0.0483	0.0846	0.1933	0.3382	0.4349	0.7610	0.773	1.353
100	124.342	77.9295	0.0503	0.0802	0.2011	0.3208	0.4524	0.7218	0.804	1.283
$\alpha = 99\%$	$\nu = n - 1$									
5	15.0863	0.5543	0.0207	0.5638	0.0829	2.2551	0.1864	5.0740	0.3314	9.0203
10	23.2093	2.5582	0.0269	0.2443	0.1077	0.9777	0.2424	2.1988	0.4309	3.9090
15	30.5779	5.2294	0.0307	0.1793	0.1226	0.7171	0.2759	1.6135	0.4906	2.8684
20	37.5662	8.2604	0.0333	0.1513	0.1331	0.6052	0.2995	1.3619	0.5324	2.4212
25	44.3141	11.5240	0.0353	0.1356	0.1410	0.5423	0.3173	1.2203	0.5642	2.1694
30	50.8922	14.9535	0.0368	0.1254	0.1474	0.5016	0.3316	1.1285	0.5895	2.0062
40	63.6907	22.1643	0.0393	0.1128	0.1570	0.4512	0.3533	1.0151	0.6280	1.8047
50	76.1539	29.7067	0.0410	0.1052	0.1641	0.4208	0.3693	0.9467	0.6566	1.6831
70	100.425	45.4418	0.0436	0.0963	0.1743	0.3851	0.3921	0.8665	0.6970	1.5404
100	135.807	70.0648	0.0460	0.0892	0.1841	0.3568	0.4142	0.8028	0.7363	1.4273

[a] Values for chi square from Pearson and Hartley, 1954, pp. 130–131.

small samples. For example, if this sample of 20 represented 20 measured lengths of quartz grains in thin section, the standard deviation 0.65 represents a moderately well sorted sediment; and as the range from well sorted to poorly sorted extends from 0.25 to 1.5, the confidence statement encompasses a large part of this range.

On this basis it is possible to recommend sampling guides for measurement of grain size in thin section where the thin section is defined as a homogeneous population and the element is a length measurement of a grain. With standard deviations of 0.25, 0.50, 0.75, and 1.0 and probability levels $\alpha = 90$ and 98 percent, the sample sizes required to achieve stipulated confidence intervals around the variance (and standard deviations) are exhibited in Table 17.1. A graphic illustration of these limits is exhibited in Fig. 17.1, and it can be seen that 30 to 50 grains (or length measurements) is ample for standard deviations below $\sigma = 0.75$ at the 90 percent probability level. Considerably more are required for higher standard deviations or higher probability levels. For example, for reasonably stable estimates of the population value, where $\sigma = 1$, at least 100 lengths (grains) should be measured.

17.4 Application of Chi Square as a Test of Goodness of Fit

The chi-square statistic was originally derived by Pearson (1900) to judge whether an observed frequency distribution could be regarded as similar to some theoretical one; in particular, the distribution of chi square is closely associated with the multinomial distribution. In this case the statistic is defined as

$$\chi^2 = \sum_{i=1}^{n} \frac{(O_i - E)^2}{E} \tag{17.5}$$

and this statistic follows the chi-square distribution with ν d.f.

In a binomial distribution, for example, this can be represented as

$$\chi^2 = \sum_{i=1}^{n} \frac{(p_i - np)^2}{npq}$$

or a ratio of n times the sum of squares of the deviations of each observed value from the mean divided by the variance, that is, ns^2/σ^2, which, as earlier defined, is a chi-square statistic with ν d.f. The correct form of the degrees of freedom varies with the particular problem and was first derived by Fisher (1922).

As an example we may take the frequency distribution of c axes of quartz grains from the Homewood quartzite, where the expected normal distribution was calculated by using the mean μ ($= 1.4278$) and variance σ^2 ($= 0.3117$) computed from the observed sample set of values (Table 13.1). The steps

TABLE 17.2
The Chi-square Test for Normality, Goodness of Fit of Normal and Observed Distribution

$\mu = 1.4278;\ \sigma^2 = 0.3117$

(1) Class, phi units	(2) x $X - \bar{X}$	(3) $\frac{x}{\sigma}$	(4) Area $\phi(t)$	(5) Difference	(6) E $N \times$ difference	(7) Observed O	(8) $O - E$	(9) $(O - E)^2$	(10) $\frac{(O - E)^2}{E}$
$-\infty$ – −0.5	−1.9278	−3.453	0.49966	0.00034	0.14 } 2.35	6	+3.65	13.3225	5.67
−0.5– 0.0	−1.4278	−2.557	0.4948	0.0049	2.21 }				
0.0– 0.5	−0.9278	−1.662	0.4515	0.0433	19.49	12	−7.49	56.1001	2.88
0.5– 1.0	−0.4278	−0.766	0.2794	0.1721	77.45	59	−18.45	340.4025	4.40
1.0– 1.5	+0.0722	+0.129	0.0517	0.3311	149.00	191	+42.00	1,764.00	11.84
1.5– 2.0	+0.5722	+1.025	0.3485	0.2968	133.56	117	−16.56	274.2336	2.05
2.0– 2.5	+1.0722	+1.920	0.4726	0.1241	55.85	55	−0.85	0.7225	0.01
2.5– 3.0	+1.5722	+2.816	0.0250	0.0250	11.25	7	−4.25	18.0625	1.61
3.0– 3.5	+2.0722	+3.712	0.4999	0.0023	1.04 } 1.09	2 } 3	+1.91	3.6481	3.35
3.5–∞				0.0001	0.05 }	1 }			
Total				1.0000	450.04	450			$\chi^2 = 31.81$

Note: d.f. = (number of classes) − 3 = 8 − 3 = 5; 1 d.f. each lost for mean, σ, and total. $P = <0.001$. The items in column 3 are found by subtracting from 0.5000 the figures found for each $-(X - \bar{X})/\sigma$ and adding to 0.5000 the figures for each $+(X - \bar{X})/\sigma$ from table of normal integral.

in the computation are illustrated in Table 17.2. It is efficacious in practice to prepare a standard form for the calculation of the required normal distribution and the subsequent steps of comparing and testing the observed and expected distributions, as in Table 17.2. Columns 1 to 6 represent the computation of the expected normal function from the observed mean and variance; column 7 contains the observed sample frequency. Deviations of observed from expected $(O - E)$ are found in column 8, and their squares are listed in column 9; the contribution of each cell to chi square is recorded in column 10 as $(O - E)^2/E$, and the sum of these values yields the appropriate total chi square. It is now necessary to find the number of degrees of freedom, and they equal the number of classes compared minus the number of restrictions imposed. In this case the mean, variance, and total area of the curve are fixed; and since there are eight classes compared, the number of degrees of freedom is $8 - 3 = 5$. Consulting tables of the chi-square distribution (Fisher and Yates, 1948, table IV, p. 33; Arkin and Colton, 1950, table 14, p. 121), we find that along the row designated $n = 5$ d.f. the value $P = 0.001$ is 20.517. The computed value of 31.81 exceeds this value; that is, $P < 0.001$. In other words, if this frequency distribution of values represents a random sample from some normal population with mean $\mu = 1.4278$, $\sigma = 0.558$, then such a sample will occur less than 1 time in 1,000. This is such an unusual event that we appear justified in rejecting the hypothesis, and either the population distribution is nonnormal or the sample is nonrandom or both. The largest contribution to chi square occurs at the tails and in the two central classes; deviations in the tails are common because of the few observations in these classes, but the deficiency in the class 0.5 to 1.00 phi and the marked excess in the modal class 1.00 to 1.5 phi are the really serious discrepancies. In this case the chi-square test agrees with that of the moment analysis and reveals exactly the departures expected in a leptokurtic curve (Fig. 13.1).

In general, the chi-square goodness-of-fit test is independent of the moment analysis and frequently yields different results; the moments are better characteristics of the shape of the curve, but the goodness-of-fit test is informative in indicating exactly where discrepancies of observed from expected occur.

The expected distribution may be binomial, Poisson, etc., and apart from adjustments in the degrees of freedom, the goodness of fit using chi square remains identical (e.g., Table 14.10). An example of the use of this test in a typical petrographic problem has been described by Chayes (in Fairbairn, 1949, p. 298), dealing with the establishment of preferred orientation in petrofabrics; in this case if the fabric is random and shows no preferred orientation, then an equal number of measurements is expected in each class. The expected distribution is, therefore, the uniform, or rectangular, distribution (see also Griffiths and Rosenfeld, 1953, p. 211), and the expected frequencies per class may be found by dividing the total frequency by the number of classes. The observed distribution may then be compared with the expected, and the goodness of fit tested by chi square. If a random distribution is rejected, some class or classes contain an excess of observations, and some degree of preferred orientation may be present in the rock sample.

Of somewhat greater interest to the paleontologist is the use of genetic ratios as expected distributions, and the test of observed individuals in a sample is frequently used to substantiate linkage, viability, etc. (Fisher, 1948, pp. 81ff.; Snedecor, 1956, pp. 188ff.; Mather, 1951).

17.5 Contingency Tables and Chi Square

The use of the chi-square statistic in a contingency table was first proposed for a geological problem by Eisenhart (1935a), and this aspect represents a very important area of application in both field and laboratory investigations in geology. Essentially the observations are classified according to certain features of interest (attributes), and the question concerns whether the features are independent or not. For example, let us suppose that a geologist had classified a set of samples at one or more outcrops into two classes (red and green) on the basis of color, and two classes on the basis of grain size (coarse and fine); he would record his observations as frequency counts, as in Table 17.3. His question would be, on the basis of this information, "Is size independent of color in the formation represented by this sample set of observations?"

It is important to realize that this 2 × 2 *contingency table* may be generalized and treated hypothetically with considerable rigor (Yule and Kendall, 1948, pp. 34ff.). Let the attributes be A (color) and B (size); then a set of geological specimens may be classified as A or not A ($\bar{A}$, or α) and as B and

TABLE 17.3
Frequency Table of 240 Geological Rock Specimens Classified According to Color and Grain Size (hypothetical data)

	Color		
Grain size	Red	Green	Total
Coarse	120	30	150
Fine	40	50	90
Total	160	80	240

TABLE 17.4
The Generalized 2 × 2 Contingency Table

	Attribute 1		
Attribute 2	B	β	Total
A	(AB)	$(A\beta)$	(A)
α	(αB)	$(\alpha\beta)$	(α)
Total	(B)	(β)	N

not B ($\bar{B}$, or β). We may designate the number frequency of A's among the entire set as (A), of B's as (B), of α's as (α), and β's as (β), as in Table 17.4.

If there is no relationship between the two attributes A and B, we expect to find the same proportion of A's among the B's as among the not-B's. We may define two such unrelated attributes as independent, and as a criterion of independence of A and B we have

$$\frac{(AB)}{(B)} = \frac{(A\beta)}{(\beta)} \tag{17.6}$$

It follows that

$$1 - \frac{(AB)}{(B)} = 1 - \frac{(A\beta)}{(\beta)}$$

$$= (B) - \frac{(AB)}{(B)} = (\beta) - \frac{(A\beta)}{(\beta)}$$

Similarly, $(\alpha B)/(B) = (\alpha\beta)/(\beta)$ and

$$\frac{(AB)}{(B)} = (AB) + \frac{(A\beta)}{(B)} + (\beta) = \frac{(A)}{N} \tag{17.7}$$

There are, of course, many other equivalent relationships among the frequencies in the table under the condition of independence (Yule and Kendall, 1948, p. 35), and perhaps the most familiar is "if the attributes A and B are independent, the proportion of AB's in the population (formation) is equal to the proportion of A's multiplied by the proportion of B's, that is,

$$\frac{(AB)}{N} = \frac{(A)}{N}\frac{(B)}{N} \tag{17.8}$$

a rule which is fundamental to the characteristic of independence throughout statistical analysis.

Suppose now that the attributes A and B are not independent but are related in some way; then we can write (Yule and Kendall, 1948, pp. 35ff.):

1 If $\quad (AB) > \dfrac{(A)(B)}{N} \quad$ A and B are positively associated (17.9*a*)

2 If $\quad (AB) < \dfrac{(A)(B)}{N} \quad$ A and B are negatively associated (17.9*b*)

It is important to note that association is expressed in terms of independence; thus A and B are associated if, and only if, they appear together in a greater number of cases than is to be expected if they were independent. Association cannot be inferred, because some A's are B's, however great the proportion.

It should be noted that in this table the classes may represent a nominal scale, e.g., red, green, or an ordinal scale, e.g., coarse > fine, and the numbers are frequencies determined by counting; this procedure is therefore applicable in all cases of measurement of "composition" wherein the elements of the population are classified in terms of attributes which may be unrelated or related; it is, of course, sometimes helpful to use this type of classification for data measured on an interval and ratio scale such as in the goodness-of-fit tests.

Returning to the test of independence, we may, from equality (17.8),

TABLE 17.5
Expected Frequencies of Rock Specimens Classified on the Basis of Color and Grain Size and the Hypothesis H_0 that These Two Classifications Are Independent

	Color		
Grain size	Red	Green	Total
Coarse	100	50	150
Fine	60	30	90
Total	160	80	240

Then $$\chi^2 = \sum_{i=1}^{4} \frac{(O - E)^2}{E}$$

yields $$\chi^2 = \frac{20^2}{100} + \frac{(-20)^2}{60} + \frac{(-20)^2}{50} + \frac{20^2}{80} = 23.66$$

find (AB) from (A), (B), and (N) if we assume that the attributes are independent. Furthermore, since $(AB) + (A) = (A)$, we can find (A) and similarly (αB) and $(\alpha\beta)$; it is obvious that once one cell frequency within the table is filled and the totals fixed, all the other cells are also fixed, which implies 1 d.f. It is also possible to calculate a table of expected cell frequencies under the hypothesis of independence, and so we have a set of expected and observed frequencies; from this set we can calculate a statistic

$$\chi^2 = \sum_{i=1}^{n} \frac{(O - E)^2}{E}$$

which follows the chi-square distribution if the hypothesis is true and if the degrees of freedom are specified. It is possible, therefore, to test the set of frequencies for independence. For the example of Table 17.3 the expected values are given in Table 17.5.

It should be noted that the sums of the deviations $(O - E)$ are zero for every row and column and over the whole table. Now we may approach the chi-square table with $\chi^2 = 23.66$ and degrees of freedom $\nu = 1$; we find at the 5 percent probability level (P_{05}) $\chi^2 = 3.841$ and at the 0.1 percent level (P_{001}) $\chi^2 = 10.827$. Since the computed chi square exceeds the latter, we can say that if the null hypothesis of independence of color and size is correct for this population, such a sample would arise by chance or by random sampling this formation less often than 1 time in 1,000. Since the event is so rare, we reject the hypothesis with a risk of being wrong in the long run less than 1 time in 1,000.

Since the frequencies imply that the attributes color and size are not independent, we may wish to examine the possible relationships by comparing (AB) and $(A)(B)/N$ as inequalities by using (17.9*a*) and (17.9*b*).

$$120 > 150 \times \tfrac{160}{240} = 100$$

Therefore A and B are positively associated, and in our example coarse beds tend to be red, and fine tend to be green. It should perhaps be emphasized that "direction" of association is really meaningful only when the attribute scales are ordinal, but the association is nevertheless characteristic of any 2×2 table.

In using chi square as a criterion to test the significance of departure from independence it should be realized that we are treating the frequencies in any one row or column as if they were binomial, and it is advisable to ensure that the number of observations is adequate: in general, the *expected* frequency in any cell should exceed 5; in 2×2 tables an improvement based on estimation requirements has been suggested by Yates (1934) and is generally known as the *Yates correction for continuity* (Fisher, 1948, p. 92; Cochran, 1954; Maxwell, 1961).

TABLE 17.6
Comparison of a Sample of Accessory Heavy Minerals with Those of a Possible Parent Population

Mineral	Zircon	Rutile	Tourma-line	Garnet	Chlori-toid	Stauro-lite	Total
Formation E	27	7	25	20	8	13	100
Sample O	25	6	28	18	4	19	100
Difference $O - E$	−2	−1	+3	−2	−4	+6	0
$(O - E)^2$	4	1	9	4	16	36	70
$(O - E)^2/E$	0.15	0.14	2.25	0.20	2.0	2.77	7.51 = χ^2

$$\text{d.f.} = (\text{number of classes compared}) - (\text{number of restrictions})$$
$$= 6 - 1 = 5$$
$$20 > P > 10$$

17.6 Extensions of the Chi-square Test to More Complex Questions

In many cases of interest there are more than two classes of one or both attributes, particularly when the data are counts of frequencies. For example, suppose we have an extensive body of data on the frequencies of a proportion of heavy minerals in a formation and a sample is taken which we wish to compare with the population "parameters" we have as in the hypothetical example of Table 17.6. Here we consider the population values as the expected, presumably based on a large number of previous estimates. In this case five of the six class frequencies may vary, but once they are fixed and the total is fixed, then the sixth class is also fixed; i.e., there are five degrees of freedom. The calculated statistic χ^2 follows the chi-square distribution with this number of degrees of freedom, and from tables of the chi-square distribution (Arkin and Colton, 1950) the value 7.51 lies between the chi-square value for $P = 20$ and that for $P = 10$; hence we may write $20 > P > 10$. In other words, if this is a random sample from this population (formation), such frequencies would be expected between 10 and 20 times on the average from "accidents of sampling." Therefore this constitutes no evidence for rejection of the hypothesis; this sample could have been drawn from this formation.

Now let us suppose that these frequencies are both from samples and we wish to know whether they are compatible, i.e., whether they could both be drawn from the same population; the calculations are exhibited in Table 17.7. First, there is obviously more opportunity for variation among two samples than when a sample is compared with a population;[1] hence the

[1] Because parameters are fixed constants.

TABLE 17.7
Comparison of the Heavy-mineral Frequencies in Two Samples to Determine Whether They Are Both from the Same Population

Mineral	Zircon	Rutile	Tourmaline	Garnet	Chloritoid	Staurolite	Total
Sample *A*	27	7	25	20	8	13	100
Sample *B*	25	6	28	18	4	19	100
Total	52	13	53	38	12	32	200
Exp. value	26	6.5	26.5	19	6	16	100
$O - E$ (A)	+1	+0.5	−1.5	+1	+2	−3	0
$O - E$ (B)	−1	−0.5	1.5	−1	−2	+3	0
Total	0	0	0	0	0	0	0
$\frac{(O - E)^2}{E}$ (A)	$\frac{1}{26}$	$\frac{0.25}{6.5}$	$\frac{2.25}{26.5}$	$\frac{1}{19}$	$\frac{4}{6}$	$\frac{9}{16}$	
(B)	$\frac{1}{26}$	$\frac{0.25}{6.5}$	$\frac{2.25}{26.5}$	$\frac{1}{19}$	$\frac{4}{6}$	$\frac{9}{16}$	
	0.038	0.038	0.085	0.053	0.667	0.563	
	0.038	0.038	0.085	0.053	0.667	0.563	
Total	0.076	0.076	0.170	0.106	1.334	1.126	$2.888 = \chi^2$

d.f. = (number of classes compared) − (number of restrictions)
$= (c - 1)(r - 1) = 5$
$80 > P > 70$

chi-square value is expected to be larger, but in this example it is actually smaller, and the probability statement becomes $80 > P > 70$. This is an event which will, on the basis of the hypothesis that these two samples are drawn from the same population, occur between 70 and 80 times in every 100 on an average. Hence these two samples are likely to have come from the same population (formation).

The degrees of freedom are calculated in this case from the general formula, which may be applied to tables of any complexity. In practice, it is often convenient to count varying numbers in the samples, and this is no disadvantage in applying chi square; indeed, the use of percentages or proportions in place of the direct counts may lead to loss of information and to a less sensitive test if the number counted is very much in excess of 100 items and the data are reported as percentages. In the contrary case, if the counts are less than the 100 index figure, computing to the base 100 inflates the "error."

Finally, the chi-square test can be applied to a more complex record, as shown in Table 17.8 which is in part the same as Table 12.2. Here there appears to be an obvious association between color and grain size, and upon examination of the criterion and testing by means of the chi-square distribution it is obvious that these two attributes are not independent in this set of data and therefore in the population they represent.

It may be somewhat more realistic to ask whether some measure of degree of association would not be more useful, but the chi-square criterion is not suitable for this purpose (Kendall, 1948, pp. 310ff.). A number of alternative criteria which measure degree of association have been proposed, and their advantages and disadvantages are discussed by Kendall (1948, vol. 1, p. 310; Goodman and Kruskal, 1954; Siegel, 1956). One such criterion is Pearson's coefficient of contingency C (Kendall, 1948, p. 319)

$$C = \sqrt{\frac{\chi^2}{N + \chi^2}} \tag{17.10}$$

In the present case

$$C = \sqrt{\frac{771.8}{1234.8}} = 0.7908$$

Unfortunately, the exact implications of the coefficient C are not very obvious; and while it is possible to use the chi-square statistic to test against independence, the degree of association is not so amenable to stringent testing. Kendall (1948) shows that in a $t \times t$ table in which the greatest degree of dependence exists

$$C = \sqrt{\frac{t - 1}{t}}$$

TABLE 17.8
Association between Color and Grain Size in Some 3-in. Well Cores from the Salt Wash Member of the Morrison Formation, Montrose County, Colorado

A = all other classes

	Texture						
Color	Cs	Ms	MsA	Fs	FsA	S + Mst	Total
(*a*) ***Observed Frequencies***							
Red						38	38
Red and green					16	116	132
Green			5	21	4	25	55
Red, white, and green	10	22	1	1	6		40
White and green	1	116	2	30	2		151
White	38	6	2	1			47
Total	49	144	10	53	28	179	463
(*b*) ***Expected Values for Part a***							
Red	4.0	11.8	0.8	4.3	2.3	14.7	38
Red and green	14.0	41.1	2.9	15.1	8.0	51.0	132
Green	5.8	17.1	1.2	6.3	3.3	21.3	55
Red, white, and green	4.2	12.4	0.9	4.6	2.4	15.5	40
White and green	16.0	47.0	3.3	17.3	9.1	58.4	151
White	5.0	14.6	1.0	5.4	2.8	18.2	47
Total	49	144	10	53	28	179	463
(*c*) ***Contributions to Criterion*** $(O-E)^2/E = \chi^2$							
Red	4.0	11.8	0.8	4.3	2.3	36.9	60.1
Red and green	14.0	41.1	2.9	15.1	8.0	82.8	163.9
Green	5.8	17.1	12.0	34.3	0.1	0.6	69.9
Red, white, and green	8.0	7.4	0.0	2.8	5.4	15.5	39.1
White and green	14.1	10.3	0.5	9.3	6.7	58.4	99.3
White	217.8	5.1	1.0	3.6	2.8	18.2	248.5
Total	263.7	92.8	17.2	69.4	25.3	212.4	680.8 = χ^2

d.f. $= (c - 1)(r - 1) = (6 - 1)(5 - 1) = 20$ $\quad P < 0.001$

Thus in a 2 × 2 table C cannot exceed 0.707, and the value in Table 17.4 for the example is

$$C = \sqrt{\frac{23.66}{263.66}} = 0.2995$$

There is an obvious association but not a very strong one. In a 3 × 3 table $C \not> 0.816$, in a 4 × 4, $C \not> 0.866$, the coefficient increasing with the number of classes in the table. It would be somewhat more in accord with general understanding if the coefficient were unity in such extreme cases of association. A thorough discussion of alternatives is given by Goodman and Kruskal (1954) and Stuart (1953).

17.7 Partitioning of Chi Square

Another important characteristic of the chi-square criterion is its additive character, which may be summarized as follows: "The sum of a number of chi-squares is itself distributed in a chi-square distribution with degrees of freedom equal to the sum of the degrees of freedom for each independent value of chi-square" (Fisher, 1948, p. 81).

TABLE 17.9
Classification of Dolomite Rock Samples From an Outcrop on Route 545 near Bellefonte, Centre County, Pennsylvania

	Grain size			
Color	Coarse	Medium	Fine	Total
(a) Observed Frequencies				
Dark	53	71	52	176
Medium	67	116	104	287
Light	14	57	121	192
Total	134	244	277	655
(b) Contribution to χ^2 from Each Cell $(O-E)^2/E$				
Dark	8.02	0.44	6.74	15.20
Medium	1.17	0.77	2.49	4.43
Light	16.29	2.94	19.51	38.74
Total	25.48	4.15	28.74	$58.37 = \chi^2$
d.f. $= (c-1)(r-1) = 4$	$P \ll 0.001$			

TABLE 17.10
Tests of Independence of Color and Grain Size in Dolomite for Each Observer

Operator	N	d.f.	χ^2	P
A	50	4	11.81	0.02–0.01
B	58	4	18.32	0.01–0.001
C	51	4	2.76	0.70–0.50
D	54	4	8.83	0.10–0.05
E	54	4	12.28	0.02–0.01
F	70	4	19.21	<0.001
G	84	4	13.70	0.02–0.01
H	52	4	4.92	0.30–0.20
I	55	4	6.94	0.20–0.10
J	67	4	3.27	0.70–0.50
K	60	4	10.80	0.05–0.02
Total	655	44	112.94	
Combined results	655	4	58.37	≪0.001

This permits the addition of a series of sample sets of observations so that much smaller departures from expected can be detected if they persist in a number of independent tests. Furthermore, this additive character allows the partitioning of chi square so that different contributory effects can be separately evaluated (Weatherburn, 1957, p. 177; Maxwell, 1961; Mather, 1951, pp. 178ff.).

A group of 11 students classified a set of hand specimens of Nittany dolomite from the formation exposed in an outcrop along Route 545 near Bellefonte, Centre County, Pennsylvania, and classified them in terms of grayness and grain size, using three classes for each attribute. The results are exhibited in Table 17.9*a*.

As a first step a set of samples covering the range of variation in the chosen attributes was classified into a 3 × 3 table on the ground in front of the outcrop, and these were subsequently used as standards for comparison. Specimens were then collected by each student, starting at one end of the outcrop at a randomly selected point and sampling at 1-ft intervals along the outcrop. Each sample was compared with the standard set, and after classification it was entered in a data table of frequencies per class.

The first test concerns whether grain size and color are related in these samples and, by inference, in the population represented by this outcrop. The contributions to χ^2 are listed in Table 17.9*b*; and under the hypothesis that these two attributes are independent, the χ^2 value follows the chi-square distribution with degrees of freedom $\nu = 4$. The high value of the criterion

TABLE 17.11
Testing Homogeneity among Observers in Classifying Dolomite Specimens

Source of variation	d.f.	Contribution to chi square	P
Total	44	112.94	<6 in 10^5
Combined	4	58.37	$\ll 0.001$
Inconsistency among operators	40	54.57	<12 in 10^2

leads to rejection of the hypothesis; i.e., size and color are related in this population. In general, dark dolomites tend to be coarse-grained, whereas light-colored dolomites are finer-grained. It is also noteworthy that the middle class (medium) for each attribute contributes little to the criterion, which indicates that perhaps these classes are redundant and are used largely as discard classes when the observers are not sure of their classification.

The results for each student observer can be arranged in a similar table, leading to 11 independent tests of the hypothesis; and we should expect that different operators would achieve different results, as exhibited in Table 17.10. First we should like to be sure that these inconsistencies are not so great as to suggest that the results are too heterogeneous to be combined; second, we should like to use the most sensitive test, i.e., the one with the largest number of degrees of freedom, if possible. By the additive property of chi square we can sum the individual results, yielding a combined value of $\chi^2 = 112.94$ with 44 d.f. Most chi-square tables terminate with the degrees of freedom $\nu = 30$, and so some other procedure must be used in this case. Generally the test proposed by Fisher may be used, in which for large values of ν the expression (see also page 342)

$$t = \sqrt{2\chi^2} - \sqrt{2\nu - 1} \tag{17.11}$$

can be used as a normal deviate with unit variance.[1]

In the present example

$$t = \sqrt{2\chi^2} - \sqrt{2\nu - 1} = \sqrt{225.88} - \sqrt{87} = 5.702 \tag{17.12}$$

The probability of exceeding such a value can be found from tables of the area of the normal curve, and since expression (17.12) refers to one-half the curve, the probability required is twice the value in the table. Actually since a t value of 5.7 standard deviations from the mean is considerably less likely than 3 in 100,000, then 6 in 100,000 represents a very unlikely event, and the hypothesis may be rejected.

[1] Alternative test procedures are critically discussed by Kendall (1948, vol. 1, pp. 294ff.).

The chi-square criterion can be further partitioned to find whether the inconsistencies among the observers are indicative of significant heterogeneity. The relevant procedure is summarized in Table 17.11, and the corresponding calculation of the probability yields

$$t = \sqrt{2\chi^2} - \sqrt{2\nu - 1} = \sqrt{109.14} - \sqrt{79} = 1.57$$

Again from tables of the area of a normal curve such a deviation yields $0.5000 - 0.4418 = 0.0582$, and for exceeding this deviation in either direction the probability is $2 \times 0.0582 = 0.1164$; in other words, such a value of the criterion is likely to arise under the hypothesis of homogeneity more than 12 times in every 100 on an average. This constitutes no evidence for rejection of the hypothesis, and the individual values of the criterion do not vary more than would be expected. In future tests it would perhaps be advisable to reduce this inconsistency by more careful standardization of technique.

Partitioning can be of much more use in isolating contributions to the criterion in appropriately designed more complex experiments (e.g., Fisher, 1948, pp. 101ff.; Snedecor, 1956, pp. 400ff.; Cochran and Cox, 1957, pp. 103ff.; Maxwell, 1961).

17.8 Other Uses of Chi Square

There are a number of other situations in which the chi-square distribution plays an important role, e.g., in the tests of homogeneity described in the discussion of the dispersion index on page 295. A useful test for homogeneity of variance derived originally by Bartlett (1937b) is described in Sec. 17.12.

A number of other situations in which the chi-square statistic is a suitable test criterion are reviewed by Cochran (1954), Kimball (1954), and Maxwell (1961), and these articles contain references to a variety of uses.

Effectively, the chi-square statistic is useful in all situations where the data occur as frequencies or counts of frequency in classes. In particular, when the scales of the variate under study are nominal or ordinal, the chi-square statistic is generally the most effective criterion for testing hypotheses; a number of examples have been described by Siegel (1956).

It should be noted that one of the advantages of the chi-square test is that it is unnecessary to state explicitly an alternative hypothesis; as a corollary, the test is not so informative as one in which the criterion is designed to test a hypothesis against a single alternative or clearly specified alternative hypotheses (cf. t test, Chap. 16).

17.9 The Variance-ratio Test Criterion

The chi-square test may be looked upon as a criterion for comparing the sum of squares of a normal deviate with its variance, and it is possible to present this test in modified form as a test of two observed variances.

Thus beginning with $\chi^2 = ns^2/\sigma^2$ and supposing we have two samples from which we can calculate the variances as s_1^2 based on n_1 observations and s_2^2 based on n_2 observations (where $n_1 = n_2$), we can express the respective relevant chi squares as

$$\chi_1^2 = \frac{n_1 s_1^2}{\sigma_1^2}$$

$$\chi_2^2 = \frac{n_2 s_2^2}{\sigma_2^2}$$

Now let us suppose we wish to test the hypothesis that these two samples were drawn from a normal population with variance σ^2. Then on the basis of the hypothesis $\sigma_1^2 = \sigma_2^2 = \sigma^2$,

$$\frac{\chi_1^2}{\chi_2^2} = \frac{n_1 s_1^2/\sigma^2}{n_2 s_2^2/\sigma^2}$$

and we may define a criterion first devised by Fisher and christened F in his honor by Snedecor (1956), where

$$F = \frac{n_1 s_1^2/(n_1 - 1)}{n_2 s_2^2/(n_2 - 1)} = \frac{\chi_1^2/\nu_1}{\chi_2^2/\nu_2} \tag{17.13}$$

Then the variance ratio F has the distribution

$$f(F) = C\,\frac{F[(\nu_1 - 2)/2]}{\nu_2 + \nu_1 F^{(\nu_1+\nu_2)/2}}$$

which has been extensively tabulated (Fisher and Yates, 1948, table V, pp. 34–43; Arkin and Colton, 1950, table 13, pp. 117–120).

It will be observed that three items of information are necessary to enter the variance-ratio table, namely, ν_1, ν_2, and F, and so the tables are generally arranged with ν_1 along the columns, ν_2 along the rows, and appropriate values of F in the body of the table. There are tables for each probability level; the two most commonly recorded are the 5 percent (P_{05}) and the 1 percent (P_{01}), but more extensive tables are given in Fisher and Yates (1948) for the additional probability levels of 10, 20, and 0.1 percent. It is important to note that the original variance ratio defined by Fisher is designated as e^{2z}, which is equivalent to Snedecor's F.

The F distribution is symmetrical; hence only one half is recorded in the tables, and by convention it is customary to insert the largest variance in the numerator and to consider this as s_1^2; hence ν_1 will accompany this variance.

17.10 Applications of the Variance-ratio Criterion

The main application of the variance-ratio criterion is in connection with analysis-of-variance procedure, which is described in Chaps. 18 to 20. A geological example which will illustrate the value of the test in a specific case is contained in testing for the presence of preferred orientation by experimental procedure advocated in Sec. 7.6.

The frequency distributions of the observations are quite generally unimodal, and the minimum variance can be found by the technique described by Chayes (1954a). Then, since we are testing against a hypothesis of random distribution which, under these conditions, yields the rectangular distribution with range ±90°, the expected variance can be calculated as $\sigma^2 = 2{,}700$. Under the experimental arrangement described, this will be the largest variance, and in all other cases the observed variance should be smaller; in addition, the degrees of freedom associated with this variance, since it is a population parameter, are effectively infinite. We may then establish a variance ratio as a test criterion defined as

$$F = \frac{\sigma^2}{s^2} \qquad \nu_1 = \infty, \nu_2 = n - 1$$

where

$$F = \frac{2{,}700}{s^2} \qquad \nu_1 = \infty, \nu_2 = n - 1$$

In practice it is suggested that a more conservative test may be used in which $\nu_1 = \nu_2 = n$, the number of observations in the sample. Choosing a value, for $n = 200$, say, we can now determine the value s^2 must attain for a significant F value. With $\nu_1 = \nu_2 = n = 200$ the 5 percent value of F is 1.26 (Arkin and Colton, 1950, p. 120), so that we can write

$$1.26 = \frac{2{,}700}{s^2} \qquad \text{or} \qquad s^2 = \frac{2{,}700}{1.26}$$

or

$$s \geq 46.29$$

Hence for the hypothesis of random orientation to be rejected the observed variance must be less than 2,143, or the standard deviation must be less than 46.3°. A number of examples are illustrated in articles in which this test procedure is described (Griffiths and Rosenfeld, 1953; Griffiths, 1952b; Secs. 7.6, 7.7). If the observed frequency distribution of orientation is normal the expected value is approximately $s \approx 30°$ (see page 153) and $s^2 = 900$; then $F = 2{,}700/900 = 3$, and the number of observations necessary should be around 20. In general, the degree of preferred orientation observed is generally less than would be expected from a normal distribution (Griffiths et al., 1956, p. 62), although such distributions do occur (Emery, 1954; Emery and Griffiths, 1954a, b).

The use of the variance ratio as a test of whether two samples are drawn from a population with variance σ^2 is, of course, quite general. For example, the standard deviation in phi units of the length of long a axes of quartzite pebbles from the Montoursville gravel pit was recorded as[1]

Sampling	s^2	n
Channel	0.707	525
Stratified	0.659	450

Then since both of these sample sets are drawn from the same population, and since the frequency distribution is likely to be normal, we can test these estimates as a variance ratio where $F = 0.707/0.659 = 1.072$ and $\nu_1 = 524$, $\nu_2 = 449$. The 5 percent probability value F_{05} for $\nu_1 = 500$, $\nu_2 = 400$ is $F = 1.16$, and $F_{01} = 1.24$. Since the value of $F = 1.072$ is less than either, there is no evidence for rejection of the hypothesis; hence these two samples are drawn from a population with the same variance.

17.11 Test of whether One Variance Exceeds the Remainder (Cochran's Test)

From more extensive experiments we have seven estimates of the variance of the long a axis of quartzite pebbles from this Montoursville gravel population, and one of them is very much larger than the remainder:

$$s_1^2 = 0.928 \qquad s_2^2 = 0.769 \qquad s_3^2 = 0.590 \qquad s_4^2 = 0.725$$

$$s_5^2 = 0.651 \qquad s_6^2 = 0.707 \qquad s_7^2 = 0.689$$

The sample sizes are not equal, but the $n_1 = 104$, and all the others exceed $n = 300$. Hence assuming approximately equal sample size because they are all "large" samples, we can use Cochran's test as an approximate test of whether this variance is too large; the test consists of

$$\frac{\text{Largest } s^2}{\sum_i s_i^2} \qquad i = 1, \ldots, 6$$

which yields

$$\frac{0.928}{4.101} = 0.2263$$

[1] Unpublished information; note that s^2 may be used for $\hat{\sigma}^2$ when $n \approx n - 1$; that is, for $n = 200$, $n/(n-1) = 1.005025$.

From special tables listing the values of Cochran's test criterion (Dixon and Massey, 1951, table 17, pp. 340–341; Eisenhart, Hastay, and Wallis, 1947, tables 15.1, 15.2), we find at the 15 percentile that for $n = 144$ and $k = 7$, the value is 0.1833 and at the 1 percentile that the value is 0.1929; hence we are led to suspect that this value is way out of line as an estimate of the variance of this population. This test suggests that considerably more than 100 measurements should be performed to obtain a stable estimate of the variance of such populations by this sampling and measurement procedure.

In performing this test the sample sizes should in general be equal, but this requirement was ignored in the above test because the sample sizes were all large.

17.12 Testing the Homogeneity of a Set of Variances

We may now reject this outlier and ask whether the remaining six sample variances can represent homogeneous estimates of the same variance; this test procedure was first devised by Bartlett (1937a) and is described in Snedecor (1956, p. 215) and Bennett and Franklin (1954, p. 196), and the general mathematical background is discussed in Hartley (1940).

The test criterion is defined as (Bennett and Franklin, 1954, p. 197)

$$B = \frac{1}{C}(\nu \ln s^2 - \sum_i \nu_i \ln s_i^2)$$

$$= \frac{2.30259}{C}(\nu \log s^2 - \sum_i \nu_i \log s_i^2) \tag{17.14}$$

where ln is the natural logarithm and log is the logarithm to the base 10. The symbol C is a correction for bias and is defined as

$$C = 1 + \frac{\sum_i (1/\nu_i) - 1/\nu}{3(k-1)} \tag{17.15}$$

where ν_i are the degrees of freedom in the individual estimates of variances and there are k such estimates. Then $\nu = \nu_1 + \nu_2 + \cdots + \nu_i$. The calculation may be usefully performed tabularly, as illustrated by Snedecor (1956, p. 251, table 10.22) and in Table 17.12.

$$C = 1 + \frac{0.016705 - 0.000445}{3(6-1)} = 1.001084$$

Then $B = 7.1977$.

TABLE 17.12
Testing the Homogeneity of Variance of the Long *a* Axes of Quartzite Pebbles from Montoursville Gravel

Sample	$\nu_i s_i^2$	d.f.	Reciprocal $1/\nu_i$	Variance	$\log s_i^2$	$\nu_i \log s_i^2$
1	244.542	318	0.003145	0.769	−0.11407	−36.27426
2	190.570	323	0.003096	0.590	−0.22915	−74.01545
3	242.875	335	0.002985	0.725	−0.13966	−46.78610
4	194.649	299	0.003344	0.651	−0.18642	−55.73958
5	370.468	524	0.001908	0.707	−0.15058	−78.90392
6	295.891	449	0.002227	0.659	−0.18111	−81.31839
$k = 6$	1,538.995	2,248	0.016705			−373.03770

$$s^2 = \sum_i \frac{\nu_i s_i^2}{\nu} = \frac{1{,}538.995}{2{,}248} = 0.684606 \qquad \log s^2 = -0.16455$$

$$\nu \log s^2 = 2{,}248 \times -0.16455 = -369.90840$$

$$\text{Then } B = \frac{1}{C}\left(\nu \ln s^2 - \sum_i \nu_i \ln s_i^2\right)$$

$$= \frac{2.30259}{C}\left(\nu \log s^2 - \sum_i \nu_i \log s_i^2\right)$$

$$= \frac{2.30259}{1.000956}(-369.90840 + 373.03770) = 7.1986$$

For values of ν_i of 5 or more the distribution of B is approximated by the chi-square distribution with $k - 1$ degrees of freedom. In this case $k - 1 = 5$, and the relevant values of chi square are $\chi^2 = 7.289$ at the 20 percent probability level and $\chi^2 = 6.064$ at the 30 percent level; hence, if these sample variances are estimates of the same population variance, such a value of B would arise between 20 and 30 times in every 100 trials on an average, and the hypothesis of homogeneity is not rejected. Evidently these variances are estimates of the same variance and may be combined to yield an improved estimate s^2 of σ^2, where $s^2 = 0.685$ based on 2,248 d.f.

18 Analysis of variance: one-way classification

The very ease with which variation can be reduced in the exact sciences and in certain fields of technology has served to obscure, on the one hand, the practical impracticability of eliminating every vestige of variability entirely and, on the other, the fact that its measurement is more important than its reduction, desirable as this may be.

H. M. Davies, 1946

18.1 Introduction

Most experiments generally encompass the comparison of more than two sets of data, and then the analysis-of-variance procedure is appropriate; it is one of the most versatile tools at the disposal of experimental analysis. It includes the simultaneous comparison of means and variances, and by adopting suitable auxiliary tests these sources of information can to a large extent be separately interrogated. Analysis of variance essentially comprises decomposition of the total variation into parts which can be assigned to the separate sources contributing to the total variation. There is literally no limit to how far this decomposition can proceed in a suitably designed experiment, and a very wide variety of designs exists, each of which is specifically arranged to suit a specific experimental situation. It is, however, necessary for the experimenter to decide before beginning the experiment what his goals are and which design is most suitable for their achievement. Usually the beginner errs by adopting too complex a design in order to achieve the highest return in information from the experiment; complex experiments are rarely successful unless there is a considerable background of information on

Quotation: H. M. Davies, The Application of Variance Analysis to Some Problems of Petroleum Technology, *J. Inst. Petrol.*, **32**(212): 466 (1946).

the nature of the variation likely to be displayed by a variate, and initially it is prudent to use simple designs and build gradually on a relatively sure background of experience.

The decomposition of total variance may be written as

$$\sum_{i=1}^{n}(X_i - \mu)^2 = \sum_{i=1}^{n}[(X_i - \bar{X}_{.g}) + (\bar{X}_{.g} - \mu)]^2 \tag{18.1}$$

indicating that the total variation of the observations around the true mean is composed of variation of the observations in each small set around the set means and, in addition, the variation of the set mean around the true mean. In order to achieve this it is necessary to collect the data in small sets; in practice, since the true mean μ is unknown, the grand mean $\bar{\bar{X}}_G$ of the entire collection is used as an estimate of the true mean, yielding

$$\sum_{i,j=1}^{n} (X_{ij} - \bar{\bar{X}}_{..})^2 = \sum_{i,j=1}^{n} [(X_{ij} - \bar{X}_{.j}) + (\bar{X}_{.j} - \bar{\bar{X}}_{..})]^2 \tag{18.2}$$

The skill of experimental design rests upon the selection of suitable sets so that the sources of variation can be segregated. Suppose each observation is written as

$$X_i = \mu + \epsilon_i \tag{18.3}$$

to indicate that each observation contains an estimate of the true mean μ plus an additional contribution from uncontrolled sources ϵ_i. For example, in weighing a specimen on an analytical balance, each weighing will differ a little from the others, and this variation from one observation to the next is represented as ϵ_i; each weighing also contains an estimate of the constant parameter μ, the true weight of the specimen. Suppose that the experimenter

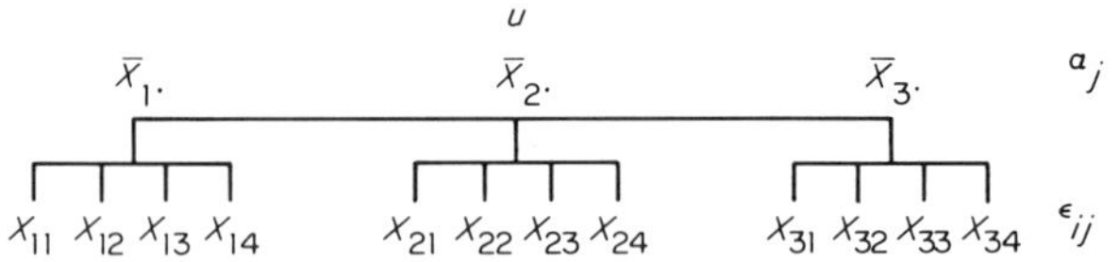

Source of variation	Degrees of freedom	Expected mean square
Between X_a	$a - 1$	$\sigma_e^2 + r\,\sigma_a^2$
Within X_a	$a(r-1)$	σ_e^2
Total	$ar - 1$	σ^2_{tot}

Figure 18.1 Experimental design and sampling; one-way classification

$$X_{ij} = \mu + \alpha_j + \epsilon_{i(j)}$$

decides that temperature variation is an important contributor to the variation of X_i around μ. Then making k observations at each of j different temperatures α, we may write

$$X_{ij} = \mu + \alpha_j + \epsilon_{i(j)} \tag{18.4}$$

or each observation X_{ij} is an estimator of μ plus a contribution from variation in temperature α_j plus a contribution from unassigned sources $\epsilon_{i(j)}$, usually called *error* (see for example, Fig. 18.1). The subscript $i(j)$ refers to the error over i within j.

It is clear that each experiment has a structure which can be written in the form of an equation relating the observations to the sources of variation around a parametric value, the true mean μ (Fig. 18.1). Furthermore, if the experimental arrangement is successful, the variation assigned to some factor, such as temperature, will reduce the error, thus leading to a more sensitive experiment. If $\epsilon_{i(j)}$ remains the same with or without allocation to temperature α_j, then temperature is not an important source of variation. Furthermore, each experiment which can be written in the form of an equation resembling (18.4) above possesses a sampling design and a corresponding analysis-of-variance table, as in Fig. 18.1.

18.2 Example of One-way-classification Analysis of Variance: Porosity Determined by Two Techniques

Suppose a new procedure is devised for the measurement of some property of materials; e.g., the variate X_i is porosity. To indicate the improvement in technique, the porosity of two sets of specimens is determined by each of two procedures. Purely for the sake of illustration we assume that the two sets of samples are chosen at random from the same population, oil-reservoir sandstones.[1] Five specimens are submitted to each technique independently, and the porosity is determined under standardized conditions. It is clear that to show the range of application of the technique a wide range in the variate is some assurance. The data are displayed in Table 18.1.

The structure of the experiment is summarized by defining each observation as follows:

$$X_{ij} = \mu + \alpha_j + \epsilon_{i(j)} \tag{18.5}$$

and there are $j = 2$ sets of data consisting of $i = 5$ items each. The sampling arrangement is illustrated in Fig. 18.1, and the observations X_{ij} are assigned at random within each set.

[1] Comparison of these techniques using two-way classification is described in Sec. 19.2.

TABLE 18.1
Porosity of Five Specimens of Oil-reservoir Sandstones Determined by Two Techniques[a]

Sample no.	Pyle and Sherborne	Coberly
1	17.1	16.92
2	20.1	20.68
3	19.3	18.05
4	10.2	10.39
5	25.9	26.00
$\sum_1^5 X_i$	92.60	92.04
$\bar{X}.$	18.52	18.408
$(\sum_1^5 X)^2$	8,574.7600	8,471.3616
$\sum_1^5 X_i^2$	1,843.7600	1,823.7034

[a] *Source*: Data taken from Pyle and Sherborne, 1939; see Table 16.4.

18.3 Computations for the Analysis of Variance

To simplify the computations it is necessary to reduce the general formula given in Eq. (18.2) by expanding the right-hand side as follows:

$$\sum_1^j \sum_1^i (X_{ij} - \bar{\bar{X}}_{..})^2 = \sum_1^j \sum_1^i [(X_{ij} - \bar{X}_{.j}) + (\bar{X}_{.j} - \bar{\bar{X}}_{..})]^2$$

$$= \sum_1^j \sum_1^i (X_{ij} - \bar{X}_{.j})^2 + \sum_1^j \sum_1^i (\bar{X}_{.j} - \bar{\bar{X}}_{..})^2$$

$$+ 2\sum_1^j \sum_1^i (X_{ij} - \bar{X}_{.j})(\bar{X}_{.j} - \bar{\bar{X}}_{..}) \qquad (18.2)$$

The third term in this expansion vanishes because the sum of the deviations around the mean is zero. Hence

$$\sum_1^j \sum_1^i (X_{ij} - \bar{\bar{X}}_{..})^2 = \sum_1^j \sum_1^i (X_{ij} - \bar{X}_{.j})^2 + i\sum_1^j (\bar{X}_{.j} - \bar{X}_{..})^2 \qquad (18.6)$$

Then by using the relationship

$$\sum(X_i - \bar{X})^2 = \sum X_i^2 - \frac{(\sum X_i)^2}{n}$$

the computing formulas become

$$\sum_{1}^{ij} X_{ij}^2 - \frac{(\sum X_{ij})^2}{ij} = \text{SS}_{\text{tot}} \qquad (18.7)$$

$$\sum_{j=1}^{2} \frac{(\sum_{i=1}^{5} X_i)^2}{5} - \frac{(\sum_{i\,j=1}^{10} X_{ij})^2}{10} = \text{SS between technique means} \qquad (18.8)$$

It is customary to label the common term $(\sum X_{ij})^2/ij$ the *correction term*, or CT for short. The computations may now start, and it is advantageous to follow a systematic pattern. First calculate the correction term; in the present case this is

$$\frac{(\sum_{1}^{ij} X_{ij})^2}{ij} = \frac{184.64^2}{10} = 3{,}409.19296 = \text{CT}$$

Then compute the total sum of squares SS_{tot} as follows [see Eq. (18.7)]:

$$\sum_{1}^{ij} X_i^2 = 3{,}667.46340$$

$$\text{CT} = 3{,}409.19296$$

$$258.27044 = \text{SS}_{\text{tot}}$$

The next step is to compute the term estimating the variation of the set means from the grand mean, or SS_j, as follows [see Eq. (18.8)]:

$$\sum_{1}^{j}(\sum_{1}^{i} X_{ij})^2 = 17{,}046.2126$$

$$\div i = 5 = 3{,}409.22452$$

$$- \text{CT} = 3{,}409.19296$$

$$0.03156 = \text{SS}_j$$

The final term is found by difference, i.e., by subtraction of the variability of sets from the total variation; it may also be calculated by summing the variability within each of the two sets around its own set mean, and this may be necessary in cases where there are unequal numbers of observations

TABLE 18.2
Comparison of Two Techniques for Measurement of Porosity; Analysis of Variance: One-way Classification

No. of items	Source of variation	d.f.	Sum of squares	Mean square	Variance ratio F	F_{05}
$g = 2$	Between techniques	$g - 1 = 1$	0.03136	0.03136	0.0009	5.32
$k = 5$	Within techniques (error)	$g(k - 1) = 8$	258.23908	32.2799		
$gk = 10$	Total	$gk - 1 = 9$	258.27044			

in the sets. It may be remarked that there are sound theoretical reasons for using equal numbers in the sets, apart from a considerable simplification in the arithmetic (Brownlee, 1960, p. 268).

These computations can now be summarized in an analysis-of-variance table (Table 18.2).

18.4 Tests of Significance in the Analysis of Variance

The analysis-of-variance table for a one-way classification with one level of subdivision, as in this example, can be written in the general form shown in Table 18.3.

The null hypothesis may be stated as "these two sets of data are random samples from the same (normal) population," in which case variation among sets will average zero, or $\sigma_j^2 = 0$, and the variance ratio $(\sigma_\epsilon^2 + i\sigma_j^2)/\sigma_\epsilon^2 = 1$. In a small set of data, such as the example under analysis, the variances are sample estimates of their expected values and will vary around their true parametric values. Since it is likely that the numerator will usually be larger than the denominator, this variance ratio will follow the F distribution[1] with degrees of freedom $j - 1$ and $j(i - 1)$, respectively. The test of significance is therefore an F test, as described in the preceding chapter (p. 359). If the means for the sets vary more than the observations within a set, this ratio will significantly exceed the tabulated F value, and the hypothesis is rejected at the appropriate level of significance.

In the example of Table 18.2 the significance test leads to an F value less than 1, and the hypothesis is accepted; i.e., the means of these two sets of data do not vary more than the variation within the sets, and therefore both techniques are yielding the same set means.

In practice, of course, it may well be questioned whether this is a suitable test of superiority of the two techniques, and so it should be decided what

[1] See Brownlee, 1960, pp. 218ff. for formal proof.

the criterion of "better technique" implies. For example, it may be claimed that one technique is more precise than the other, which would lead directly to the variance-ratio test among the two groups. The individual variances are

$$\hat{\sigma}_1^2 = 32.2020 \qquad \hat{\sigma}_2^2 = 32.3578$$

and the variance-ratio test F_{α,ν_1,ν_2} yields the variance ratio $\hat{\sigma}_2^2/\hat{\sigma}_1^2 = 1.0048$; the tabulated value of F at the 95 percent level with 4 and 4 d.f. is $F_{95,4,4} = 6.39$; hence there is no evidence for increased precision in one test over the other (see Sec. 19.2 for continuation of the comparison of these two techniques).

18.5 An Experiment Leading to One-way Classification with Two Levels of Subdivision

It is customary, in the absence of knowledge about the outcome of an experiment, to use the simplest possible experimental arrangement so that subsequent investigations can be designed more efficiently and on a sound basis. In some cases when the sampling plan is hierarchical or multistage, there is little choice; the experimental design is predetermined by the sampling plan. Both these restrictions occur in many investigations in geology.

In such a situation it is essential to run a preliminary experiment to determine the range of variation involved and particularly to determine that the range of variation is such that unknown sources of variation, usefully classified as experimental error, represent only a small part of the total variation. As an example of this type of experiment we take the analysis of bulk density of sediments by mercury balance (Griffiths, 1948).

Five rock samples t were chosen to cover a range of sedimentary rock types. A core sample (5516) of Middle Miocene sediments from Trinidad

TABLE 18.3
Analysis of Variance: One-way Classification, One Level of Subdivision

Source of variation	d.f.	Sum of squares	Mean square	Expected mean square
Among sets	$j - 1 = 1$	SS_j	$\dfrac{SS_j}{j-1}$	$\sigma_\epsilon^2 + i\sigma_j^2$
Within sets (error)	$j(i - 1) = 8$	SS_i	$\dfrac{SS_i}{j(i-1)}$	σ_ϵ^2
Total	$ji - 1 = 9$	SS_{tot}		

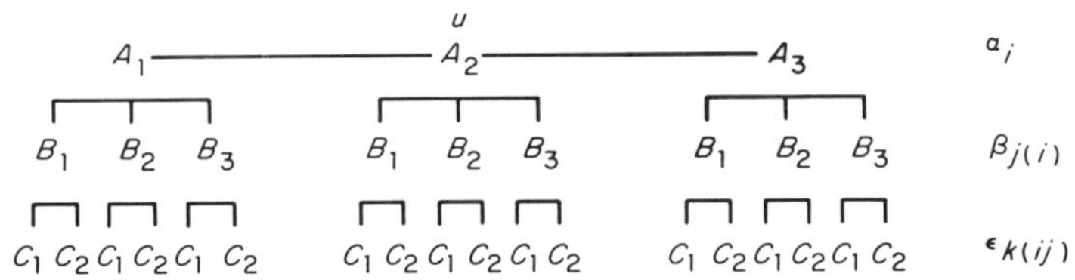

Source of variation	Degrees of freedom	Expected mean square
Among A: α_i	$a - 1$	$\sigma_e^2 + k\sigma_\beta^2 + kb\sigma_a^2$
Among B : $1\beta_{j(i)}$	$a(b-1)$	$\sigma_e^2 + k\sigma_\beta^2$
Within B :$\epsilon_{k(ij)}$	$ab(k-1)$	σ_e^2
Total	$abk - 1$	σ^2_{tot}

Figure 18.2 Experimental design and sampling; one-way classification with two levels of subdivision.

$$X_{ijk} = \mu + \alpha_i + \beta_{j(i)} + \epsilon_{k(ij)}$$

was selected to represent the low-rank graywackes;[1] another core sample (B-8) from the Upper Devonian Bradford sand of Pennsylvania represents a more quartzose type of low-rank graywacke (Krynine, 1940). A core from the Upper Devonian Venango formation (H-13) represents the protoquartzites, and an outcrop sample of Oriskany orthoquartzite (5292) from the Lower Devonian of West Virginia completes the compositional range from graywacke to quartzite. By way of contrast an arkose core sample (5293) from the Oligocene Sespe formation of California was also included.

Each sample was subdivided into three subsamples m; variation among subsamples yields some measure of uniformity of a rock sample. Finally, each 8- to 10-g subsample was analyzed five times k; variation among these replicates represents differences arising from unassigned sources and yields a suitable measure of experimental error. The sampling plan and experimental arrangement are summarized in Table 18.4, and Fig. 18.2 represents a multistage or hierarchical sampling pattern. Now each determination of bulk density for each subsample is assigned to the replicates at random; i.e., replicate 1 of subsample 1 of sample 5292 has no relationship to any other replicate 1, and the analysis is, therefore, of the completely randomized type. If each replicate 1 represented the first run on each subsample, differences between replicates would also include a systematic variation arising from order of performing the determinations. In such a case the experimental arrangement would be entirely different, namely, a two-way classification.

The objectives of this reconnaissance experiment are twofold. The main interest lies in the degree of variation at the different levels, so that the

[1] All petrographic terms follow the general usage prescribed by Krynine (1948).

TABLE 18.4
Multistage Sampling Design for a One-way Classification with Two Levels of Subdivision

Rock type	Oriskany orthoquartzite			Venango protoquartzite			Bradford graywacke			Tertiary graywacke			Sespe arkose		
Age	Lower Devonian			Upper Devonian			Upper Devonian			Miocene (Middle)			Oligocene		
Locality	West Virginia			Pennsylvania			Pennsylvania			Trinidad			California		
Sample no.	5292			H-13			B-8			5516			5293		
Subsample	*A*	*B*	*C*	*A*	*B*	*C*	*A*	*B*	*C*	*A*	*B*	*C*	*A*	*B*	*C*
	1	1	1	1	1	1	1	1	1	1	1	1	1	1	1
	2	2	2	2	2	2	2	2	2	2	2	2	2	2	2
Replicate runs k	3	3	3	3	3	3	3	3	3	3	3	3	3	3	3
	4	4	4	4	4	4	4	4	4	4	4	4	4	4	4
	5	5	5	5	5	5	5	5	5	5	5	5	5	5	5
Subsample m	$\bar{X}_m$	$\bar{X}_m$	$\bar{X}_m$	$\bar{X}_m$	$\bar{X}_m$	$\bar{X}_m$	$\bar{X}_m$	$\bar{X}_m$	$\bar{X}_m$	$\bar{X}_m$	$\bar{X}_m$	$\bar{X}_m$	$\bar{X}_m$	$\bar{X}_m$	$\bar{X}_m$
Rock type t		$\bar{X}_t$			$\bar{X}_t$			$\bar{X}_t$			$\bar{X}_t$			$\bar{X}_t$	

Note: $t = 5$ rock samples; $m = 3$ subsamples; $k = 5$ replicates.

variances are the important features of the experiment. In order that these variances may be suitable estimators, the samples (rock types and subsamples) should be random samples of the population. Since bulk density ranges widely within any one petrographic class, these specimens represent a wide range; whether they are random and representative may be determined only by more elaborate experimentation. A secondary objective of an entirely different type concerns the question of whether the bulk density for an arkose lies outside the range for the quartzite–low-rank graywacke sequence; this question concerns means and represents a contrast among means for the rock types.

The important feature here is that these two questions are not really answerable by the same set of data; the contrasts among means are related to estimators of population values, and there are only four detrital rock types: the quartzites, low-rank and high-rank graywacke, and arkoses, and three of these are included in the experiment. These are not, therefore, random samples from an infinite population of rock types. Questions of this kind lead to a model 1 type of analysis (e.g., Eisenhart, 1947; Bennett and Franklin, 1954, pp. 348ff.; Brownlee, 1960, p. 245); they are essentially concerned with differences among means, and the variances are incidental. On the other hand, when the questions at issue are concerned with variation, the sampling is random sampling from infinite populations, and this is known as a model 2 type of analysis. Model 1 is sometimes called fixed and model 2, random; there are, of course, mixed models, of which the fixed and random are extreme cases. It is always possible to look at the data through a model 1, or fixed, type of approach or a model 2, or random, type of approach for the sake of illustration, but in practice an experiment is usually performed with some definite objective, and one approach will be appropriate and the other not because the sampling and populations are different in the two cases. Examples of fixed, random, and mixed models will be given in subsequent chapters; as the designs become more complex and more specific, there is less latitude in applying a model 1 or 2 approach, often only one being suitable to the experiment under analysis. Choice of which model and design to use should be made before the experiment is started and is usually implicit in the kind of question the experiment is chosen to answer.

18.6 Analysis of the Data

In the present experiment consisting of five replicate determinations on three subsamples of each of five rock types the data consist of 75 values of bulk density in grams per cubic centimeter. The experimental procedure is described in Sec. 11.2.

The computations may be summarized in the form of a generalized analysis-of-variance table for the one-way classification with two levels of subdivision (Table 18.5); there are t $(= 5)$ rock types, m $(= 3)$ subsamples,

TABLE 18.5
Generalized Computing form for Analysis of Variance of a One-way Classification with Two Levels of Subdivision

Source of variation	d.f.	Sums of squares	Mean squares	Expected mean square
Among rock-type means	$t - 1 = 4$	$\dfrac{\sum_1^t (\sum_1^{mk} X)^2}{mk} - \text{CT} = \text{SS}_t$	$\dfrac{\text{SS}_t}{t - 1}$	$\sigma_\epsilon^2 + k\sigma_m^2 + mk\sigma_t^2$
Among subsample means	$t(m - 1) = 10$	$\dfrac{\sum_1^{tm} (\sum_1^{k} X)^2}{k} - \text{CT} - \text{SS}_t = \text{SS}_m$	$\dfrac{\text{SS}_m}{t(m - 1)}$	$\sigma_\epsilon^2 + k\sigma_m^2$
Among replicates (experimental error)	$tm(k - 1) = 60$	$\sum_1^{tmk} X^2 - \text{CT} - \text{SS}_\epsilon - \text{SS}_m = \text{SS}_k$	$\dfrac{\text{SS}_k}{tm(k - 1)}$	σ_ϵ^2
Total	$tmk - 1 = 74$	$\sum_1^{tmk} X^2 - \text{CT} = \text{SS}_{\text{tot}}$		

Note: $\text{CT} = \dfrac{(\sum_1^{tmk} X)^2}{tmk}$

and r ($= 5$) replicates, and the computational program starts with the correction term (see Table 18.5 footnote), followed by the total sum of squares SS_{tot} and then by the sum of squares for the highest level SS_t, and for successively lower levels, the "error term" SS_ϵ ($= SS_k$) being determined by difference. The sums and means used in this analysis are illustrated in Table 18.6, and the analysis of variance is summarized in Table 18.7.

The steps in significance testing may be determined from the expected mean squares of Table 18.5; the first step is to compare variation among subsample means with that among replicates within subsamples. The null hypothesis H_0 may be stated as "these subsamples are random samples from the same normal population."

It follows from H_0 that the expected value of $\sigma_m^2 = 0$ and therefore the variance ratio

$$F_{\alpha,\nu_1,\nu_2} = \frac{\sigma_\epsilon^2 + k\sigma_m^2}{\sigma_\epsilon^2} = 1$$

The data furnish estimates of these expected mean squares, and we have

$$F_{95,10,60} = \frac{\hat\sigma_\epsilon^2 + k\hat\sigma_m^2}{\hat\sigma_\epsilon^2} = \frac{0.0012230}{0.00060996} = 2.005$$

The tabulated 95 percent value of F for degrees of freedom 10 and 60 is 1.99, and our value just exceeds it. Hence the null hypothesis is rejected at the 5 percent level and $\sigma_m^2 \neq 0$.

We may now proceed to test variation among rock-type means by using the variance ratio

$$F_{95,4,10} = \frac{\hat\sigma_\epsilon^2 + k\hat\sigma_m^2 + km\hat\sigma_t^2}{\hat\sigma_\epsilon^2 + k\hat\sigma_m^2} = \frac{0.7717395}{0.0012230} = 631.0217$$

The tabulated F value at a probability level of 0.001 is 11.28, so this is an unexpectedly large value on the basis of the hypothesis. H_0 is rejected, and $\sigma_t^2 \neq 0$.

We may conclude that variation among subsample means exceeds variation among replicates within subsamples and that variation among rock-type means exceeds variation among subsample means and therefore exceeds variation among replicates within all classes. It is evident that this experimental procedure, using this design, is able to detect differences among mean bulk densities of rock types. This is all we can conclude from the analysis of variance; auxiliary tests are necessary for finding where the differences lie and what degree of variation exists at each level.

18.7 Isolating Estimates of Components of Variance

The first objective of this experiment is, by examining the magnitude of variation at different levels, to decide upon the most efficient sampling design for use in future experiments of similar kind. If we could reduce the number of replicates, we could perhaps increase the number of rock specimens, and this is the information level we most wish to extend. In this case we are using the model 2, or infinite-population, approach; i.e., all samples are a few from an infinite number that could conceivably have been taken. This is reasonable if we consider rock specimens and not the few petrographic classes of rock types that exist.

For the purpose of segregating variation at different levels or making point estimates of the variances, the source of information is the components of variance. Thus the average variance of replicates throughout the experiment is σ_ϵ^2, of which we have the estimator $\hat{\sigma}_\epsilon^2 = 0.000610$; a point estimate of the variance at the subsample level may be found as

$$\frac{\hat{\sigma}_\epsilon^2 + k\hat{\sigma}_m^2 - \hat{\sigma}_\epsilon^2}{k} = \frac{0.001223 - 0.000610}{5} = 0.000123$$

and, finally, at the rock-type level the estimate is

$$\frac{(\hat{\sigma}_\epsilon^2 + k\hat{\sigma}_m^2 + mk\hat{\sigma}_t^2) - (\hat{\sigma}_\epsilon^2 + k\hat{\sigma}_m^2)}{mk} = \frac{0.7717395 - 0.0012230}{15} = 0.051368$$

These estimates, together with their appropriate standard deviations, are summarized in Table 18.8. It seems clear that variation among rock-specimen means is so much larger than all other sources that this technique could be very much simplified and would still suffice to diagnose differences at that level. On the other hand, variation among subsamples of the same specimen is only one-fifth as large as replicate variation, so that if it is important to detect differences among subsamples, five replicates is a minimum.

These estimates may be combined to obtain an estimate of the variance around the grand mean as follows:

$$\text{Var}\,(\bar{\bar{X}}_{..}) = \frac{\sigma_t^2}{t} + \frac{\sigma_m^2}{tm} + \frac{\sigma_\epsilon^2}{tmk} \tag{18.9}$$

$$\text{Var}\,(\bar{\bar{X}}_{..}) \approx \frac{\hat{\sigma}_\epsilon^2 + k\hat{\sigma}_m^2 + mk\hat{\sigma}_t^2}{tmk}$$

$$= \frac{0.000610 + 5 \times 0.000123 + 15 \times 0.051368}{75}$$

$$= 0.010290 \tag{18.10}$$

The equivalent standard deviation is $\hat{\sigma}_{\bar{X}_{..}} = 0.1$. From (18.9) it would seem appropriate to increase r to reduce the error variance if differences among subsample means are of interest; whereas if all we are interested in is specimen mean differences, it may be more efficient to dispense with subsamples altogether.

TABLE 18.6
Computations of Sums, Means, Sums of Squares for Analysis of Variance;

	Rock					
	Oriskany orthoquartzite			Venango protoquartzite		
Age Locality	Lower Devonian West Virginia			Upper Devonian Pennsylvania		
Sample no.	5292A	5292B	5292C	H-13-A	H-13-B	H-13-C
Bulk density:						
Run 1	2.461	2.487	2.459	2.235	2.242	2.222
2	2.465	2.505	2.471	2.219	2.228	2.221
3	2.440	2.466	2.453	2.206	2.233	2.222
4	2.442	2.472	2.450	2.210	2.230	2.211
5	2.447	2.469	2.496	2.221	2.225	2.243
$\sum_1^k X = \sum_1^5 X$	12.255	12.399	12.329	11.091	11.158	11.119
$\bar{X}$	2.4510	2.4798	2.4658	2.2182	2.2316	2.2238
$(\sum_1^k X)^2 = (\sum_1^5 X)^2$	150.185025	153.735201	152.004241	123.010281	124.500964	123.632161
$\sum_1^{mk} = \sum_1^3 \sum_1^5 X$		36.983			33.368	
$\bar{X}$		2.4655			2.2245	
$(\sum_1^{mk} X)^2 = (\sum_1^{15} X)^2$		1,367.742289			1,113.423424	
$\sum_1^t (\sum_1^{mk} X)^2$						
$\div (mk) = 15$						
$\sum_1^t \sum_1^m (\sum_1^k X)^2$						
$\div 5$						
$SS_{tot} = \sum_1^{tmk} X^2 = \sum_1^{75} X^2$						
$CT = \sum_1^{tmk} X = \sum_1^{75} X$	161.771					
$(\sum_1^{tmk} X)^2 = (\sum_1^{75} X)^2$	26,169.856441					
$\div tmk \div 75$	348.931419					
$\sum_1^k X^2$	30.037519	30.748095	30.402347	24.602563	24.900362	24.726979
$\dfrac{(\sum_1^k X)^2}{k}$	30.037005	30.747040	30.400848	24.602056	24.900193	24.726432

type

Bradford graywacke			Tertiary graywacke			Sespe Arkose		
Upper Devonian Pennsylvania			Miocene (Middle) Trinidad			Oligocene California		
B-8-A	B-8-B	B-8-C	5516A	5516B	5516C	5293A	5293B	5293C
2.182	2.193	2.181	1.816	1.994	1.871	2.097	2.096	2.109
2.146	2.135	2.108	1.798	1.832	1.847	2.076	2.096	2.091
2.176	2.163	2.147	1.813	1.812	1.852	2.087	2.098	2.092
2.179	2.148	2.167	1.802	1.822	1.855	2.089	2.083	2.108
2.179	2.177	2.171	1.790	1.813	1.848	2.091	2.093	2.097
10.862	10.816	10.774	9.019	9.273	9.273	10.440	10.466	10.497
2.1724	2.1632	2.1548	1.8038	1.8546	1.8546	2.0880	2.0932	2.0994
117.983044	116.985856	116.079076	81.342361	85.988529	85.988529	108.993600	109.537156	110.187009
	32.452			27.565			31.403	
	2.16346			1.8376			2.09353	
	1,053.132304			759.829225			986.148409	
5,280.27565								
352.018377								
1,760.155033								
352.030607								
352.067205								
$\bar{\bar{X}}_G = 2.156946$								
23.597498	23.399276	23.219164	16.268933	17.222257	17.198083	21.798956	21.907574	22.037699
23.596609	23.377171	23.215815	16.268472	17.197706	17.197706	21.798720	21.907431	22.037402

TABLE 18.7
Analysis of Variance of Bulk Densities of Sedimentary Rock Samples

Source of variation	d.f.	Sum of squares	Mean square	F	F_{05}	F_{01}	F_{001}
Among rock type means	4	3.086958	0.7717395	631.0217***	3.48	5.99	11.28
Among subsample means	10	0.012230	0.0012230	2.0050*	1.99	2.63	3.87
Among replicates (error)	60	0.036598	0.00060996				
Total	74	3.135786					

TABLE 18.8
Point Estimates of Variance from the Different Sampling Levels

Source of variation	Variance	Standard deviation
Rock-type means t	0.051368	0.227
Subsample means m	0.000123	0.011
Replicates (error) r	0.000610	0.025

18.8 Differences among Means

The means for the different levels can be calculated, and they are listed in Table 18.9. With estimates of variances at different levels we may now examine the differences among the means. One comparison between the Sespe arkose and the remainder was contemplated while the experiment was being designed, and if we assume constant variance[1] and random samples, we may proceed to calculate confidence limits by using the variance around, say, rock-type means as follows:

$$\text{Var}\ (\bar{X}_t) = \frac{k\sigma_m^2 + \sigma_\epsilon^2}{mk} = \frac{0.00615 + 0.000610}{15} = 0.0000816$$

$$\hat{\sigma}_{\bar{X}_t} = 0.009$$

Then the confidence limits around a rock-type mean are

$$\bar{X}_t - t_{\alpha,\nu}\hat{\sigma}_{\bar{X}_t} < \mu < \bar{X}_t + t_{\alpha,\nu}\hat{\sigma}_{\bar{X}_t}$$

where t is Student's t with selected probability level α, say 95 percent (or 0.05 level), and degrees of freedom $\nu = 14$, whence $t_{05,14} = 2.14$. For the mean of the Sespe arkose this yields

$$2.0935 - 2.14 \times 0.009 < \mu < 2.0935 + 2.14 \times 0.009$$

or

$$2.075 < \mu < 2.113$$

Evidently the mean bulk density of the Sespe arkose differs at the 5 percent level from the means of all the other specimens (see Fig. 18.3).

If we wished to make general contrasts which were not envisaged prior to the experiment, it would be necessary to use a different approach first described by Scheffé (1959, p. 66). In this case we define the contrasts as θ, where

$$\theta = \sum_1^k c_i\bar{X}_i \qquad \text{and} \qquad \sum_1^k c_i = 0 \tag{18.11}$$

[1] An assumption later demonstrated to be invalid; see Sec. 18.9.

TABLE 18.9
Mean Bulk Densities for Rock Types and Subsamples

Rock type	Rock-type mean	Subsample mean		
		A	*B*	*C*
Oriskany (5292)	2.4655	2.4510	2.4793	2.4658
Venango (H-13)	2.2245	2.2182	2.2316	2.2238
Bradford (B-8)	2.1635	2.1724	2.1632	2.1548
Miocene (5516)	1.8376	1.8038	1.8546	1.8546
Sespe (5293)	2.0935	2.0880	2.0932	2.0994

The specific estimate of θ is H, and

$$H = \sum_{i}^{k} c_i \bar{X}_i$$

with variance

$$\text{Var}\,(H) = \sum_{i}^{k} c_i^2 \frac{\sigma^2}{n_i} \tag{18.12}$$

The estimated variance of H is

$$\hat{V}(H) = \sum_{1}^{k} c_i^2 \frac{\hat{\sigma}^2}{n_i} = \hat{\sigma}^2 \sum_{1}^{k} \frac{c_i^2}{n_i}$$

In the present example the contrast is

$$H = (\bar{X}_1 + \bar{X}_2 + \bar{X}_3 + \bar{X}_4 - 4\bar{X}_5)$$

and the c_i are 1, 1, 1, 1, −4, which sum to zero. The $\bar{X}_i$ are rock-type means which yield

$$H = 2.4655 + 2.2245 + 2.1635 + 1.8376 - 4(2.0935) = 0.317$$

We can now calculate confidence limits of the same type as above but with

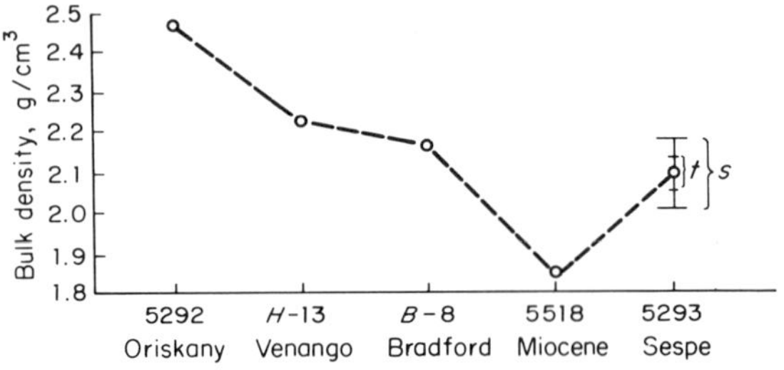

Figure 18.3 Mean bulk density for rock types.

the coefficient Student's t replaced by a coefficient due to Scheffé, namely, S:

$$H - S\sqrt{\hat{V}(H)} < \theta < H + S\sqrt{\hat{V}(H)} \tag{18.13}$$

and S is obtained from

$$S^2 = (k - 1)F_\alpha(k - 1, f) \tag{18.14}$$

where k is the number of means, α is the selected probability level, and f is the number of degrees of freedom associated with the estimate of the appropriate $\hat{\sigma}^2$. For the mean bulk densities of rock types

$$S^2 = (5 - 1)F_{001}(4,14)$$

$$= 4 \times 8.62 = 34.48$$

$$S = 5.87$$

The necessary confidence limits are

$$H - 5.87 \times 0.009 < \theta < H + 5.87 \times 0.009$$

For the Sespe arkose specimen this leads to

$$2.041 < \theta < 2.147$$

which can be written as $H \pm 0.053$, where $H = 0.317$; then the confidence interval becomes 0.264 to 0.370, which, since the interval does not contain zero, indicates that the Sespe differs from the other four specimens in mean bulk density.

A number of other tests for contrasts among means meet different requirements. For example, such a test, proposed by Tukey, requires that the variances around each mean be equal and that the numbers in each class be equal (see Brownlee, 1960, p. 253). Dunnett (1955) has proposed a test for comparison of k means with a control. These tests have different properties and lead to different intervals (see Brownlee, 1960, p. 253, Scheffé, 1959, pp. 66ff.).

18.9 Homogeneity of Variances

The basis of significance testing in analysis of variance is that it is appropriate to combine the variance estimates for the different groups or that the variances at the different levels are homogeneous. If this assumption fails, and variances at some level, particularly the error level, are heterogeneous, conclusions from the analysis may be vitiated; the analysis simultaneously compares means and variances, and heterogeneous variance tends to confuse the comparison. One possible symptom of heterogeneity is the presence of an unusually large variance ratio, or F value. If an experiment is constructed to detect a difference between means, one does not usually expect the difference to be extreme; hence an unusually large F value, such

TABLE 18.10
Test of Homogeneity of Error Variance in Bulk Density: Bartlett's Test for Homogeneity of Variance; Equal Numbers in Samples

$n = 15;\ k = 5$

Sample	Variance	$\log \hat{\sigma}^2$
1	0.000514	$\bar{4}.71096$
2	0.001055	$\bar{3}.02325$
3	0.001399	$\bar{3}.14582$
4	0.000507	$\bar{4}.70501$
5	0.000169	$\bar{4}.22789$
6	0.000547	$\bar{4}.73799$
7	0.000889	$\bar{4}.94890$
8	0.002105	$\bar{3}.32325$
9	0.004349	$\bar{3}.63839$
10	0.000461	$\bar{4}.66370$
11	0.024551	$\bar{2}.39007$
12	0.000377	$\bar{4}.57634$
13	0.000236	$\bar{4}.37291$
14	0.000143	$\bar{4}.15534$
15	0.000297	$\bar{4}.47276$
Total		$\overline{54} + 7.09258 = \overline{47}.092580 = \sum_1^{15} \log \hat{\sigma}_i^2$

$$\sum_1^{15} \hat{\sigma}_i^2 = 0.037599 \qquad \bar{\hat{\sigma}}^2 = 0.0025066$$

$$\log \bar{\hat{\sigma}}^2 = \bar{3}.399088 \qquad n \log \bar{\hat{\sigma}}^2 = \overline{40}.986320$$

$$n \log \bar{\hat{\sigma}}^2 - \sum \log \hat{\sigma}_i^2 = \overline{40}.986320 - \overline{47}.092580$$
$$= 7.89344$$

$$\chi^2 = 2.3026\,(k-1)\,(n \log \bar{\hat{\sigma}}^2 - \sum \log \hat{\sigma}_i^2)$$
$$= 2.3026 \times 4 \times 7.89344$$
$$= 72.701740$$

$$C = 1 + \frac{1}{3(n-1)}\left[\sum \frac{1}{k-1} - \frac{1}{\sum (k-1)}\right]$$
$$= 1 + \frac{1}{3(15-1)}\left[\sum \frac{1}{5-1} - \frac{1}{\sum (5-1)}\right]$$
$$= 1 + \tfrac{1}{42}\left(\tfrac{225}{60} - \tfrac{1}{60}\right)$$
$$= 1 + \tfrac{4}{45}$$
$$= 1.0889$$

Corrected

$$\chi^2 = \frac{\chi^2}{C} = \frac{72.70174}{1.0889} = 66.762$$

With d.f. $= n - 1 = 14$, $\chi^2_{001,14} = 36.123$, therefore the hypothesis of homogeneous variance is rejected.

TABLE 18.11
Independent Analysis of Variance of Bulk Density for Each Rock Type

Source of variation	d.f.	Sum of squares	Mean square	F	F_{05}	F_{01}
Oriskany:						
Between samples	2	0.002074	0.001037	4.19*	3.88	6.93
Between runs	12	0.002968	0.0002473···			
Total	14	0.005042				
Venango:						
Between samples	2	0.000453	0.0002265	2.22	3.88	6.93
Between runs	12	0.001223	0.000101916···			
Total	14	0.001676				
Bradford:						
Between samples	2	0.000775	0.0003875	0.73	3.88	6.93
Between runs	12	0.006343	0.000528583···			
Total	14	0.007118				
Trinidad, Miocene:						
Between samples	2	0.008602	0.004301	2.03	3.88	6.93
Between runs	12	0.025389	0.00211575			
Total	14	0.033991				
Sespe:						
Between samples	2	0.000326	0.000163	2.89	3.88	6.93
Between runs	12	0.000676	0.0000563···			
Total	14	0.001002				

as that at the rock-type level in Table 18.7, is grounds for suspecting inhomogeneity in the data. It would be prudent in such a case to test the homogeneity of variance; a test for this purpose has been devised by Bartlett (1937a), which, with equal numbers in the samples, as in this example, can be described as a comparison of the logarithm of the average variance, multiplied by the number of variances, with the sum of the logarithms of each variance. Specifically, the test comprises calculating a statistic B as follows (see also Sec. 17.12):[1]

$$B = \frac{1}{C}\left(n \ln \hat{\sigma}^2 - \sum_{i}^{n} k_i \ln \hat{\sigma}_i^2\right) \tag{18.15}$$

[1] The more general formula for unequal-sized samples is given in Bennett and Franklin (1954, p. 197) and Brownlee (1960, p. 225).

TABLE 18.12
Variances and Standard Deviations of Subsample Means and Error and Coefficients of Variation for Each Rock Type

Rock type	Subsample $\hat{\sigma}_m^2$	Subsample $\hat{\sigma}_m$	Error $\hat{\sigma}_\epsilon^2$	Error $\hat{\sigma}_\epsilon$	CV, %
Oriskany	0.000263	0.016	0.000247	0.016	0.64
Venango	0.000041	0.006	0.000102	0.010	0.45
Bradford	Negative	0.000	0.000529	0.023	1.38
Miocene	0.000728	0.027	0.002116	0.046	2.50
Sespe	0.000036	0.006	0.000056	0.008	0.36

or, equivalently,

$$B = \frac{2.3026}{C}\left(n \log \hat{\sigma}^2 - \sum_1^n k_i \log \hat{\sigma}_i^2\right)$$

and

$$C = 1 + \frac{\sum_1^n (1/k) - 1/k}{3(n - 1)}$$

where there are n estimates of the variance, each one based on k observations, and $\hat{\sigma}^2$ is the mean of the variances. In the present example we calculate $n = 15$ variances for the three subsamples in each of five rock types based on $k = 5$ observations in each. The work is carried out tabularly as in Table 18.10. The calculated ratio B/C under the null hypothesis that all the variances are calculated from random samples from the same population approximates the chi-square distribution with $n - 1$ d.f. In the present case the value exceeds the tabulated chi square at the 1 percent level; hence the hypothesis is rejected, and the variances are heterogeneous.

Under these conditions it is necessary to consider each rock type separately, and the analysis of variance for individual specimens is exhibited in Table 18.11. From this analysis only the variance for subsamples in the Oriskany significantly exceeds the replication or error variation (see also Fig. 18.4).

The components of variance, standard deviation, and coefficients of variation CV [ratio of standard deviation to the mean, in percent, that is, $CV\% = 100(\hat{\sigma}/\bar{X})$] for each rock type are displayed in Table 18.12. The error variation is quite large and variable in comparison with the subsample variation. Presumably the error variation in the Miocene specimen is large because the specimen is so friable.

The variation among subspecimen means in the Oriskany rock type could be a reflection of the patchy distribution of carbonate cement.

18.10 Conclusions and Recommendations

On the basis of the experiment it seems clear that the technique of determining the bulk density of small specimens (≈ 8 g) of rocks by the mercury balance is adequate to detect differences among rock-type means. If subsample variation is of interest, adequate replication is necessary to reduce the error variation in order to detect differences among subsample means from the same sample. In adjusting the sampling design to achieve this reduction it is necessary to obtain, by experiment, an estimate of the magnitude of the error variation, and in friable rock specimens this may be quite large because of losses while handling the specimens. On the other hand, in some consolidated rocks the patchy distribution of carbonate cement leads to inflated variation at the subsample level. It is necessary to evaluate these sources of variation before attempting an elaborate investigation of some sedimentary-rock population (e.g., Cochran, 1960; Griffiths et al., 1953a, b; and Sec. 11.3).

18.11 Examples of the Use of One-way-classification Analysis of Variance in Geological Problems

Multistage, or hierarchical, sampling is a common sampling design in the analysis of rock populations, and, as a result, one-way-classification analysis of variance with one or more levels of subdivisions is a common experimental design and analysis in the study of rocks. Furthermore, this completely randomized design is most suitable where little or no preliminary information exists about the relative magnitude of variability at different levels; the experimental arrangement is so simple that it can accommodate unequal numbers in the subdivisions which may arise either because equal numbers are not available or because of loss during the laboratory procedure. This design is therefore excellent for preliminary reconnaissance and is par-

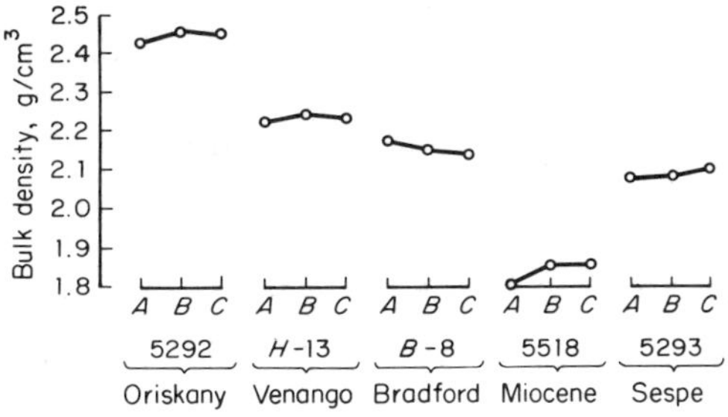

Figure 18.4 Mean bulk density for subsamples within rock types.

ticularly suitable for controlling unassigned or error sources of variation, because this lowest level possesses the largest number of degrees of freedom.

Various examples ranging from simple to sophisticated designs exist in the geological literature, encompassing a wide variety of subject-matter fields (e.g., Miller and Kahn, 1962, pp. 178–181). The single level of subdivision is extremely useful for isolating the contribution of experimental error; the particular sampling design illustrated in Fig. 2.2, first proposed by Krumbein (1934a) and subsequently adapted to analysis of variance (Griffiths, 1953b), is particularly attractive and efficient for comparing sampling variation with all other sources of variation in attempting to distinguish a pattern of variation over and above experimental error (estimated by channel sampling in this design).

A detailed analysis of components of variance and their use in determining the most efficient sampling allocation by using cost estimates is described by Olson and Potter (1954). A nested (multistage) design with reduction in the number of subunits at specific levels, useful in regional studies, is described by Krumbein and Slack (1956).

19 Analysis of variance: two-way classification

There is always an unexplained component . . . which arises at least in part from the influence of factors not included in the model. These are called "residual" or "chance" variables. If these variables account for more variation in outcome than those which are treated explicitly in our models, we have not taken a broad enough view of the system. The objective, of course, is to be able to reduce this unexplained variation as much as possible.

R. L. Ackoff, 1961

19.1 Introduction

The preceding chapter dealt with classification into hierarchies in which each subclass is nested within the preceding class. Suppose that it is possible to subdivide the population on the basis of two criteria of classification; then the subclasses in one set correspond to the subclasses of the other. This is *stratified* sampling, and stratification has obvious geological connotations. The principle of stratification, or blocking, however, is of very wide application, and whereas geological stratification is generally related to order of superimposition and often represents a temporal sequence, statistical stratification does not necessarily imply an ordinal arrangement. For example, suppose a series of specimens is classified into different color classes and at the same time into different rock types, and the random variable is the proportion of carbonate or grain size. The color classes and the rock types are not necessarily ordered in any way; in other words, the scale underlying the classifications may be nominal. On the other hand, of course, the classifications may possess ordinal, interval, or ratio scaling. Generally, if

Quotation: R. L. Ackoff, The Meaning, Scope and Methods of Operations Research, chap. 1 in R. L. Ackoff (ed.), "Progress in Operations Research," p. 29, ORSA publ. 5, John Wiley & Sons, Inc., 1961.

the scaling is ordinal, interval, or ratio, the designs are thought of as *factorial*, whereas if they are nominal, they are merely two-way classifications without qualification. In the absence of order the questions asked concern the existence of differences among the class means (main effects), on either or both bases of classification, and the independence or association among the classifications (interaction). If the classes are ordered or possess a higher scaling, additional questions may be resolved concerning the relationship of variation in the variable to the order among the classes on either or both classifications, and, again, tests can be made for consistency of the relationship in one dimension (equivalent to one basis of classification) against the other dimension.

The structure of the simplest two-way-classification design, the associated sampling pattern, and the accompanying analysis-of-variance table are illustrated in Fig. 19.1; each observation X_{ij} may be expressed as

$$X_{ij} = \mu + \alpha_i + \beta_j + (\alpha\beta_{ij} + \epsilon_{ij}) \tag{19.1}$$

indicating that each observation contains an estimate of the true mean μ, plus a contribution from one main effect α_i, plus a contribution from the second main effect β_j, plus contributions from all other (unassigned) sources. Equation (19.1) indicates that the error term is composed of a contribution $\alpha\beta_{ij}$ from interaction plus variation ϵ_{ij} from all other sources. In this experimental arrangement the effect of α can be separately considered from the effect of β only if $\alpha\beta_{ij}$ is zero; otherwise, the effect of α in, say, the first class compared with the second may not be the same in each class. It is therefore necessary to ensure that $\alpha\beta_{ij} = 0$, and this requirement is incorporated as a part of the model assumptions. In practice it is advisable to determine whether this requirement is fulfilled, and so a separate estimate of the contribution of $\alpha\beta_{ij}$ is necessary, together with an appropriate error term to test

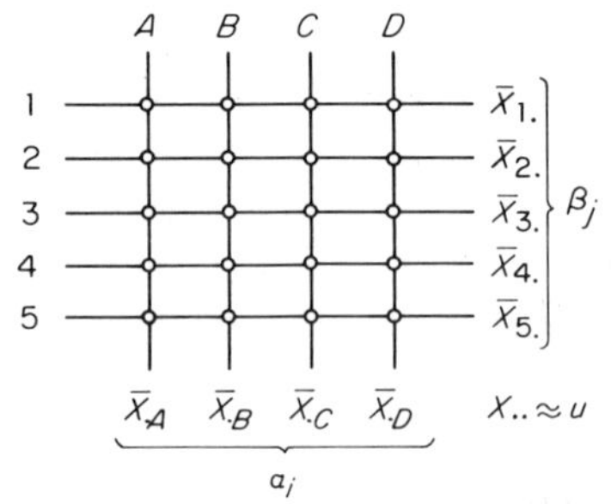

$$X_{ij} = u + \alpha_i + \beta_j + [\alpha\beta_{ij} + \epsilon_{ij}]$$

Source of variation	Degrees of freedom	Expected mean square
Among a_i	$a - 1$	$\sigma_e^2 + b\sigma_a^2$
Among β_j	$b - 1$	$\sigma_e^2 + a\sigma_\beta^2$
Error $\alpha\beta_{ij} + \epsilon_{ij}$	$(a-1)(b-1)$	σ_e^2
Total	$ab - 1$	σ^2_{tot}

Figure 19.1 Experimental design and sampling; two-way classification.
$X_{ij} = \mu + \alpha_i + \beta_j + (\alpha\beta_{ij} + \epsilon_{ij})$

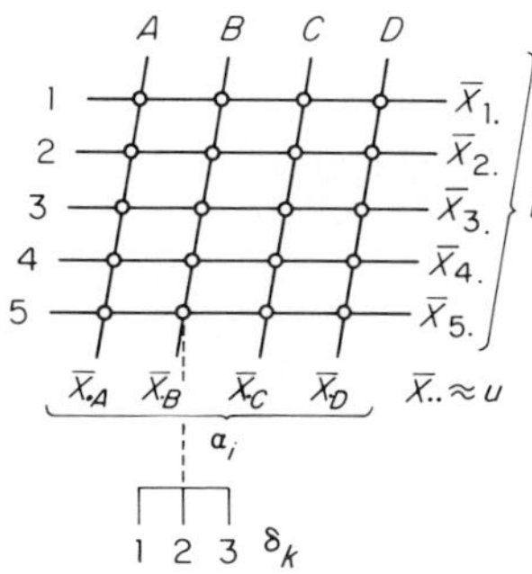

Source of variation	Degrees of freedom	Expected mean square
Among α	$a-1$	$\sigma_e^2 + r\sigma_{\alpha\beta}^2 + br\sigma_\alpha^2$
Among β	$b-1$	$\sigma_e^2 + r\sigma_{\alpha\beta}^2 + ar\sigma_\beta^2$
α and β	$(a-1)(b-1)$	$\sigma_e^2 + r\sigma_{\alpha\beta}^2$
Error $\delta+\epsilon$	$ab(r-1)$	σ_e^2
Total	$abr-1$	$\sigma_{\text{tot.}}^2$

Figure 19.2 Experimental design and sampling; two-way classification with replicates.

$$X_{ijk} = \mu + \alpha_i + \beta_j + \alpha\beta_{ij} + (\delta_{k(ij)} + \epsilon_{(ijk)})$$

the hypothesis $\alpha\beta_{ij} = 0$. For this purpose it is mandatory to collect more than one observation in each cell, and these *replicates* are nested within each cell of the two-way classification. The structure of the experiment now becomes (see Fig. 19.2)

$$X_{ijk} = \mu + \alpha_i + \beta_j + \alpha\beta_{ij} + (\delta_{k(ij)} + \epsilon_{(ijk)}) \tag{19.2}$$

wherein each observation X_{ijk} contains an estimate of the true mean μ, plus a contribution α_i from the first main effect, plus a contribution β_j from the second main effect, plus a contribution $\alpha\beta_{ij}$ from the inconsistency of α_i across β_j, plus variation from all unassigned sources. This last term is composed of variation from replication $\delta_{k(ij)}$ plus variation from all other sources $\epsilon_{(ijk)}$, and the (ij) indicates nesting of δ_k within the α_i and β_j cells. The experimental structure, sampling arrangement, and accompanying analysis-of-variance table are illustrated in Fig. 19.2. The numbers are assigned at random to the replicate determinations. The two-way-classification design arose from an experimental arrangement used in agronomic field experiments called a *randomized block design*. In this case the agronomist was interested in comparing the yield of several varieties of wheat, and a number of plants of each variety were arranged in a block (Cochran and Cox, 1957, p. 106); differences among yields of varieties represent the main effect of interest, and the blocking or stratification is used to remove the effect of different locations, soil heterogeneity, or climatic zones. No attempt is made to study differences among blocks, and the essential value of the design relies on the fact that behavior within a block is more similar than that among blocks for any single variety of wheat. This arrangement removes the block source of variation from the error term and so, if effective, reduces the magnitude of variation from unassigned sources and leads to a more sensitive experiment, the smaller error permitting the investigator to detect smaller differences among variety means. In most geological experiments, on the other hand, the second classification, corresponding to blocks in the agronomy case, is also of interest, and so the investigator intends to evaluate variation from this source; such an experiment is a two-way classification and should not be confused with a randomized block design.

19.2 Example of a Two-way Classification: Porosity Determined by Two Techniques

In the experiment described in Sec. 18.2 the objective is to compare two procedures for the determination of porosity of some oil-reservoir sandstones. The data were classified into the two techniques with several determinations performed on the different rock specimens by each technique; the specimens were selected at random. In practice, it is known that different specimens of rocks show widely different properties, and it would therefore be prudent to make use of this information in designing the experiment. A wide range in porosity is required, so five rocks were randomly selected covering this range, and each specimen was subdivided into two parts; in this way both procedures were applied to subsamples of the same rock specimens, and so the data may now be represented in a two-way classification in which the main effects are differences in porosity between procedures and among rock specimens. The data are recorded in terms of the two-way classification in Table 19.1 (cf. Table 18.1).

TABLE 19.1
Porosity of Five Rock Specimens Determined by Each of Two Techniques

Specimen, k	Percent g		$\sum_1^2$	Average, %
	Technique 1	Technique 2		
1	17.1	16.92	34.02	17.01
2	20.1	20.68	40.78	20.39
3	19.3	18.05	37.35	18.675
4	10.2	10.39	20.59	10.29
5	25.9	26.00	51.90	25.95
$\sum_1^5$	92.60	92.04	184.64	
$\bar{X}_j$	18.52	18.408		$\bar{X}_G = 18.464$

The analysis of the data follows exactly the same procedure as that described in Sec. 18.3 for the total sum of squares and the sum of squares arising from differences between techniques; but now we can also compute the sum of squares due to differences among k specimen means SS_k as follows:

$$SS_k = \frac{\sum_1^k(\sum_1^g X)^2}{g} - CT$$

$$\sum_1^5(\sum_1^2 X)^2 = 7{,}332.9494$$

$$\div\ 2 = 3{,}666.4747$$

$$-\ CT = 3{,}409.19296$$

$$= 257.28174 = SS_k$$

The error term becomes

$$\text{Error } (SS_\epsilon) = SS_{tot} - (SS_g + SS_k)$$

$$= 258.27044 - (0.03156 + 257.28174) = 0.95714$$

The analysis of variance is displayed in Table 19.2. The appropriate null hypotheses are:

1 H_0: The specimen means are means of random samples from the same homogeneous population; therefore $\sigma_k{}^2 = 0$.
2 H_0: The technique means are means of random samples from the same homogeneous population; $\sigma_g{}^2 = 0$.

Under the first null hypothesis the variance ratio $F_{\alpha,\nu_1,\nu_2} = (\sigma_\epsilon{}^2 + g\sigma_k{}^2)/\sigma_\epsilon{}^2$ follows an F distribution with $\nu_1 = 4$ and $\nu_2 = 4$ d.f. The tabulated $\alpha = 0.05$ level yields $F_{05,4,4} = 6.39$ (Arkin and Colton, 1950, p. 117), and the observed F far exceeds this; hence the hypothesis is rejected at the 5 percent level (and even at the 0.001 percent level). Evidently the rock specimens differ in mean porosity. On the other hand, under the second null hypothesis the variance ratio $F_{05,\nu_1,\nu_2} = (\sigma_\epsilon{}^2 + k\sigma_g{}^2)/\sigma_\epsilon{}^2$ follows an F distribution with $\nu_1 = 1$ and $\nu_2 = 4$ d.f., and the tabulated value is $F_{05,1,4} = 7.71$ (Arkin and Colton, 1950, p. 117). The second null hypothesis is not rejected, and apparently the two techniques do not yield a significant difference among means.

This example illustrates the importance of experimental design wherein, by appropriate arrangement of the sampling, a marked reduction in the magnitude of variation associated with unassigned (error) sources of variation is achieved. Stratification or blocking (in this case paired sampling) enables the experimenter to extract a contribution to variation from the error term and thus achieve a more sensitive experiment for the same number of obser-

TABLE 19.2
Comparison of Two Techniques for Measurement of Porosity—Analysis of Variance: Two-way Classification

No. of items	Source of variation	d.f.	Sum of squares	Mean square	Variance ratio F	F_{05}
$g = 2$	Between techniques	$g - 1 = 1$	0.03156	0.03156	0.131	7.71
$k = 5$	Among specimens	$k - 1 = 4$	257.28174	64.32043	268.808***	6.39
	Error	$(g - 1)(k - 1) = 4$	0.95714	0.23928		
$gk = 10$	Total	$gk - 1 = 9$	258.27044			

vations and therefore at no extra cost. It should be observed that a reduction in the number of degrees of freedom for error accompanies this rearrangement; and unless the stratification achieves a reduction in variation equivalent to this loss of degrees of freedom, the experiment is less efficient. Since stratification requires less variation within than among the strata to achieve this reduction, such a procedure can be used to test whether stratification actually exists.

19.3 An Example of Two-way Classification with Replicates: Bulk-density Variation in a Pocono Sandstone

The objective of this experiment is to obtain a preliminary estimate of the variability in bulk density of a single sandstone layer and is essentially concerned with sampling variability. The sandstone bed is 10 ft thick and is exposed in a road cut 6 miles northwest of Port Matilda, Pennsylvania, on Route 322. The sandstone is a slumped layer and contains large 2- to 3-in.-diameter concretions of pyrite. Otherwise it appears homogeneous in texture, i.e., massive.

Six specimens were taken from this layer, four equally spaced on a vertical traverse across the sandstone and two equally spaced on either side of the vertical traverse. The interval between specimens was 2 ft.

Subsamples of 8 to 10 g were used to measure the bulk density of these specimens by means of a mercury balance (Chap. 11), and each of five operators[1] ran each of the six specimens four times, yielding 120 determinations.

19.4 Experimental Design

The specimens may be considered random samples from an infinite population; and since the experiment is aimed at an estimate of sampling variability as a basis for future experimentation, the operators are six randomly selected from an infinite number. Similarly, the replicates are four of an infinite number which could be run. This therefore represents the infinite population model, model 2 of Bennett and Franklin (1954, p. 348), or the components-of-variance model of Eisenhart (1947, p. 15), or the random-effects model of Scheffé (1959, chaps. 7 et seq., particularly sec. 7.4).

The experimental design is a two-way classification (specimens and operators) with replicates, where the replicate determinations are randomly assigned within each specimen and operator class. Each observation can be written as

$$X_{ijk} = \mu + \alpha_i + \beta_j + \gamma_{ij} + (\delta_{k(ij)} + \epsilon_{ijk})$$

expressing the structure of the experiment. Each observation is composed

[1] The experiment was performed by R. Cadigan, R. E. Cobb, J. A. Cochran, A. D. Leonard, and R. C. Pearson in a graduate class under the author's tutelage in 1952.

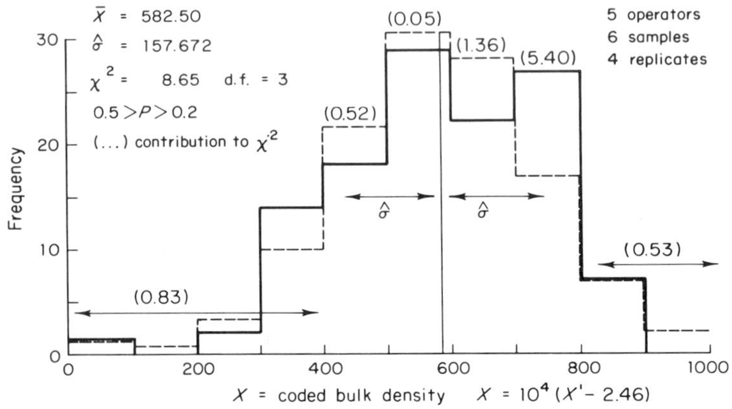

Figure 19.3 Frequency distribution of bulk-density observations on six specimens of Pocono graywacke.

of a constant (the overall mean μ), a contribution from the specimen α_i, from the operator β_j, and from "discrepancy" or differential effect of operators over specimens γ_{ij}, plus a final contribution from replicates $\delta_{k(ij)}$ and from unspecified (unassigned) sources, otherwise known as error $\epsilon_{k(ij)}$.[1]

The assumptions necessary to the analysis of these data by the variance ratio or F test are:

1 The observations X_{ijk} are random variables.
2 The variables are sums of the component variables α_i and β_j; that is, $\sum \alpha_i = \sum \beta_j = \sum \gamma_{ij} = 0$.
3 The random variables are distributed with variances σ_α^2, σ_β^2, and σ_ϵ^2, and all covariances are zero.

19.5 Preliminary Analysis of the Data

The 120 observations X_{ijk} are coded by subtracting a constant (= 2.46) from each value and multiplying the remainder by 10^4 to remove decimals. The coded data are listed in Table 19.3 and are expressed as a histogram in Fig. 19.3.

The frequency distribution yields the summary statistics

$$\bar{X} = 2.518 \qquad \hat{\sigma} = 0.016 \qquad \sqrt{b_1} = -0.367^* \qquad b_2 = -3.686^{**}$$

The frequency distribution of these 120 observations is skewed ($\sqrt{b_1}$ significant at the 5 percent level) and platykurtic (b_2 significantly low at the

[1] Note that the replicate runs are nested within operators and specimens indicated by parentheses around the ij. This error term includes variation from differences among replicate determinations plus variation from all other unassigned sources.

TABLE 19.3
Bulk Density of Pocono Sandstone Data Coded BD = 10^{-4} (X_{ij}–2.46)

Operator	Replicate	1	2	3	4	5	6	$\sum_{1}^{24} X_i$
A	1	379	567	619	861	669	843	
	2	468	377	557	752	670	797	
	3	485	402	647	780	648	776	
	4	343	375	599	728	626	793	
	Subtotal	1,675	1,721	2,422	3,121	2,613	3,209	14,761
B	1	409	635	564	710	561	770	
	2	500	536	562	677	604	713	
	3	456	528	482	566	336	646	
	4	870	570	576	705	459	782	
	Subtotal	2,235	2,269	2,184	2,658	1,960	2,911	14,217
C	1	338	27	438	794	660	719	
	2	397	549	529	807	677	791	
	3	405	554	498	849	681	763	
	4	441	510	568	809	755	790	
	Subtotal	1,581	1,640	2,033	3,259	2,773	3,063	14,349
D	1	294	442	525	630	695	524	
	2	333	518	603	611	735	516	
	3	442	516	632	772	743	724	
	4	308	512	641	710	736	724	
	Subtotal	1,377	1,988	2,401	2,723	2,909	2,488	13,886
E	1	364	350	477	406	515	810	
	2	361	577	473	603	512	787	
	3	502	526	603	402	466	769	
	4	338	317	621	586	268	784	
	Subtotal	1,565	1,770	2,174	1,997	1,761	3,150	12,417
Total		8,433	9,388	11,214	13,758	12,016	14,821	69,630 $\sum_{1}^{120} X_i$

1 percent level). When a class-by-class comparison is performed and the deviations of the observed frequencies compared with those expected from a normal distribution with the same mean and variance as this sample of observations, the resulting $\chi^2 = 8.65$; under the null hypothesis H_0 that these 120 observations are a random sample from a normal population with

mean $\mu = 2.518$ and standard deviation $\sigma = 0.016$, the statistic χ^2 follows the chi-square distribution with degrees of freedom equal to number of classes compared minus number of restrictions imposed; that is, $c - r = 6 - 3 = 3$ and $\chi^2_{05,3} = 7.815$, whereas $\chi^2_{02,3} = 9.837$. The null hypothesis leads to the conclusion that if these are indeed a random sample from a normal distribution with the stated parameters, such a sample would occur between 2 and 5 times in every 100 such cases. The hypothesis is rejected; either the population distribution is not normal, or these observations are not random. Since 6 specimens and 20 determinations on each are unlikely to be random samples from the infinite set of replicates on an infinite number of specimens from the sandstone, the rejection of the hypothesis really throws doubt on the randomness of the sampling rather than on the form of the population frequency distribution.

As a further illustration of this feature the data are plotted in the form of a control chart in Fig. 19.4. We should expect 3 observations per 1,000 to lie beyond the three standard-deviation limits (even relaxing the requirement of normality); the number out of 120 which lie beyond $\pm 3\hat{\sigma}$ is 5, which is certainly indicative of nonrandom variation around the population mean.

Variation around each specimen mean is large and of the same order of magnitude. The scatter of observations tends to be systematically arranged around each specimen mean, as would be expected, and it seems likely that more specimens would be needed to represent adequately the bulk density of this sandstone. It also appears likely that the replications show much greater

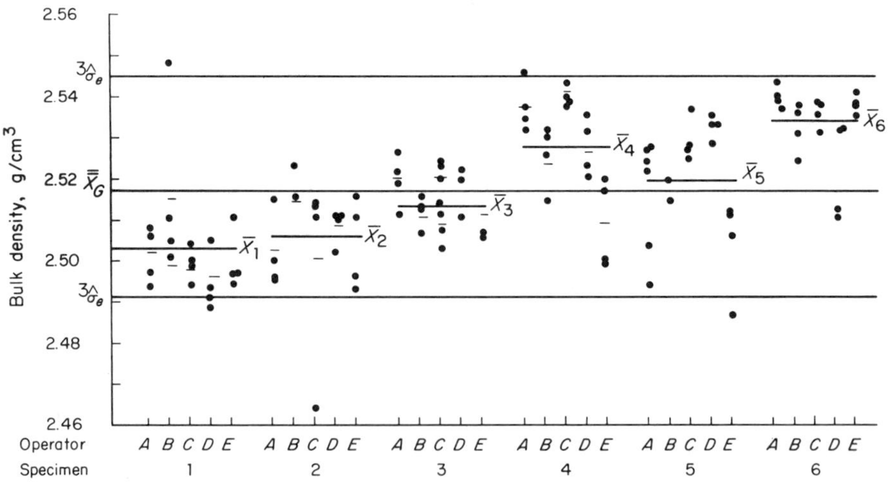

Figure 19.4 Quality control chart to show variation in bulk density of six specimens of Pocono sandstone.

TABLE 19.4
Totals for Operators and Specimens

Sample	Operator A	B	C	D	E	Total
1	1,675	2,235	1,581	1,377	1,565	8,433
2	1,721	2,269	1,640	1,988	1,770	9,388
3	2,422	2,184	2,033	2,401	2,174	11,214
4	3,121	2,658	3,259	2,723	1,997	13,758
5	2,613	1,960	2,773	2,909	1,761	12,016
6	3,209	2,911	3,063	2,488	3,150	14,821
Total	14,761	14,217	14,349	13,886	12,417	69,630

spread than we have been led to expect from experienced operators demonstrably in control.

In other words, in future experiments, the operators should be better trained and their behavior should be tested before performing the experiment. Second, to estimate the mean bulk density of this layer will need a larger number of specimens.

The extremely aberrant observation at 2.4627 is likely to represent a gross mistake and possibly should be replaced in the form of a missing value.

A graphic plot of operator means and specimen means against their respective standard deviations shows no trend; hence additivity is a reasonable assumption.

19.6 Analysis of Variance

Computations for the sums of squares follow; the correction term is first computed, followed by the total sum of squares, both taken from Table 19.3. The sums of squares for both main effects and the interaction are found from the totals of Table 19.4. The analysis of variance is summarized in Table 19.6.

Bulk Density of Pocono Graywacke: Computations for Sums of Squares

1. Correction term CT

$$\sum_{i=1}^{120} X_i = 69.630; \qquad \frac{\left(\sum_{i=1}^{120} X_i\right)^2}{120} = 40{,}402{,}807.50$$

$$\bar{X} = 580.250$$

2. Total sum of squares (SS_{tot})

$$\sum_{i=1}^{120} X_i^2 = 43{,}444{,}456.00$$

$$-\ CT = 40{,}402{,}807.50$$

$$= 3{,}041{,}648.50 = SS_{tot}$$

3. Rock-specimen sum of squares SS_m

$$\sum_{i=1}^{6} (\sum_{jk=1}^{20} X_i)^2 = 838{,}332{,}690.0$$

$$\div\ 20 = 41{,}916{,}634.5$$

$$-\ CT = 40{,}402{,}807.5$$

$$= 1{,}513{,}827.0 = SS_m$$

4. Operator's sum of squares SS_p

$$\sum_{j=1}^{5} (\sum_{ik=1}^{24} X_i)^2 = 972{,}906{,}896.00$$

$$\div\ 24 = 40{,}537{,}787.33$$

$$-\ CT = 40{,}402{,}807.50$$

$$= 134{,}979.83 = SS_p$$

5. Specimens by operators sums of squares SS_{mp}

$$\sum_{ij}^{30} (\sum_{k=1}^{4} X_i)^2 = 170{,}808{,}656.0$$

$$\div\ 4 = 42{,}702{,}164.0$$

$$-\ CT = 40{,}402{,}807.5$$

$$= 2{,}299{,}356.5$$

$$-\ (SS_m + SS_p) = 1{,}648{,}806.83$$

$$= 650{,}549.67 = SS_{mp}$$

6. Error sum of squares SS_ϵ

$$SS_{tot} = 3{,}041{,}648.50$$

$$-\ (SS_m + SS_p + SS_{mp}) = 2{,}299{,}356.50$$

$$= 742{,}292.00 = SS_\epsilon$$

The experiment should appear as summarized in Table 19.7. Significance tests start with a test of the hypothesis that $\sigma_{mp}^2 = 0$, an integral part of the assumptions already listed; this test is performed by means of the variance ratio

$$\frac{\sigma_\epsilon^2 + r\sigma_{mp}^2}{\sigma_\epsilon^2} = F_{\alpha,20,90}$$

or from Table 19.6, 32,527.48/8,247.69 = 3.94. The 5 percent level for

TABLE 19.5
Operator and Sample Means; Bulk Density of Pocono Sandstone

	Sample						
Operator	1	2	3	4	5	6	$\bar{X}$
A	418.75	430.25	605.5	780.25	653.25	802.25	615.042
B	558.75	567.25	546.0	664.5	490.0	727.75	592.375
C	395.25	410.0	508.25	814.75	693.25	765.75	597.875
D	344.25	497.0	600.25	680.75	727.25	622.0	578.583
E	391.25	442.5	543.50	499.25	440.25	787.5	517.375
Op. $\bar{X}$	421.65	469.40	560.70	687.90	600.80	741.05	580.25 = $\bar{\bar{X}}_G$

$F_{05,20,90}$ is 1.69 and 1 percent $F_{01,20,90} = 2.09$. Therefore, since the observed F exceeds that expected, the hypothesis is rejected, and $\sigma_{mp}{}^2 \neq 0$; the source of variation operators by specimens is contributing more than the error source of variation.

At this stage the analysis breaks down, and the significant interaction clearly indicates that operators are discrepant from specimen to specimen. This is presumably another display of the lack of control due to lack of experience. In subsequent better-arranged experiments this source of variation was demonstrably removed (Griffiths et al., 1954a, pp. 47ff.).

When an interaction is significant, it implies that no general recommendation can be made; in fact, the significant interaction offsets the usefulness of the summary table. The sources of variation are not exhibiting additivity of variation at this level; and so in order to compare operators each specimen must be taken individually, or to compare specimens each operator must be taken one at a time. There appears little purpose in attempting to draw conclusions about the differences in the bulk density among specimens if they are not consistent across operators. For illustration we can continue testing and assume that the operator-by-specimen interaction is the error for testing the main effects. First, we can test the operator mean square against the interaction mean square as follows:

$$\frac{\sigma_\epsilon{}^2 + r\sigma_{mp}{}^2 + rm\sigma_p{}^2}{\sigma_\epsilon{}^2 + r\sigma_{mp}{}^2} = \frac{33{,}744.96}{32{,}527.48} = 1.037$$

$F_{05,4,20} = 2.87$, and so this is not significant; i.e., operator means are not varying more than the interaction—not much comfort since the interaction is so large.

TABLE 19.6
Analysis of Variance; Bulk Density of Pocono Sandstones

Source of variation	d.f.	Sum of squares	Mean square	Variance ratio F	F_{05}	F_{01}	F_{001}
Six specimens	$m - 1 = 5$	1,513,827.0	302,765.4	9.31***	2.71		6.46
Five operators	$p - 1 = 4$	134,979.83	33,744.96	1.037	2.87	4.43	
Operators × specimens	$(p - 1)(m - 1) = 20$	650,549.67	32,527.48	3.94**	1.69	2.09	
Four error	$pm(r - 1) = 90$	742,292.00	8,247.69				
Total 120	$pmr - 1 = 119$	3,041,648.50					

$\hat{\sigma}_\epsilon = 90.817$ decoded, $\hat{\sigma}_\epsilon = 0.0091$

$\overline{\overline{X}}_G = 2.518025$

TABLE 19.7
Model Analysis-of-variance Table for Two-way Classification with Replicates

Source of variation	d.f.	Expected mean square
Operators	$p - 1 = 4$	$\sigma_\epsilon^2 + r\sigma_{mp}^2 + mr\sigma_p^2$
Specimens	$m - 1 = 5$	$\sigma_\epsilon^2 + r\sigma_{mp}^2 + pr\sigma_m^2$
Operators × specimens	$(m - 1)(p - 1) = 20$	$\sigma_\epsilon^2 + r\sigma_{mp}^2$
Replicates (error)	$mp(r - 1) = 90$	σ_ϵ^2
Total	$mpr - 1 = 119$	

Suppose the interaction were nonsignificant, i.e., effectively zero; then we could test operator variation against error on the variance ratio:

$$\frac{\sigma_\epsilon^2 + r\sigma_{mp}^2 + rm\sigma_p^2}{\sigma_\epsilon^2} = \frac{33{,}744.96}{8{,}247.69} = 4.091$$

The expected $F_{01,4,80} = 3.56$ and $F_{01,4,100} = 3.51$; hence the operator differences are significantly larger than unassigned or error source of variation.

Last, in testing differences among specimen means against the interaction as an error term we have

$$\frac{\sigma_\epsilon^2 + r\sigma_{mp}^2 + rp\sigma_m^2}{\sigma_\epsilon^2 + r\sigma_{mp}^2} = \frac{302{,}765.4}{32{,}527.48} = 9.31.$$

$F_{001,5,60} = 4.76$; hence this F value is significant beyond the 001 level. The hypothesis of no differences among the specimen means is rejected despite the significant interaction.

It is of considerable interest to note here that if there had been no replicates, the interaction term could not have been separately evaluated, and this mean square would have been used as an error term. We should not, however, be justified in deciding that because operator means show variation not significantly larger than this error they were "uniform." Similarly, in a two-way classification the absence of differences among specimen means does not imply that the rock is "uniform," i.e., shows little difference among specimen means (Scheffé, 1959, p. 94). In fact, to achieve proof of this contention it is necessary first to ensure that the interaction is zero; therefore, replicates are required to serve as an error term for testing the interaction. Second, it is necessary to specify, *before* starting the experiment, what magnitude of difference among specimen means can be tolerated while maintaining the hypothesis that the rock is uniform (cf. Chayes and Fairbairn, 1951).

In general, there is no way of testing uniformity in the form of a null hypothesis; it is necessary to state initially that if differences are less in magnitude than some specified amount, the population sampled will be considered uniform (cf. Krumbein, 1953b).

19.7 Graphic Illustration of Differences among Means

This experiment was aimed at evaluating variability rather than estimating differences among means, but for illustration it is of some interest to examine the significant differences and what they imply.

Differences among means for both main effects are plotted in Fig. 19.5 (see also Table 19.5), and the much greater spread of specimen means compared with the operator means is clearly exhibited.

Two graphs can be drawn for the interaction, using either specimen or operator means as the abscissa. Specimen means are used in Fig. 19.6 and operator means in Fig. 19.7.

From Fig. 19.6 it can be seen that there is a general trend for operators to find increasing bulk density from specimen 1 to 6 with a slight break at specimen 5. The order of treating the specimens is of course arbitrary but serves to establish a basis for comparison. For example, contrary to all other operators, operator *B* crosses this trend in the region of specimens 1 to 4. Similarly, operator *D* crosses the general trend for specimens 5 and 6. This is a graphic portrayal of interaction. The crossovers in Fig. 19.7 illustrate the same feature with operators along the abscissa.

Discrepancy or interaction among main effects can also be represented by comparing the means in each cell of Table 19.5; an example is given in Table 19.8.

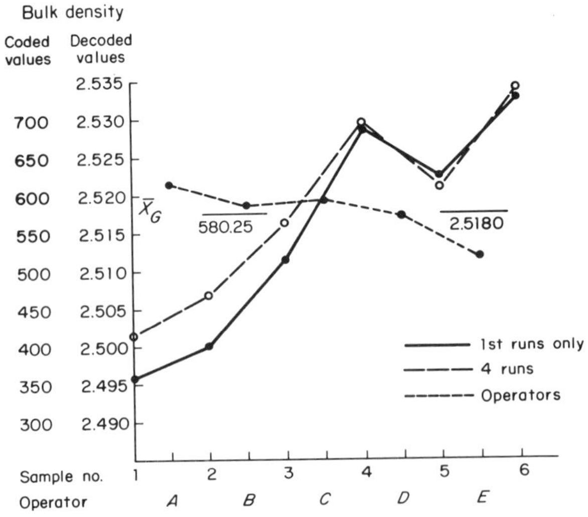

Figure 19.5 Differences among means for main effects, operators and specimens.

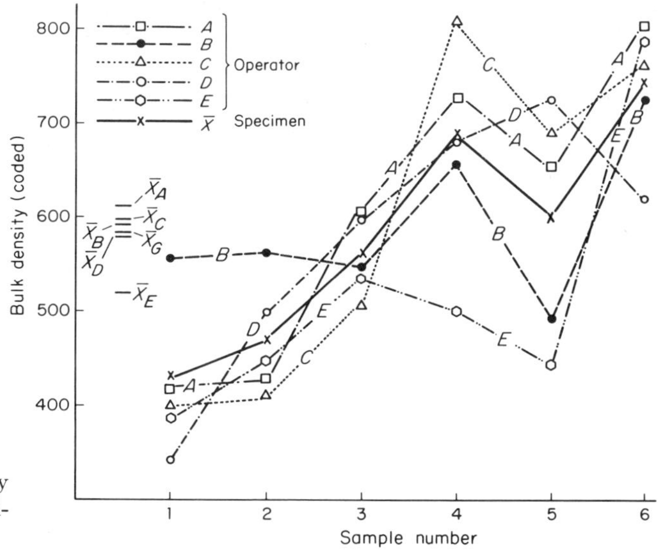

Figure 19.6 Discrepancy of operators over specimens.

If the interaction term were indeed zero, there would be no discrepancies, and the differences among a pair of operators over the samples would be consistent; i.e., when one operator obtained higher values than another on one specimen, he would do so on all the specimens. Operator A finds that specimen 3 yields a greater value for bulk density than specimen 2 (+255.25), whereas operator B finds specimen 3 to be less in bulk density than specimen 2 (−21.25). There does not have to be a change in sign to induce discrepancy; if one operator finds a small positive difference between two specimen means such as that found by operators A, B, and C in specimens 1 and 2 in Table

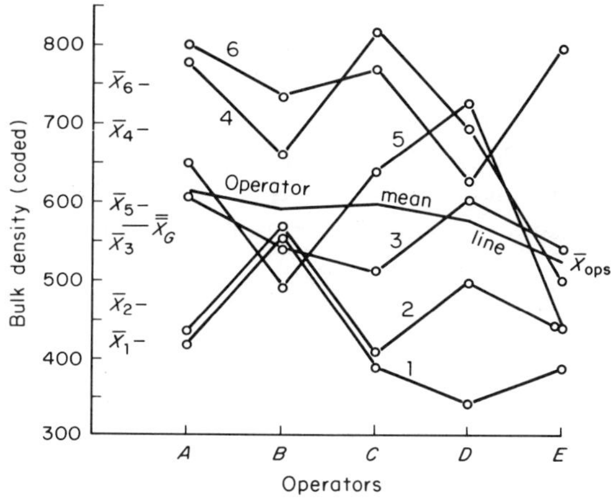

Figure 19.7 Discrepancy of specimens over operators.

TABLE 19.8
Differences among Sample Means over Operators

	Operator				
	A	*B*	*C*	*D*	*E*
Mean of specimen 2 − mean of specimen 1	+11.50	+8.50	+14.75	+52.75	+51.25
Mean of specimen 3 − mean of specimen 2	+255.25	−21.25	+98.25	+103.25	+101.0

19.8 whereas other operators find a large positive difference like that found by operators D and E on the same pair, this too will add to discrepancy.

Evidently one of the most desirable features of an experiment is the reduction of discrepancy to zero. Inconsistency among operators performing the same tasks generally reduces the sensitivity of an experiment, and considerable effort should be expended to ensure the model requirements $\gamma_{ij} = 0$. It is also necessary in most cases to demonstrate, by testing against a suitable error term, that this requirement is fulfilled. After a suitable period of testing has demonstrated that under the conditions of an experiment $\gamma_{ij} = 0$ is a reasonable representation, it may be possible to dispense with the actual testing. In other words, the two-way classification with replicates is necessary, at least at first, and a two-way classification without replicates and without fulfillment of the requirement $\gamma_{ij} = 0$ is likely to be very misleading.

19.8 Estimates of Components of Variance

The main purpose of the present experiment is to estimate magnitudes of variability among the sources tested and to make recommendations about future experiments. It should be remarked again that the populations from which the α_i, β_j, and $\delta_{k(ij)}$ are sampled are infinite, so that the infinite population model is the basis of this analysis. Under such a model the components of variance for each source tested may be estimated from the observed mean squares. More precisely, the observed mean squares are unbiased estimates of their corresponding expected mean squares (Scheffé, 1959, pp. 243ff.).

The computations are straightforward; thus the mean square for error $\hat{\sigma}_\epsilon^2$ is 8,247.69 from Table 19.6, and under H_0, $\hat{\sigma}_\epsilon^2$ is an unbiased estimator of unassigned sources of variation σ_ϵ^2. Similarly, $\hat{\sigma}_\epsilon^2 + r\hat{\sigma}_{mp}^2$ estimates the expected interaction mean square of Table 19.7 and is listed in Table 19.6 as

32,527.48. Then $\sigma_\epsilon^2 + r\sigma_{mp}^2 - \sigma_\epsilon^2 = r\sigma_{mp}^2$, and dividing by r yields an estimate of σ_{mp}^2; the computation follows:

$$\hat{\sigma}_{mp}^2 = \frac{32{,}527.48 - 8{,}247.69}{4} = \frac{24{,}279.79}{4} = 6{,}069.95$$

Similarly, estimates of the expected mean squares for the two main effects are

$$\hat{\sigma}_p^2 = \frac{\text{MS}_p - \text{MS}_{mp}}{mr}$$

$$= \frac{33{,}744.96 - 32{,}527.48}{24} = 50.728$$

and

$$\hat{\sigma}_m^2 = \frac{\text{MS}_m - \text{MS}_{mp}}{pr}$$

$$= \frac{302{,}765.4 - 32{,}527.48}{20} = 13{,}511.896$$

Differences among specimen means are obviously larger than any other source; error variation is the next largest, and interaction is almost as large. Operator difference is relatively small.

It seems quite obvious that for an adequate recommendation of optimum sample allocation an experiment in which interaction is in much better control is necessary. In a suitably controlled experiment, operator difference is independent of specimen difference by design; and if the interaction is zero,

TABLE 19.9
Point Estimates[a] of Expected Mean Squares and Standard Deviations

Source of variation	Estimate of mean squares	Standard deviation	
		Coded	Decoded
Operators	$\hat{\sigma}_p^2 = 50.728$	$\hat{\sigma}_p = 7.12$	0.0007
Specimens	$\hat{\sigma}_m^2 = 13{,}511.896$	$\hat{\sigma}_m = 116.24$	0.0116
Interaction	$\hat{\sigma}_{mp}^2 = 6{,}069.95$	$\hat{\sigma}_{mp} = 77.91$	0.0078
Error	$\hat{\sigma}_\epsilon^2 = 8{,}247.69$	$\hat{\sigma}_\epsilon = 90.82$	0.0091

[a] For use of this term see Scheffe, 1959, p. 228.

the whole investigation can easily be arranged to reduce the error (variation from unassigned sources) by adequate replication.

It should be observed that the variance of the grand mean may be written as

$$\text{Var}\ (\overline{X}_G) = \frac{\sigma_p^{\ 2}}{p} + \frac{\sigma_m^{\ 2}}{m} + \frac{\sigma_{mp}^{\ 2}}{mp} + \frac{\sigma_\epsilon^{\ 2}}{mpr}$$

and it can then be seen that increase in replicates merely serves to reduce the error term, whereas increase in the number of specimens contributes to the denominator of three sources of variance.

It is therefore advisable and necessary to increase the number of specimens in any future experiment as well as to reduce variation due to interaction by appropriate preparation of operators. Operator differences will not prove troublesome if they can be made consistent, i.e., if $\gamma_{ij} = 0$.

19.9 Other Aspects of the Two-way Layout

While the objective of the present experiment, determining the bulk density of sandstone from the Pocono formation, has been strictly specified, the experiment may be used to illustrate some of the implications in similar but differently structured experiments. Thus an experimenter may perform a reconnaissance investigation of this kind with the intention of repeating the experiment on these specimens and with these operators, and he would then be analyzing the total population of specimens and operators. He would, of course, have ultimate interest in population means and not in variances. Thus if all the items from a population are present, the sum of the deviations is equal to zero. No estimate of variability is possible; since the entire population is analyzed, the results are parameters, not statistics.

Again, these six specimens may be all that exist, but the operators may be a random sample from an infinite population; alternatively, the same operators are to be used again and again, but the specimens are random samples from an infinite population. This is generally known as the *mixed model* (Bennett and Franklin, 1954, p. 348).

Finally, any combination of these may arise in practice. Each of these situations alters the structure of the experiment, as would be expected because the predictions made from the analyses are to apply to different physical situations and different populations.

If we may generalize the experimental possibilities and write some alternative structures, the nature of the different arguments accompanying the different arrangements can be better illustrated; the structure of these different alternatives is represented in Table 19.10. In all cases the replicates are assumed to be a random sample from an infinite population. Adjustments at this level can be made by inserting the coefficient $1 - r/R$ for r and pursuing

TABLE 19.10
Analysis of Variance for Random, Fixed, and Mixed Models

$$X_{ijk} = \mu + \alpha_i + \beta_j + \alpha\beta_{ij} + [\delta_{k(ij)} + \epsilon_{ijk}]$$

Each observation = mean + operators + specimens + interaction + replicates + others

No. of items	Source of variation	d.f.	Sums of squares	Mean squares	Variance ratio F	Expected mean squares: Random[a]	Expected mean squares: Fixed[b]	Expected mean squares: Mixed[c]
5	Operators	$p - 1 = 4$	134,979.83	33,744.96	1.037†	$\sigma_\epsilon^2 + r\sigma_{mp}^2 + mr\sigma_p^2$	$\sigma_\epsilon^2 + 0 + mr\theta_p$	$\sigma_\epsilon^2 + r\sigma_{mp}^2 + mr\theta_p$
6	Specimens	$m - 1 = 5$	1,513,827.0	302,765.4	9.31***	$\sigma_\epsilon^2 + r\sigma_{mp}^2 + pr\sigma_m^2$	$\sigma_\epsilon^2 + 0 + pr\theta_m$	$\sigma_\epsilon^2 + 0 + pr\sigma_m^2$
	Operators × specimens	$(p - 1)(m - 1) = 20$	650,549.67	32,527.48	3.94**	$\sigma_\epsilon^2 + r\sigma_{mp}^2$	$\sigma_\epsilon^2 + r\sigma_{mp}^2$	$\sigma_\epsilon^2 + r\sigma_{mp}^2$
4	Error (replicates)	$pm(r - 1) = 90$	742,292.0	8,247.69		σ_ϵ^2	σ_ϵ^2	σ_ϵ^2
120	Total	$pmr - 1 = 119$	3,041,648.5			σ_{tot}^2	σ_{tot}^2	σ_{tot}^2

[a] All sampling levels represent random samples from infinite populations.
[b] Main-effects levels are fixed; random variation only at replicate level.
[c] Operators fixed effect; specimens and replicates random effects.

the argument outlined below for $1 - m/M$, etc. In all cases the replicates are required to test for the presence of interaction; in most cases it is expected that the interaction will be found not significant, that is, $\gamma_{ij} = 0$. Failure to meet this requirement is more serious for the finite model and mixed model; it is not quite so serious but is still very restrictive in the infinite model.

Let us now assume that these are all the operators and specimens that exist and we wish to draw conclusions about contrasts among the means from the main effects. In such a case we can examine the finite-population correction factor (Bennett and Franklin, 1954) and see what happens to the expected mean squares. Thus if $m = M$, the coefficient $1 - m/M = 1 - 1 = 0$, and the interaction term drops out of the expected mean squares in the finite model. Supposing the experiment is properly performed and $\gamma_{ij} = 0$; then the variance ratio testing the interaction against the error will be nonsignificant, that is, $\sigma_\epsilon^2 + r\sigma_{mp}^2/\sigma_\epsilon^2 < F_{05,(p-1)(m-1),pm(r-1)}$.

The appropriate error term for testing the main effects is, therefore, σ_ϵ^2, and sometimes the expected values of the mean squares for the main effects are written as $\sigma_\epsilon^2 + rm\theta_p$ and $\sigma_\epsilon^2 + rp\theta_m$, respectively, to emphasize that they are not estimates of variability such as we get in the infinite model.

Now let us consider the infinite-population model, model 2; since m is now a negligibly small number compared with M (and p with P, respectively), the finite-population correction factor becomes

$$1 - \frac{m}{M} = 1 - 0 = 1$$

and similarly for $1 - p/P$; under such circumstances the interaction remains as a contributor to the variance in the expected mean square of each main effect. If the interaction proves nonsignificant, then either the error σ_ϵ^2 or the interaction may be used to test the main effects. Sometimes the two estimates are combined as estimates of the same source of variability if σ_{mp}^2 is effectively zero, i.e., nonsignificant, or not greater than σ_ϵ^2, and the two are combined to obtain a "better" estimate of error. However, since the estimate of interaction variance is subject to sampling error and may not be a "stable" estimate of σ_{mp}^2, its population value, because of the small number of degrees of freedom, combining may or may not be an advantage (Brownlee, p. 414).

If, on the other hand, the interaction is significantly larger than error ($\sigma_\epsilon^2 + r\sigma_{mp}^2/\sigma_\epsilon^2$ significant), then the interaction becomes the basis for testing the main effects. Two features should be mentioned. If $\gamma_{ij} \neq 0$, then recommendations are not so exact as if $\gamma_{ij} = 0$. As pointed out by Scheffé (1959, pp. 92ff.) and exemplified in the previously described experiment, if the main effect is significantly larger than a significant interaction, it does not necessarily imply that each determination of bulk density by each operator is consistently larger or smaller than that determined on another specimen but only that on average the differences are larger. In any specific comparison the reverse may be true.

Second, in a two-way layout without replicates there is no estimate of the interaction, and it is tacitly assumed that $\gamma_{ij} = 0$;[1] if this assumption is likely to be unfulfilled, the testing of main effects is liable to yield misleading conclusions.

Let us now suppose that these operators are to be used again but the new specimens are to be randomly selected from the infinite population. Then $1 - p/P = 0$ and $1 - m/M = 1$; in such circumstances the interaction does not contribute to the main effect of differences among specimen means, but it may contribute to the main effect representing differences among operator means. The appropriate error term for testing the specimen mean squares is the error term σ_ϵ^2, whereas the appropriate "error" for testing operator difference is the interaction. Significant interaction is again a confusing influence and renders it difficult to obtain general conclusions.

These proposals may be modified if the number of operators are considered to be drawn from a pool of operators of finite size when p/P is appreciably larger than zero but not equal to unity. The coefficients multiplying the variance estimate from the interaction is then fractional; and unless the interaction is zero, tests of the main effects are clumsy to perform (see Bennett and Franklin, 1954, p. 373).

19.10 Assumptions Included in the Various Experimental Structures of a Two-way Classification

It seems clear from the above discussion that the assumptions contained in the various forms of analysis of variance are not the same, although computational procedure and the summary analysis-of-variance table are identical in appearance, and so it is advisable to specify explicitly the assumptions included in each experiment as part of its structure. These are discussed at length by Eisenhart (1947), Bennett and Franklin (1954), and Scheffé (1959). Effects of failures in the assumptions are also described by Cochran (1947) and in the texts cited.

The model may be written generally as

$$X_{ijk} = \mu + \alpha_i + \beta_j + \gamma_{ij} + \epsilon_{k(ij)}$$

Let us first of all consider a two-way arrangement of observations (one observation per cell) in rows and columns (Eisenhart, 1947, table 1, p. 5). Then if the "analysis of variance" is performed and the resulting summary table of analysis is compiled solely as a summary of these data, the relationships among the sums of squares, etc., are algebraic properties of the array, and no assumptions are required. As soon as the data are regarded as a

[1] If the interaction is present, it inflates the estimate of error.

sample drawn from a population and are to be used to argue that conclusions based on the analysis of the data are to be applied to a population from which the data are drawn, the problem becomes one of statistical inference, and earlier assumptions must be fulfilled for the argument to be valid.

At this stage it is necessary to be explicit about the kinds of conclusions to be drawn and, therefore, the kind of model which the experiment represents. Let us first assume that a *fixed-effects* model[1] represents the experimental situation. The first assumption is that the observations are random variables distributed around true mean values which are fixed constants.

From this assumption it follows that the observations X_{ij} of the two-way array are distributed around true mean values m_{ij} and that $E(X_{ij}) = m_{ij}$ or, alternatively, that the $\sum(X_{ij} - m_{ij}) = 0$.

The second assumption concerns the additivity of the effects of the factors in the experiment. Thus if the hypothetical true means m_{ij} for each cell may be written as

$$m_{ij} = \mu + (\alpha_i - \alpha_.) + (\beta_j - \beta_.)$$

where μ is a fixed constant, the true mean $\alpha_.$ is the true mean of the ith row (mean of the ith effect), and $\beta_.$ is the true mean of the jth column, or jth-column effect, then the observations will permit us to estimate the differences among means due to any of the specified effects of the factors. In terms of the two-way layout of operators and specimens, differences among means of specimens, say $\alpha_{1.} - \alpha_{2.}$, are estimated by the difference among means of the first and second rows of the array, and this observed difference is an unbiased estimator of the true difference in bulk density that exists between the main effects due to specimens 1 and 2.

Again in this case the variation among different specimen means does not affect the variation among different operator means, i.e., interaction is zero ($\gamma_{ij} = 0$).

When additivity does not hold, the difference between $\alpha_{1.} - \alpha_{2.}$ applies only to the specific operator (column) under examination.

The third assumption is that the random variables X_{ij} have common variance σ^2 and all covariances are zero; this assumption ensures that the variances are additive and that the use of error variances averaged over the entire experiment is valid.

To complete the analysis and use tests of significance, a fourth assumption is required: the X_{ij} are jointly distributed in a normal distribution.

Relaxation of any of these assumptions results in inexact conclusions based on the comparison of means and the accompanying significance tests. In order to understand what is involved in such relaxations, each case should be carefully and individually considered.

[1] The fixed-effects model is called class 1 by Eisenhart (1947).

19.11 Differences among Means

Quite frequently the experimenter wishes to go further than the demonstration that there exist significant differences among the means in one classification or another. It is then necessary to be able to compare the means and rank them in some order; a number of alternatives exist, and, in particular, procedures differ when the comparison is built into the experiment prior to its completion and when the comparison is inspired by the results of the experiment (compare Sec. 18.8). It is possible to partition the contrasts among means into pairs accompanied by 1 d.f. for each comparison (Brownlee, 1960, pp. 417ff.) and, of course, to consider those comparisons of specific interest. The requirement of orthogonality must be fulfilled among the several comparisons.

In general, for comparisons or contrasts among means specified prior to the experiment, the error term consists of the variance of the difference among two means or $2\sigma_i^2$, where σ_i^2 is the appropriate error term; on the other hand, when comparisons are performed after examination of the experimental results, the Scheffé or Tukey confidence intervals are appropriate, depending on whether the number of items in each mean are unequal or equal, respectively (Scheffé, 1959, pp. 68ff.). It is necessary to differentiate between the finite (model 1), the infinite (model 2), and the mixed model in such cases, and the significance tests become quite complex[1] (Brownlee, 1960, pp. 383ff.).

Methods for ranking means are also described by Bennett and Franklin (1954, pp. 339ff.), Bechofer (1954), and Bechofer, Dunnett, and Sobel (1954).

19.12 Examples of Two-way Classifications

A number of examples of the application of two-way-classification analysis of variance in geological problems are described by Miller and Kahn (1962, pp. 411–416), Krumbein and Miller (1953), and Griffiths and Rosenfeld (1954). A detailed exposition of the two-way classification with replicates applied to measurement of axial-ratio shape and preferred orientation of quartz grains in sediments is given in Griffiths and Rosenfeld (1953).

It seems evident that this design is an effective tool in a wide variety of experimental situations which may arise in geological problems. It is worthwhile, therefore, to emphasize the advantages and disadvantages entailed in the use of this design.

The first requirement is prior knowledge that the stratifying or blocking of the population is effective; this means that the degrees of freedom associated with the stratification which are lost from "error" do represent a sizable

[1] Note that the arithmetic is simple enough but the reasoning underlying the appropriate inference is complex.

proportion of the variance. If stratification is performed and no strata exist, this basis of classification reduces the degrees of freedom for "error" and reduces the sensitivity of the experiment.

Second, it is most important to decide which model (fixed, random, or mixed) is involved, so that the correct inferences can be drawn from the experimental results.

Third, the experiment should be so controlled that the interaction term is indeed zero; and if there is any doubt, it is advisable to use replicates to test this assumption. Quite frequently, proof of the presence of interaction will enable future experiments to be redesigned to remove or at least reduce this effect.

Finally, the use of unequal numbers of replicates in the classes can be easily accommodated in the design; this should not be taken as advocating the use of unequal numbers, because this serves to complicate both point estimates of variance and comparisons among means and, in the presence of significant interaction variance, may vitiate the analysis.

20 Other experimental designs

If the numerous factors interact in a puzzling manner, prolonged study of the results and further experimentation may be needed before the facts are mastered. The trouble in this case is that the phenomena are complex, not that the experimentation is faulty.

W. G. Cochran and G. M. Cox, 1957

20.1 Introduction

As the experimenter becomes more familiar with the potentialities of experimental design, it will be found that a wide variety of procedures exist in which analysis of variance is the basic form. The next stage is to consider three-way classifications, and these have obvious extensions to four-way and higher classifications. However, as the complexity of classification increases, the number of observations increases very rapidly; and so the experimenter, with considerable knowledge about outcome, can afford to ignore some sources of variation and use balanced incomplete-block and unbalanced designs to reduce the number of observations required. Examples of balanced incomplete-block and unbalanced designs will be described.

It has already been mentioned that if the basis of classification represents an ordinal or higher measurement scale, it is possible to use factorial designs; but, so far, at least, there are few examples of these designs in the geological literature. Similarly, it is possible to combine analysis of variance with regression and perform covariance analyses where necessary. Examples are discussed in Miller and Kahn (1962, pp. 214–227, 240–242).

Quotation: W. G. Cochran and G. M. Cox, "Experimental Designs," 2d ed., p. 153, John Wiley & Sons, Inc., 1957.

At this stage it is advisable to consult reference texts on experimental design and choose suitable sampling arrangements to match the design requirements; excellent sources for this are Cochran and Cox (1957), Bennett and Franklin (1954), Brownlee (1960), and Kempthorne (1952). Factorial experiments are well represented in industrial experimentation (Davies, 1956), and this does not by any means exhaust the literature on experimental design. It is as well, when choosing a design, to realize that in some of the more complicated arrangements the arithmetic calculations become a sizable problem, and then automatic electronic computing is of material benefit. It is perhaps most necessary to realize that collecting a set of data and subsequently searching for a design to fit the haphazard arrangement is least desirable. In general, the simpler the design, the easier it is to extract all the information; and, presumably, if an experimenter knows what he is doing, he will be able to reduce his problem to a reasonably simple design.

20.2 Three-way-classification Analysis of Variance

The three-way classification has an intuitive appeal for the geologist because he sees the three bases of classification as one stratigraphic and two geographic dimensions, and this is a typical framework in the geological area. In preparation for the use of this design the experimenter should write down explicitly the populations of interest and whether the aim of the experimenter is to compare means (fixed effects, model 1), to compare variances (random effects, model 2), or some of both (mixed model, model 3). These two features will suffice to decide upon a suitable sampling plan. If this is the first time this volume of rock has been tackled, it is advisable to include replications as a means for testing the second-order interaction; if it is known, from previous work, that some or all of the interactions are absent or negligible, a more efficient design such as a Latin square can be adopted (Sec. 20.4).

Suppose the problem concerns sampling a scree and the variate is the long *a* axis of quartzite boulders in the scree; a number of operators will be

TABLE 20.1
Sources of Variation in a Five-way Classification

Main effects	5
First-order interactions	10
Second-order interactions	10
Third-order interactions	5
Fourth-order interactions	1
Total	31

used to reduce the number of measurements per operator to a reasonable task. The main objective is to study variability in long axis of the quartzite pebbles so that the infinite-population model appears appropriate. On the other hand, to establish whether a size gradient is present, differences among means of rows and means of columns across the scree are also of interest. The sampling pattern consists of subdivision of the scree into small areas and sampling these; in order to maintain the infinite-population model, since the scree is not infinite in extent, the sampling will be considered to be *with replacement* and will be arranged in rows and columns across the scree. The distance between the rows and columns could be taken as a random variable, but for an initial investigation the distance will be some integer times the largest boulder length. Sampling with replacement would permit the repetition of some rows and/or columns if the sampling is extensive compared with the area of scree.

At each intersection of the rows-by-columns grid, each of the p operators measures q boulders; and, on the basis of previous experience, at least eight boulders are measured at each site, that is, $pq \geq 8$. In practice each operator chooses a diagonal of a circle of 12 ft or more and measures the boulder at each end of the diameter. The diameters are arranged symmetrically to yield the $pq \geq 8$ measurements, and the operators rotate $\theta°$ from sampling site to sampling site to avoid any positional bias, i.e., to randomize the selection of boulders at each site.

This sampling arrangement leads to three-way classification with the main effects operators, rows, and columns. There are three *first-order* interactions, namely, operators by rows, operators by columns, and rows by columns; and there is one *second-order* interaction of operators by rows by columns. Since each operator measured q boulders at each row-by-column intersection, there are q replicates nested within each operator-by-row-by-column class. This design is considerably more complex than a two-way classification, and four-way and higher-order classifications become even more complex. For example, in a four-way design there are four main effects, six first-order interactions, four second-order interactions, and one third-order interaction; this design may be required to analyze contributions to variation of long axes of quartz grains measured in thin section, for example (Chap. 3). The five fundamental properties (Chap. 3) will require a five-way classification with the arrangement shown in Table 20.1.

This will require 32 observations with only one observation per cell; and if q replicates are taken, then $32q$ observations are required for a completely balanced design. This preview of what is entailed is one of the advantages of applying the scientific-method approach.

To return to the three-way classification, the total number of observations is $N = prcq$, and the experimental structure may be written as

$$X_{ijkl} = \mu + \alpha_i + \beta_j + \gamma_k + \alpha\beta_{ij} + \alpha\gamma_{ik} + \beta\gamma_{jk} + \alpha\beta\gamma_{ijk} + (\delta_{l(ijk)} + \epsilon_{ijkl})$$

$$i = 1, \ldots, p; j = 1, \ldots, r; k = 1, \ldots, c; q = 1, \ldots, q \qquad (20.1)$$

where α is the operator effect, β and γ are the respective row and column effects, and δ is the contribution from replicates which is confounded with all other source of variation ϵ_{ijkl}. Table 20.2 is the corresponding analysis-of-variance table. An examination of the expected mean squares will indicate the appropriate errors for testing the significance of the contributions to variance at the various levels. Ideally, if all the interactions are zero, the main effects can be tested against error. All other outcomes lead to difficulty in one way or another. Suppose the second-order interaction is effectively zero; then each first-order interaction can be tested against error, and if any one is significant, this becomes an error term for some main effect. If two first-order interactions are significantly larger than error, there is no simple test for at least two of the main effects. Such considerations suggest that unless the contributions from interactions are well known to be zero, they should be examined with care; and this design is obviously difficult to handle unless the nature and the order of magnitude of variation are known before the experiment is performed. In addition, this discussion refers to the infinite-population model; the finite model and certainly the mixed model are likely to lead to greater complexity (see Bennett and Franklin, 1954, pp. 385ff.). At the present stage in most geological research, if this design is adopted, the investigator should be prepared to repeat the experiment and modify the measurement procedure and sampling balance before useful conclusions can be drawn.

As an example, the analysis of the lengths (long a axis) of boulders of quartzite in a Tuscarora scree is given in Table 20.3. There are seven operators, four rows, and five columns; and two boulders are measured by each operator at each column-row intersection, leading to a total of 280 measurements of long axes. The data are part of a series of continuing sampling experiments on geological populations, but this discussion is restricted to Table 20.3 as an example of procedure for a three-way-classification analysis of variance.

Calculation begins by finding, preferably on an electronic computer, the sum of the 280 observations and the sum of the squares of the 280 observations; from these the correction term CT and total sum of squares are calculated.

To facilitate computation three tables are constructed containing the totals for operators by rows, operators by columns, and rows by columns; from these the sum of squares for each main effect, operators, rows, and columns, may be calculated from the respective marginal totals. Each first-order-interaction sum of squares is then obtained from the totals in the body of each table. There are four tables, one for each row of operator-by-column totals, which will yield the second-order interaction, operators by rows by columns. The error term is found by difference. The results of these several calculations appear in column 4 of Table 20.3. The mean squares (column 5) are found by dividing each sum of squares by its appropriate degrees of freedom (Table 20.3, column 3).

TABLE 20.2
Generalized Analysis-of-variance Table for Three-way Classification (Infinite Population = Random-effects Model)

No. of items	Source of variation	d.f.	Expected mean square
p	Operators	$p - 1$	$\sigma_\epsilon^2 + q\sigma_{prc}^2 + rq\sigma_{pc}^2 + cq\sigma_{pr}^2 + rcq\sigma_p^2$
r	Rows	$r - 1$	$\sigma_\epsilon^2 + q\sigma_{prc}^2 + pq\sigma_{rc}^2 + cq\sigma_{pr}^2 + pcq\sigma_r^2$
c	Columns	$c - 1$	$\sigma_\epsilon^2 + q\sigma_{prc2} + pq\sigma_{rc}^2 + rq\sigma_{pc}^2 + prq\sigma_c^2$
	Operators × rows	$(p - 1)(r - 1)$	$\sigma_\epsilon^2 + q\sigma_{prc}^2 + cq\sigma_{pr}^2$
	Operators × columns	$(p - 1)(c - 1)$	$\sigma_\epsilon^2 + q\sigma_{prc}^2 + rq\sigma_{pc}^2$
	Rows × columns	$(r - 1)(c - 1)$	$\sigma_\epsilon^2 + q\sigma_{prc}^2 + pq\sigma_{rc}^2$
	Operators × rows × columns	$(p - 1)(r - 1)(c - 1)$	$\sigma_\epsilon^2 + q\sigma_{prc}^2$
q	Replicates (error)	$prc(q - 1)$	σ_ϵ^2
N	Total	$prcq - 1$	σ_{tot}^2

TABLE 20.3
Three-way-classification Analysis of Variance of Long *a* Axes of Quartzite Boulders from Tuscarora Scree, Lamar, Pennsylvania

+ = present; − = absent; 0 = zero

No. of items	Source of variation	d.f.	Sum of squares	Mean square	Variance ratio F	Expected mean square							
						σ_ϵ^2	$q\sigma_{prc}^2$	$pq\sigma_{rc}^2$	$rq\sigma_{pc}^2$	$cq\sigma_{pr}^2$	$prq\sigma_c^2$	$pcq\sigma_r^2$	$rcq\sigma_p^2$
7	Operators	6	8.4321	1.2046	2.699*	+	0	−	0	0	−	−	+
5	Columns	4	18.1763	4.5441	3.731*	+	0	+	−	0	−	+	
4	Rows	3	13.2860	4.4287	3.636*	+	0	+	0	−	+		
	Operators × columns	24	10.9410	0.4559	1.022†	+	0	−	−	0			
	Operators × rows	18	8.3769	0.4654	1.043†	+	0	−	0				
	Rows × columns	12	14.6146	1.2179	2.729**	+	0	+					
	Operators × rows × columns	72	17.6500	0.2451	0.549†	+	0						
2	Replicates (error)	140	62.4757	0.4463	1.820**	+							
280	Total	279	153.9525	0.5518									

The variance (mean square) associated with the total sum of squares is included and would be used if all sources of variation proved nonsignificant, when it becomes the *best unbiased estimator* of the population variance; on the other hand, if any other source of variation is significant, this estimate is biased.

The first step in testing hypotheses is to compare the mean squares for the second-order interaction against the error term in a variance ratio $F_{\alpha,\nu_1,\nu_2} = (\sigma_\epsilon^2 + q\sigma_{prc}^2)/\sigma_\epsilon^2$, or $F_{05,72,140} = 0.2451/0.4463 = 0.549$ when the respective components of variance are used as estimators of their corresponding mean squares. The hypothesis H_0 comprises the statement that the 140 means for operators by columns by rows are all means of random samples from the same (normal) population; it follows that $\sigma_{prc}^2 = 0$ if H_0 is true; and from tables (Arkin and Colton, 1950, p. 120) the calculated F value would exceed $1.39 = F_{05,75,125}$ by chance 5 times in 100. Since $F = 0.549$ is less than the tabulated F, there is no evidence against the hypothesis, and $\sigma_{prc}^2 = 0$ is accepted. If this variance ratio is reversed, the expected $F_{05,100,70} = 1.45$ and the observed value, $F = 0.4463/0.2451 = 1.820$, is significantly larger at the 1 percent level; evidently these means are more homogeneous than would be expected from random sampling the same (normal) population. In some experiments this may imply that the sampling is biased, i.e., the means are too much alike. Here it reflects the fact that variation in long axes among pairs of boulders is very much larger than the means of sets of two, a feature which emphasizes the extreme variability among more or less unsorted materials (stage 1 population of the sampling model, Fig. 2.3).

Since σ_{prc}^2 is effectively zero, these two estimates can be regarded as two independent estimates of the same variance and so can be pooled to yield an estimate based on 212 d.f.; but because the means of small sets vary so much less than those of the single boulders, it seems more realistic (geologically) to use the error without pooling for subsequent testing. Actually, the conclusions and significance or nonsignificance would not be affected by either procedure in this case.

The next step is to test each first-order interaction against error, and both operators by rows and operators by columns prove nonsignificant, whereas rows by columns is significant beyond the 1 percent level. In effect, then, this suggests that σ_{pc}^2 and σ_{pr}^2 are effectively zero; the operators are not inconsistent from row to row or column to column. This is some assurance that the measurement procedure is in control. The fact that the rows-by-columns interaction is significant means that σ_{rc}^2 is not zero, and therefore differences among row means are not consistent over the columns, and vice versa. If there is a consistent size gradient in the scree, e.g., when the lengths of the boulders increase toward the foot of the scree, the interaction should be zero; the gradient among row means is similar along each column. On the other hand, if there is a consistent size gradient from side to side of the scree, the column differences would be significant and the interaction zero. Where the interaction rows by columns is significant, the "gradient" is inconsistent, and subsequent conclusions based on testing of the main effects of row differences and column differences are obscure.

Since the two first- and second-order interactions are zero, the error term for testing operator differences is the contribution from differences among replicates; and upon testing the appropriate null hypothesis, $\sigma_p^2 = 0$, the differences among operator means are significant at the 5 percent level, i.e., $\sigma_p^2 = 0$ is rejected, and operators differ. This is a commonly observed feature in such experiments, but because the operator interactions are zero, these differences are orthogonal to differences among columns means and differences among rows means and so do not seriously affect these comparisons. It is most important to achieve this result because if operators are inconsistent, i.e., some operator interaction is significant, no general conclusions across operators can be made. Such a result suggests that if the experiment were repeated by other investigators, different results would be found.

Because the row-by-column interaction is significant, that is, $\sigma_{rc}^2 \neq 0$, it becomes the error term for testing row and column main effects; for both rows and columns, $\sigma_r^2 = 0$ and $\sigma_c^2 = 0$ are rejected, and the differences among row means (and column means) significantly exceed the contribution from row-by-columns interaction (at the 5 percent level). No general gradient emerges, however, because the significant interaction implies that some column means exceed others in some rows and do not in other rows, and the same for row mean differences. In practice it would now be advisable to draw graphs of the row-by-column interaction and insert the row means per column and column means per row to examine the nature of the differences. Similarly, by calculating the Scheffé confidence limits for contrasts, using the interaction variance, the scree could be contoured on lengths of a axes to elucidate any size patterns.

Of course the experiment was primarily concerned with point estimates of variance components as a basis of recommendations for future sampling so that the variance components would be isolated to examine the magnitude of the contribution from each source.

20.3 Other Examples of Three-way-classification Analysis of Variance

The first example of analysis of variance in the geological literature is a three-way classification, mixed model, in which different sieving techniques are compared (Swineford and Swineford, 1946). Another example, in which interactions prove somewhat troublesome, concerns the comparison of visual estimation of the sphericity and roundness of quartz grains from a beach sand (Rosenfeld and Griffiths, 1953); the absence of replicates and the relatively large magnitude of the contribution from the second-order interaction leads to some ambiguity in the results.

Examples of four-way classification and mixed models are described by Bennett and Franklin (1954, pp. 410ff.) and Brownlee (1960, chap. 14). The experimenter interested in adopting such designs should follow through these examples with considerable care before performing the experiment.

20.4 Latin Squares

Latin square experiments are a form of three-way classification which makes use of confounding to reduce the number of observations. For example, in a completely balanced three-way classification without replicates the structure is

$$X_{ijkl} = \mu + \alpha_i + \beta_j + \gamma_k + \alpha\beta_{ij} + \alpha\gamma_{ik} + \beta\gamma_{jk} + (\alpha\beta\gamma_{ijk} + \epsilon_{ijkl}) \qquad (20.2)$$

wherein the error term contains a contribution from the second-order interaction together with all unassigned sources (ϵ_{ijkl}). The structure of an analogous experiment arranged in a Latin square is

$$X_{ijk} = \mu + \alpha_i + \beta_j + \gamma_k + \epsilon_{ijk} \qquad (20.3)$$

and it can be seen that unassigned sources embrace all contributions to variance from the three first-order interactions and the second-order interaction. If any interaction is contributing significantly, this will inflate the error term and reduce the sensitivity of the experiment; because there is no way of testing for these effects, the conclusions in such cases may be woefully misleading. This design was evolved for a very specific purpose, again for field experiments in agronomy; the main objective is usually a comparison among mean yields of varieties (or treatments, such as the effect of different fertilizers) and the rows and columns effects are merely used as a two-way stratification to reduce the contribution of heterogeneous gradients in soil fertility in the blocks in which the varieties are planted. Again, in general, no attempt is made to interpret the "variances" among rows or columns; indeed, these are looked upon as fixed effects, and the experiment is a model 1 or fixed-effect type (Kempthorne, 1952, p. 195). It should be emphasized that the sampling arrangements in agronomic field experiments are based on extensive sampling experiments, called *uniformity trials,* which were performed on many common plants before these designs were adopted and it was well known, on the basis of these experiments, that the effect of interactions could be ignored. In geological experiments this extensive study of sampling variation has not yet been performed, and so the assumption of absence of interaction effects is liable to vitiate the experimental results. Furthermore, the Latin square design is not aimed at estimates of components of variance but only at comparisons among means; to emphasize this aspect the model is written with θ's in the place of variances in the expected-mean-square columns (Table 20.4).

The attraction of the Latin square arrangement is that it leads to a substantial reduction in the number of observations—a feature which is quite

TABLE 20.4
Generalized Analysis-of-variance Model for Latin Square Experiments

No. of items	Source of variation	d.f.	Expected mean squares
4	Rows	$t - 1 = 3$	$\sigma_\epsilon^2 + t\theta_r$
4	Columns	$t - 1 = 3$	$\sigma_\epsilon^2 + t\theta_c$
4	Treatments	$t - 1 = 3$	$\sigma_\epsilon^2 + t\theta_p$
	Error	$(t - 1)(t - 2) = 6$	σ_ϵ^2
	Total	$t^2 - 1 = 15$	

desirable when the necessary knowledge exists and which suggests that the reduction can be achieved without loss. For example, if a three-way-classification experiment is performed without replicates and using four classes in each main effect, there are $4 \times 4 \times 4 = 64$ observations; in a similar Latin square experiment with four rows, four columns, and four classes in the third main effect (the arrangement is given in Table 20.5), there are a total of 16 observations. The degrees of freedom for main effects are the same in both designs, but the error term contains 6 d.f. in the Latin square design, and for the same contributions to variation in a three-way classification the degrees of freedom are 54. The "error" in both cases contains contributions from all four interaction terms plus unassigned sources. It is evident that effective use of Latin square designs requires that the interactions all be zero.

The arrangement of the four letters in Table 20.5, once in each row and once in each column, is the typical feature of a Latin square. In performing a Latin square experiment the rows and columns should be permuted in successive blocks, or an arrangement should be selected which will serve effectively as a randomization device; examples are given in Kempthorne (1952) and Cochran and Cox (1957), and a choice of Latin square arrangements is contained in the tables of Fisher and Yates (1948). An interesting and popular discussion of Latin and Greco-Latin squares and their combinatorial properties is given by Gardner (1959).

20.5 Examples of the Use of Latin Squares in Geological Problems

Chayes and Fairbairn (1951) were the first to use a Latin square design in which the main effects were six operators, six thin sections, and six days. The basic arrangement of the 6×6 Latin square is given in their table 1 (1951, p. 705); the experiment, a test of the point-count technique, was used to determine whether operators are consistent from day to day and whether six slides from "a single strip of Westerly granite" are relatively uniform in proportion of mineral constituents. The results were quite favorable; it should be observed, however, that if any interactions are suspected to be contributors to variance, it would be more informative, although considerably more expensive, to isolate the interaction contributions by using a three-way classification design. Furthermore, tests of uniformity are not always so straightforward as expected. Hypothesis testing, in general, attempts to disprove uniformity; and if the experiment fails in this respect, the conclusion that the material is uniform is not a very strong claim unless the basis for comparison, the error term, can be shown to be small. In the experiment quoted a theoretical binomial error is used to judge the relative magnitude of the variance from unassigned sources, and the error term is close to this theoretical error, so the writers conclude that the technique is successful in estimating the proportions of mineral constituents in this granite. Similar tests of operator variation in point counting thin sections of sediments using a two-way classification with replicates imply that considerable practice is necessary to reduce operator inconsistency to a desirable level (Griffiths and Rosenfeld, 1954; see also Griffiths, 1960d; and Sec. 9.4 and Fig. 15.5).

An extensive series of examples of Latin square experiments is described by Krumbein (1953b) and Krumbein and Miller (1953); in these experiments it is emphasized that the use of the Latin square design is preliminary to a more extensive three-way-classification design (Krumbein, 1953b, p. 280) in evaluating the various factors. Again the problem of homogeneity is raised; for example, Krumbein's third experiment (1953b) is an attempt to show that the Negaumee iron formation is homogeneous in thickness of jasper and hematite beds over an exposure of about 60 by 20 ft. The design is a 4×4 Latin square including the three main effects along beds, across beds, and operators. The conclusion is that the variation from each of the three main effects is not significant, and so it is presumed that the bed thickness is uni-

TABLE 20.5
Arrangement of the Classes *A*, *B*, *C*, *D* across Rows and Columns in a Latin Square

	Column			
Row	1	2	3	4
1	*A*	*B*	*C*	*D*
2	*D*	*A*	*B*	*C*
3	*C*	*D*	*A*	*B*
4	*B*	*C*	*D*	*A*

form. Since the error variance is relatively large (about half the total sum of squares is associated with error), this conclusion is not very informative and may be misleading.

In attempting to prove homogeneity it is first necessary to establish a specific critical limit which represents the breakpoint between variation which may be accepted as relatively small (homogeneous) and variation which is not considered small (heterogeneous). Then it is necessary to evaluate this variability among beds against variation within beds *after* interfering sources of variation are removed. In particular, it would be most advisable to estimate variation from interactions among main effects, e.g., operator inconsistency, and (after demonstrating that their contributions may be safely neglected) to compare variation among beds with an appropriate measure of within-bed variation; otherwise, conclusions regarding homogeneity may be woefully misleading.

At present the Latin square design has limited usefulness in geological problems; but when enough has been learned about operator control and other interfering sources of variation to achieve the requirement of zero contribution from interaction, this design is an efficient tool for establishing differences among means for a main effect of specific interest.

A special advantage of the Latin square design occurs when the block sizes are fixed, such as when the experimental unit is an automobile and the object of interest is tire wear; these experiments usually entail the use of more than one square to obtain replicates for control of the error term (see, for example, Cochran and Cox, 1957, p. 517).

20.6 Incomplete Blocks: Balanced and Unbalanced

When a very large number of specimens must be analyzed and replication is required as an estimate of error, the number of determinations becomes very large and the experiment is time-consuming and expensive. In agronomic experimentation a large number of varieties, i.e., specimens, requires very large blocks of land, and it is well known that large blocks vary more than small ones, so that increasing heterogeneity of soil fertility limits the block size. The same feature will occur in geological experiments: where the block, i.e., stratum or area, is large, a greater heterogeneity will generally exist. In nearly all experiments large numbers of specimens require increased time to perform the analyses, and variation over time will contribute its share to the error term. Intricate experimental arrangements have been devised to reduce block size by using incomplete replication and by confounding some effects believed to be unimportant, e.g., the Latin square designs earlier described.

An extensive study of variation in bulk density of $3 \times 1\frac{5}{8}$-in. cores of the Salt Wash member of the Morrison formation on the Colorado Plateau led to the analysis of some 161 core specimens from a single well representing 155 ft of the formation (Griffiths, Groff, and Kahn, 1953, pp. 4–26). The well core was first stratified into four zones of mudstone and sandstone on the basis of lithology; one of the two mudstone zones, zone 3, was further subdivided into two lithological subzones of cemented sandstones and mudstones. This kind of stratification was used to increase homogeneity in bulk density and thus reduce the number of samples within zones. It only partially succeeded in its aim, because within the two sandstone zones the bulk density varied widely in association with carbonate cement (see Cochran, 1960; Griffiths, Groff, and Kahn, 1953, fig. 1, p. 29). Even with this stratification, the number of specimens, together with repeated measurements on each, required large numbers of bulk-density determinations for each zone;[1] to reduce the time involved, a number of operators, varying from three to nine, were used to perform the experimental work (see Griffiths, Groff, and Kahn, 1953, table 4, p. 11). It was found that the variability among operators decreased as the number of operators was increased from three to five and then tended to increase beyond six or seven; apparently the team size of four or five is near optimum. Various experimental designs were adopted, including a randomized balanced complete block and balanced incomplete blocks such as square and rectangular lattices.

As a specific example of a design and corresponding analysis, consider zone 3*a*; there are 16 samples of core, and three replicates yields 48 determinations of bulk density. Using a balanced complete-block design, i.e., a randomized block design, with four operators would require $16 \times 4 \times 3 = 192$ bulk density determinations (Fig. 20.1*a*), whereas by adopting a balanced incomplete-block design such as a square lattice this may be reduced to $4 \times 4 \times 3 = 48$ determinations (Fig. 20.1*b*). The corresponding analysis of variance is summarized in Table 20.5, and it will readily be appreciated that the analysis is much more complicated than in the symmetrical randomized complete-block design. In calculating the sums of squares it is advisable to follow the steps very carefully, and almost always the analysis should be repeated by several persons if mistakes in arithmetic are to be avoided; it is best to use automatic electronic data processing for these analyses. However, to find the appropriate error term it is necessary to understand the intricacies of the design, so that it is as well to work through a published example completely before attempting the analysis of a set of data; examples are described along with recommended procedures for analysis in Cochran and Cox (1957, chaps. 9–13).

It will be observed from Table 20.6 that the error from the incomplete-block design may be compared with the complete-block arrangement, and it

[1] It was known from previous experiments on the bulk density of sediments determined by mercury balance (Chap. 11) that three replicates are necessary to reduce error when the operators are sufficiently in control for the operator interaction to be ignored.

is apparent that despite the reduction in the number of observations, the error term is less and the experiment therefore more efficient. Furthermore, an observation was lost during the experiment, a not infrequent occurrence, and a missing value was calculated to replace it (see Cochran and Cox, 1957, p. 413) with the loss of 1 d.f. from the error term.

In this series of bulk-density experiments each operator first analyzed a set of standard specimens until his procedure yielded results in control, in particular to ensure that the inconsistency (interaction) was negligible; for balanced and unbalanced incomplete-block arrangements this is an essential prerequisite. The replication error ranges around 0.01 to 0.02 bulk density unit, and sample-to-sample variation far exceeds this value. In general, then, it is advisable, before adopting an incomplete block design, to make a thorough study of the main sources of variation and to arrange the experimental procedure very carefully to ensure that saving on the number of observations does not result in failure to achieve the experimental objective.

Incomplete-block designs, both balanced and unbalanced, increase the variety of arrangements available to the experimenter and permit a wide choice in sampling arrangement. For the most part these designs are essentially concerned with comparisons among means; i.e., they are generally considered to belong to the fixed-effects model.

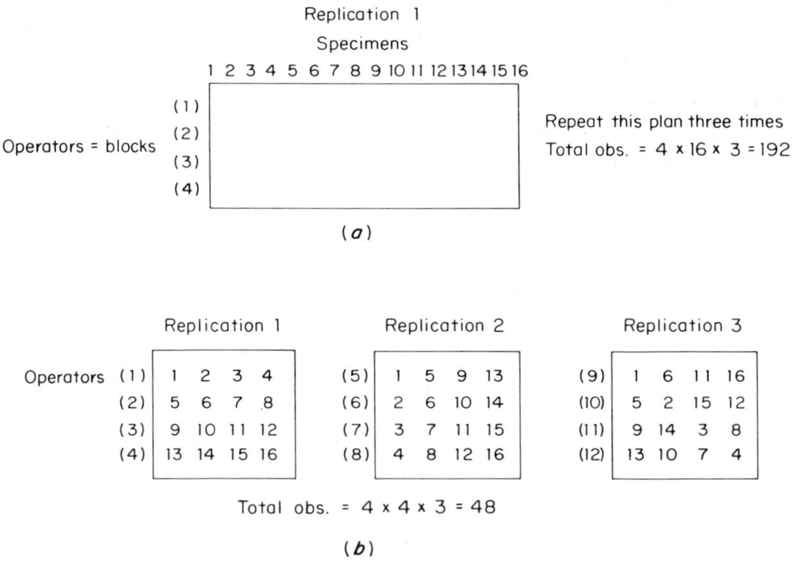

Figure 20.1 Comparison of experimental arrangement for equivalent balanced complete and incomplete blocks. (*a*) Randomized complete blocks; (*b*) randomized incomplete blocks (balanced square lattice).

TABLE 20.6
Square Lattice, Incomplete-block-design Analysis of Variance: Bulk Density of 16 Cores from Zone 3*a*, Well 155*C*, of the Salt Wash Member of Morrison Formation, Bull Canyon, Colorado

Data coded: $(BD - 2) \times 10^3 = X_i$

Source of variation	d.f.	Sum of squares	Mean square
Replications	3	1,062.67	354.22
Component *a*	(6)	4,279.09	713.18
Component *b*	(6)	1,749.60	291.60
Blocks (eliminating samples)	12	6,028.69	502.39
Samples (ignoring blocks)	15	14,792.86	986.19
Samples (eliminating blocks)	(15)	13,275.86	885.07
Error (intrablock)	32[a]	4,579.39	143.11
Error (randomized block)	44[a]	10,608.08	241.09
Total	62[a]	26,463.61	

[a] One degree of freedom lost for calculating missing value.

20.7 The Chain Block: An Example of Unbalanced Incomplete-block Design

The chain-block design was first proposed as a special sampling arrangement for a particular problem in quantitative spectrographic analysis; the plate used in this procedure will not take more than 10 to 12 determinations. Again, in the initial experiment, the precision of the determination was known to be high, and therefore few replicates were required to achieve a small error; and since only small mean differences were expected, a sensitive test was a prerequisite (Youden and Connor, 1953). In quantitative spectrographic analysis for trace elements one of the main contributions to "error" arises from differences among plates, both in terms of variation in emulsion and in variation in development after exposure; by using the plate as a "block" this source of variation can be removed from the error term. However, since trace-element analysis usually entails a large number of samples and the plate is limited to 10 to 12 exposures (samples), the required replication leads to incomplete-block design. The two characteristics, a fixed block size and high degree of precision, are quite general in quantitative spectrographic analysis, so that incomplete-block designs, such as the chain block, are the most suitable sampling arrangement for this kind of analysis. The original chain block has been extended so that two-way blocking can be used in suitable sampling arrangements (Mandel, 1954).

TABLE 20.7
Sampling Arrangement for a Chain-block Design with 30 Treatments

Plates = blocks

Number of specimens on a plate	Block 1	2	3	4	5	Class
	A	*E*	*I*	*M*	*Q*	
	B	*F*	*J*	*N*	*R*	
	C	*G*	*K*	*O*	*S*	
	D	*H*	*L*	*P*	*T*	
$v_2 = 20$	*E*	*I*	*M*	*Q*	*A*	C_1
	F	*J*	*N*	*R*	*B*	
	G	*K*	*O*	*S*	*C*	
	H	*L*	*P*	*T*	*D*	
$v_1 = 10$	*a*	*c*	*e*	*j*	*l*	C_2
	b	*d*	*f*	*k*	*m*	

$v_1 + v_2 = v = 30$ treatments = specimens $v_1 + 2v_2 = N = 50$

As an example, a reconnaissance experiment aimed at isolating contributions to replication error in determining boron in samples of sedimentary rocks is described;[1] the problem arose because it was found that subsampling vials of powdered specimens showed unexpectedly large variation. Proportion of boron was chosen as the variate of interest and may be represented as the yield X_{ijuz} in parts per million.

There are 30 specimens arranged as 10 determinations on each of 5 plates, i.e., a total of 50 determinations. From the arrangement of the specimens in Table 20.7 it will be noted that there are two classes C_i of letters, the 20 capital letters *A* to *T* (C_1), which occur twice, and the 10 small letters *a* to *m* (C_2), which occur once. Let t_{iju} denote the effect of the *u*th treatment of the *j*th group of C_i, and let b_z denote the effect of the *z*th block. Then the structure of the experiment is (Youden and Connor, 1953; Cochran and Cox, 1957, pp. 463ff.)

$$X_{iju} = t_{iju} + b_z + \epsilon_{iju} \tag{20.4}$$

The requirements are that the t_{iju} be random variables with mean 0 and variance σ^2, and ϵ_{iju} be independent of the random variable associated

[1] These data were released through the courtesy of M. L. Keith, Director, Mineral Conservation Section, College of Mineral Industries, The Pennsylvania State University.

with yield. Now, if

$$X_{2jz} = \sum_{u=1}^{n_2} X_{2juz} \qquad \text{and} \qquad D_j = X_{2jj} = X_{2j(j+1)}$$

then iuj is defined as

$$i = 1, 2 \qquad j = 1, \ldots, 5 \qquad u = 1, \ldots, n_{ij}$$

$$n_{ij} = \begin{cases} 20 & i = 1 \\ 10 & i = 2 \end{cases}$$

and denotes the uth treatment of the jth group of the ith class.

The observations are summarized in Table 20.8 together with their individual labels and the block totals. The analysis-of-variance table is summarized in Table 20.9, and the calculations for the sums of squares are as follows:

1. Sum of squares (SS) due to block differences

$$\text{SS}_b = \sum_{j=1}^{b} \frac{B_j^2}{k_j} - \frac{G^2}{N}$$

$$= \sum_{j=1}^{5} \frac{B_j^2}{10} - \frac{G^2}{50}$$

$$= \frac{4{,}845{,}955}{10} - \frac{24{,}176{,}889}{50} = 1{,}057.72$$

2. Total sum of squares SS_{tot}

$$\text{SS}_{\text{tot}} = \sum X_{ijuz}^2 - \frac{G^2}{N}$$

$$= 508{,}081 - 483{,}537.78$$

$$= 24{,}543.22$$

3. Sum of squares due to error SS_ϵ. For the jth group of C_2 find the differences d_{ju}:

$$d_{jv} = (X_{2juj} - X_{2ju(j+1)}) \qquad u = 1, \cdots, n_2$$

For example,

$$\left.\begin{aligned} d_{11} &= X_{2111} - X_{2112} = 138 - 100 = 38 \\ d_{12} &= X_{2121} - X_{2122} = 81 - 98 = -17 \\ d_{13} &= X_{2123} - X_{2132} = 79 - 74 = 5 \\ d_{14} &= X_{2141} - X_{2142} = 73 - 93 = -20 \end{aligned}\right\} \sum_{u=1}^{4} d_{1u} = 6$$

then

$$\sum_{u=1}^{4} d_{1u} = 6 \qquad \sum_{1}^{4} d_{2u} = 9 \qquad \sum_{1}^{4} d_{3u} = 53 \qquad \sum_{u=1}^{4} d_{4u} = -114 \qquad \sum_{u=1}^{4} d_{5u} = 10$$

TABLE 20.8
Proportions of Boron (ppm) in 30 Specimens of Powdered Sedimentary Rock

Block	ppm	Block	ppm	Block	ppm	Block	ppm	Block	ppm
2511	107	2112	100	2213	97	2314	97	2415	112
2521	77	2122	98	2223	79	2324	95	2425	62
2531	97	2132	74	2233	105	2334	102	2435	107
2541	73	2142	93	2243	120	2344	87	2445	195
2111	138	2212	100	2313	105	2414	130	2515	95
2121	81	2222	93	2323	79	2424	71	2525	74
2131	79	2232	100	2333	130	2434	87	2535	110
2141	73	2242	117	2343	120	2444	74	2545	85
1111	115	1212	117	1313	85	1414	110	1515	102
1121	67	1222	100	1323	89	1424	112	1525	102
$\sum_1^{10}$	907		992		1,009		965		1,044

Observation = X_{ijuz} $i = 1, 2$ $j = 1, \ldots, 5$ $u = 1$ to n_{ij} $z = j$ or $j - 1$

TABLE 20.9
Analysis-of-variance Table for Chain-block Design

Source of variation	d.f.	Sum of squares	Mean square
Treatments (corrected for blocks)	$v - 1$	SS_t	$\overline{SS}_t = \dfrac{SS_t}{v-1}$
Blocks (uncorrected)	$b - 1$	SS_b	$\overline{SS}_b = \dfrac{SS_b}{b-1}$
Error	$N - v - b + 1$	SS_ϵ	$\overline{SS}_\epsilon = \dfrac{SS_\epsilon}{N-v-b+1}$
Total	$N - 1$	SS_{tot}	

also

$$\sum_{u=1}^{4} d_{1u}^2 = 2{,}158 \qquad \sum_{u=1}^{4} d_{2u}^2 = 239 \qquad \sum_{u=1}^{4} d_{3u}^2 = 2{,}193$$

$$\sum_{1}^{4} d_{4u}^2 = 15{,}446 \qquad \sum_{1}^{5} d_{5u}^2 = 466$$

Then the sum of squares due to error is SS_ϵ, where

$$SS_{\epsilon,j} = \frac{1}{2}\sum_{u=1}^{n_2}(d_{ju} - \bar{d}_{ju})^2 = \frac{1}{2}\left[\sum_{1}^{n_2} d_{ju}^2 - \frac{1}{n_2}\left(\sum_{1}^{n_2} d_{ju}\right)^2\right]$$

For example,

$$SS_{\epsilon,1} = \frac{1}{2}\left[\sum_{u=1}^{4} d_{ju}^2 - \frac{1}{4}\left(\sum_{u=1}^{4} d_{ju}\right)^2\right]$$

$$= \tfrac{1}{2}\,[2{,}158 - \tfrac{1}{4}\,(6^2)] = 1{,}074.5$$

and, similarly, $SS_{\epsilon,2} = 109.375$; $SS_{\epsilon,3} = 745.375$; $SS_{\epsilon,4} = 6{,}098.5$, and $SS_{\epsilon,5} = 220.5$ or a total $SS_{\epsilon,j} = 8{,}248.50$.

Now $SS_{\epsilon,j}$ is an estimate of the sum of squares due to error based on $n_2 - 1$ d.f. By computing a similar quantity for each group of C_2 we find the sum of squares due to error based on $b(n_2 - 1) = 15$ d.f. The remaining degree of freedom is given by

$$Y = \frac{1}{2bn_2}\left(\sum_{z=1}^{b} X_{2z(z+1)} - \sum_{z=1}^{b} X_{2zz}\right)$$

$$= \frac{1}{2 \times 5 \times 4}\left(\sum_{z=1}^{5} X_{2z(z+1)} - \sum_{z=1}^{5} X_{2zz}\right)$$

$$= \tfrac{1}{40}\,[(1{,}977 - 1{,}941)]^2 = 32.4$$

TABLE 20.10
Analysis of Variance for Boron (ppm)

Source of variation	d.f.	Sum of squares	Mean square
Treatments (corrected for blocks)	$v - 1 = 29$	15,204.85	524.305
Blocks (uncorrected)	$b - 1 = 4$	1,057.72	264.48
Error	$N - v - b + 1 = 16$	8,280.65	517.54
Total	$N - 1 = 49$	24,543.22	

$$\frac{SS_t}{SS_\epsilon} = \frac{524.305}{517.540} = 1.01, \text{ with } F_{v_2-1,N-v-b+1}$$

Then

$$SS_\epsilon = \sum_{j=1}^{b} SS_{\epsilon,j} + Y$$

$$= 8{,}280.650$$

4. Treatment sum of squares SS_t

$$SS_t = SS_{\text{tot}} - SS_b - SS_\epsilon = 16{,}279.35$$

The resulting analysis of variance is summarized in Table 20.10. The test of treatments against error is the comparison of interest, and the expected F value under the null hypothesis H_0 of no differences among treatment means is $F_{05,29,16} = 2.28$; the observed F value is 1.01, so that there is no evidence against the hypothesis, and therefore no difference among treatment means can be detected. This result, with its relatively large error term, indicates that replicate analyses are not very reproducible, and so steps were taken to ensure that the powdered samples were homogeneous in future experiments, and the reproducibility was then checked by a similar experiment.

When the reproducibility achieves a desirable precision level, the aim of the experiment is to compare means of specimens (treatments), and then it is necessary to calculate appropriate error terms. Three kinds of comparisons may be performed, and each requires its own error:

1. Comparisons among any two means in C_1, say t_{1ju} and t_{1sw}, where

$$\text{Var}\,(\hat{t}_{1uj} - \hat{t}_{1sw}) = \left[f(i) + \left(1 + \frac{1}{n_2}\right)\right]\sigma^2 \qquad i = j - s \pmod b$$

2. Comparisons of two means in C_2, say t_{2ju} and t_{2sw}, where

$$\text{Var}\ (\hat{t}_{2ju} - \hat{t}_{2sw}) = \begin{cases} \sigma^2 & j = s \\ [f(i)]\sigma^2 & j \neq s \end{cases}$$

3. Comparisons of two means in C_1 and C_2, respectively, say t_{1ju} and t_{2sw}, where

$$\text{Var}\ (\hat{t}_{1ju} - \hat{t}_{2sw}) = \begin{cases} \frac{1}{2}\left(3 + \dfrac{b-1}{n_2 b}\right)\sigma^2 & j = s \\ \left[f(i') + \frac{1}{2}\left(1 - \dfrac{b+1-4i'}{n_2 b}\right)\right]\sigma^2 & j \neq s \\ & i' = j - s[\text{mod}\,(b+1)] \end{cases}$$

$$f(z) = 1 + \frac{(2z-1)b - 2z^2}{n_2 b}$$

It should be evident that comparisons are not so easily performed as with symmetrical complete designs, e.g., randomized blocks; and unless the error term is indeed small, differences among means may not emerge. In using special designs of this kind it is essential to ensure that the requirements are fulfilled, or the experiment will not be very informative.

21 Measuremnet of association among variates

If we can find a mathematical form for the population which adequately represents the data, and then calculate from the data the best possible estimates of the required parameters, then it would seem that there is little, or nothing, more that the data can tell us; we shall have extracted from it all the available relevant information.

R. A. Fisher, 1948

21.1 Introduction

Most of the analytical procedures described in the preceding chapters deal with variation in a single variate, but frequently more than one variate is measured on each of a number of specimens, and then questions arise concerning the form and intensity of the relationship among the variables. Two kinds of model may be distinguished, the first dealing essentially with the form of the relationship and the second dealing essentially with the degree of intensity of association among the variates. In the first case, where we are concerned with the form of the relationship, a set of values of one variate (X_i; $i = 1, 2, \ldots, k$) is chosen and the value of a second variate Y measured with the X variate at a specified level; this is the regression model. If both variates are permitted to vary at random, we are concerned with the degree of association, or the second model, wherein a measure of correlation is more meaningful. Both cases may be extended, for example, from simple linear regression or correlation to multiple regression and correlation and from linear to curvilinear relationships.

Quotation: R. A. Fisher, "Statistical Methods for Research Workers," 10th ed., p. 7, Oliver & Boyd Ltd., 1948.

Because many nonlinear relationships may, by suitable transformations (see below), be reduced to linear form, this kind of regression will be described in detail; for extension to multiple and curvilinear relationships the reader is referred to standard texts (Brownlee, 1960; Bennett and Franklin, 1954; Quenouille, 1952) and for geological examples to Griffiths (1957, 1958b, 1960b, 1962a) and Miller and Kahn (1962).

21.2 Some Relationships among Variates and Their Linear Representation

Linear relationship among variates may be expressed as

$$Y = \alpha + \beta X \tag{21.1}$$

where β is a measure of how X changes per unit change in Y and α represents the intercept or distance intersected by the regression line along the Y axis. The linear regressions of Fig. 21.1*a* illustrate variation in β while α is constant at zero; the regression line passes through the zero coordinates. When β remains constant and α changes, the regression lines remain parallel (see Fig. 21.1*b*, lines 2, 3, 4), and when β changes but α remains fixed, the regression changes slope (Fig. 21.1*b*, lines 1, 2).

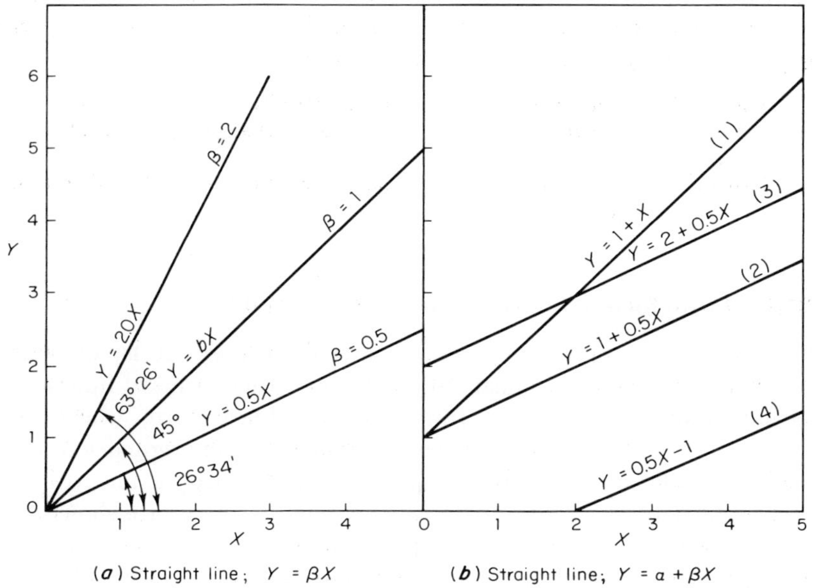

Figure 21.1 Graphic representations of the relationship $Y = \alpha + \beta X$. (*a*) Straight line, $Y = \beta X$; (*b*) straight line not through the origin, $Y = \alpha + \beta X$.

Any regression relationship of the form

$$Y = \alpha + \beta X$$

can be reduced to $Y = \beta X$ by translating the origin to the location $(\bar{Y},\bar{X})$, and then the equation is usually written $y = \beta x$ to indicate that the variates are expressed as deviations from their respective means.

21.3 Nonlinear, Nonperiodic Curves

This group of relationships includes the parabola, hyperbola, exponential, and potential series curves as special cases. In each case the relationship expressed in terms of measuring units of the variate contains a power.

Thus for the parabola and hyperbola the general equation is

$$Y = \alpha X^{\beta} \tag{21.2}$$

When β is positive, the curve is a parabola; when it is negative the curve is a hyperbola. Two simple cases are illustrated, a parabola in Fig. 21.2*a*, where $Y = X^2$, that is, $\alpha = 1$, $\beta = 2$, and a hyperbola in Fig. 21.2*b*, where $Y = X^{-1}$ (or $Y = 1/X$ or $XY = 1$) when $\alpha = 1$ and $\beta = -1$.

Similarly, for the exponential curve the general equation is

$$Y = \alpha\beta^{X} \tag{21.3}$$

An example $Y = 2^X$, where $\alpha = 1$ and $\beta = 2$, is illustrated in Fig. 21.2*c*.

The relationships between the straight line, parabola, hyperbola, and exponential may be summarized as follows:

Linear: When one variate Y changes arithmetically, so does the other X.

Parabola and hyperbola: When one variate Y changes geometrically, so does the other X.

Exponential: When one variate Y changes geometrically, the other changes arithmetically X.

21.4 Transformation of Scale of the Variable

For both graphic and computational purposes it is, of course, much easier to manipulate a linear function than other forms of function; hence, it is advantageous to express functions in a linear form whenever possible. To accomplish this it is necessary to transform the original variate to a new scale of units in which changes in the variates are on an arithmetic scale. Logarithms are designed for this purpose, and any variate showing geometric change will yield logarithms which change in arithmetic intervals. The changes of variable or transformation may be summarized as follows:

Parabola and hyperbola:

$$Y = \alpha X^{\beta}$$

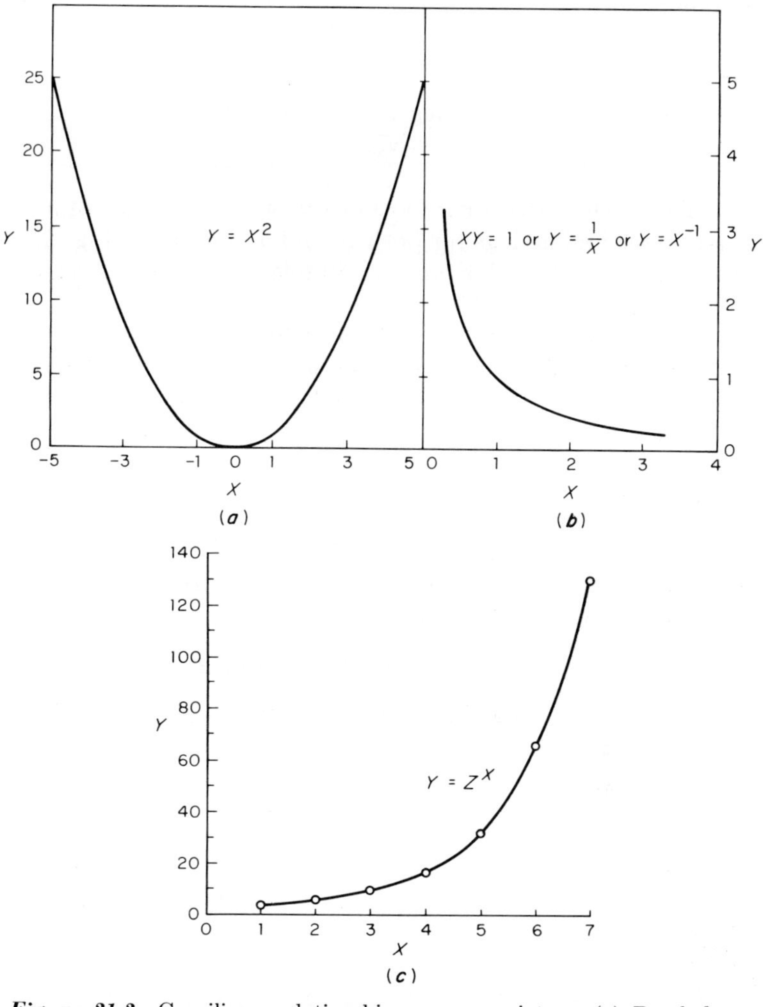

Figure 21.2 Curvilinear relationships among variates. (*a*) Parabola on arithmetic scale; (*b*) equilateral hyperbola on arithmetic scale; (*c*) exponential curve on arithmetic scale.

becomes

$$\log Y = \log \alpha + \beta \log X$$

Whence putting

$$Y' = \log Y \qquad \alpha' = \log \alpha \qquad \text{and} \qquad X' = \log X$$

yields

$$Y' = \alpha' + \beta X'$$

For an exponential:

$$Y = \alpha\beta^X$$

becomes

$$\log Y = \log \alpha + X \log \beta$$

or

$$Y' = \alpha' + \beta' X$$

Thus to linearize a parabola or hyperbola the logarithmic transformation necessitates finding logs for both X and Y variables (Fig. 21.3*a*), but for an exponential the log of Y and a suitable change in intercept and coefficient are sufficient (Fig. 21.3*b*).

Hence, graphically, a parabola or hyperbola may be linearized by using logarithmic graph paper (Fig. 21.3*a*); similarly, an exponential may be expressed in linear form by taking logarithms of the Y variate or plotting the values on semilog paper with the Y scale logarithmic (Fig. 21.3*b*).

A typical example of this arrangement in sedimentology is Krumbein's phi scale (1938), expressed as

$$d = 2^{-\phi}$$

where d is diameter in millimeters; this relationship is an example of an exponential and may be linearized by transformation as follows:

$$\log d = -\phi \log 2$$

(Krumbein and Pettijohn, 1938, fig. 110, p. 235; Griffiths and McIntyre, 1958). A potential series with more than one power ($Y = \alpha + \beta_1 X_1 +$

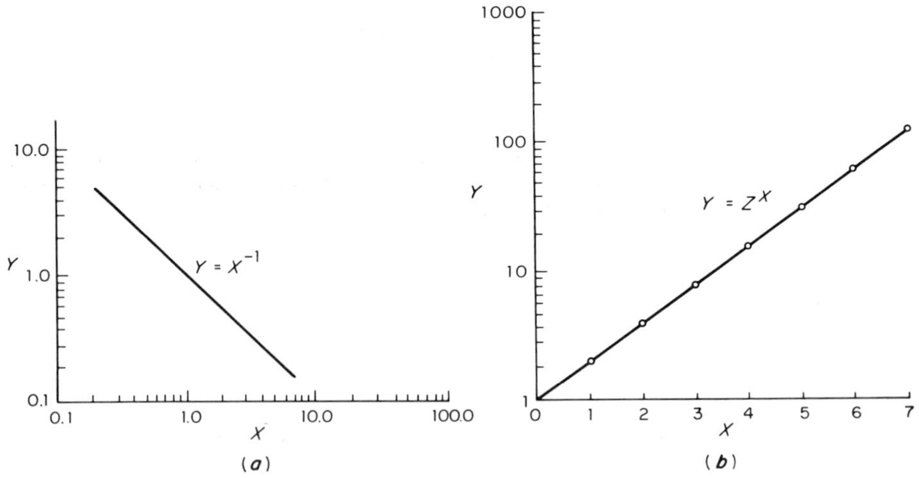

Figure 21.3 Linearized hyperbola and exponential curves. (*a*) Hyperbola, $Y = X^{-1}$, on log-log graph paper; (*b*) exponential $Y = Z^X$ on arithmetic-log (semilog) graph paper.

$\beta_2 X_2^2 + \beta_3 X_3^3 + \cdots$) and periodic curves of the type $Y = \sin X$ are illustrated in standard texts (Ezekiel, 1950; Arkin and Colton, 1950, p. 90, fig. 23).

Relationships of the type described are usually termed *functional relationships* and occur in the field of pure mathematics; in experimental research a functional relationship is rarely known on a priori grounds, and one of the problems in research of this kind is that of estimating the kind of functional relationship which may exist between variates, e.g., pressure, temperature, and volume of a gas, depth and density for rock samples, change in roundness with time or distance, and so on. In attempting to discover these relationships certain values of one variate are chosen, and the values of the second are determined experimentally; thus, a petrographer will estimate roundness of pebbles on a beach at say 1, 2, and 3 miles distance from some base line. The relationship is plotted on a graph, and an attempt is made to find a functional relationship which will represent the variation. Now in such empirical determinations there are, of course, errors in the determined values; i.e., every pebble which is measured at the third mile will not possess the same roundness. In a functional relationship, on the other hand, both variables are presumed to be without error; in the experimental example one variable is chosen and is considered to be measured without error, and this is called the independent variable X, whereas the second is a random variable and yields different values at one and the same value of the independent variable; the second variable becomes the dependent variable Y.

Given two variables measured on a set of specimens, the next step is to graph the values of these variables and examine the possible form of the function. In general, it may be linear or be represented by one of the functions described earlier, in which case suitable transformation will reduce it to linearity. The next step is to attempt to fit a curve to the variation represented by the observations in the graphs; the general procedure is to use the technique known as least squares (originally described by Legendre and Gauss; see Whittaker and Robinson, 1949, pp. 209ff.).

21.5 Computation of Line of "Best Fit" by the Method of Least Squares

Given that there are two characteristics, or properties, measured on n specimens, the one (X) selected without error and a number of determinations of the other (Y_i) being available at each X value, we may consider the population of Y's at each selected value of X and calculate their respective means $\mu_{y \cdot x}$; there is a linear relationship between the expected Y value $= \mu_{Y \cdot X}$ and the value of X, as shown in Fig. 21.4. The populations Y_i are normally and independently distributed around their respective $\bar{Y}$'s ($= \mu_{Y \cdot X}$) with common variance $= \sigma^2$, that is, $\sigma^2_{Y \cdot X} = \sigma^2$, or the variances are homoscedastic; then

$$\mu_{Y \cdot X} = \alpha + \beta X \qquad (21.4)$$

Suppose specimens are taken at at least two given X's and the Y_i measured on each set at each X value; then the Y_i may be defined as

$$Y_i = \hat{\alpha} + \hat{\beta}X_i + \epsilon_i \tag{21.5}$$

where ϵ_i is a measure of "error," or the amount by which Y_i departs from $\mu_{Y \cdot X}$ for all Y_i at given X_i. Then the least-squares procedure is a means for estimating the "best-fitting" straight line to the observed points and leads to estimates $\hat{\alpha}$ and $\hat{\beta}$ of α and β. The line of "best fit" is defined as that line which minimizes the ϵ_i^2 or minimizes the squared deviation of each Y_i from its appropriate $\mu_{Y \cdot X}$. Now

$$\epsilon_i = Y_i - (\hat{\alpha} + \hat{\beta}X_i)$$

and we minimize

$$\epsilon_i^2 = [Y_i - (\hat{\alpha} + \hat{\beta}X_i)]^2$$

or summing over all values of X_i,

$$\sum_{i=1}^{n} \epsilon_i^2 = \sum_{i=1}^{n} [Y_i - (\hat{\alpha} + \hat{\beta}X_i)]^2 \tag{21.6}$$

Hence by minimizing this expression we find the best estimates $\hat{\alpha}$ of α and $\hat{\beta}$ of β. To minimize, put

$$Q = \sum (Y_i - \hat{\alpha} - \hat{\beta}X_i)^2$$

and find the partial differentials of Q for $\hat{\alpha}$ and $\hat{\beta}$ respectively (where the derivative of a sum is the sum of the derivatives), and equate the derivatives to zero as follows:

$$\frac{\partial Q}{\partial \hat{\alpha}} = -2\sum (Y_i - \hat{\alpha} - \hat{\beta}X_i) = 0$$

and

$$\frac{\partial Q}{\partial \hat{\beta}} = -2\sum X_i(Y_i - \hat{\alpha} - \hat{\beta}X_i) = 0$$

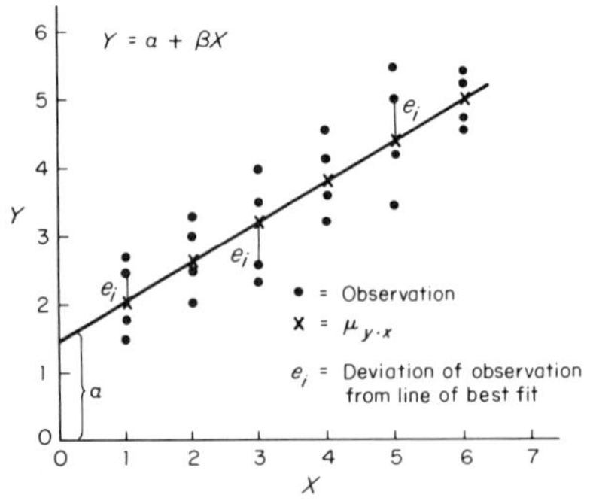

Figure 21.4 The generalized linear relationship.

which yields the "normal equations"

$$\sum_{i=1}^{n} Y_i = n\hat{\alpha} + \hat{\beta} \sum_{i=1}^{n} X_i \tag{21.7}$$

$$\sum_{i=1}^{n} X_iY_i = \hat{\alpha} \sum_{i=1}^{n} X_i + \hat{\beta} \sum_{i=1}^{n} X_i^2 \tag{21.8}$$

The solution to these simultaneous equations yields the required best estimates of α and β.

Calculate $\sum X_i$, $\sum X_i^2$, $\sum Y_i$, $\sum Y_i^2$, $\sum X_iY_i$ where the summation extends from $i = 1$ to n, insert the appropriate values in Eqs. (21.7) and (21.8), and solve for $\hat{\alpha}$ and $\hat{\beta}$. There are several procedures for solving these simultaneous equations; see, for example, Bennett and Franklin (1954, pp. 225ff.) or Brownlee (1960, pp. 282ff.). Suppose we convert the observations to deviations around their respective means as follows:

$$x_i = X_i - \bar{X}$$

whence

$$\sum x_i = 0$$

and

$$y_i = Y_i - \bar{Y}$$

whence

$$\sum y_i = 0$$

then

$$Sx_i^2 = \sum (X_i - \bar{X})^2 = \sum X_i^2 - \frac{(\sum X_i)^2}{n} \tag{21.9}$$

$$Sy_i^2 = \sum (Y_i - \bar{Y})^2 = \sum Y_i^2 - \frac{(\sum Y_i)^2}{n} \tag{21.10}$$

$$Sx_iy_i = \sum (X_i - \bar{X})(Y_i - Y) = \sum X_iY_i - \frac{\sum X_i \sum Y_i}{n} \tag{21.11}$$

then

$$\hat{\beta} = \frac{Sx_iy_i}{Sx_i^2} \tag{21.12}$$

and

$$\hat{\alpha} = \frac{Sy_i - \hat{\beta} Sx_i}{n} \tag{21.13}$$

or, alternatively,

$$Y - \bar{Y} = \hat{\beta}(X - \bar{X}) \tag{21.14}$$

which leads to

$$Y = \bar{Y} + \hat{\beta}(X - \bar{X})$$

and, upon substituting the numerical values for $\bar{Y}$ and $\bar{X}$, this yields

$$Y = \hat{\alpha} + \hat{\beta}X \tag{21.15}$$

The variance of the regression values of Y is given by

$$\sigma^2_{Y \cdot X} = \frac{\sum (Y_i - Y_c)^2}{n}$$

where Y_i is an observed value of Y and Y_c is the calculated value of Y ($= \mu_{Y \cdot X}$) for a given value of X_i. The estimate is

$$\hat{\sigma}^2_{Y \cdot X} = \frac{\sum (Y_i - Y_c)^2}{n - 2} \tag{21.16}$$

where 2 d.f. are lost in estimating $\hat{\alpha}$ and $\hat{\beta}$ from the observations. In order to test the variance for significance we have $(n - 2)\hat{\sigma}^2_{Y \cdot X}/\sigma^2$ has a chi-square distribution with $n - 2$ d.f. (Bennett and Franklin, 1954, p. 226; Brownlee, 1960, p. 282; Williams, 1959, p. 13). The most convenient computing formula for finding the variance is found as follows. Since

$$\begin{aligned} \sum (Y_i - Y_c)^2 &= \sum (Y_i - \hat{\alpha} - \hat{\beta}X_i)^2 \\ &= \sum (Y_i - \hat{\alpha} - \hat{\beta}X_i)(Y_i - \hat{\alpha} - \hat{\beta}X_i) \\ &= \sum Y_i(Y_i - \hat{\alpha} - \hat{\beta}X_i) - \hat{\alpha}\sum (Y_i - \hat{\alpha} - \hat{\beta}X_i) \\ &\qquad - \hat{\beta}\sum (Y_i - \hat{\alpha} - \hat{\beta}X_i) \end{aligned}$$

and because in finding the line of best fit $\sum (Y_i - \hat{\alpha} - \hat{\beta}X_i)$ is placed equal to zero, the last two terms vanish; therefore

$$\begin{aligned} \sum (Y_i - \hat{\alpha} - \hat{\beta}X_i)^2 &= \sum Y_i(Y_i - \hat{\alpha} - \hat{\beta}X_i) \\ &= \sum Y_i^2 - \hat{\alpha}\sum Y_i - \hat{\beta}\sum X_iY_i \end{aligned}$$

then

$$\hat{\sigma}^2_{Y \cdot X} = \frac{\sum Y_i^2 - \hat{\alpha}\sum Y_i - \hat{\beta}\sum X_iY_i}{n - 2} \tag{21.17}$$

21.6 Relationship between Carbonate Content and Porosity in Cow Run Sandstone

The pore space of an oil sand, estimated as its porosity in percent, is often strongly affected by the amount of carbonate cement present in the sand and, in the Cow Run sand, a Pennsylvanian oil-bearing sandstone from West Virginia (Griffiths, 1958b), there is a linear relationship over part of the range. Where the content of carbonate is less than 5 percent, there appears to be no relationship between carbonate and porosity (see Griffiths, 1958b, fig. 8, p. 28), and in these specimens the carbonate is usually present as fragments of fossil shells; the maximum porosity of this sandstone is

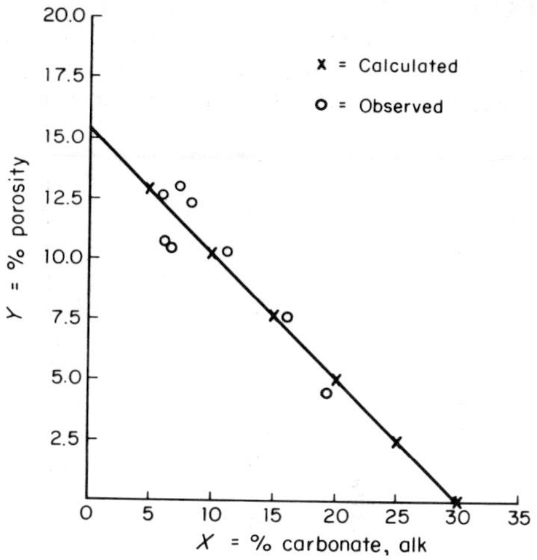

Figure 21.5 Percent porosity Y vs. percent carbonate X Cow Run sandstone.

around 25 percent, and when the carbonate content exceeds this value, there is again no obvious relationship between carbonate content and porosity (see figure cited above). In the range from 5 to 25 percent, however, the relationship appears distinct and linear, and as the carbonate in these specimens is crystalline, it tends to occupy pore space (Fig. 21.5). In attempting to fit a regression line to this set of observations (Table 21.1) it should be noted that the carbonate content[1] is the independent variable X_i and the porosity at various selected values of carbonate content is the dependent variable Y_i. It is assumed that the porosity determinations are normally and independently distributed around their respective $\mu_{Y \cdot X}$ in the range 5 to 25 percent, although this requirement is not necessary for a least-squares calculation of the line of "best fit" but only for the subsequent tests of significance. The data are summarized in Table 21.1.

From the calculated sums and sums of squares and products in the table the corrected sums of squares and products using Eqs. (21.9) to (21.11) become

$$Sx_i^2 = 164.6845 \qquad Sy_i^2 = 52.80 \qquad Sx_iy_i = -87.8425$$

and hence

$$\hat{\beta} = \frac{Sx_iy_i}{Sx_i^2} = \frac{-84.8425}{164.6845} = -0.5152$$

$$\hat{\alpha} = \frac{\sum Y_i - \hat{\beta} \sum X_i}{n} = \frac{80.4 + 0.5152 \times 82.89}{8} = 15.388$$

and the regression equation is

$$Y = 15.388 - 0.515X$$

[1] The carbonate content was chemically determined by absorption of evolved CO_2 in an alkalimeter; the effective porosity was determined by saturation.

TABLE 21.1
Carbonate Content and Percent Porosity of Eight Core Specimens of Cow Run Sandstone, Pennsylvanian, West Virginia

Specimen no. i	Carbonate X_i, %	Porosity Y_i, %	
1	7.68	12.7	
2	7.09	10.2	
3	19.43	4.6	
4	6.71	10.4	
5	15.93	7.6	
6	10.97	10.4	
7	8.94	12.0	
8	6.14	12.5	
$\sum_{i=1}^{8} X_i$	82.89	80.40	$\sum_{i=1}^{8} Y_i$
$\bar{X}$	10.36	10.05	$\bar{Y}$
$\sum X_i^2$	1,023.5285	860.82	$\sum Y_i^2$
$\sum X_i Y_i$	748.2020		

Note: For description of sediments see Griffiths, 1958b.

and this line is inserted in Fig. 21.5. It implies that for every half-unit decrease in carbonate content, porosity increases 1 unit.

The variance of Y_i around the regression line is, from Eq. (21.17),

$$\hat{\sigma}^2_{Y \cdot X} = \frac{860.82 - 15.388 \times 80.40 + 0.515 \times 748.202}{6}$$

$$= 1.515$$

or $$\hat{\sigma}_{Y \cdot X} = 1.231$$

21.7 Testing the Regression Coefficient and Intercept for Significance

From this value of $\hat{\sigma}_{Y \cdot X}$ the variance of the estimated regression coefficient ($\hat{\beta}$) is

$$\text{Var}(\hat{\beta}) = \hat{\sigma}_{\hat{\beta}}^2 = \frac{\hat{\sigma}^2_{Y \cdot x}}{Sx_i^2} \tag{21.18}$$

$$= \frac{1.515}{164.6845}$$

$$= 0.00920$$

and we can test whether this estimate reflects the fact that the true population value $\beta = 0$, by using the statistic t as follows:

$$t_{\alpha,\nu} = t_{\alpha,n-2} = \frac{\hat{\beta} - \beta}{\sqrt{\text{Var}(\hat{\beta})}}$$

$$= \frac{-0.515 - 0}{0.0959} = -5.370 \qquad (21.19)$$

Under the null hypothesis that $\hat{\beta}$ is a sample estimate of $\beta = 0$ the statistic t is distributed as Student's t with degrees of freedom $\nu = n - 2$; therefore at the 1 percent level (P_{01}) for $\nu = n - 2 = 6$ d.f., $t = 3.71$ and, since our calculated value exceeds this, we reject the null hypothesis (H_0) and agree that the $\hat{\beta} = -0.515$ is not from a population in which $\beta = 0$; the confidence limits for β at the 1 percent level are

$$\hat{\beta} - t_{01,n-2}\hat{\sigma}_{\hat{\beta}} < \beta < \hat{\beta} + t_{01,n-2}\hat{\sigma}_{\hat{\beta}} \qquad (21.20)$$

or

$$-0.515 - 3.71 \times 0.0959 < \beta < -0.515 + 3.17 \times 0.0959$$

$$-0.870 < \beta < -0.159$$

and as this interval does not include zero, the regression coefficient β differs significantly from zero.

Similarly, the variance for the intercept value is

$$\text{Var}(\hat{\alpha}) = \frac{\hat{\sigma}^2_{Y \cdot X}}{n - 2} \qquad (21.21)$$

In the present example this yields

$$\text{Var}(\hat{\alpha}) = \frac{1.515}{6} = 0.2525$$

and

$$\hat{\sigma}_{\alpha} = 0.5025$$

Confidence limits may then be calculated for α as follows:

$$\hat{\alpha} - t_{\alpha,n-2}\hat{\sigma}_{\hat{\alpha}} > \alpha > \hat{\alpha} + t_{\alpha,n-2}\hat{\sigma}_{\hat{\alpha}} \qquad (21.22)$$

or, for probability level $P = 0.01$ when $t_{01,6} = 3.71$

$$15.388 - 3.71 \times 0.503 < \alpha < 15.388 + 3.71 \times 0.503$$

$$13.522 < \alpha < 17.254$$

From the nature of the regression line in Fig. 21.6, it is quite clear that $\alpha = 0$ is not a very meaningful test in this case.

21.8 Errors of Estimation in Predicted Values

The value of $\sigma^2_{Y \cdot X}$ refers essentially to the variance around $\bar{Y}$, the mean of all the Y_i at the mean $(\bar{X})$ value, and the confidence intervals for this variance are inserted as solid lines $\bar{Y} \pm \hat{\sigma}_{Y \cdot X}$ in Fig. 21.6. It is sometimes of interest to consider the precision of prediction of the $\bar{Y}_o$ at a specific value of X_o which includes errors in estimates of α and β, that is,

$$\text{Ave}\ (\bar{Y}_o) = \alpha + \beta X_o \tag{21.23}$$

$$\text{Var}\ (\bar{Y}_o) = \frac{\sigma^2}{n} + \frac{\sigma^2 (X_o - \bar{X})^2}{\sum (X_i - \bar{X})^2}$$

$$= \sigma^2 \left[\frac{1}{n} + \frac{(X_o - \bar{X})^2}{\sum (X_i - \bar{X})^2} \right] \tag{21.24}$$

and substituting the sample estimate $\hat{\sigma}^2_{Y \cdot X}$ for σ^2 the statistic

$$t_{\alpha, n-2} = \frac{\bar{Y}_o - (\hat{\alpha} + \hat{\beta} X_o)}{\hat{\sigma}_{Y \cdot X} \sqrt{\dfrac{1}{n} + \dfrac{(X_o - \bar{X})^2}{\sum (X_o - \bar{X})^2}}} \tag{21.25}$$

follows Student's t distribution with $\nu = n - 2$ degrees of freedom and may be used to test hypotheses concerning the mean $\alpha + \beta X_o$ of $\bar{Y}_o$ for a given X_o. The appropriate $100(1 - \alpha)$ percent confidence intervals for $\alpha + \beta X_o$ are

$$\bar{Y}_o - t_{\alpha, n-2} \hat{\sigma}_{Y \cdot X} \left[\frac{1}{n} + \frac{(X_o - \bar{X})^2}{\sum (X_i - \bar{X})^2} \right]^{1/2} < \alpha + \beta X_o$$

$$< \bar{Y}_o + t_{\alpha, n-2} \hat{\sigma}_{Y \cdot X} \left[\frac{1}{n} + \frac{(X_o - \bar{X})^2}{\sum (X_i - \bar{X})^2} \right]^{1/2} \tag{21.26}$$

The confidence limits around the regression line in Fig. 21.6 labeled $\bar{Y}_o$ exemplify these limits, and it can be seen that prediction falls off markedly away from the region of the mean $\bar{Y}$ at $\bar{X}$.

Again, it may be necessary to specify the confidence limits for a new observation Y_o at some selected X_o in which the variance for individual observations $\sigma^2_{Y \cdot X}$ must be added to the expression as follows:

$$\text{Var}\ (Y_o) = \sigma^2 \left[1 + \frac{1}{n} + \frac{(X_o - \bar{X})^2}{\sum (X_1 - \bar{X})^2} \right] \tag{21.27}$$

and the appropriate confidence interval is

$$Y_o \pm t_{\alpha, n-2} \hat{\sigma}_{Y \cdot X} \left[1 + \frac{1}{n} + \frac{(X_o - \bar{X})^2}{\sum (X_1 - \bar{X})^2} \right]^{1/2} \tag{21.28}$$

The corresponding intervals are inserted in Fig. 21.6 and labeled Y_o; single observations are not predicted with any great precision unless the number of observations is fairly large. Again the precision falls off away

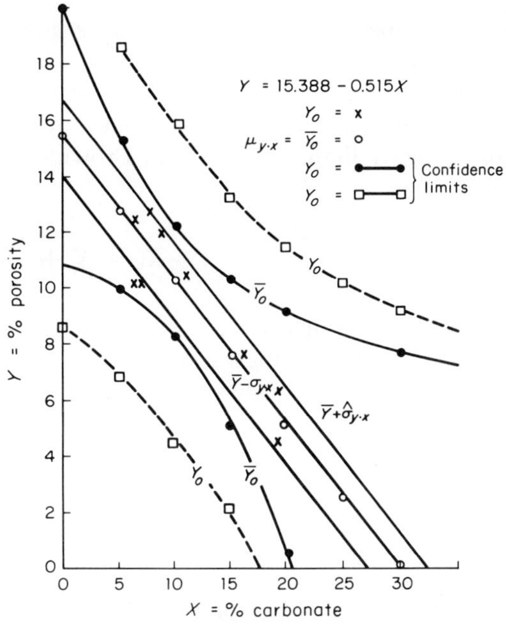

Figure 21.6 Confidence limits for predicted values of Y from the regression line of porosity Y on carbonate X content in Cow Run sandstone.

from the mean $\bar{Y}$ at mean $\bar{X}$. It should be clear that to use $\hat{\sigma}_{Y \cdot X}$ as estimate of error may be very misleading if the measures of interest are individual observations of Y_i.

21.9 Estimation of Errors in "Predicted X"

Quite frequently an experimenter uses a set of standards, determined essentially without error, to establish the relationship between some property of the standard and the response to an experimental test, e.g., the density of a line on a spectrograph plate to estimate the proportion of an element in a compound, the height of a peak on an x-ray diffraction pattern to estimate the proportion of some mineral in a mixture, or the weight of carbon dioxide gas lost in heating a carbonate-containing sample in a combustion train to estimate the carbonate content. The objective is to set up a standard curve, and in doing so the requirements for using a regression model are fulfilled. However, the experimenter then wishes to reverse his estimation procedure and, from the response measured upon an unknown, to predict, via the standard curve, the concentration of some element or mineral in the specimen—an example of inverse estimation in which X is to be predicted from an observed Y and the experimenter wishes to state confidence limits around "predicted X." Since X was initially selected, it is not a random variable in a probability sense, and because X does not have an error distribution, estimation is not strictly applicable; a number of possible alternatives exist and are discussed in the statistical literature (e.g., Williams, 1959, p. 95; Acton, 1959, p. 43; Brownlee, 1960, pp. 284ff.). The experimenter should be familiar with the possible alternatives because each model is designed for a specific situation (Acton, 1959, pp. 43–47). In the present instance, following Williams (1959) and Brownlee (1960) the problem may be stated as

$$X_o = \bar{X} + \frac{\bar{Y}'_o - \alpha}{\beta} \tag{21.29}$$

where X_o is the predicted X corresponding to the new set of Y_j $(j = 1, \ldots, n)$ yielding $\bar{Y}'_o$ and the regression line based on n pairs of Y_{ij}, X_j is $Y_o = \hat{\alpha} + \hat{\beta}X_o$. Given the expected value of $\bar{Y}'_o = \mu_{Y \cdot X}$, the true value of X from the regression is ϵ and

$$\epsilon = \bar{X} + \frac{\mu_{Y \cdot X} - \alpha}{\beta} \tag{21.30}$$

or

$$\mu_{Y \cdot X} - \alpha - \beta(\epsilon - \bar{X}) = 0$$

Then the variance of $\bar{Y}'_o - \alpha - \beta(\epsilon - \bar{X})$ is (Brownlee, 1960, p. 285)

$$P_2 - P_1 = P(t_1 < t < t_2) \tag{21.31}$$

where P_2, P_1 are the probability limits and t_1, t_2 are the appropriate values of Student's t; then the range in ϵ yields a quadratic expression in ϵ for both upper and lower confidence bounds. This may be solved for values of ϵ, say ϵ_1 and ϵ_2, yielding expressions of the form

$$\epsilon_2 = \bar{X} + \frac{\hat{\beta}(Y'_o - \hat{\alpha})}{\hat{\beta}^2 - p^2} - \frac{t_2 \hat{\sigma}_{Y \cdot X}}{\hat{\beta}^2 - p^2} \left[(\hat{\beta}^2 - p^2)\left(\frac{1}{m} + \frac{1}{k}\right) + \frac{(\bar{Y}'_o - \hat{\alpha})^2}{\sum (X_i - \bar{X})^2} \right]^{1/2}$$

and

$$p^2 = \frac{t_2^2 \hat{\sigma}^2_{Y \cdot X}}{\sum (X_1 - \bar{X})^2} \tag{21.32}$$

and a similar expression for ϵ_1 containing t_1 in place of t_2. The variance for the regression coefficient $\hat{\beta}$ enters this expression in the form

$$\hat{\beta}^2 - \frac{t_j^2 \hat{\sigma}^2_{Y \cdot X}}{\sum (X_1 - \bar{X})^2} \qquad j = 1, 2 \tag{21.33}$$

If this expression approaches zero, no realistic confidence limits exist. The relationship between $\hat{\beta}^2$ and the right-hand term of (21.33) appears critical, so define

$$g = \frac{t_j^2 \hat{\sigma}^2_{Y \cdot X}}{\hat{\beta}^2 \sum (X_i - \bar{X})^2} \tag{21.34}$$

then if $0 < g \ll 1$, we may use

$$\epsilon_2 \approx \bar{X} + \frac{\bar{Y}'_o - \hat{\alpha}}{\hat{\beta}} - \frac{t_2\hat{\sigma}_{Y\cdot X}}{|\hat{\beta}|}\left[\left(\frac{1}{m} + \frac{1}{n}\right) + \frac{(\bar{Y}'_o - \hat{\alpha})^2}{\hat{\beta}^2 \sum (X_i - \bar{X})^2}\right]^{1/2} \qquad (21.35)$$

$$\epsilon_1 \approx \bar{X} + \frac{\bar{Y}'_o - \hat{\alpha}}{\hat{\beta}} - \frac{t_1\hat{\sigma}_{Y\cdot X}}{|\hat{\beta}|}\left[\left(\frac{1}{m} + \frac{1}{n}\right) + \frac{(\bar{Y}'_o - \hat{\alpha})^2}{\hat{\beta}^2 \sum (X_1 - \bar{X})^2}\right]$$

as approximate limits for ϵ, where $g < 0.1$ (Brownlee, 1960, p. 287). Note that if there is only one observation at the new Y_o, then $m = 1$, and the equation is reduced to $1 + 1/n$ in place of $1/m + 1/n$.

In the present example suppose that an observation of porosity is made on a new specimen from the same population; what is the predicted carbonate content and its confidence limits? The new value of Y'_o is 11.0 percent porosity, which, substituted in the regression equation $Y = 15.388 - 0.515X$, yields a value of 8.52 percent carbonate. The necessary quantities for calculating the confidence limits are

$$\bar{X} = 10.36 \qquad \bar{Y} = 10.05 \qquad \hat{\sigma}_{Y\cdot X} = 1.231 \qquad Sx_1^2 = 164.6845$$

$$m = 1 \qquad n = 8 \qquad Y'_o = 11.0$$

Then the criterion quantity, with $t = 2.447$ for 6 d.f. and probability level P_{05},

$$g = \frac{t^2_{k-2}\hat{\sigma}^2_{Y\cdot X}}{\hat{\beta}^2 Sx_i^2} = \frac{2.447^2 \times 1.231^2}{(-0.515)^2 \times 164.6845}$$

$$= \frac{9.073692}{43.678447}$$

$$= 0.208$$

suggests the approximation is not adequate, but for purposes of illustration we calculate

$$\epsilon_2 \approx 10.36 + \left(11.0 - \frac{15.388}{-0.515}\right) - \left(2.447\,\frac{1.231}{0.515}\right) \times \sqrt{1 + \tfrac{1}{8} + \frac{(4.388^2)}{(-0.515)^2 \times 164.6845}}$$

$$\epsilon_1 \approx 10.36 + \left(11.0 - \frac{15.4}{-0.515}\right) + \left(2.447\,\frac{1.231}{0.515}\right) \times \sqrt{1 + \tfrac{1}{8} + \frac{(4.388^2)}{(-0.515)^2 \times 164.6845}}$$

and the range at the 95 percent level is 11.563 to 26.197 percent carbonate.

The confidence limits calculated from the full equations are 12.599 to 29.629. Both these intervals are, of course, quite unrealistic, reflecting the fact that the criterion quantity g is outside its range for adequate estimation. In order to reduce the magnitude of g it would be advisable to increase the number of observations and their range as far as possible, thereby decreasing the value of t and perhaps $\hat{\sigma}_{Y \cdot X}$ and increasing the quantity Sx_i^2.

21.10 Testing the Regression Line by Analysis of Variance

When the regression equation is written in terms of its estimated values as

$$\hat{Y}_i = \hat{\alpha} + \hat{\beta} X_i + \epsilon_i$$

it resembles the structural equation (18.4) described on page 367 in terms of analysis of variance, and it is possible, of course, to test the significance of the regression coefficient by means of analysis of variance where the total variance of the Y observations is

$$\sum_{i=1}^{k} (Y_o - \bar{Y})^2$$

This may be written

$$\sum_{i=1}^{k} (Y_o - \bar{Y})^2 = \sum_{1}^{k} (Y_o - Y_c)^2 + \sum_{1}^{k} (Y_c - \bar{Y})^2 \qquad (21.36)$$

which states that the variance of the observations around the mean is composed of the variation of the observations around the regression value Y_c plus the variation of the regression values around the mean, the cross products vanishing as usual. If we consider the relationship in terms of variations

TABLE 21.2
Testing the Regression Coefficient by Analysis of Variance: $\hat{\beta} = \beta = 0$

Source of variation	d.f.	Sum of squares	Mean square	Expected mean squares
Regression	1	$\sum (Y_c - \bar{Y})^2$	$\dfrac{SS_R}{1}$	$\sigma_\epsilon^2 + \beta^2 \sum (X_i - \bar{X})^2$
Residual	$k - 2$	$\sum (Y_o - Y_c)^2$	$\dfrac{SS_\epsilon}{k-2}$	σ_ϵ^2
Total	$k - 1$	$\sum (Y_o - \bar{Y})^2$		

TABLE 21.3
Testing the Regression of Porosity on Carbonate Content of Cow Run Sandstone

Source of variation	d.f.	Sum of squares	Mean square	Variance ratio F
Regression	1	43.7093	43.7093	28.849
Residual	$k - 2 = 6$	9.0907	1.5151	
Total	$k - 1 = 7$	52.80		

Note: It will be observed that $F = 28.849 = t^2$ of Eq. (21.19).

from means, that is, x_i and y_i, then $\alpha = 0$, and the analysis may be written tabularly as shown in Table 21.2.

The procedure for calculation comprises finding the total sum of squares Sy_i^2 and the sum of squares due to regression $\hat{\beta}_i^2 Sx_i^2$; the residual is found by difference.

In the example relating porosity Y_i to carbonate content this yields the analysis of variance of Table 21.3.

21.11 Comparing Two or More Regression Lines

When two or more sets of data are collected, possibly from the same population or from a population in which the functional relationship between two variates is likely to be similar, it becomes necessary to test whether the regression lines calculated from the observations differ or not. Several possible questions may arise; e.g., are the lines parallel, i.e., of equal slope ($\beta_1 = \beta_2 = \beta$) irrespective of intercept values; alternatively, are they coincident ($\beta_1 = \beta_2 = \beta$ and $\alpha_1 = \alpha_2 = \alpha$)? Generally, if slopes differ ($\beta_1 \neq \beta_2$), questions about similarity of intercept value do not arise.

If we consider the comparison of slopes first, we can remove the effect of intercept by using the mean values of the two variates as origin of the coordinates and express the observations as deviations from their respective means (Tippett, 1952, p. 261). Since by definition each least-squares line is fitted to the data by minimizing the sums of squares of the deviations, we may state that if the two sets of data are from a population with common regression line the residual sum of squares from a common regression line will equal the sum of the residual sum of squares from individual lines apart from sampling errors. On the other hand, if $\beta_1 \neq \beta_2 \neq \beta$, the residual sums of squares of the common line will exceed the sum of the residuals for the individual regressions.

TABLE 21.4
Testing the Significance of Differences among Regression Lines

Sources of variation	d.f.	Sums of squares	Mean square
Residual from common line	$n - k - 1$	SS_{tr}	
Residual from individual lines	$n - 2k$	SS_1	$\dfrac{SS_1}{n - 2k}$
Difference	$k - 1$	$SS_{tr} - SS_1 = SS_d$	$\dfrac{SS_d}{k - 1}$

It is customary to calculate the residual sums of squares for each individual regression line and sum them, and (where there are k sets of data) this would appear as follows:

$$\left(S_1y_i^2 - \frac{S_1x_iy_i}{S_1x_i^2}\right) + \left(S_2y_i^2 - \frac{S_2x_iy_i}{S_2x_i^2}\right) + \cdots + \left(S_ky_i^2 - \frac{S_kx_iy_i}{S_kx_i^2}\right) \qquad (21.37)$$

with degrees of freedom

$$(n_1 - 2) + (n_2 - 2) + \cdots + (n_k - 2) = \sum_{i=1}^{k} k_i - 2k$$

On the other hand, the residual sum of squares obtained from fitting a common regression line is

$$S_1y_i^2 + S_2y_i^2 + \cdots + S_ky_i^2 - \frac{S_1x_iy_i + S_2x_iy_i + \cdots + S_1x_iy_i}{S_1x_i^2 + \cdots + S_kx_i^2} \qquad (21.38)$$

with degrees of freedom

$$\sum_{i=1}^{k} n_i - k - 1$$

that is, $n - 1$ from each set of data minus 1 from the common regression coefficient. Unfortunately, these two residual variances are not independent, but by finding their difference, along with the difference in degrees of freedom, we can compare the variance of (21.37) with the difference in residuals in the form of a variance ratio as in Table 21.4.

Then the variance ratio becomes

$$F_{\alpha,\nu_1,\nu_2} = \frac{SS_d/(k - 1)}{SS_1/(n - 2k)}$$

where $\qquad \nu_1 = k - 1 \qquad$ and $\qquad \nu_2 = n - 2k$

For example, in studying the relationships between percent porosity X and logarithm of permeability in millidarcys Y in Cow Run sandstone, West Virginia, core-analysis data from four wells are available, and it is of interest to determine whether the relationship between permeability and porosity is the same in cores from all four wells. The necessary data are summarized in Table 21.5, and the regression lines and their respective sums of squares are illustrated in Table 21.6. The analysis of variance, testing whether the common regression line is adequate, is summarized in Table 21.7; the variance ratio testing the differences is $F_{05,3,62} = 1.874/0.3138 = 3.784$, whereas the tabulated value is $F_{05,3,62} = 2.76$, which leads to rejection of the hypothesis of a common regression line. From the individual lines of Fig. 21.7, it appears as if the relationship between porosity and permeability is similar in wells 1 to 3 but differs in well 4, and this is borne out by further testing. Furthermore, since the values which lead to the difference in slope of the regression line representing conditions in well 4 is largely in the region of 0.1 millidarcy, it seems likely that difficulty in determining permeability in cores with such low values has led to this discrepancy. It would appear, therefore, that the relationship between porosity and permeability within these cores of Cow Run sandstone is likely to be represented by a common regression line.

21.12 Degree of Association among Variates

Previous sections in this chapter have considered a model in which Y is a random variate and X is selected (or essentially measured without error), but on occasion it is not possible to designate a particular variate as selected because both variates include random variation; in this case correlation is a more suitable analytical tool than regression. The arithmetic procedures leading to both statistical estimators are very similar, and, indeed, regression and correlation are closely related concepts and are frequently used together; nevertheless, it should be remembered that the populations to which the estimators are applied are of different character, there being only one random variable in regression and two in correlation. Again, the main emphasis will be given to the simple linear model, although correlation, like regression, may be extended to include more than a pair of variables, and multiple linear correlation bears the same relationship to multiple linear regression as simple correlation does to simple regression; furthermore, this concept may also be extended to encompass curvilinear relationships by using the correlation ratio (Tippett, 1952, p. 310). In fact, the concepts of correlation, regression, and analysis of variance may be looked upon as complementary tools for the analysis of a set of data, and they are generally used together as the problem becomes multivariate in character.

Suppose we wish to estimate from some variable Y the "best value." We shall, in general, find $\bar{Y}$ and, as a measure of reliability, evaluate the standard deviation of Y, that is, σ_Y. Now, if we know the relationship of

TABLE 21.5
Sums and Sums of Squares from Simple Linear Regression of Log Permeability Y on Porosity X

$X = 10(P - 7)$; $Y = \log(10k)$, and P = Porosity, Percent, and k = Permeability, Millidarcys

Well	N	$\sum X_i$	$\bar{X}$	$\sum Y_i$	$\bar{Y}$	$\sum X_i^2$	$\sum Y_i^2$	$\sum X_i Y_i$	Sx_i^2	Sy_i^2	Sx_iy_i	$\frac{Sx_iy_i}{Sx_i^2}$
17	19	1,146	60.316	21.131	1.112	75,422	33.056819	1,451.494	6,300.105	9.555811	176.961	0.028089
18	19	1,228	64.632	31.713	1.669	86,514	62.782879	2,233.367	7,146.421	9.850544	183.706	0.025706
19	18	1,168	64.889	27.312	1.517	90,012	68.760454	2,326.508	14,221.778	27.319046	554.263	0.038973
20	14	936	66.857	21.550	1.539	67,492	39.015764	1,535.284	4,913.714	5.844157	94.513	0.019235
Four wells	70	4,478	63.971	101.706	1.453	319,440	203.615916	7,546.653	32,975.943	55.842910	1,040.374886	0.031550
Within all 4 wells									32,582.018	52.569558	1,009.443	0.030982

TABLE 21.6
Testing Regression Relationships between Permeability and Porosity in Cow Run Sandstone from St. Marys, West Virginia

Well	N	Regression relationship	d.f.	Sum of squares due to regression	d.f.	Residual sum of squares	Variance ratio F	Significance level
17	19	$Y_1 = 0.028X_1 - 0.582$	1	4.971	17	4.585	18.4	$P < 0.001$
18	19	$Y_2 = 0.025X_2 + 0.146$	1	4.722	17	5.128	15.68	$P < 0.01$
19	18	$Y_3 = 0.039X_3 - 1.012$	1	21.601	16	5.718	60.43	$P < 0.001$
20	14	$Y_4 = 0.019X_4 + 0.253$	1	1.818	12	4.026	5.419	$P < 0.05$
All 4 wells	70	$Y = 0.032X - 0.565$	1	32.824	68	23.019	96.83	$P < 0.001$

Note: $r^2 = \dfrac{\text{SS regression}}{\text{SS}_{\text{tot}}}$

Y to some other variable X, that is, $Y = \alpha + \beta X$, we may make a second independent estimate of Y from the regression equation, and the reliability of this mean will be expressed by its standard error of estimate $\sigma_{Y \cdot X}$. In other words,

$$\sigma_Y{}^2 = \frac{(Y_o - \bar{Y})^2}{N}$$

$$\sigma^2_{Y \cdot X} = \frac{(Y_o - Y_c)^2}{N}$$

We may now compare the reliability of these two estimates by comparing $\sigma^2_{Y \cdot X}$ with $\sigma_Y{}^2$; if $\sigma^2_{Y \cdot X}$ is as large as $\sigma_Y{}^2$, then regression has added nothing to

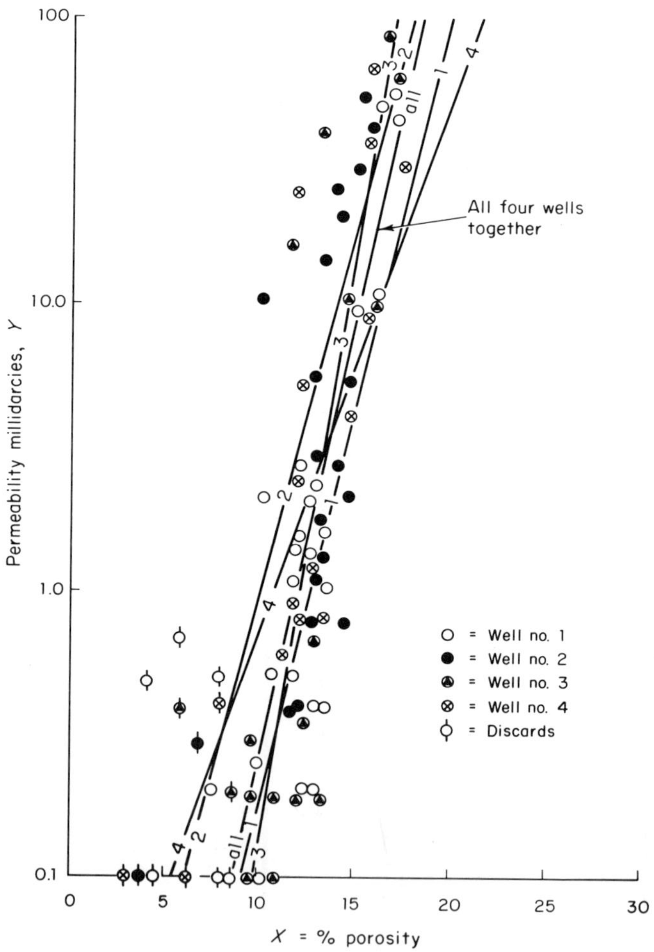

Figure 21.7 Relationship between porosity and permeability in cores from four wells in Cow Run sandstone, St. Marys, W.Va.

TABLE 21.7
Analysis of Variance among and within Regression Lines of Porosity and Permeability of Cow Run Sandstone Cores from Four Wells

Source of variation	d.f.	Sum of squares	Mean squares
Residual variances from common line	65	23.0196	
Residual variances from individual lines	62	19.4575	0.3138
Difference	3	3.5621	1.1874

our information. If $\sigma^2_{Y \cdot X}$ is less than $\sigma_Y{}^2$, then regression gives a better estimate than the variable Y alone. The comparison may be carried out as follows:

$$\frac{\sigma^2_{Y \cdot X}}{\sigma_Y{}^2} = \frac{\sum(Y_o - Y_c)^2}{\sum(Y_o - \bar{Y})^2} \tag{21.39}$$

The ratio $\sigma^2_{YX}/\sigma_Y{}^2$ is, of course, dimensionless like the coefficient of variation CV.

As an example three hypothetical sets of data are plotted in Fig. 21.8 to illustrate graphically the meaning of the relationship between these two variances. For example, in the figures, the deviation of each observation Y_o from the mean $\bar{Y}$ of the Y's is marked by a solid line, whereas the deviation of each observation Y_o from the regression line Y_c is marked by a broken line; it can then be seen that $\sum(Y_o - Y_c)^2$ can be equal to $\sum(Y_o - \bar{Y})^2$, as it

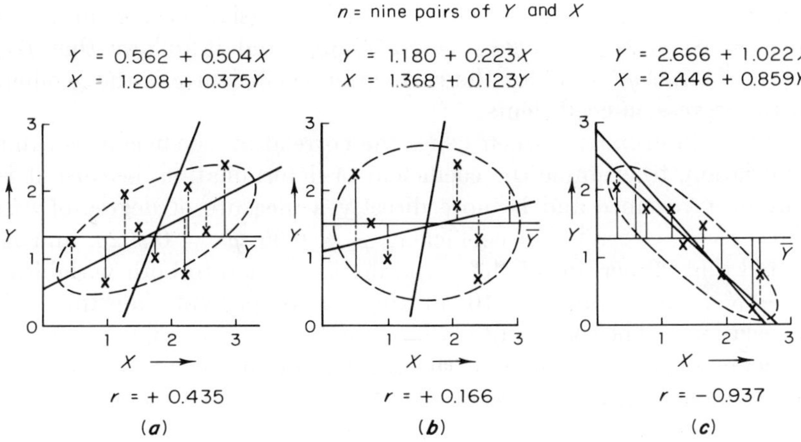

Figure 21.8 Correlation, or the degree of association among variates.

very nearly is in Fig. 21.8*b*, or less than $\sum(Y_o - \bar{Y})^2$ as it is in Fig. 21.8*a* and *c*, but never greater than $\sum(Y_o - \bar{Y})^2$. Hence, as the observations scatter less around the regression line, the ratio

$$\frac{\sigma^2_{Y \cdot X}}{\sigma_Y{}^2} = \frac{\sum(Y_o - Y_c)^2}{\sum(Y_o - \bar{Y})^2}$$

becomes smaller, and the arrangement of the observations tends to vary from circular (Fig. 21.8*b*), in cases where Y and X are not linearly associated, to elliptical, in cases where variation in Y is closely associated with variation in X (Fig. 21.8*c*). It is somewhat more convenient to use a parameter which increases as the association improves, so we define the coefficient of determination ρ^2 as

$$\rho^2 = 1 - \frac{\sigma^2_{Y \cdot X}}{\sigma_Y{}^2} = 1 - \frac{\sum(Y_o - Y_c)^2}{\sum(Y_o - \bar{Y})^2} \tag{21.40}$$

and the coefficient of correlation is then $\rho = \sqrt{\rho^2}$. The statistical estimators are r^2 and r respectively, where $r = (1 - \hat{\sigma}^2_{Y \cdot X}/\hat{\sigma}_Y{}^2)^{1/2}$. It follows that when $\sigma^2_{Y \cdot X} = \sigma_Y{}^2$, then $\rho^2 = 1 - 1 = 0$, whereas when the observations approach closely to the regression line $\sigma^2_{Y \cdot X}$ is very small compared with $\sigma_Y{}^2$, and ρ^2 approaches unity. From Fig. 21.8*a* to *c* the sign of ρ may be positive or negative, ranging from -1 through 0 to $+1$, and the correlation coefficient $\rho_{Y \cdot X}$ has the same sign as the corresponding regression coefficient $\beta_{Y \cdot X}$.

Another feature worthy of note in Fig. 21.8 is the relationship between the regression coefficient and the correlation coefficient; thus when there is a high correlation, for example, $\rho = -0.937$ in Fig. 21.8*c*, the regression lines are close together and the angle θ between them is small, whereas when the correlation is low, for example, $\rho = 0.166$ in Fig. 21.8*b*, the lines tend to be widely spread, and the angle θ between them approaches 90°. When the regression lines are perpendicular, $\rho = 0$, and the observations are linearly independent; in fact $\rho = \cos\theta$. The regression coefficient is defined as $\beta_{Y \cdot X} = Sx_iy_i/Sx_i{}^2$ and $\beta_{X \cdot Y} = Sx_iy_i/Sy_i{}^2$, and it follows that $\beta_{Y \cdot X}\beta_{X \cdot Y} = (Sx_iy_i)^2/Sx_i{}^2Sy_i{}^2 = \rho^2$, or the correlation coefficient ρ is the geometric mean of the regression coefficients.

It is informative, when using the correlation coefficient as a measure of association, to examine the coefficient of determination, because it is equivalent to a variance and is more directly a measure of degree of association; thus, given a correlation coefficient, $\rho = 0.95$, $\rho^2 = 0.9025$, and it may be stated that 90 percent of the variation is common to both variables, whereas $1 - \rho^2$ is approximately 10, or there is 10 percent "unexplained" in the association. The coefficient of determination is a much more realistic measure of degree of association than the correlation coefficient, and when $\rho = 0.6$, for example, only 36 percent $(100\rho^2)$ is common to both variables while $100(1 - \rho^2)$, or 64 percent, is "unexplained"; it needs a very high value of ρ

to establish a close relationship. Of course, the same thing holds for the sample estimator r of the population correlation coefficient (ρ).

21.13 Calculation of the Correlation Coefficient

Since the correlation coefficient is so closely related to regression, it is not surprising that the arithmetic procedures leading to r are very similar to those leading to $\beta_{Y \cdot X}$; thus it is necessary to calculate the sums, $\sum X_i$, $\sum Y_i$; means; sums of squares, $\sum X_i^2$, $\sum Y_i^2$; and products, $\sum X_iY_i$, for both procedures, and it is convenient to use the corrected sums of squares and products, Sx_i^2, Sy_i^2, Sx_iy_i. The correlation coefficient may then be calculated directly as

$$\rho = \frac{Sx_iy_i}{(Sx_i^2Sy_i^2)^{1/2}} \tag{21.41}$$

or as the square root of the coefficient of determination ρ^2, where

$$\rho^2 = \frac{(Sx_iy_i)^2}{Sx_i^2Sy_i^2} \tag{21.42}$$

In both cases, of course, using a set of observations which contain random variation, the calculations lead to the statistical estimators r and r^2 respectively.

21.14 Tests of Significance of the Correlation Coefficient

Since most experiments leading to sets of observations are samples from some possible population, the calculated statistical estimators should be tested for significance before using them as a basis for inference. Thus, given $r = 0.166$ (see Fig. 21.8*b*), on the basis of nine pairs of observations Y, X we may inquire whether this sample is a random sample from a population in which $\rho = 0$. If we set up the null hypothesis that these nine pairs are random samples from a population in which $\rho = 0$, we can use tables in which, for $\rho = 0$ and given sample size, the probability of r's of known magnitude are listed (see, for example, Arkin and Colton, 1950, table 21, p. 140). The degrees of freedom for r are similar to those for the regression coefficient, $\nu = n - 2$, or in the present instance $\nu = 9 - 2 = 7$, whence from the table $r = 0.666$ or greater is expected 5 percent of the time or $r = 0.798$ or greater is expected 1 percent of the time. It seems quite clear that in the example of Fig. 21.8*b* this sample could be drawn from a population with $\rho = 0$. Similarly, $r = 0.435$ (Fig. 21.8*a*) contributes no definite evidence for correlation different from zero in the population; indeed, for $r = 0.435$ to be significant at the 5 percent level at least 21 pairs would be required. Now in the example illustrated in Fig. 21.8*c*, where $r = -0.937$ and $\nu = 7$, the

TABLE 21.8
Comparing Two Correlation Coefficients

Well	r	n	$\nu = n - 3$	$\frac{1}{n-3}$	$\frac{1+r}{1-r}$	$\log \frac{1+r}{1-r}$	z
C	0.5220	29	26	0.0370	3.1841	0.50298	0.5791
F	0.9207	50	47	0.0280	24.2207	1.38418	1.5936

hypothesis of $\rho = 0$ is rejected at the 1 percent level; or, if this set of nine pairs represents a random sample from a population in which $\rho = 0$, we should expect a value of $r = -0.937$ much less than 1 time in 100 such sample sets.

This test against zero correlation ($\rho = 0$) is based on the fact that the sample r's from a population with $\rho = 0$ are symmetrically and approximately normally distributed around mean $\rho = 0$, so that a Student's t test is appropriate with $\nu = n - 2$ degrees of freedom (Fisher, 1948, fig. 7, p. 200) as follows:

$$t = r\sqrt{\frac{n-2}{1-r^2}} \tag{21.43}$$

In all other cases the distributions of sample r's around their respective population values of ρ are highly skewed, and the skewness increases as ρ increases. Fisher (1948, p. 201) found a transformation from ρ to z which leads to approximate normal distributions of z at all values of ρ, and the variance of z is solely a function of the degrees of freedom. In other words, if the sample value of r is transformed to z, then z is normally distributed around its appropriate mean $\bar{z}$ with variance $\sigma_{\bar{z}}^2 = 1/n - 3$, where n is the number of pairs in the sample. This may then be used as a basis for significance tests and for setting up confidence limits.

The transformation is

$$z = \tfrac{1}{2} \ln \frac{1+r}{1-r}$$

or, in common logarithms,

$$z = 1.1573 \log \frac{1+r}{1-r} \tag{21.44}$$

A table facilitating this transformation is included as Table 21.14.

As an example we may first compare two correlation coefficients resulting from a study of the relationship between size (phi median, Md_ϕ) and sorting (phi percentile deviation, PD_ϕ) in core samples from two wells, C and F, in the Miocene sediments of South Trinidad (Griffiths, 1951a, p. 220). The null hypothesis presumes that these two sample sets of data are random samples from the same population; they yield $r = 0.5220$ and $r = 0.9207$ based on 29 and 50 pairs of observations respectively. The calculations may be arranged as in Table 21.8.

The standard error of the difference between the two z values is

$$\hat{\sigma}^2_{z_1-z_2} = \frac{1}{n_1 - 3} + \frac{1}{n_2 - 3} \tag{21.45}$$

or

$$\hat{\sigma}^2_{z_1-z_2} = 0.0370 + 0.0208 = 0.2404$$

then

$$t = \frac{z_1 - z_2}{\hat{\sigma}_{z_1-z_2}} = \frac{1.0145}{0.4904} = 2.06$$

and t is 2.06 times its standard error, which is of course significant (t = twice its standard error is conventionally chosen as the limit beyond which a difference is regarded as significantly larger than its standard error).

By the same procedure we could set up confidence limits for a correlation coefficient; thus from a sample of 531 pairs of phi medians and phi percentile deviations we obtain $r = 0.7585$ (Griffiths, 1951*a*, p. 220).

In this case

$$z = 1.1513 \log \frac{1 + z}{1 - z} = 1.1513 \log \frac{1.7585}{0.2415}$$

$$= 1.1513 \log 7.2816 = 0.9927$$

and

$$\hat{\sigma}_z = \frac{1}{528^{1/2}} = 0.0435$$

then 95 percent confidence limits are

$$z \pm 2\hat{\sigma}_z = 0.9927 \pm 0.0870$$

or

$$1.0797 > \zeta > 0.9057 \qquad 0.7932 > \rho > 0.7196$$

where ζ and ρ are the respective population values. The 99 percent limits are

$$1.2537 > \zeta > 0.7317 \qquad 0.8493 > \rho > 0.6241$$

Any value of r beyond these limits differs from the estimate found at the appropriate probability level. In comparing a second estimate of r use the degrees of freedom associated with second r and the first one as ρ.

Finally, in cases where sets of r have been obtained, it is often necessary to test whether these values of r could all have arisen from sampling a single homogeneous parent population and, if so, how they can best be combined

TABLE 21.9
Test of Homogeneity of Several Correlation Coefficients and Determination of Best Estimate of Mean Value of r

No. of samples	$n - 3$	r	z	Weighted z $(n - 3)z$	Weighted square $(n - 3)z^2$
15	12	0.4524	0.4877	5.8524	2.8542
20	17	0.5077	0.5596	9.5132	5.3236
25	22	0.4320	0.4623	10.1706	4.7019
30	27	0.4054	0.4301	11.6127	4.9946
90	78			37.1489	17.8743
			$\bar{z} = 0.4763$	$\bar{r} = 0.4432$	

into a single value of r for a more efficient estimate of the population correlation coefficient.

Once again the z transformation may be used, and the different values are weighted by the inverse of their variance and the weighted mean z transformed to r. The work is most efficiently performed tabularly (Table 21.9). Four sets of sample pairs were randomly selected from 69 to yield the values in Table 21.9 of the size-sorting study (Griffiths, 1951a, p. 217); are these values such that they could have been drawn from a single population?

Remembering that χ^2 is the distribution of the sum of squares of normal deviates with mean zero and unit standard deviation (see Chap. 17), we may note that z is a normal deviate with standard deviation $1/\sqrt{n - 3}$, whence $z\sqrt{n - 3}$ fulfills the requirements for such a normal deviate.

Suppose we now wish to determine whether, despite the fact that each correlation is small, the aggregate tends to show evidence for a significant association; we may use chi square where

$$\chi^2 = z^2 \sum (n - 3) \tag{21.46}$$

with degrees of freedom equal to the number of samples. In the present case d.f. $= 4$ and $\chi^2 = 17.87$; then $0.01 > P > 0.001$, and there is clear evidence for association in this set of figures.

We can also test whether these sample correlation coefficients are homogeneous, i.e., all drawn from the same population, by means of chi square where chi square is defined as[1]

$$\begin{aligned} \chi^2 &= \sum [(z - \bar{z})^2 (n - 3)] \\ &= \sum (n - 3) z^2 - [\sum (n - 3) z] \bar{z} \end{aligned} \tag{21.47}$$

[1] See Tippett, 1952, p. 257, and Snedecor, 1956, p. 154.

with degrees of freedom one less than the number of sample r's. From Table 21.9

$$\chi^2 = 17.8743 - 37.1489 \times 0.4763$$

$$= 17.8743 - 17.6940 = 0.1803 \qquad \text{d.f.} = 3$$

$$0.99 > P > 0.98$$

These sample values are considerably more homogeneous than we should expect on average in draws from even a single population.

As a further example we analyze some of the conclusions regarding the correlation coefficients of Table 21.10 (Griffiths, 1951a, p. 220). It is claimed that trends C, E, F, K are similar associations; are they homogeneous?

Therefore, these values are not consistent but contain significant differences, and the sensitivity is of course due to the large number of pairs in the samples. It is important to note how the weights $(n - 3)$ affect the importance of z $(\equiv r)$ as well as the magnitude of z (or r) itself.

In this same set of values wells D and H yield correlation coefficients which are not incompatible with their being samples from a population in which $\rho = 0$. Suppose they are from the same population; do they reinforce each other? From Table 21.11,

$$\chi^2 = \sum[z^2(n - 3)] = 3.7803 \qquad \text{d.f.} = 2$$

$$0.20 > P > 0.10$$

There is still no evidence to support association in the combined figures. In testing whether there is any evidence for incompatibility $\chi^2 = 0.5926$ with d.f. $= 1$, then $50 > P > 30$.

TABLE 21.10
Tests of Homogeneity of a Set of Correlation Coefficients of Size and Size Sorting

Well	No. of samples	$n - 3$	r	z	$(n - 3)z$	$(n - 3)z^2$
C	29	26	0.5220	0.5791	15.0566	8.7193
E	50	47	0.7423	0.9556	44.9132	42.9191
F	50	47	0.9207	1.5936	74.8992	119.3594
K	37	34	0.6264	0.7355	25.0070	18.3926
	166	154			159.8760	189.3904
					$\bar{z} = 1.0382$	

$$\chi^2 = 189.3904 - 159.876 \times 1.0382$$

$$= 189.3904 - 165.9833 = 23.4071 \qquad \text{d.f.} = 3$$

$$p < 0.001$$

TABLE 21.11
Comparison of Correlation Coefficients Which May Be Drawn from the Same Population

Well	No. of samples	$n-3$	r	z	$n-3$	$(n-3)z^2$
D	36	33	0.0405	0.0405	1.3365	0.0541
H	31	28	0.3494	0.3648	10.2144	3.7262
	61		$\bar{z} = 0.1894$	$\bar{r} = 0.1913$		

According to the principle of hydraulic equivalence (Rittenhouse, 1943b; and page 216), minerals of like density and habit (shape) should behave in a similar manner during transport in a fluid; in general, one would therefore expect variations in proportion of amphibole and pyroxene to behave more similarly than those of amphibole and garnet. Sixty-two specimens of Mississippi River sand were analyzed for their heavy residue by Russell (1937) yielding the correlation coefficients of Table 21.12 between proportions of amphibole and pyroxene and amphibole and garnet respectively. Tested against a population value of $\rho = 0$, both are highly significant ($P < 0.001$), so the correlations are real. We may now test whether they are different by following the steps in the lower two rows of Table 21.12.

The standard error of the difference $z_1 - z_2$ is

$$\hat{\sigma}_{z1-z2} = \left(\frac{1}{n_1 - 3} + \frac{1}{n_2 - 3}\right)^{1/2} = 0.0338^{1/2} = 0.1838$$

then $t = z_1 - z_2/\hat{\sigma}_{z1-z2} = 0.5944/0.1838 = 3.23$. Since such a value of t would arise about 14 times in 10,000 in random sampling the same population, we reject the hypothesis that these two sample r's are based on random samples from the same population. Evidently variation in the proportion of amphibole is more closely associated with variation in proportion of pyroxene ($r^2 = 62$ percent in common) than with variation in proportion of garnet ($r^2 = 19$ percent in common), as would be expected on the basis of hydraulic equivalence. Actually such a test is very approximate, because variation in percentage data tends to be related to the amount present, small percentages varying less than large, so that the variances tend to be heteroscedastic and thus violate one of the requirements, homogeneous variance, underlying the use of simple correlation.

21.15 The Correlation Matrix

When a quantitative analysis of some rock is undertaken, it is customary to assess the composition by point counting the several elements of which the rock is composed and to estimate the variation of proportion in, say, four or five classes, and then the size and the size sorting of the quartz grains can be measured. When two axes, the long a and short b, are used, this will yield a measure of shape of the quartz grains. In addition, it is also customary to measure the orientation and packing of the grains and perhaps the bulk density, porosity, and permeability on the rock. If these measurements are performed on n specimens, there results a $p \times n$ matrix of observations consisting of p estimates of the properties on n specimens of rock. One of the first steps in reduction of these data is to convert the matrix of observations (which, if $n \neq p$, is rectangular) to the matrix of correlation coefficients, which is a square $p \times p$ matrix (see, for example, Griffiths, 1958b, 1960b, 1962b). For example, the Maxton sand, a Mississippian low-rank graywacke from West Virginia, was analyzed by measuring 11 properties on 33 core specimens, and this 11 × 33 matrix of observations was condensed to the 11 × 11 matrix of correlation coefficients displayed in Table 21.13. The matrix of correlation coefficients is both square and symmetrical, and therefore only half the correlation coefficients need be recorded.

Another important feature of this correlation matrix is that the values of the variates have been transformed so that each mean is zero ($\bar{X}_i = 0$) and each variance is unity ($\hat{\sigma}^2 = 1$); this permits comparison between variables of different kinds, and the correlation matrix is the starting point for many excursions into multivariate analysis (Rao, 1952; Griffiths, 1958b, 1960b, 1962b).

Some features of this table essentially resulting from the petrographic analysis follow. First, the correlations among the variates estimating variation in composition are likely to be, at least in part, induced by the fact that all these variates sum to 100; Chayes has discussed the effects of this restraint and exhibited the pseudocorrelation which may arise among such

TABLE 21.12
Association in Variation of Proportions of Amphibole, Pyroxene, and Garnet in Mississippi River Sands

Amphibole X		Pyroxene Y_1 r_{y_1x} 0.7882		Garnet Y_2 r_{y_2x} 0.4410	
n	$n-3$	r	z	$z_1 - z_2$	$\frac{1}{n-3}$
62	59	0.7882	1.0667		0.0169
62	59	0.4110	0.4723	0.5944	0.0169

Source: Data from Russell, 1937.

TABLE 21.13
Correlation Coefficients r Estimating Association among Petrographic Properties and Reservoir Characteristics in Maxton Sandstone, West Virginia

Property	p	(1)	(2)	(3)	(4)	(5)	(6)	(7)	(8)	(9)	(10)	(11)
Quartz	(1)	1.0	−0.2303	−0.2558	−0.4406*	−0.1864	−0.0481	−0.1473	0.0492	−0.2249	0.0442	0.0961
Rock fragments	(2)		1.0	−0.5054**	−0.5117**	−0.5802**	0.5549**	−0.5815*	0.4267*	0.5340**	0.7023**	0.5751**
Silica	(3)			1.0	0.3815	0.3417	−0.3224	0.3362	−0.3452*	−0.1903	−0.6064**	−0.6784**
Matrix	(4)				1.0	0.8113**	−0.3465*	0.7808**	−0.2742	−0.4710**	−0.6671	−0.6174**
Size $\bar{X}_{a\phi}$	(5)					1.0	0.2625	0.9931**	−0.2550	−0.5580**	−0.7094**	−0.6167**
Sorting $\hat{\sigma}_{a\phi}$	(6)						1.0	−0.2762	−0.7236**	0.2412	0.4156*	0.3858*
Size $\bar{X}_{b\phi}$	(7)							1.0	−0.2798	−0.5885**	−0.7004**	−0.6045**
Sorting $\hat{\sigma}_{b\phi}$	(8)								1.0	0.1723	0.1848	0.2171
Orientation $\hat{\sigma}^{\circ}$	(9)									1.0	0.4773**	0.3061
Porosity	(10)										1.0	0.8496**
Log permeability	(11)											1.0

Note: $n = 33$, therefore significant r at $P_{05} = 0.344$; $P_{01} = 0.441$.

variates (Chayes, 1949b, 1960; Vistelius and Sarmanov, 1961). Second, we know from previous work (see Chap. 7, for example) that the axial measures of the quartz grains are generally very highly correlated, particularly so when expressed as mean values. Similarly, the standard deviations of these axes are correlated, yielding coefficients which achieve 0.6 to 0.7 unit; on the other hand, the means are seldom related to standard deviations (Griffiths et al., 1956; Griffiths 1958b, 1959b; Stanonis, 1956; Wood, 1960; Pachman, 1961; Siegrist, 1961). In a number of cases too, it has been found that the porosity is related to log permeability with correlation varying between 0.65 to 0.90 ($r^2 = 40$ to 80 percent).

Of course, it is essential to add that correlation coefficients of zero order such as these are guides that should be interpreted with considerable care. Quite frequently, because relationships are complex, measurements of properties are interdependent, and several may be interrelated; the relationships displayed by the zero-order correlation coefficients are anything but simple and obvious. Thus, if proportion of quartz is related to size of quartz as measured by the long a axis, it will also be related to the short b axis because a and b are closely related. Similarly, proportion of matrix and/or cement may be inversely related to packing (quartz-to-quartz grain contacts), because as matrix increases, the quartz proportion drops, and so contacts are less frequent (see, for example, Emery, 1954; Griffiths et al., 1956). Within the compositional variables, increase in one is quite likely to induce changes in the others because they sum to constant value. Nevertheless, the matrix of zero-order correlation coefficients is an important step in the analysis of variation in sets of data and may be used quite successfully as a guide to concomitant variation.

21.16 Partial Correlation

Some improvement in the measurement of association may be achieved by the use of partial correlation coefficients in which association between two variates can be assessed independently of a third, essentially by holding the third constant at its mean value. The partial correlation coefficient may be computed from the formula (Fisher, 1948, p. 188)

$$r_{12.3} = \frac{r_{12} - r_{13}r_{23}}{[(1 - r_{13}^2)(1 - r_{23}^2)]^{1/2}} \tag{21.48}$$

By permuting the subscripts, three new partial correlation coefficients can be calculated. These can be tested for significance against zero correlation by using the same table as before for two variables with degrees of freedom $\nu = n - 3$. Furthermore, partial correlation coefficients can also be transformed to z and several tested for homogeneity, etc.

As an example it may be noted from Table 21.13 that grain size as measured by the long a axis in thin section X_5 is negatively associated with porosity X_{10}, and again proportion of matrix X_4 is positively related to grain size X_5, while variation in proportion of matrix is negatively related to variation in porosity. It would be of interest to determine whether grain size is related to porosity independently of change in matrix proportion. Label grain size as X_1, matrix as X_2, and porosity as X_3; then the required correlations are

$$r_{12} = +0.8113 \qquad r_{13} = -0.7094 \qquad r_{23} = -0.6671$$

and we calculate the association between grain size X_1 and porosity X_3 holding proportion of matrix X_2 constant

$$r_{13.2} = \frac{r_{13} - r_{12}r_{23}}{[(1 - r_{12}^2)(1 - r_{23}^2)]^{1/2}}$$

or

$$r_{13.2} = \frac{-0.7094 + 0.8113 \times 0.6671}{[1 - 0.8113^2 \times 1 - (-0.6671)^2]^{1/2}}$$

$$= -0.386$$

For $n = 33$, $\nu = n - 3 = 30$, and the tabulated value of $r_{13.2}$ equals 0.349 at the 5 percent level (Arkin and Colton, 1950, table 21, p. 140). Evidently variation in grain size is still significantly associated with variation in porosity, but the intensity is much less if proportion of matrix remains constant. If we consider the removal of the effect of a third variable in each of the other two in turn, we have

$$r_{12} = +0.8113 \qquad r_{13} = -0.7094 \qquad r_{23} = -0.6671$$

$$r_{12.3} = +0.6439 \qquad r_{13.2} = -0.3862 \qquad r_{23.1} = -0.2222$$

Here the association is lowered in each case, but that between grain size and matrix is less affected than either grain size and porosity or matrix and porosity. The relationship between variation in proportion of matrix and variation in porosity, with grain size held constant ($r_{23.1}$), is reduced below the level of significance, and therefore when this relationship is assessed, with the effect of grain size removed, variation in porosity is independent of variation in proportion of matrix.

It is possible, of course, to extend the concept of partial correlation to the removal of more than one variable, but it is then better to consider multiple correlation and multivariate analysis using principal components or factor analysis (Holzinger and Harman, 1941; Griffiths, 1962b; Miller and Kahn, 1962). It is also obvious that many other combinations of variables could be evaluated, but it is necessary to avoid wholesale particulate analysis of this kind, first because a limited set of data can contain only a limited amount of information and probably much redundancy; and second, because consistently testing pairs and triplets of interdependent variates contradicts the requirements of independence necessitated for successive testing; as a result, the probability levels of the various tests may be strongly affected. It is advisable therefore to test only those features which are of prime interest and which were envisioned prior to the examination of the correlation matrix.

21.17 Extension from Simple to Multiple and from Linear to Curvilinear Relationships

It is only a step from the measurement of one variate to the measurement of two and so on to three, four, or more; and, in each case, it is possible to express the relationship among the variates in some form such as, for example, the zero-order correlation coefficients of the previous section or linear regression with more than one variate. This may be formalized by writing the general formula for the polynomial as

$$Y = \beta_0 X_0 + \beta_1 X_1 + \beta_2 X_2 + \cdots + \beta_k X_k \qquad (21.49)$$

in its simplest form $X_0 = 1$, and β_0 is equivalent to α of the simple linear regression formula; Eq. (21.49) relates variation in Y to several X variates, thus yielding multiple linear regression, which may be analyzed by a procedure analogous to that for simple linear regression by extending the treatment directly to several variables (Rao, 1952; Quenouille, 1952); similarly, if we set $X_1 = X_1$, $X_2 = X_2$, $X_3 = X_1^2$, $X_4 = X_2^2$, $X_5 = X_1X_2$, and so on, we can encompass curvilinear relationships, and again the procedures, while leading to more complex arithmetic, do not require any very new concepts. Again, in estimating degree of association, the correlation of Y with several X_i leads to the multiple (linear) correlation coefficient $R^2_{Y \cdot X_1 X_2 \cdots X_k}$ and so on. The resolution of problems of this kind is described in standard texts such as Rao (1952), Quenouille (1952), Williams (1959), Bennett and Franklin (1954), and Brownlee (1960). A number of examples of the use of these techniques in geological problems also exist (Griffiths et al., 1956; Griffiths, 1958b, 1960b; Krumbein, 1959; Miller and Kahn, 1962, pp. 242ff.).

21.18 Examples of the Use of Simple Regression and Correlation in Geoscience Problems

A number of examples of the use of simple regression and correlation in geological problems is referred to in the preceding text, and a wide variety of examples is discussed in Miller and Kahn (1962, chaps. 8 to 11).[1] These writers also discuss (1962, pp. 203ff.) alternative procedures for defining the "best fitting" line and its application to problems in paleobiometrics. An excellent article on exponential functions in sedimentology is also available (Krumbein, 1937).

[1] Particularly useful are their annotated references on pp. 199–200, 210–213, and 226–227.

Some specific relationships which appear to be of general application in the geosciences are those between bulk density and porosity, which is inverse linear and exhibits a high degree of association, over 90 percent (Davis, 1954); between size and size sorting when the measurements are performed by indirect procedure (Griffiths, 1951a, b; Inman, 1949); between porosity and log permeability in oil-reservoir sandstones and carbonates (Ryder, 1948; Griffiths, 1958b; Headlee and Joseph, 1945; Law, 1944); and between the long and short axes of quartz grains and quartzite rock fragments, which is linear in phi units and possesses a very high degree of association when the measurements are performed in thin section (Griffiths et al., 1956; Griffiths, 1958b, 1959a).

Similarly of general interest is the development of pseudocorrelation among proportions of constituents induced by the restraint that all constituents sum to a fixed value (Chayes, 1949b, 1960, 1962; Vistelius and Sarmanov, 1961); this feature pervades the whole analysis of composition, whether bulk chemistry, thin section, or accessory heavy minerals are the items of interest, and until some effective transformation is found, analysis of this kind of data will suffer from a largely undefined degree of ambiguity. A related pseudocorrelation occurs among long and short axes of constituents measured in thin section or the long, intermediate, and short axes measured on loose particles, because the axes are ordered by definition as $a > b$, or $a > b > c$, respectively. The degree of association induced by these constraints can be estimated, and unless the sample correlations observed significantly exceed these values, they may be assumed to estimate a population $\rho = 0$ (Smith, 1964).

An unusual use of the correlation coefficient reported by Dryden (1935) led to comment both pro and con (Eisenhart, 1935b; Helson, 1936). Dryden used the correlation coefficient to measure "likeness" between two heavy mineral suites, and Eisenhart suggested that the chi-square statistic was more appropriate to Dryden's purpose. Helson, on the other hand, could see no harm in Dryden's procedure. This example serves to emphasize two different uses of "statistics"; when a statistic is used as a summary number to represent data, as long as the algebraic requirements are met, no further justification is needed, but if the statistic is used as a sample estimator of a population parameter, then a number of additional requirements must be met. Thus frequencies in discrete classes imply the multinomial frequency distribution and nominal or at best ordinal scaling, and the chi-square statistic is the appropriate test statistic; whereas if the two variates are random variables, a continuous (interval or ratio) scale is implied, and parametric statistics such as the correlation coefficient are the more efficient estimators.

Extensive use of the correlation coefficient in paleobiometry is illustrated in Olson and Miller (1958), and here the correlation matrix is used as a basis of extended rigorous analysis by multivariate techniques.

TABLE 21.14
Values of z for Values of r

0.0 to 0.66

R	0	1	2	3	4	5	6	7	8	9
0.0	0.0	0.0010	0.0020	0.0030	0.0040	0.0050	0.0060	0.0070	0.0080	0.0090
0.01	0.0100	0.0110	0.0120	0.0130	0.0140	0.0150	0.0160	0.0170	0.0180	0.0190
0.02	0.0200	0.0210	0.0220	0.0230	0.0240	0.0250	0.0260	0.0270	0.0280	0.0290
0.03	0.0300	0.0310	0.0320	0.0330	0.0340	0.0350	0.0360	0.0370	0.0380	0.0390
0.04	0.0400	0.0410	0.0420	0.0430	0.0440	0.0450	0.0460	0.0470	0.0480	0.0490
0.05	0.0500	0.0510	0.0520	0.0530	0.0541	0.0551	0.0561	0.0571	0.0581	0.0591
0.06	0.0601	0.0611	0.0621	0.0631	0.0641	0.0651	0.0661	0.0671	0.0681	0.0691
0.07	0.0701	0.0711	0.0721	0.0731	0.0741	0.0751	0.0761	0.0772	0.0782	0.0792
0.08	0.0802	0.0812	0.0822	0.0832	0.0842	0.0852	0.0862	0.0872	0.0882	0.0892
0.09	0.0902	0.0913	0.0923	0.0933	0.0943	0.0953	0.0963	0.0973	0.0983	0.0993
0.10	0.1003	0.1013	0.1024	0.1034	0.1044	0.1054	0.1064	0.1074	0.1084	0.1094
0.11	0.1104	0.1115	0.1125	0.1135	0.1145	0.1155	0.1165	0.1175	0.1186	0.1196
0.12	0.1206	0.1216	0.1226	0.1236	0.1246	0.1257	0.1267	0.1277	0.1287	0.1297
0.13	0.1307	0.1318	0.1328	0.1338	0.1348	0.1358	0.1368	0.1379	0.1389	0.1399
0.14	0.1409	0.1419	0.1430	0.1440	0.1450	0.1460	0.1471	0.1481	0.1491	0.1501
0.15	0.1511	0.1522	0.1532	0.1542	0.1552	0.1563	0.1573	0.1583	0.1593	0.1604
0.16	0.1614	0.1624	0.1634	0.1645	0.1655	0.1665	0.1676	0.1686	0.1696	0.1706
0.17	0.1717	0.1727	0.1737	0.1748	0.1758	0.1768	0.1779	0.1789	0.1799	0.1809
0.18	0.1820	0.1830	0.1841	0.1851	0.1861	0.1872	0.1882	0.1892	0.1903	0.1913
0.19	0.1923	0.1934	0.1944	0.1955	0.1965	0.1975	0.1986	0.1996	0.2007	0.2017
0.20	0.2027	0.2038	0.2048	0.2059	0.2069	0.2079	0.2090	0.2100	0.2111	0.2121
0.21	0.2132	0.2142	0.2153	0.2163	0.2174	0.2184	0.2195	0.2205	0.2216	0.2226
0.22	0.2237	0.2247	0.2258	0.2268	0.2279	0.2289	0.2300	0.2310	0.2321	0.2331
0.23	0.2342	0.2352	0.2363	0.2374	0.2384	0.2395	0.2405	0.2416	0.2427	0.2437
0.24	0.2448	0.2458	0.2469	0.2480	0.2490	0.2501	0.2512	0.2522	0.2533	0.2543
0.25	0.2554	0.2565	0.2575	0.2586	0.2597	0.2608	0.2618	0.2629	0.2640	0.2650
0.26	0.2661	0.2672	0.2683	0.2693	0.2704	0.2715	0.2726	0.2736	0.2747	0.2758
0.27	0.2769	0.2779	0.2790	0.2801	0.2812	0.2823	0.2833	0.2844	0.2855	0.2866
0.28	0.2877	0.2888	0.2899	0.2909	0.2920	0.2931	0.2942	0.2953	0.2964	0.2975
0.29	0.2986	0.2997	0.3008	0.3018	0.3029	0.3040	0.3051	0.3062	0.3073	0.3084
0.30	0.3095	0.3106	0.3117	0.3128	0.3139	0.3150	0.3161	0.3172	0.3183	0.3194
0.31	0.3205	0.3217	0.3228	0.3239	0.3250	0.3261	0.3272	0.3283	0.3294	0.3305
0.32	0.3316	0.3328	0.3339	0.3350	0.3361	0.3372	0.3383	0.3395	0.3406	0.3417
0.33	0.3428	0.3440	0.3451	0.3462	0.3473	0.3484	0.3496	0.3507	0.3518	0.3530
0.34	0.3541	0.3552	0.3564	0.3575	0.3586	0.3598	0.3609	0.3620	0.3632	0.3643
0.35	0.3654	0.3666	0.3677	0.3689	0.3700	0.3712	0.3723	0.3734	0.3746	0.3757
0.36	0.3769	0.3780	0.3792	0.3803	0.3815	0.3826	0.3838	0.3850	0.3861	0.3873
0.37	0.3884	0.3896	0.3907	0.3919	0.3931	0.3942	0.3954	0.3966	0.3977	0.3989
0.38	0.4001	0.4012	0.4024	0.4036	0.4047	0.4059	0.4071	0.4083	0.4094	0.4106
0.39	0.4118	0.4130	0.4142	0.4153	0.4165	0.4177	0.4189	0.4201	0.4213	0.4225
0.40	0.4236	0.4248	0.4260	0.4272	0.4284	0.4296	0.4308	0.4320	0.4332	0.4344
0.41	0.4356	0.4368	0.4380	0.4392	0.4404	0.4416	0.4428	0.4441	0.4453	0.4465
0.42	0.4477	0.4489	0.4501	0.4513	0.4526	0.4538	0.4550	0.4562	0.4574	0.4587
0.43	0.4599	0.4611	0.4624	0.4636	0.4648	0.4660	0.4673	0.4685	0.4698	0.4710
0.44	0.4722	0.4735	0.4747	0.4760	0.4772	0.4784	0.4797	0.4809	0.4822	0.4834
0.45	0.4847	0.4860	0.4872	0.4885	0.4897	0.4910	0.4922	0.4935	0.4948	0.4960
0.46	0.4973	0.4986	0.4999	0.5011	0.5024	0.5037	0.5049	0.5062	0.5075	0.5088
0.47	0.5101	0.5114	0.5126	0.5139	0.5152	0.5165	0.5178	0.5191	0.5204	0.5217
0.48	0.5230	0.5243	0.5256	0.5269	0.5282	0.5295	0.5308	0.5321	0.5334	0.5347
0.49	0.5361	0.5374	0.5387	0.5400	0.5413	0.5427	0.5440	0.5453	0.5466	0.5480
0.50	0.5493	0.5506	0.5520	0.5533	0.5547	0.5560	0.5573	0.5587	0.5600	0.5614
0.51	0.5627	0.5641	0.5654	0.5668	0.5682	0.5695	0.5709	0.5722	0.5736	0.5750
0.52	0.5763	0.5777	0.5791	0.5805	0.5818	0.5832	0.5846	0.5860	0.5874	0.5888
0.53	0.5901	0.5915	0.5929	0.5943	0.5957	0.5971	0.5985	0.5999	0.6013	0.6027
0.54	0.6042	0.6056	0.6070	0.6084	0.6098	0.6112	0.6127	0.6141	0.6155	0.6169
0.55	0.6184	0.6198	0.6213	0.6227	0.6241	0.6256	0.6270	0.6285	0.6299	0.6314
0.56	0.6328	0.6343	0.6358	0.6372	0.6387	0.6401	0.6416	0.6431	0.6446	0.6460
0.57	0.6475	0.6490	0.6505	0.6520	0.6535	0.6550	0.6565	0.6580	0.6595	0.6610
0.58	0.6625	0.6640	0.6655	0.6670	0.6685	0.6700	0.6716	0.6731	0.6746	0.6761
0.59	0.6777	0.6792	0.6807	0.6823	0.6838	0.6854	0.6869	0.6885	0.6900	0.6916
0.60	0.6931	0.6947	0.6963	0.6978	0.6994	0.7010	0.7026	0.7042	0.7057	0.7073
0.61	0.7089	0.7105	0.7121	0.7137	0.7153	0.7169	0.7185	0.7201	0.7218	0.7234
0.62	0.7250	0.7266	0.7283	0.7299	0.7315	0.7332	0.7348	0.7365	0.7381	0.7398
0.63	0.7414	0.7431	0.7447	0.7464	0.7481	0.7498	0.7514	0.7531	0.7548	0.7565
0.64	0.7582	0.7599	0.7616	0.7633	0.7650	0.7667	0.7684	0.7701	0.7718	0.7736
0.65	0.7753	0.7770	0.7788	0.7805	0.7823	0.7840	0.7858	0.7875	0.7893	0.7910
0.66	0.7928	0.7946	0.7964	0.7981	0.7999	0.8017	0.8035	0.8053	0.8071	0.8089

TABLE 21.14
(continued)

0.67 to 0.99

R	0	1	2	3	4	5	6	7	8	9
0.67	0.8107	0.8126	0.8144	0.8162	0.8180	0.8199	0.8217	0.8236	0.8254	0.8273
0.68	0.8291	0.8310	0.8328	0.8347	0.8366	0.8385	0.8404	0.8423	0.8441	0.8460
0.69	0.8480	0.8499	0.8518	0.8537	0.8556	0.8576	0.8595	0.8614	0.8634	0.8653
0.70	0.8673	0.8693	0.8712	0.8732	0.8752	0.8772	0.8792	0.8812	0.8832	0.8852
0.71	0.8872	0.8892	0.8912	0.8933	0.8953	0.8973	0.8994	0.9014	0.9035	0.9056
0.72	0.9076	0.9097	0.9118	0.9139	0.9160	0.9181	0.9202	0.9223	0.9245	0.9266
0.73	0.9287	0.9309	0.9330	0.9352	0.9373	0.9395	0.9417	0.9439	0.9461	0.9483
0.74	0.9505	0.9527	0.9549	0.9571	0.9594	0.9616	0.9639	0.9661	0.9684	0.9707
0.75	0.9730	0.9752	0.9775	0.9798	0.9822	0.9845	0.9868	0.9892	0.9915	0.9939
0.76	0.9962	0.9986	1.0010	1.0034	1.0058	1.0082	1.0106	1.0130	1.0154	1.0179
0.77	1.0203	1.0228	1.0253	1.0277	1.0302	1.0327	1.0352	1.0378	1.0403	1.0428
0.78	1.0454	1.0479	1.0505	1.0531	1.0557	1.0583	1.0609	1.0635	1.0661	1.0688
0.79	1.0714	1.0741	1.0768	1.0795	1.0822	1.0849	1.0876	1.0903	1.0931	1.0958
0.80	1.0986	1.1014	1.1042	1.1070	1.1098	1.1127	1.1155	1.1184	1.1212	1.1241
0.81	1.1270	1.1299	1.1329	1.1358	1.1388	1.1417	1.1447	1.1477	1.1507	1.1538
0.82	1.1568	1.1599	1.1630	1.1660	1.1692	1.1723	1.1754	1.1786	1.1817	1.1849
0.83	1.1881	1.1914	1.1946	1.1979	1.2011	1.2044	1.2077	1.2111	1.2144	1.2178
0.84	1.2212	1.2246	1.2280	1.2315	1.2349	1.2384	1.2419	1.2454	1.2490	1.2526
0.85	1.2562	1.2598	1.2634	1.2671	1.2707	1.2745	1.2782	1.2819	1.2857	1.2895
0.86	1.2933	1.2972	1.3011	1.3050	1.3089	1.3129	1.3169	1.3209	1.3249	1.3290
0.87	1.3331	1.3372	1.3414	1.3456	1.3498	1.3540	1.3583	1.3626	1.3670	1.3714
0.88	1.3758	1.3802	1.3847	1.3892	1.3938	1.3984	1.4030	1.4077	1.4124	1.4171
0.89	1.4219	1.4268	1.4316	1.4365	1.4415	1.4465	1.4516	1.4566	1.4618	1.4670
0.90	1.4722	1.4775	1.4828	1.4882	1.4937	1.4992	1.5047	1.5103	1.5160	1.5217
0.91	1.5275	1.5334	1.5393	1.5453	1.5513	1.5574	1.5636	1.5698	1.5762	1.5826
0.92	1.5890	1.5956	1.6022	1.6089	1.6157	1.6226	1.6296	1.6366	1.6438	1.6510
0.93	1.6584	1.6658	1.6734	1.6811	1.6888	1.6967	1.7047	1.7129	1.7211	1.7295
0.94	1.7380	1.7467	1.7555	1.7645	1.7736	1.7828	1.7923	1.8019	1.8117	1.8216
0.95	1.8318	1.8421	1.8527	1.8635	1.8745	1.8857	1.8972	1.9090	1.9210	1.9333
0.96	1.9459	1.9588	1.9721	1.9857	1.9996	2.0139	2.0287	2.0439	2.0595	2.0756
0.97	2.0923	2.1095	2.1273	2.1457	2.1649	2.1847	2.2054	2.2269	2.2494	2.2729
0.98	2.2976	2.3235	2.3507	2.3796	2.4101	2.4427	2.4774	2.5147	2.5550	2.5987
0.99	2.6466	2.6996	2.7587	2.8257	2.9031	2.9945	3.1063	3.2504	3.4534	3.8002

This table relating r to z is a table of inverse hypertangent and was supplied through the courtesy of C. M. Smith and the Computer Laboratory at The Pennsylvania State University.

22 Some principles of experimental inquiry

. . . although indisputably one purpose of experimentation is to reach decisions, another purpose is to gain knowledge about the state of nature (that is, about the parameters) without having specific actions in mind.

D. V. Lindley, 1956

22.1 Introduction

Any investigation begins with an objective, and it is important that the objective be made as explicit as possible. It is also clear that the objective eventually becomes formulated as a hypothesis to be tested, and this in turn encompasses a philosophy of method, a methodology. The scientific method, advocated in the first chapter, represents a consistent approach to this aspect of experimental investigation. Subsequently the steps include the design of an experiment to test the hypothesis; a sampling plan is proposed to accompany the design, and, for the detrital sediments, a sampling model is described which forms the basis of the sampling arrangement (Chap. 2). Random samples are required to apply probability theory to the results of the experiment.

The procedures for data collection are next discussed in terms of an axiomatic treatment of the analysis of a population of elements, wherein a set of properties is defined as fundamental on the basis of the necessary and sufficient conditions to achieve a unique description of the population. The fundamental properties selected are five: proportions of elements of different

Quotation: D. V. Lindley, On a Measure of the Information Provided by an Experiment, *Ann. Math. Statistics,* **27:** 987 (1956).

kinds, their sizes, shapes, orientation, and packing. All other properties may be considered as derived, i.e., dependent upon, the particular combination of the values of the five fundamental properties (Chap. 3).

With this axiomatic treatment as a basis, the selection of appropriate techniques with which to generate the data for descriptions of rocks (populations) is simplified; simple procedures comprising direct measurement performed on sets of specific elements of one kind are preferable, and the data are ultimately expressed as summary statistics representing common frequency distributions (Chaps. 4 to 11). The statistics are sample estimators of population parameters, and by appropriate sampling design leading to random samples, the statistics fulfill the requirements of unbiased, consistent, efficient, and often sufficient estimators. In many cases the information or scale of the resulting measurements achieves a ratio level or at least an interval level, so that parametric statistical analysis can be used for subsequent data reduction (Chap. 12).

The properties of the frequency distributions are described along with tests for departures from these ideal models, and the departures are assigned largely to nonrandomness in the sampling arrangement (Chaps. 13 to 15). Generally when random sampling is achieved, by suitable sampling design, the observed frequencies fit the theoretical distributions, and since these distributions all fall within the central-limit law, they tend asymptotically, i.e., stochastically, to normal distributions. Thus the subsequent statistical testing may be applied with confidence to the measured data and examples of comparisons among pairs, as in Student's t testing (Chap. 16), and among more complex comparisons by analysis of variance are described (Chaps. 18 to 20). Finally, measurement of association between two variates is described, again using the parametric form of analysis (Chap. 21).

Extension to multivariate analysis is mentioned, and as long as the requirements, earlier prescribed, are fulfilled, the collected data are amenable to the most stringent testing by these more sophisticated techniques. However, it is most necessary that the fulfillment of these minimal requirements be achieved, or more comprehensive analysis will not yield the required information for answering the more complex and interesting questions. It is most unfortunate that many of the challenging problems in the earth sciences appear complex and almost certainly require multivariate analysis for their resolution; this may well be because no suitable models have been postulated which are comprehensive enough to lead to adequate objective formulation of the usual geological questions in sufficiently simple terms, so that geological problem solving is still largely in the heuristic stage.

In fact the procedures advocated in the preceding chapters are largely of use only when the problem of interest has been adequately formulated so that the question can be reduced to one of the simple forms in which univariate, or at most bivariate, analysis is adequate to provide an answer. Nevertheless, if the data are not collected according to suitable design and by suitable procedures, which produce observations of known information content, no amount of complexity in the statistical analysis will suffice to provide a consistent and useful answer. The rather low level of predictability based on answers to the usual earth-science questions is evidenced by the low success ratio in exploration for natural resources, and until more adequate formulation of geological problems is achieved, very little improvement can be expected in the search for new natural resources (Griffiths, 1962c, 1963a, b).

22.2 The Basis of Inference or Decision Making in Testing Hypotheses

Throughout the book the selection of a probability level for accepting or rejecting a hypothesis upon the basis of a statistical test has been intuitive or conventional, e.g., the 5 percent level ($P < 0.05$) is often chosen. Such an approach is quite frequently challenged, and the usual recourse is, where possible, empirical demonstration of the effectiveness of the procedure. Actually an extensive body of literature exists in which the procedures for selection are discussed in detail, and a consistent basis of choice is developed using rigorous logical analysis. The most recent and exhaustive treatment in the context of the geosciences is given by Miller and Kahn (1962, chaps. 4–7), and equivalent discussions occur in many recent statistical textbooks e.g., Bennett and Franklin, 1954, pp. 141ff.; Brownlee, 1960, pp. 74ff.; Dixon and Massey, 1951, pp. 89–99).

In general, given a specific null hypothesis, analysis of an appropriate set of data leads to its acceptance or rejection; two different kinds of error may occur:

1 The null hypothesis may be rejected when it is true.
2 The null hypothesis may be accepted when it is false.

The former is called an α error, or error of the first kind, and the latter is known as a β error, or error of the second kind. It is, of course, desirable

TABLE 22.1
Sources of False Conclusions in Hypothesis Testing

	True state of nature	
Decision based on experiment	H_0 true H_1 false	H_0 false H_1 true
H_0 true H_1 false	Correct	False (β error)
H_0 false H_1 true	False (α error)	Correct

to avoid these errors as far as possible, by suitable adjustment of the statistical test, but achievement of these requirements is not by any means easy.

First, consider a "simple" test based on the comparison of two means $\bar{X}_1$ and $\bar{X}_2$, in which the null hypothesis H_0 states that these are two means of random samples from the same normal population; the appropriate test is Student's t test, and the basis of comparison of the difference among the means is an estimate of the variance provided by the two sets of data (see Sec. 16.4). If the hypothesis is true, the expected value of the mean difference in an extended series of trials is zero, and the mean of the population of differences is therefore $\mu = 0$. Alternatives to equality are that $\mu_1 > \mu_2$ or $\mu_2 > \mu_1$, and the two samples are drawn from populations with different means. To maintain the simplicity of the test it is assumed that either the two estimates of variance are estimates of the same variance, that is, $\hat{\sigma}_1^2 = \hat{\sigma}_2^2 = \sigma^2$, or that if these samples represent two populations, they possess homogeneous variances. It appears quite evident that even this "simple" test is difficult to refine, but for the resolution of a question it is necessary to decide what the appropriate question is, i.e., the objective must be specified exactly. Even when the exact question is defined, it is still possible to err in accepting or rejecting the hypothesis, and the less well-defined the question, the more ambiguous the answer. In the case where the exact question is formulated at the beginning of the investigation the possible outcomes are shown in Table 22.1.

In other words, given a hypothesis H_0 and an alternative H_1 it is desirable to accept H_0 when H_0 is true and H_1 false or to reject H_0 when it is false and H_1 is true. It is possible to commit two kinds of error; to reject H_0 when it is true (α error) or to accept H_0 when H_1 is true (β error). Suppose H_0 is clearly defined and H_1 represents a number of possible alternatives; then even the correct rejection of H_0 when false leaves the experimenter very little better off than before the experiment.

This discussion serves to emphasize the necessity for a very clear statement of the problem before performing the experiment and should encourage the investigator to list the possible alternatives and examine all possible outcomes (as recommended by Chamberlin, 1897, and Fisher, 1942) before selecting the most suitable test. Student's t test is very specifically designed and requires an explicit alternative, whereas the chi-square test, for example, requires a well-defined hypothesis but alternative hypotheses are generally left undefined.

Consider now the problem of differentiating a beach sand from its associated dune sand; the hypothesis H_0 may be well defined in the sense that the variance (sorting) of long a axes of the quartz grains in the beach sand (σ_1^2) is expected to be larger (sorting poorer) than that (σ_2^2) of the a axes of quartz grains in the dune sand, that is, H_1: $\sigma_1^2 > \sigma_2^2$.[1] The test, a

[1] The null hypothesis H_0 is generally phrased "these two samples are randomly drawn from a population with common variance ($\sigma_1^2 = \sigma_2^2 = \sigma^2$)."

variance ratio or F test (see Chap. 17.9), is performed with the expected larger variance in the numerator, that is, $F_{\alpha,\nu_1,\nu_2} = \sigma_1^2/\sigma_2^2$. Alternative hypotheses are that the variances are either equal (H_0) or the dune is more poorly sorted than the beach, that is, H_2: $\sigma_1^2 < \sigma_2^2$. Given adequate representative sampling, all possible outcomes may be listed, and the results of the entire experiment may be made explicit prior to the experimental analysis. Ideally H_1 is accepted, and the experimenter will be satisfied. In actual examples (Hulbe, 1957; Biederman, 1959) the results were much less simple and required a multivariate approach, and a discriminant test (the multivariate analog of the univariate t test) proved satisfactory. However, in practice, when investigating geological problems of this kind, the alternative hypotheses are much less obvious. For example, let us suppose that dune sands can be differentiated from their associated beach sands quite generally by this approach; what will happen if the sand sampled is other than a beach or a dune sand? Quite frequently geological investigations are of this kind, and while the hypothesis H_1 is simple enough, the alternatives are wide open. If H_1 is rejected, no satisfactorily definite conclusions can then be drawn. It is necessary, therefore, before attempting to consider the α and β errors, to be quite explicit about the question the experimenter has in mind. When this has been achieved, it is then possible to consider levels of probability much more exactly in certain specific cases.

Given explicit definition of the hypothesis H_0 and the alternative hypothesis H_1, we wish to ensure that H_0 is accepted when true and H_1 is accepted when true: in other words, we should like to ensure a high probability of being correct or, alternatively, a low probability of committing both α and β errors. In terms of Table 22.2 we wish α and β to be small and $1 - \alpha$ and $1 - \beta$ to be large. In a comparison of variances by means of an F test, as in the example above described, for given values of the variances σ_1^2, σ_2^2, we can select a small α and β, and then the degrees of freedom (or the sample size) will be determined by the α and β levels chosen. For most of the tests in this book α and the sample size (or degrees of freedom) were selected, and this automatically fixes the level of β (or the power of the test, as $1 - \beta$ is called).

It is at present extremely rare to find a geological question sufficiently well formulated so that the more rigorous treatment of the α and β errors and the power of the statistical test of hypothesis are adequately prescribed to lead to precise levels of probability; and so, in general, the probability level is selected in an apparently arbitrary fashion although in fact this automatically leads to implicit values for α and β and the sample size. As experience with hypothesis testing grows in the geosciences, the tendency toward more rigorous formulation of the hypothesis will lead to much more explicit consideration of these important characteristics of experimentation. For those interested in pursuing these ideas, recourse to the articles by Neyman and Pearson (e.g., 1928, 1936, 1938) and to discussions of decision theory is

TABLE 22.2
Diagrammatic Representation of α and β Errors

Experimental outcome		True state of nature: H_0 true, H_1 false	True state of nature: H_0 false, H_1 true	
H_0 true, H_1 false	(Large)	$1-\alpha$	β	Error (small)
H_0 false, H_1 true	Error (small)	α	$1-\beta$	Power (large)

recommended (Chernoff and Moses, 1959). An example of hypothesis formulation and testing by sequential analysis using decision criteria is illustrated by McIntyre (1961).

22.3 Parametric Versus Nonparametric Tests

A number of examples of the use of statistical tests which are categorized rather loosely as *nonparametric* tests are mentioned throughout the preceding text, e.g., the use of order statistics (Sec. 5.3), the sign test (Sec. 16.6), and the phi medial correlation coefficient (Sec. 15.4). These so-called "quick and dirty" procedures are attractive because they are simple to perform, and for preliminary rapid testing they certainly have their place; an excellent presentation of the available tests and examples of their application are given by Siegel (1956). The fundamentals of nonparametric statistical analysis are described by Fraser (1957), and some applications to multivariate analysis are given by Roy (1957), while applications to geoscience problems are illustrated in Miller and Kahn (1962, e.g., p. 182, chap. 14, and appendix G, pp. 464–470) and by McCammon (1962a).

In general, when the scale of the experimental observations fails to achieve interval level, it is necessary to use less efficient test procedures; for the greater part the use of parametric analysis is recommended where it is justified.

Nonparametric statistical tests are used with data when the scale level is ordinal or nominal, and the use of these tests with data on interval or ratio scales certainly implies a loss of information; in the text, extended criticism of size determination by indirect measurement procedures is based, at least in part, on the fact that nonparametric or order statistics are the only recourse with such data. Similarly, percentage data generally require less rigorous test procedures than parametric analysis; thus Garner and McGill (1956), in comparing the relationships between uncertainty and variance analysis, emphasize that "the uncertainty measure has greater generality" whereas use of the "variance measure is more specific and retains information about the metric." They suggest that "the decision as to which to use depends not only upon the properties of the criterion variable but also upon the gain expected from being more sensitive instead of more general." (Garner and McGill, 1956, p. 228). There is in fact no easy way to compare parametric with nonparametric analysis; in most cases the "efficiency" of the nonparametric procedure is compared with the equivalent parametric analysis, using the normal distribution as a common base. It is often demonstrated that the loss in efficiency is small, and therefore the gain in simplicity from using the equivalent nonparametric procedure leads to its selection. Actually, if one can assume a normal distribution as the basic frequency function of the variate, there appears no obvious reason for resorting to nonparametric analysis, and this may be particularly detrimental if the data are to be subsequently submitted to multivariate analysis. When the frequency distribution of the variate is not known, the estimate of relative efficiency of nonparametric against the normal model does not necessarily apply, and where nonparametric procedure is adopted, the level of information achieved is quite often unknown and impossible to specify.

In general, then, the use of parametric and nonparametric procedures cannot be usefully compared; in many cases, there is little choice. If the variate achieves interval or ratio level, information on the metric should be preserved in the subsequent analysis. On the other hand, if the variate fails to achieve interval level, nonparametric testing is mandatory, and the results must be lived with unless, or until, some improvement in the procedure for data gathering can be achieved.

There seems little question that in the analysis of geological problems, because of varying degrees of interdependence among the properties measured, much of the information gathered is redundant, and we are now at the stage where the matrix of observations should be tested for this redundance (Griffiths, 1962b, 1963b). We can then eliminate some of the data-gathering procedures and either add new properties or analyze more specimens to achieve better prediction and control.

22.4 The Nature of Geoscience Problems

Certain characteristics are common to all experimental analysis, question answering, and problem solving; other characteristics differ or at least appear to. Recognition of these similarities and differences is important in achieving "better" results.

Any approach requires an explicit statement of the objective or formulation of the problem. At first sight this step appears to be simpler to accomplish in some cases than others, and it seems evident that where the formulation is complete, the solutions are more reliable and meaningful than when the formulation is vague. It is necessary therefore to achieve this formal statement of the problem and to recognize that failure to achieve it may defeat the entire purpose of the experiment.

In this respect it is interesting to note the similarity in statement of the nature of problems by Chamberlin (1897) and by Fisher (1935, see reference 1942); the latter may make statements in more formal terms, but the requirements are materially the same: list all possible outcomes (the multiplicity of hypotheses), devise a test procedure which will eliminate all but one of the alternatives, or accept the one it is impossible to reject. The fact that Chamberlin recognized the negative nature of this approach, i.e., the possibility of rejection but acceptance only by default, is quite striking, and it is regrettable that the prolific development of subsequent experimental design flowed not from work in the geosciences but from a combination of mathematics and agriculture (Fisher, 1942). Fisher also quotes Lyell's research in establishing Tertiary stratigraphy and taxonomy as an outstanding early example of the use of statistical analysis (Fisher, 1953), and again it has taken a very long time for this pioneering effort to be more widely adopted in paleontology although in biometry it is well established as common practice.

It would be rather strange if a procedure which has proved overwhelmingly successful in such varied fields as physics, chemistry (particularly industrial practice), biology, psychology, sociology, agriculture, medicine, and industry should be inapplicable to, say, the earth sciences, or for that matter to the space sciences;[1] it becomes even stranger when it turns out that this same methodological procedure is effective in some of the earth sciences, such as meteorology, and even some aspects of geophysics.

Certain features which are common to research in the natural sciences do not appear in many phases of the precise sciences, e.g., physics and chemistry, just as there are differences between controlled laboratory experiments and investigation of real-world situations. Essentially research in the natural sciences has required the study of variation in populations or aggregates, and here, by its very nature, the statistical method is extremely effective. Controlled experiments are difficult to perform in such cases, and so continued observation of events becomes necessary to extract the common factors and, by replication, to estimate the error or magnitude of uncontrolled (unassigned) variation; only when variation arising from the effects of the factors exceeds the variation from the effects of the uncontrolled, or error, sources of variation can assignable causes be adduced. Since it was precisely in such environments that the design of experiments (Fisher, 1942) developed, it would seem obvious that these procedures are quite generally applicable to the natural sciences including the geosciences. Just as the areas of biometrics, econometrics,

[1] This plea for the adoption of the philosophy of statistics, especially the inclusion of uncertainty in the basic model of the universe, is very eloquently presented by Neyman and Scott (1958).

psychometrics, and sociometrics evolved to adapt the statistical method to biology, economics, psychology, and sociology, respectively, so the area of *geometrics* could be used to embrace the quantification and application of statistical method to the field of the geosciences (Griffiths, 1958a); this book may be looked upon as an introduction to the area of geometrics. In this context it may be worthwhile to emphasize that statistical analysis and, indeed, the scientific method, of which it forms an important part, are in the process of evolution and are not by any means static, final, or complete.

22.5 Possible Future Developments of Importance in the Geosciences

Whereas, initially, hypothesis testing as mentioned above concerned a purely logical and mathematical approach, largely developed by Neyman and Pearson (1928, 1936, 1938), more recent work by Wald (1950), Savage (1951), and others has led to the decision-making aspects of hypothesis testing, and even newer facets are envisaged in value theory, wherein each outcome is scored and decisions are made by optimizing expected outcomes on the bases of ordered value judgments (Churchman, 1961; Ackoff, Gupta, and Minas, 1962). Thus the scientific method evolves and adapts to increasing knowledge about outcome. There seems to be an increasing awareness of the human factor in experimental research, and attempts are made to allow for this factor in the more recent work; an early example in the geoscience area is described by Griffiths and Rosenfeld (1954), and in research projects using a multiplicity of observers it is essential to design the experiment to "take care of" the contribution due to variation among and within operators.

As research becomes more complex and interdisciplinary, the program includes contributions from many experts, and problems of organization of, and communication among, the various facets of the research become more and more important. Indeed, in industrial research, this has become a well-recognized problem and permeates the whole operation of the firm. The completion of a piece of research is not achieved until the results are at least communicated, and often, particularly in "useful" research, the results must be implemented before the research can be considered complete. These features introduce an entirely different view of research programs and of the part played by the experimenter, and with decreasing lead times between discoveries and their applications, these aspects are likely to become more and more important (Churchman and Ratoosh, 1960). This will become increasingly true of geoscience research in the future and will require a reorientation of research effort. In fact, it is becoming increasingly clear that the "scientist" can no longer afford to subdivide his living habits into those exercised within the laboratory and those exercised in the "outside" world, because the results of his laboratory performance directly and immediately affect what he does—or what is done to him—in the outside world. Just as decision theory is pervading hypothesis testing, an accompanying value scale is adopted by, or forced upon, the researcher to enable him to reach decisions. Here it would be most dangerous to permit the adoption of a value scale to remain implicit; hence, in geoscience, as elsewhere, the future of research will require an entirely new look if the researcher is to adapt to the dynamic new environment which continually interacts with, and impinges more intimately upon, his research. Indeed, one may now claim with some candor that the "scientific method," when taken in its broadest sense, becomes a (formally defined) way of life.

Concomitant with these changes, new areas of a research activity are opened up to the geosciences, e.g., the use of operations research, management science, systems analysis; and, with the growing use of computers, these relatively new weapons of the research arsenal are almost forced upon the experimenter. For example, one way of looking at a research program in the natural sciences is as "a game against nature," and a very difficult game it turns out to be; game theory, in which the opponent may be looked upon as responding in a human fashion, presents a number of possible solutions to problems in which conflict of interest plays a dominant role (e.g., Luce and Raiffa, 1957, von Neumann and Morgenstern, 1953). However, nature does not necessarily respond in the expected human fashion but often appears to act as if the experimenter did not exist; there appear to be many difficulties and few advantages in this kind of response, and our approach to scientific problems in game-theoretic terms is woefully naïve at present.

It seems then that scientific methodology is as dynamic in its evolution as any other scientific or human activity. There are a plethora of opportunities awaiting the experimenter in geoscience, all of which need serious attention, and some of which must surely help to advance us toward our ultimate goals. The least that such approaches can achieve is the clarification of our goals, and this is certainly one of the important aspects, if not the most important, of our life and work.

References

ACKOFF, R. L. (ed.), 1961, The Meaning, Scope and Methods of Operations Research, chap. 1, in "Progress in Operations Research," ORSA publ. 5, John Wiley & Sons, Inc., New York, 505 pp.

———, S. K. GUPTA, and L. S. MINAS, 1962, "Scientific Method: Optimizing Applied Research Decisions," John Wiley & Sons, Inc., New York, 464 pp.

ACTON, F. S., 1959, "Analysis of Straight Line Data," John Wiley & Sons, Inc., New York, 267 pp.

AHRENS, L., 1954a, b, 1957, The Lognormal Distribution of the Elements, I, II, III, *Geochim. Cosmochim. Acta,* **5:** 49–73, **6:** 121–131, **11:** 205–212.

———, 1954c, "Quantitative Spectrochemical Analysis of Silicates," Pergamon Press, New York, 122 pp.

AITCHISON, J., and J. A. C. BROWN, 1957, "The Lognormal Distribution," Cambridge University Press, New York, 176 pp.

AITKEN, A. C., 1957, "Statistical Mathematics," 8th ed., Oliver & Boyd Ltd., Edinburgh and London, 153 pp.

ALLEN, P., 1945a, Lognormal Distributions, *Nature,* **156:** 746.

———, 1945b, Sedimentary Variation: Some New Facts and Theories, *J. Sediment Petrol.,* **15:** 75–83.

———, 1947, Correlation between Allogenic Grade Size and Allogenic Frequency in Sediments, *J. Sediment. Petrol.,* **17:** 3–7.

———, 1948, Wealden Petrology: The Top Ashdown Pebble Bed and the Top Ashdown Sandstone, *Quart. J. Geol. Soc. London,* **104:** 257–321.

AMES, D. P., R. R. IRANI, and C. F. CALLIS, 1959, The Agreement between Independent Methods for Particle Size Distribution Measurements on Finely Divided Powders Including Phosphates, *J. Phys. Chem.,* **63:** 531–534.

ANDERSON, R. L., and T. A. BANCROFT, 1952, "Statistical Theory in Research," McGraw-Hill Book Co., Inc., New York, 399 pp.

APFEL, E. T., 1938, Phase Sampling of Sediments, *J. Sediment. Petrol.,* **8:** 67–68.

ARKIN, H., and R. R. COLTON, 1950, "Tables for Statisticians," Barnes & Noble, Inc., New York, 152 pp.

ARTINI, E., 1898, Intorno alla composizione mineralogica della sabbie di alcune fiumi del Veneto, con applicazione terreni di trasporte, *Riv. Mineral. Crist. Ital. Padova,* **19:** 33–94.

ATHY, L. F., 1930, Density, Porosity, and Compaction of Sedimentary Rocks, *Bull. Am. Assoc. Petrol. Geologists,* **14:** 1–35.

BAGNOLD, R. A., 1941, "Physics of Blown Sand and Desert Dunes," Methuen & Co., Ltd., London, 265 pp.

BAKER, H. A., 1920, On the Investigation of the Mechanical Constitution of Loose Arenaceous Sediments by the Method of Elutriation . . . , *Geol. Mag.,* **57:** 366–370.

BARTLETT, M. S., 1937a, Properties of Sufficiency and Statistical Tests, *Proc. Roy. Soc. Ser. A,* **160:** 268–282.

———, 1937b, *J. Roy. Statist. Soc. Suppl.,* **4:** 137.

———, 1947, The Use of Transformations, *Biometrics,* **3:** 39–52.

BECHOFER, R. E., 1954, A Single-sample Multiple Decision Procedure for Ranking Means of Normal Populations with Known Variances, *Ann. Math. Statistics,* **25:** 16–39.

———, C. W. DUNNETT, and M. SOBEL, 1954, A Two-sample Multiple Decision Procedure for Ranking Means of Normal Populations with a Common Unknown Variance, *Biometrika,* **41:** 170–176.

BECKER, R. M., and S. W. HAZEN, JR., 1961, Particle Statistics of Infinite Populations as Applied to Mine Sampling, *U.S. Bur. Mines Rept. Invest.* 5669, 79 pp.

BEERS, Y., 1957, "Introduction to the Theory of Error," 2d ed., Addison-Wesley Publishing Company, Inc., Reading, Mass., 66 pp.

BENNETT, C. A., and N. L. FRANKLIN, 1954, "Statistical Analysis in Chemistry and the Chemical Industry," John Wiley & Sons, Inc., New York, 724 pp.

BIEDERMAN, E. W., JR., 1959, Shoreline Sedimentation in New Jersey, unpublished Ph.D. thesis, Dept. Mineralogy, Pennsylvania State University, University Park, Pa., 270 pp.

BLISS, C. I., and R. A. FISHER, 1953, Fitting the Negative Binomial Distribution to Biological Data and Note on Efficient Fitting of the Negative Binomial, *Biometrics*, **9:** 176–200.

BOKMAN, J., 1952, Clastic Quartz Particles as Indices of Provenance, *J. Sediment. Petrol.*, **22:** 17–24.

———, 1953, Lithology and Petrology of the Stanley and Jackfork Formations, *J. Geol.*, **61:** 152–170.

———, 1957, Comparison of Two and Three Dimensional Sphericity of Sand Grains, *Bull. Geol. Soc. Am.*, **68:** 1689–1692.

BOSWELL, P. G. H., 1916a, The Stratigraphy and Petrology of the Lower Eocene Deposits of the North-eastern Part of the London Basin, *Quart. J., Geol. Soc. London*, **71:** 536–588.

———, 1916b, "A Memoir on British Resources of Sands Suitable for Glass-making," Longmans, Green & Co., Ltd., London, 92 pp.

———, 1927, The Rarer Detrital Minerals of British Sedimentary Rocks, *Trans. Geol. Soc. Glasgow*, **18:** 129–148.

———, 1933, "On the Mineralogy of Sedimentary Rocks, Thomas Murby Co., London, 393 pp.

BRAMMALL, A., 1928, Dartmoor Detritals: A Study in Provenance, *Proc. Geologists' Assoc. Engl.*, **39:** 27–48.

——— and J. G. C. LEECH, 1943, The Petrology of Coal Measures Shales and Sandstones, part E, in *Brit. Med. Res. Council, Spec. Rept.* 244, H. M. Stationery Office, London, 222 pp.

BRESSLER, C. T., 1951, The Petrology of the Roslyn Arkose, Central Washington, unpublished Ph.D. thesis, Dept. Mineralogy, Pennsylvania State University, University Park, Pa., 147 pp.

BRIDGMAN, P. W., 1946, "The Logic of Modern Physics," The Macmillan Company, New York, 228 pp.

———, 1950, "Reflections of a Physicist," Philosophical Library, Inc., New York, 392 pp.

BROTHERHOOD, G. R., and J. C. GRIFFITHS, 1947, Mathematical Derivation of the Unique Frequency Curve, *J. Sediment. Petrol.*, **17:** 77–82.

BROWNLEE, K. A., 1948, "Industrial Experimentation," H. M. Stationery Office, London, 169 pp.

———, 1960, "Statistical Theory and Methodology in Science and Engineering," John Wiley & Sons, Inc., New York, 570 pp.

BRUCE, W. A., and H. J. WELGE, 1947, The Restored State Method for Determination of Oil in Place and Connate Water, American Petroleum Institute, New York, in "Drilling and Production Practice," pp. 166ff.

BUSH, V., 1945, As We May Think, *Atlantic Monthly*, **176:** 101–108.

BUTTERFIELD, J. A., 1936, Outgrowths on Zircon, *Geol. Mag.*, **73:** 511.

CAILLEUX, A., 1938, La Disposition individuelle des galets dans les formations détritiques," *Rev. Geograph. Phys. Geol. Dynamique*, **11:** 171–198.

CAMPBELL, N. R., 1952, "What Is Science," Dover Publications, Inc., New York, 186 pp.

———, 1957, "Foundations of Science," Dover Publications, Inc., New York, 565 pp.

CASEY, R. S., C. F. BAILEY, and G. D. COX, 1946, Punched Card Techniques and Applications, *J. Chem. Educ.*, **23:** 495–499.

———, J. W. PERRY, M. M. BERRY, and A. KENT, 1958, "Punched Cards: Their Application to Science and Industry," 2d ed., Reinhold Publishing Corporation, New York, 506 pp.

CAYEUX, L., 1897, Contribution à l'étude micrographie des terrains sédimentaires, *Mém. Soc. Geol. Nord*, **4:** 1–589.

CHAMBERLIN, T. C., 1897, The Method of Multiple Working Hypotheses, *J. Geol.*, **5:** 837–848, reprinted in *J. Geol.*, **39:** 155–165 (1931).
CHAYES, F., 1949a, A Simple Point-counter for Thin Section Analysis, *Am. Mineralogist*, **34:** 1–11.
———, 1949b, On Correlation in Petrography, *J. Geol.*, **57:** 239–254.
———, 1950a, Measurement of Intercept Distances in Thin Section, *Trans. Am. Geophys. Union*, **31:** 870–872.
———, 1950b, On the Bias of Grain-size Measurements in Thin Section, *J. Geol.*, **58:** 156–160.
——— and H. W. FAIRBAIRN, 1951, A Test of Precision of Thin Section Analysis by Point-counter, *Am. Mineralogist*, **36:** 704–712.
———, 1952, Notes on the Staining of Potash Feldspar with Sodium Cobaltinitrite in Thin Section, *Am. Mineralogist*, **37:** 337–340.
———, 1954a, Effect of Change of Origin on Mean and Variance of Two Dimensional Fabrics, *Am. J. Sci.*, **252:** 567–570.
———, 1954b, The Lognormal Distribution of the Elements: A Discussion, *Geochim. Cosmochim. Acta*, **6:** 119–120.
———, 1956a, "Petrographic Modal Analysis," John Wiley & Sons, Inc., New York, 113 pp.
———, 1956b, The Holmes' Effect and the Lower Limit of Modal Analysis, *Mineral. Mag.*, **31:** 276–281.
———, 1960, On Correlation between Variables of Constant Sum, *J. Geophys. Res.*, **65:** 4185–4193.
———, 1961, Correlation in Closed Tables, *Carnegie Inst. Wash. Ann. Rept. Director Geophys. Lab.*, 1959–60, 1340, pp. 165–168.
———, 1962, Numerical Correlation and Petrographic Variation, *J. Geol.*, **70:** 440–452.
CHERNOFF, H., and E. L. MOSES, 1959, "Elementary Decision Theory," John Wiley & Sons, Inc., New York, 364 pp.
CHURCHMAN, C. W., 1948, "Theory of Experimental Inference," The Macmillan Company, New York, 292 pp.
——— and P. RATOOSH (eds.), 1959, "Measurement: Definitions and Theories," John Wiley & Sons, Inc., New York, 274 pp.
——— and ———, 1960, Innovation in Group Behavior, in "Proceedings of the Second International Conference on Operational Research," pp. 122–128, English Universities Press, London.
———, 1961, "Prediction and Optimal Decision," Prentice-Hall, Inc., Englewood Cliffs, N.J., 349 pp.
CLOUD, W. F., 1941, Effects of Sand Grain Size Distribution upon Porosity and Permeability, *Oil Weekly*, **103:** 25–32.
COCHRAN, J. A., 1960, The Sampling Problem in Sedimentary Petrography: A Contribution, unpublished M.S. thesis, Dept. Mineralogy, Pennsylvania State University, University Park, Pa., 94 pp.
COCHRAN, W. G., 1947, Some Consequences when the Assumptions for the Analysis of Variance Are Not Satisfied, *Biometrics*, **3:** 22–38.
———, 1953, "Sampling Techniques," John Wiley & Sons, Inc., New York, 330 pp.
———, 1954, Some Methods for Strengthening the Common χ^2 Tests, *Biometrics*, **10:** 417–451.
———, F. MOSTELLER, and J. W. TUKEY, 1954, Appendix G, Principles of Sampling, pp. 309–331, in "Statistical Problems of the Kinsey Report," American Statistical Association, Washington, D.C.
——— and G. M. COX, 1957, "Experimental Designs," John Wiley & Sons, Inc., New York, 611 pp. (1st ed., 1950, 454 pp.).
COOMBS, C. H., A Theory of Psychological Scaling, *Eng., Res. Inst. Bull.*, **34:** 94.
———, 1959, Inconsistency of Preferences as a Measure of Psychological Distance, pp. 221–232 in Churchman and Ratoosh, 1959.

CURRAY, J. R., 1949, An Analysis of the Sphericity and Roundness of Quartz Grains, unpublished M.S. thesis, Dept. Mineralogy, Pennsylvania State University, University Park, Pa., 100 pp.

——— and J. C. GRIFFITHS, 1955, Sphericity and Roundness of Quartz Grains in Sediments, *Bull. Geol. Soc. Am.*, **66:** 1075–1096.

———, 1956, Dimensional Grain Orientation Studies of Recent Coastal Sands, *Bull. Am. Assoc. Petrol. Geologists*, **40:** 2440–2456.

CURTISS, J. H., 1941, On the Distribution of the Quotient of Two Chance Variables, *Ann. Math. Statistics*, **12:** 409–421.

DALLA VALLE, J. M., 1943, "Micromeritics: The Technology of Fine Particles," Pitman Publishing Corporation, New York, 428 pp.

DAMOUR, A., 1856, Nouvelles recherches sur les sables diamantifères, *Bull. Soc. Geol. France*, **13:** 542–554.

DAPPLES, E. C., and J. F. ROMINGER, 1945, Orientation Analysis of Finegrained Clastic Sediments, *J. Geol.*, **53:** 246–261.

DAUBRÉE, A., 1879, "Études synthétiques de géologie expérimentale," Dunod, Paris, 2 vols.

DAVIES, G. M., 1919, Chromite in Beer Stone, *Geol. Mag.*, **56:** 506–507.

DAVIES, H. M., 1946, The Application of Variance Analysis to Some Problems of Petroleum Technology, *J. Inst. Petrol.*, **32:** 465–491.

DAVIES, O. L. (ed.), 1956, "The Design and Analysis of Industrial Experiments," Hafner Publishing Company, Inc., New York, 636 pp.

DAVIS, G. H., 1954, Estimating Porosity of Sedimentary Rocks from Bulk Density, *J. Geol.*, **62:** 102–107.

DELESSE, M. A., 1848, Procédé mécanique pour déterminer la composition des roches, *Ann. Mines*, **13:** 379–388.

DEMING, W. W., 1950, "Some Theory of Sampling," John Wiley & Sons, Inc., New York, 602 pp.

DÉVERIN, L., 1923, Étude lithologique des roches crétacées des Alpes-maritimes et de topographies souterraines, *Bull. Serv. Carte Geol. France*, **26:** 120–150.

DIXON, W. J., and F. J. MASSEY, 1951, "Introduction to Statistical Analysis," McGraw-Hill Book Company, New York, 370 pp. (2d ed., 1957).

DOEGLAS, D. J., 1940, The Importance of Heavy Mineral Analysis for Regional Sedimentary Petrology, *Natl. Res. Council Rept. Comm. Sedimentation* 1939–1940, exhibit G, pp. 102–121.

——— and N. C. B. SMITHUYZEN, 1941, Die Interpretatie van de Resultaten van Korrelgroote-analysen, *Geol. Mijnbouw*, **8:** 273–296.

———, 1946, Interpretation of the Results of Mechanical Analysis, *J. Sediment Petrol.*, **16:** 19–40.

DOLLAR, A. T. J., 1937, An Integrating Micrometer for the Geometrical Analysis of Rocks, *Mineral. Mag.*, **24:** 577–594.

DONNAY, J. D. H., 1938, A Small Set of Guides for the Determination of Non-opaque Minerals, *Am. Mineralogist*, **23:** 91–100.

DRYDEN, A. L., 1931, Accuracy in Percentage Representation of Heavy Mineral Frequencies, *Proc. Natl. Acad. Sci. U.S.*, **17:** 233–238.

———, 1935, A Statistical Method for the Comparison of Heavy Mineral Suites, *Am. J. Sci.*, **29:** 393–408.

DUNNETT, C. W., 1955, A Multiple Comparison Procedure for Comparing Several Treatments with a Control, *J. Am. Statist. Assoc.*, **50:** 1096–1121.

EDELMAN, C. H., 1931, Diagenetische Umwandlungserscheinungen an detritischen Pyroxen und Amphibolen, *Fortschr. Minerol.*, **16:** 323–324.

EDSON, F. C., 1929, Pre-Mississippian Sediments in Central Kansas, *Bull. Am. Assoc. Petrol. Geologists*, **13:** 441–458.

EISENHART, C., 1935a, A Test of Significance of Lithological Variations, *J. Sediment. Petrol.*, **5:** 137–145.

———, 1935b, A Note on "A Statistical Method for the Comparison of Heavy Mineral Suites," *Am. J. Sci.*, **130:** 549–553.

———, 1947, The Assumptions Underlying the Analysis of Variance, *Biometrics*, **3:** 1–21.

———, M. W. HASTAY, and W. A. WALLIS, 1947, "Techniques of Statistical Analysis," McGraw-Hill Book Company, Inc., New York, 473 pp.

ELDERTON, W. P., 1938, "Frequency Curves and Correlation," Harren Press, Washington, D.C. (4th ed., 1953, 272 pp.).

ELLIOTT, R. B., 1952, The "Superposition Error" in Micrometric Analysis of Rocks, *Mineral. Mag.*, **29:** 833–837.

EMERY, J. R., 1954, The Application of a Discriminant Function to a Problem in Petrography, unpublished M.S. thesis, Dept. Mineralogy, Pennsylvania State University, University Park, Pa., 120 pp.

——— and J. C. GRIFFITHS, 1954a, Reconnaissance Investigation into Relationships between Behavior and Petrographic Properties of Some Mississippian Sediments, *Bull. Mineral Ind. Expt. Sta. Pennsylvania State University*, **62:** 67–80.

——— and ———, 1954b, Differentiation of Oil-bearing from Barren Sediments by Quantitative Petrographic Analysis, *Bull. Mineral Ind. Expt. Sta. Penn. State Univ.*, **64:** 63–68, and *Producers Monthly*, **19:** 33–37.

EPSTEIN, B., 1947, The Mathematical Description of Certain Breakage Mechanisms Leading to the Logarithmico-normal Distribution, *J. Franklin Inst.*, **224:** 471.

EVANS, P., R. J. HAYMAN, and M. A. MAJEED, 1933, The Graphical Representation of Heavy Mineral Analysis, *World Petrol. Congr.*, **1:** 251–256.

EZEKIEL, M., 1941, "Methods of Correlation Analysis," John Wiley & Sons, Inc., New York, 531 pp.

FAIRBAIRN, H. W., 1943, Packing in Ionic Minerals, *Bull. Geol. Soc. Am.*, **54:** 1305–1374.

———, 1949, "Structural Petrology of Deformed Rocks," Addison-Wesley Publishing Company, Inc., Reading, Mass., 344 pp.

——— et al., 1951, A Co-operative Investigation of Precision and Accuracy in Chemical, Spectrochemical and Modal Analysis of Silicate Rocks, *U.S. Geol. Surv. Bull.* 980, 71 pp.

———, 1953, Precision and Accuracy of Chemical Analysis in Silicate Rocks, *Geochim. Cosmochim. Acta*, **4:** 143–156.

FATT, I., 1957, Compressibility of a Sphere Pack: Comparison of Theory and Experiment, *J. Appl. Mech.*, **24:** 148–149.

———, 1958, Pore Structure in Sandstones by Compressible Sphere-pack Models, *Bull. Am. Assoc. Petrol. Geologists*, **42:** 1914–1923.

FISHER, R. A., 1922, On the Interpretation of Chi-square from Contingency Tables and the Calculation of *P*, *J. Roy. Statist. Soc.*, **85:** 87–94.

———, 1942, "The Design of Experiments," 3d ed., Oliver & Boyd Ltd., Edinburgh and London, 263 pp. (1st ed., 1935).

———, 1948, "Statistical Methods for Research Workers," 10th ed., Oliver & Boyd Ltd., Edinburgh and London, 354 pp. (1st ed., 1925).

——— and F. YATES, 1948, "Statistical Tables for Biological, Agricultural and Medical Research," 3d ed., Oliver & Boyd Ltd., Edinburgh and London, 112 pp.

———, 1956, "Statistical Methods and Scientific Inference," Oliver & Boyd Ltd., Edinburgh and London, 175 pp.

———, 1953, The Expansion of Statistics, *J. Roy. Statist. Soc.*, **A116**, pt. 1, reprinted in *Am. Scientist*, **42:** 275–282 (1954).

FLEET, W. F., 1926, Petrological Notes on the Old Red Sandstone of the West Midlands, *Geol. Mag.*, **63:** 505–516.

FOLK, R. L., 1951, Stages of Textural Maturity in Sedimentary Rocks," *J. Sediment. Petrol.*, **21:** 127–130.

——— and W. C. WARD, 1957, Brazos River Bar: A Study in the Significance of Grain Size Parameters, *J. Sediment. Petrol.*, **27:** 3–26.

FRASER, D. A. S., 1957, "Nonparametric Methods in Statistics," John Wiley & Sons, Inc., New York, 299 pp.

FRASER, H. J., 1935, Experimental Study of the Porosity and Permeability of Clastic Sediments, *J. Geol.*, **43:** 910–1010.

FRIEDMAN, G. M., 1958, Determinations of Sieve-size Distribution from the Thin-section Data for Sedimentary Petrological Studies, *J. Geol.*, **66:** 394–416.

———, 1962, On Sorting Coefficients and the Lognormality of the Grain Size Distribution of Sandstones, *J. Geol.*, **70:** 737–753.

GADDUM, J. H., 1945, Lognormal Distributions, *Nature,* **156:** 463–466.

GARDNER, M., 1959, Mathematical Games, *Sci. Am.*, **301:** 181–188.

GARNER, W. R., and W. J. MCGILL, 1956, The Relation between Information and Variance Analysis, *Psychometrika,* **21:** 219–228.

GILBERT, G. K., 1914, The Transportation of Debris by Running Water, *U.S. Geol. Surv., Profess. Paper* 86, 263 pp.

GLAGOLEV, A. A., 1933, On the Geometrical Methods of Quantitative Mineralogical Analysis of Rocks, *Trans. Inst. Econ. Mineral. U.S.S.R.*, **59:** 1–47 (English summary, p. 47).

———, 1934, Quantitative Analysis with the Microscope by the Point Method, *Eng. Mining J.*, **135:** 399.

GOODMAN, L. A., and W. H. KRUSKAL, 1954, Measures of Association for Cross Classifications, *J. Am. Statist. Assoc.*, **49:** 732–764.

GRATON, L. C., and H. J. FRASER, 1935, Systematic Packing of Spheres with Particular Relation to Porosity and Permeability, *J. Geol.*, **43:** 785–909.

GRIFFITHS, J. C., 1940, The Glacial Deposits West of the Taff, South Wales, unpublished Ph.D. thesis, Imperial College of Science and Technology, London, 187 pp.

——— and A. STUART, 1940, An Occurrence of Detrital Diaspore in South Wales, *Geol. Mag.*, **77:** 74–76.

———, 1948, Petrographic Evaluation of Porosity Measurement, *Bull. Mineral Ind. Expt. Sta. Penn. State Univ.*, **52:** 6–30.

——— and M. A. ROSENFELD, 1950, Progress in Measurement of Orientation in Bradford Sand, *Bull. Mineral Ind. Expt. Sta. Penn. State Univ.*, **56:** 202–236.

———, 1951a, Size versus Sorting in Some Caribbean Sediments, *J. Geol.*, **59:** 211–243.

———, 1951b, Size and Sorting in Sediments, *Bull. Geol. Soc. Am. Abstr.*, **62:** 1551.

———, 1952a, Grain Size Distribution and Reservoir Rock Characteristics, *Bull. Am. Assoc. Petrol. Geologists,* **36:** 205–229.

———, 1952b, A Review of Dimensional Orientation of Quartz Grains in Sediments, *Bull. Mineral Ind. Expt. Sta. Penn. State Univ.*, **60:** 47–55.

———, 1953a, *Producers Monthly,* **17:** 14–20.

———, 1953b, Estimation of Error in Grain Size Analysis, *J. Sediment. Petrol.*, **23:** 75–84.

——— and M. A. ROSENFELD, 1953, A Further Test of Dimensional Orientation of Quartz Grains in Bradford Sand, *Am. J. Sci.*, **251:** 192–214.

———, D. W. GROFF, and J. S. KAHN, 1953, Petrographical Investigations of the Salt Wash Sediments, *U.S. At. Energy Comm. RME* 3054, 56 pp.

———, J. A. COCHRAN, D. W. GROFF, and J. S. KAHN, 1953, Petrographical Investigations of the Salt Wash Sediments, *U.S. At. Energy Comm. RME* 3079, 94 pp.

———, ———, ———, and ———, 1954a, Petrographical Investigations of the Salt Wash Sediments, *U.S. At. Energy Comm. RME* 3097, 63 pp.

———, ———, ———, and ———, 1954b, Petrographical Investigations of the Salt Wash Sediments, *U.S. At. Energy Comm. RME* 3106, 98 pp.

——— and M. A. ROSENFELD, 1954, Operator Variation in Experimental Research, *J. Geol.*, **62:** 74–91.

———, 1955a, Statistics for the Description of Frequency Distributions Generated by the Measurement of Petrographic Properties of Sediments, *The Compass* (of Sigma Gamma Epsilon), **32:** 329–346.

———, 1955b, Sampling Sediments for Measurement of Grain Size and Shape, *Bull. Geol. Soc. Am. Abstr.*, **66:** 1568.

———, J. A. COCHRAN, J. J. HUTTA, and R. STEINMETZ, 1956, Petrographical Investigations of the Salt Wash Sediments, *U.S. At. Energy Comm. RME* 3122, pt. 1, 66 pp., pt. 2, 84 pp.

———, 1957, Petrographical Investigations of Salt Wash Sediments, *U.S. At. Energy Comm. RME* 3151, 37 pp.

———, 1958a, Geometrics in Petroleum Petrography, *Producers Monthly,* **22:** 40.

———, 1958b, Petrography and Porosity of the Cow Run Sandstone, St. Marys, W.Va., *J. Sediment. Petrol.,* **28:** 15–30.

——— and D. D. MCINTYRE, 1958, A Table for Conversion of Millimeters to Phi Units, Department of Mineralogy, Pennsylvania State University, University Park, Pa., 56 pp.

———, 1959a, Size and Shape of Rock-fragments in Tuscarora Scree, Fishing Creek, Lamar, Central Pennsylvania, *J. Sediment. Petrol.,* **29:** 391–401.

———, 1959b, Sampling Pebbles of "Quartzite" from Gravel at Montoursville, Penna., *Bull. Geol. Soc. Am. Abstr.,* **70:** 1612.

———, 1960a, Frequency Distributions in Accessory Mineral Analysis, *J. Geol.,* **68:** 353–365.

———, 1960b, Relationships between Reservoir Petrography and Reservoir Behavior of Some Appalachian Oil Sands, *Tulsa Geol. Soc. Dig.,* **28:** 43–58.

———, 1960c, Some Aspects of Measurement in the Geosciences, *Mineral Ind.,* **29:** 1, 4, 5, 8.

———, 1960d, Modal Analysis of Sediments, *Rev. Geograph. Phys. Geol. Dynamique,* **3:** 29–48.

———, 1961, Measurement of the Properties of Sediments, *J. Geol.,* **69:** 487–498.

———, 1962a, Statistical Methods in Sedimentary Petrography, pp. 565–617, in Milner, 1962, vol. 1, 4th ed.

———, 1962b, Translating Sedimentary Petrography into Petrology, *Geol. Soc. Am. Spec. Papers* **68** (**1961**): 186.

———, 1962c, Uses of Computers and Statistics in Exploration and Development of Mineral Resources, in "Mathematical Techniques and Computer Applications in Mining and Exploration," vol. 1, pp. E1-1–E1-19, College of Mines, University of Arizona, Tucson, Ariz.

———, 1963a, Frequency Distributions of Some Natural Resources, *Penn. State Univ. Mineral Ind. Expt. Sta. Circ.,* **63:** 174–198.

———, 1963b, Statistical Approach to the Study of Potential Oil Reservoir Sandstones, pp. 637–668, in G. A. Parks (ed.), "Computers in the Mineral Industries," School of the Earth Sciences, Stanford University, Stanford, Calif.

——— and C. M. SMITH, 1964, Relationship between Volume and Axes of Some Quartzite Pebbles from the Olean Conglomerate, Rock City, N.Y., *Am. J. Sci.,* **262:** 497–512.

GROVES, A. W., 1931, The Unroofing of the Dartmoor Granite and the Distribution of Its Detritus in the Sediments of Southern England, *Quart. J. Geol. Soc. London,* **137:** 62–96.

HANSEN, M. H., W. N. HURWITZ, and W. G. MADOW, 1953, "Sample Survey Methods and Theory," John Wiley & Sons, Inc., New York, vol. 1, 638 pp.

HARKER, A., 1923, "Petrology for Students, 6th ed., Cambridge University Press, 302 pp.

———, 1939, "Metamorphism: A Study of the Transformations of Rock-masses," 2d ed., Methuen & Co., Ltd., London, 302 pp.

HARTLEY, H. O., 1940, Testing the Homogeneity of a Set of Variances, *Biometrika,* **31:** 249–255.

HATCH, F. H., R. H. RASTALL, and M. BLACK, 1938, "The Petrology of the Sedimentary Rocks," 3d ed., George Allen & Unwin, Ltd., London, 383 pp.

HATCH, T., and S. CHOATE, 1929, Statistical Description of the Size Properties of Non-uniform Particulate Substances, *J. Franklin Inst.,* **207:** 369–387.

———, 1933, Determination of "Average Particle Size" from the Screen Analysis of Non-uniform Particulate Substances, *J. Franklin Inst.,* **215:** 227–237.

HAZEN, W., JR., 1958, A Comparative Study of Statistical Analysis and Other Methods of Computing Ore Reserves, Utilizing Analytical Data from Maggie Canyon Manganese Deposit, Artillery Mountains Region, Mohave Co., Arizona, *U.S. Bur. Mines Rept. Invest.* 5375, 188 pp.

———, 1961, Statistical Analysis of Sample Data for Estimating Ore, *U.S. Bur. Mines Rept. Invest.* 5835, 27 pp.

HEADLEE, A. J. W., and J. S. JOSEPH, 1945, Permeability, Porosity and Oil and Water Content of Natural Reservoirs; Kanawha-Jackson Campbells Creek Oriskany Fields, *West Va. Geol. Econ. Surv. Bull.* 8.

HEDBERG, H. D., 1926, The Effect of Gravitational Compaction on the Structure of Sedimentary Rocks, *Bull. Am. Assoc. Petrol. Geologists,* **10:** 1035–1072.

———, 1928, Some Aspects of Sedimentary Petrography in Relation to Stratigraphy in the Bolivia Coast Fields of the Maracaibo Basin, Venez., *J. Paleontol.,* **2:** 32–42.

HEINRICH, E. W., 1956, "Microscopic Petrography," McGraw-Hill Book Company, New York, 296 pp.

HELSON, H., 1936, On Statistical Methods of Comparing Heavy Mineral Suites: An Answer to Mr. Churchill Eisenhart, *Am. J. Sci.,* **232:** 392–395.

HEMPEL, C. G., 1952, "Fundamentals of Concept Formation in Empirical Science," The University of Chicago Press, Chicago, 93 pp.

HOEL, P. G., 1947, "Introduction to Mathematical Statistics," John Wiley & Sons, Inc., New York (3d ed., 1962).

HOLMES, A., 1930, "Petrographic Methods and Calculations," rev. ed., Thomas Murby Co., London, 515 pp.

HOLMES, C. D., 1941, Till Fabric, *Bull. Geol. Soc. Am.,* **52:** 1299–1354.

HOLZINGER, K. J., and H. H. HARMAN, 1941, "Factor Analysis," The University of Chicago Press, Chicago, 417 pp.

HUBERT, J. F., 1958, Petrology of the Fountain and Lyons Formations Along the Colorado Front Range, unpublished Ph.D. thesis, Dept. Mineralogy, Pennsylvania State University, University Park, Pa., 443 pp.

HULBE, C. W. H., 1955, Mounting Technique for Grain Size and Shape Measurement, *J. Sediment. Petrol.,* **25:** 302–303.

———, 1957, An Investigation of the Size and Shape of Quartz Grains, Pedro Beach, California, unpublished M.S. thesis, Dept. Mineralogy, Pennsylvania State University, University Park, Pa., 69 pp.

HUNT, W. F., 1924, An Improved Wentworth Recording Micrometer, *Am. Mineralogist,* **9:** 190–193.

HURLBUT, C. S., JR., 1939, An Electric Counter for Thin-section Analysis, *Am. J. Sci.,* **237:** 253–261.

———, 1948, Mineral Determination by Means of Punched Cards, *Am. Mineralogist,* **33:** 508–512.

HUTTA, J. J., 1956, Relation of Dimensional Orientation of Quartz Grains to Directional Permeability in Sandstones, unpublished M.S. thesis, Dept. Mineralogy, Pennsylvania State University, University Park, Pa., 97 pp.

HUTTON, C. O., 1959, Mineralogy of the Beach Sands between Halfmoon and Monterey Bays, California, *Calif. Div. Mines Spec. Rept.* 59, 32 pp.

ILLING, V. C., 1916, The Oilfields of Trinidad, *Proc. Geol. Assoc. Engl.,* **27:** 115.

———, 1939, Some Factors in Oil Accumulation, *J. Inst. Petrol.,* **25:** 201–225.

INGERSON, E., and J. L. RAMISCH, 1942, Origin of the Shapes of Quartz Sand Grains, *Am. Mineralogist,* **27:** 595–606.

INMAN, D. L., 1949, Sorting of Sediments in the Light of Fluid Mechanics, *J. Sediment. Petrol.,* **19:** 51–70.

———, 1952, Measures for Describing the Size Distribution of Sediments, *J. Sediment. Petrol.,* **22:** 125–145.

IRANI, R. R., 1959, The Interpretation of Abnormalities in the Log-normal Distribution of Particle Size, *J. Phys. Chem.*, **63:** 1603–1607.

———, 1961, How Various Techniques Contribute to Particle Size Distribution Data, *Cereal Sci. Today*, **6:** 35–39.

JEFFREYS, H., 1948, "Theory of Probability," 2d ed., Oxford University Press, Fair Lawn, N.J., 411 pp.

JEFFRIES, C. D., 1941, A Double Centrifuge Tube for the Separation of Soil Minerals by Means of Heavy Liquids, *Soil Sci.*, **52:** 167–171.

JIZBA, Z. V., 1953, Mean and Standard Deviation of Certain Geologic Data: A Discussion, *Am. J. Sci.*, **251:** 899–906.

JOHNSON, N. L., and H. TETLEY, 1950, 1951, "Statistics: An Intermediate Textbook," Cambridge University Press, New York, vol. 1, 294 pp., vol. 2, 318 pp.

JOHNSON, P. O., 1949, "Statistical Methods in Research," Prentice-Hall, Inc., Englewood Cliffs, N.J., 377 pp.

JOHNSON, W. E., and J. N. BRESTON, 1951, Directional Permeability Measurements on Oil Sandstones from Various States, *Producers Monthly*, **15:** 10–19.

JONES, G. K., 1947, Measurement of Rock Densities with the Mercury Balance, private report, United British Oilfields of Trinidad, now Shell Trinidad Ltd.

KAHN, J. S., 1954, The Measurement of Packing in Sandstones, unpublished M.S. thesis, Dept. Mineralogy, Pennsylvania State University, University Park, Pa., 135 pp.

———, 1956, The Analysis and Distribution of the Properties of Packing in Sand-size Sediments, 1. On the Measurement of Packing in Sandstones, 2. The Distribution of the Packing Measurements and an Example of Packing Analysis, *J. Geol.*, **64:** 385–395, 578–606.

———, 1959, Anisotropic Sedimentary Parameters, *Trans. N.Y. Acad. Sci.* (II), **21:** 373–386.

KAPTEYN, J. C., 1916, Skew Frequency Curves in Biology and Statistics, *Rec. Trav. Botan. Néerl.*, **13:** 105–157.

KEITH, M. L., 1939, Selective Staining to Facilitate Rosiwal Analysis, *Am. Mineralogist*, **24:** 561–565.

KELLAGHER, R. C., and F. J. FLANAGAN, 1956, The Multiple-cone Sample Splitter, *J. Sediment. Petrol.*, **26:** 213–221.

KELLER, W. D., 1939, Differential Packing as a Possible Display of Bedding, *J. Sediment. Petrol.*, **9:** 131–133.

KEMENY, J. G., H. MIRKIL, J. L. SNELL, and G. L. THOMPSON, 1959, "Finite Mathematical Structures," Prentice-Hall, Inc., Englewood Cliffs, N.J., 487 pp.

KEMPTHORNE, O., 1952, "The Design and Analysis of Experiments," John Wiley & Sons, Inc., New York, 631 pp.

KENDALL, M. G., 1948, "The Advanced Theory of Statistics," vol. 1, 4th ed., 1948, 457 pp., vol. 2, 3d ed., 1951, 521 pp.

———, 1957, "A Course in Multivariate Analysis," Charles Griffin & Company, Ltd., London, 185 pp.

KIMBALL, A. W., 1954, Short-cut Formulas for the Exact Partition of χ^2 in Contingency Tables, *Biometrics*, **10:** 452–458.

KINSEY, A. C., W. B. POMEROY, and C. E. MARTIN, 1948, "Sexual Behavior in the Human Male," W. B. Saunders Company, Philadelphia, 804 pp.

KIRKPATRICK, A. F., 1948, An Application of Punch Cards; Filing of Optical Properties of Crystals, *Anal. Chem.*, **20:** 847–849.

KRIGE, D. G., 1951, A Statistical Approach to Some Basic Mine Valuation Problems on the Witwatersrand, *J. Chem. Soc. S. Africa*, **52:** 119–139.

KRUMBEIN, W. C., 1934a, The Probable Error of Sampling Sediments for Mechanical Analysis, *Am. J. Sci.*, **27:** 204–214.

———, 1934b, Size Frequency Distribution of Sediments, *J. Sediment. Petrol.*, **4:** 65–77.

———, 1935, Thin Section Mechanical Analysis of Indurated Sediments, *J. Geol.*, **43:** 482–496.

———, 1936a, Application of Logarithmic Moments to Size-frequency Distribution of Sediments, *J. Sediment. Petrol.*, **6:** 35–47.

———, 1936b, Rejoinder to Wentworth's Discussion of the Method of Moments, *J. Sediment. Petrol.*, **6:** 159–160.

———, 1936c, The Use of Quartile Measures in Describing and Comparing Sediments, *Am. J. Sci.*, **32:** 98–111.

———, 1937, Sediments and Exponential Curves, *J. Geol.*, **45:** 577–601.

——— and F. ABERDEEN, 1937, The Sediments of Barataria Bay, *J. Sediment. Petrol.*, **7:** 3–17.

———, 1938, Size Frequency Distribution of Sediments and the Normal Phi Curve, *J. Sediment. Petrol.*, **8:** 84–90.

——— and F. J. PETTIJOHN, 1938, "Manual of Sedimentary Petrography," Appleton-Century-Crofts, Inc., New York, 549 pp.

———, 1939a, Preferred Orientation of Pebbles in Sedimentary Deposits, *J. Geol.*, **19:** 675–706.

———, 1939b, Graphic Presentation and Statistical Analysis of Sedimentary Data, pp. 558–591, in P. D. Trask (ed.), "Recent Marine Sediments," Thomas Murby Co., London, 736 pp.

———, 1941a, The Effect of Abrasion on the Size, Shape and Roundness of Rock-fragments, *J. Geol.*, **49:** 482–520.

———, 1941b, Measurement and Geological Significance of Shape and Roundness of Sedimentary Particles, *J. Sediment. Petrol.*, **11:** 64–72.

——— and W. C. RASMUSSEN, 1941, The Probable Error of Sampling Beach Sand for Heavy Mineral Analysis, *J. Sediment. Petrol.*, **11:** 10–20.

——— and G. D. MONK, 1943, Permeability as a Function of the Size Parameters of Unconsolidated Sand, *Trans. AIME*, **151:** 153–163.

———, 1953a, Statistical Designs for Sampling Beach Sand, *Trans. Am. Geophys. Union*, **34:** 857–868.

———, 1953b, Latin Square Experiments in Sedimentary Petrology, *J. Sediment. Petrol.*, **23:** 280–283.

——— and R. L. MILLER, 1953, Design of Experiments for Statistical Analysis of Geological Data, *J. Geol.*, **61:** 510–532.

———, 1954a, Statistical Problems of Sample Size and Spacing on Lake Michigan Beaches, pp. 147–162, *Proc. Conf. Coastal Eng. Council on Wave Research, 4th Conf., Berkeley, Calif.*

———, 1954b, Applications of Statistical Methods to Sedimentary Rocks, *J. Am. Statist. Assoc.*, **49:** 51–66.

——— and R. L. MILLER, 1954, A Note on Transformations of Data for Analysis of Variance, *J. Geol.*, **62:** 192–193.

———, 1955, Experimental Designs in the Earth Sciences, *Trans. Am. Geophys. Union*, **36:** 1–11.

——— and H. A. SLACK, 1956, Statistical Analysis of Low-level Radioactivity of Pennsylvanian Black Fissile Shale in Illinois, *Bull. Geol. Soc. Am.*, **67:** 739–762.

——— and J. W. TUKEY, 1956, Multivariate Analysis of Mineralogic, Lithologic and Chemical Composition of Rock-bodies, *J. Sediment. Petrol.*, **26:** 322–337.

——— and L. L. SLOSS, 1958, High-speed Digital Computers in Stratigraphic and Facies Analysis, *Bull. Am. Assoc. Petrol. Geologists*, **42:** 2650–2669.

———, 1959, The Sorting Out of Geological Variables Illustrated by Regression Analysis, *J. Sediment. Petrol.*, **29:** 575–587.

———, 1960a, The "Geological Population" as a Framework for Analyzing Data in Geology, *Liverpool and Manchester Geol. J.*, **2:** 341–368.

———, 1960b, Some Problems in Applying Statistics to Geology, *Appl. Statistics*, **14:** 82–91.

KRYNINE, P. D., 1940, Petrology and Genesis of Third Bradford Sand, *Bull. Mineral Ind. Expt. Sta. Penn. State Univ.*, **29:** 130–165.

———, 1946, The Tourmaline Group in Sediments, *J. Geol.*, **54:** 65–87.

———, 1948, The Megascopic Study and Field Classification of Sedimentary Rocks, *J. Geol.*, **56:** 130–165.

KUENEN, P. H., and W. G. PERDOK, 1962, Experimental Abrasion, 5. Frosting and Defrosting of Quartz Grains, *J. Geol.*, **70:** 648–658.

KULLBACK, S., 1959, "Information Theory and Statistics," John Wiley & Sons, Inc., New York, 395 pp.

LANDY, R. L., 1956, The Genetic Pottsville-Tuscarora Relationship in Eastern Pennsylvania, unpublished M.S. thesis, Dept. Mineralogy Pennsylvania State University, University Park, Pa., 142 pp.

———, 1961, Variation in Chemical Composition of Rock Bodies: Metabasalts in the Iron Springs Quadrangle, South Mountain, Pennsylvania, unpublished Ph.D. thesis, Dept. Mineralogy, Pennsylvania State University, University Park, Pa., 185 pp.

LARSEN, E. S., and H. BERMAN, 1934, The Microscopic Determination of the Non-opaque Minerals, 2d ed., *U.S. Geol. Surv. Bull.* 848, 266 pp.

LAW, J., 1944, A Statistical Approach to the Interstitial Heterogeneity of Sand Reservoirs, *AIME Petrol. Technol. Tech. Paper* 1732.

LE CORBEILLER, P., 1954, The Curvature of Space, *Sci. Am.*, **191:** 80–86.

LEDLEY, R. S., and L. B. LUSTED, 1959, Reasoning Foundations of Medical Diagnosis, *Sci.*, **130:** 9–21.

LICASTRO, P., and J. C. GRIFFITHS, 1953, The Effect of Solvent Extraction on the Determination of Porosity by Saturation, *Bull. Mineral Ind. Expt. Sta. Penn. State Univ.*, **62:** 41–45.

LINDLEY, D. V., 1956, On a Measure of the Information Provided by an Experiment, *Ann. Math. Statistics*, **27:** 986–1005.

LOEWINSON-LESSING, F. Y., 1954, "A Historical Survey of Petrology," transl. S. I. Tomkeieff, Oliver & Boyd Ltd., Edinburgh and London, 112 pp.

LUCE, R. D., and H. RAIFFA, 1957, "Games and Decisions," John Wiley & Sons, Inc., New York, 509 pp.

MACDONALD, D. K. C., 1952, Information Theory and Its Applications to Taxonomy, *J. Appl. Phys.*, **23:** 529–531.

MACKIE, W., 1897, The Sands and Sandstones of Eastern Moray, *Trans. Edinburgh Geol. Soc.*, **7:** 148–172.

———, 1897, On the Laws That Govern the Rounding of Particles of Sand, *Trans. Edinburgh Geol. Soc.*, **7:** 298–311.

———, 1923, The Source of the Purple Zircon in the Sedimentary Rocks of Scotland, *Trans. Edinburgh Geol. Soc.*, **11:** 200–213.

MANDEL, J., 1954, Chain-block Designs with Two-way Elimination of Heterogeneity, *Biometrics*, **10:** 251–272.

MASSON, P., 1951, Measurement of Grain Packing in Sandstone, unpublished paper given at the Annual Meeting, Am. Assoc. Petrol. Geologists, St. Louis, Mo., 14 pp.

MATHER, K., 1951, "Statistical Analysis in Biology," Methuen & Co., Ltd., London, 267 pp.

MAXWELL, A. E., 1961, "Analyzing Qualitative Data," Methuen & Co., Ltd., London, 163 pp.

MCCAMMON, R. B., 1962a, Efficiencies of Percentile Measures for Describing the Mean Size and Sorting of Sedimentary Particles, *J. Geol.*, **70:** 453–465.

———, 1962b, Moment Measures and the Shape of Size-frequency Distributions, *J. Geol.*, **70:** 89–91.

MCCARTHY, P. J., 1957, "Introduction to Statistical Reasoning," McGraw-Hill Book Company, New York, 402 pp.

MCINTYRE, D. D., 1959, The Hydraulic Equivalence and Size Distributions of Some Mineral Grains from a Beach, *J. Geol.*, **67:** 278–301.

———, 1961, A Comparison of Three Associated Environments, Glacial Till, Fluvio-glacial Delta and Beach Sand, in Terms of the Shapes of Their Quartz, Garnet and Hornblende Grains, unpublished API Rept., Dept. Mineralogy, Pennsylvania State University, University Park, Pa., 78 pp.

MIDDLETON, G. M., 1962, On Sorting Coefficients and the Lognormality of the Grain Size Distribution of Sandstones: A Discussion, *J. Geol.*, **70:** 754–756.

MILLER, R. L., and E. D. GOLDBERG, 1955, The Normal Distribution in Geochemistry, *Geochim. Cosmochim. Acta*, **8:** 53–62.

——— and E. C. OLSON, 1955, The Statistical Stability of Quantitative Properties as a Fundamental Criterion for the Study of Environments, *J. Geol.*, **63:** 376–387.

——— and J. S. KAHN, 1962, "Statistical Analysis in the Geological Sciences," John Wiley & Sons, Inc., New York, 483 pp.

MILNER, H. B., 1962, "Sedimentary Petrography," George Allen & Unwin, Ltd., London, 4th ed., vol. 1, 643 pp., vol. 2, 715 pp. (3d ed., 1940, Thomas Murby Co., London, 666 pp.).

MINER, N. A., 1934, Talus Slopes of the Gaspé Peninsula, *Science*, **79:** 229–230.

MISES, R. VON, 1957, "Probability, Statistics and Truth," 2d ed., George Allen & Unwin, Ltd., London, 244 pp.

MOLINA, E. C., 1942, "Poisson's Exponential Binomial Limit," D. Van Nostrand Company, Inc., Princeton, N.J., 47 pp.

MOORHOUSE, W. W., 1959, "The Study of Rocks in Thin Section," Harper & Row, Publishers, Incorporated, New York, 514 pp.

MUSKAT, M., 1937, "The Flow of Homogeneous Fluids through Porous Media," McGraw-Hill Book Company, New York, 763 pp.

NATIONAL BUREAU OF STANDARDS, 1950, Tables of the Binomial Probability Distribution, *Natl. Bur. Std. U.S. Appl. Math. Ser.* 6, 387 pp.

NEUMANN, J. VON, and O. MORGENSTERN, 1953, "Theory of Games and Economic Behavior," 3d ed., Princeton University Press, Princeton, N.J., 641 pp.

NEVIN, C., 1946, Competency of Running Water to Transport Debris, *Bull. Geol. Soc. Am.*, **57:** 651–674.

NEYMAN, J., and E. S. PEARSON, 1928, On the Use and Interpretation of Certain Test Criteria for Purposes of Statistical Inference, I, II, *Biometrika*, **20A:** 175–240, 263–294.

——— and ———, 1936, 1938, Contributions to the Theory of Testing Statistical Hypotheses, *Statist. Res. Mem.*, **1:** 1–37, **2:** 25–57.

——— and E. L. SCOTT, 1958, Statistical Approach to Problems of Cosmology, *J. Roy. Statist. Soc., Ser. B*, **20:** 1–43.

NIGGLI, P., 1935, Die Charakterisierung der klastischen Sedimente nach der Kornzusammensetzung, *Schweiz. Mineral. Petrog. Mitt.*, **15:** 31–38.

OLSON, E. C., and R. L. MILLER, 1958, "Morphological Integration," The University of Chicago Press, Chicago, 317 pp.

OLSON, J. S., and P. E. POTTER, 1954, Variance Components of Cross-bedding Direction in Some Basal Pennsylvanian Sandstones of the Eastern Interior Basin: Statistical Methods, *J. Geol.*, **62:** 26–49.

OTTO, G. H., 1933, Comparative Tests of Several Methods of Sampling Heavy Mineral Concentrates, *J. Sediment. Petrol.*, **3:** 30–39.

———, 1937, The Use of Statistical Methods in Effecting Improvements on a Jones' Sample Splitter, *J. Sediment. Petrol.*, **7:** 110–132.

———, 1938, The Sedimentation Unit and Its Use in Field Sampling, *J. Geol.*, **46:** 569–582.

———, 1939, A Modified Logarithmic Probability Graph for the Interpretation of the Mechanical Analysis of Sediments, *J. Sediment. Petrol.*, **9:** 62–76.

PACHMAN, J. M., 1961, Inter-relationships among Petrographic, Textural and Oil Reservoir Properties in the Chipmunk Sandstone, unpublished Ph.D. thesis, Dept. Mineralogy, Pennsylvania State University, University Park, Pa., 95 pp.

PARKER, M. A., 1952, Punched-card Techniques Speed Map-making, *Bull. Geol. Soc. Am. Abstr.*, **63:** 1288.

———, 1957, Application of Punched-card Analysis to Limestone Exploration, *Bull. Geol. Soc. Am. Abstr.*, **68:** 1777.

PEARSON, E. S., and J. WISHART, 1942, "Student's Collected Papers," Cambridge University Press, New York, 224 pp.

——— and H. O. HARTLEY, 1954, "Biometrika Tables for Statisticians," vol. 1, Cambridge University Press, New York, 238 pp.

———, 1948, "Karl Pearson's Early Statistical Papers," Cambridge University Press, New York, 557 pp.

PEARSON, K., 1895, Contributions to the Mathematical Theory of Evolution, II. Skew Variation in Homogeneous Material, *Phil. Trans. Roy. Soc. London Ser. A*, **186:** 343–414.

———, 1900, On the Criterion that a Given System of Deviations from the Probable in the Case of a Correlated System of Variables Is Such that It Can Reasonably Be Supposed to Have Risen from Random Sampling, *Phil. Mag.*, (5)**50:** 157–175.

———, 1902, On the Mathematical Theory of Errors of Judgement with Special Reference to the Personal Equation, *Phil. Trans. Roy. Soc. London Ser. A*, **198:** 235–299.

PELTO, C. R., 1954, Mapping of Multicomponent Systems, *J. Geol.*, **62:** 501–511.

PETTIJOHN, F. J., 1957, "Sedimentary Rocks," 2d ed., Harper & Row, Publishers, Incorporated, 718 pp.

PIRSON, S. J., 1950, "Oil Reservoir Engineering," 2d ed., McGraw-Hill Book Company, New York, 441 pp.

PLUMLEY, W. J., 1948, Black Hills Terrace Gravels: A Study in Sediment Transport, *J. Geol.*, **56:** 526–577.

POINCARÉ, H., 1892, "Thermodynamique," quoted in J. Mellor, 1955, "Higher Mathematics for Students of Chemistry and Physics," Dover Publications, Inc., New York, 641 pp.

POPPER, K., 1959, "The Logic of Scientific Discovery," Science Editions, Inc., New York, 480 pp.

POTTER, P. E., and J. S. OLSON, 1954, Variance Components of Cross-bedding Directions in Some Basal Pennsylvanian Sandstones in the Eastern Interior Basin: Geological Applications, *J. Geol.*, **62:** 50–73.

——— and R. SIEVER, 1956, Sources of Basal Pennsylvanian Sediments in the Eastern Interior Basin, I. Cross-bedding, *J. Geol.*, **64:** 225–244.

——— and R. F. MAST, 1963, Sedimentary Structures, Sand Shape Fabrics and Permeability, *J. Geol.*, **71:** 441–471.

——— and F. J. PETTIJOHN, 1963, "Palaeocurrents and Basin Analysis," Academic Press Inc., New York, 296 pp.

PYE, W. D., 1943, Rapid Methods of Making Sedimentational Analyses of Arenaceous Sediments, *J. Sediment. Petrol.*, **13:** 85–104.

PYLE, H. C., and J. E. SHERBORNE, 1939, Core Analysis, *Trans. AIME, Petrol. Tech.*, **132:** 33–61.

QUENOUILLE, M. H., 1950, "Introductory Statistics," Pergamon Press, New York, 248 pp.

———, 1952, "Associated Measurements," Butterworth & Co. (Publishers), Ltd., London, 254 pp.

RAEBURN, C., and H. B. MILNER, 1927, "Alluvial Prospecting," Thomas Murby Co., London.

RAND Corporation, 1955, "A Million Random Digits," The Free Press of Glencoe, New York, 200 pp.

RAO, C. R., 1952, "Advanced Statistical Methods in Biometric Research," John Wiley & Sons, Inc., New York, 390 pp.

RASTALL, R. H., and W. H. WILCOCKSON, 1917, The Accessory Minerals of the Granitic Rocks of the English Lake District, *Quart. J. Geol. Soc. London*, **71:** 592–620.

RÉAUMUR, R. A. F. DE, 1718, Essais de l'histoire des rivières et des ruisseaux du royaume qui roulent des paillettes d'or, *Mem. l'Acad. Roy.*, pp. 68–88.

REED, R. D., and J. P. BAILEY, 1927, Subsurface Correlation by Means of Heavy Minerals, *Bull. Am. Assoc. Petrol. Geologists*, **11:** 359–368.

RETGERS, J. W., 1895, Ueber die mineralogische und chemische Zusammensetzung der Dunensands Hollands, *Neues Jahrb. Mineral*, **1:** 16–74.

RICHTER, K., 1932, Die Bewegungsrichtung des Inlandeises, rekonstruiert aus den Kritzen und Langsachsen der Geschiebe, *Z. Geschiebeforschung*, **8:** 62–66.

RIDER, P. R., 1939, "An Introduction to Modern Statistical Methods," John Wiley & Sons, Inc., New York, 220 pp.

RIEMANN, G., 1868, On the Hypotheses Which Lie at the Foundations of Geometry, pp. 411–425, in D. E. Smith, "A Source Book in Mathematics," McGraw-Hill Book Company, New York, 1929.

RILEY, N. A., 1941, Projection Sphericity, *J. Sediment. Petrol.*, **11:** 94–97.

RITTENHOUSE, G., 1939, The Pipette Method Modified for Mass Production, *Natl. Res. Council Rept. Comm. Sedimentation*, for 1938–1939, exhibit G, pp. 88–102.

———, 1940, Curves for Determining Probable Errors in Heavy Mineral Studies, *Natl. Res. Council Rept. Comm. Sedimentation 1939–1940*, exhibit F, pp. 97–101.

———, 1941, Size of Sand as Measured by Sieving and Microprojection, *Bull. Geol. Soc. Am. Abstr.*, **52:** 1930.

——— and W. E. BERTHOLF, JR., 1942, Gravity versus Centrifuge Separation of Heavy Minerals from Sand, *J. Sediment. Petrol.*, **12:** 85–89.

———, 1943a, A Visual Method of Estimating Two-dimensional Sphericity, *J. Sediment. Petrol.*, **13:** 79–81.

———, 1943b, Transportation and Deposition of Heavy Minerals, *Bull. Geol. Soc. Am.*, **54:** 1725–1750.

———, 1944, Sources of Modern Sands in the Middle Rio Grande Valley, New Mexico, *J. Geol.*, **11:** 145–183.

ROMIG, H., 1947, "50–100 Binomial Tables," John Wiley & Sons, Inc., New York, 172 pp. (2d ed., 1953).

ROSENFELD, M. A., 1950, Porosity, 1. A Survey of the Problem, 2. Some Statistical Techniques Applied to Porosity Data, unpublished M.S. thesis, Dept. Mineralogy, Pennsylvania State University, University Park, Pa., 149 pp.

——— and J. C. GRIFFITHS, 1950, A New Approach to the Problem of Porosity Measurement, *Bull. Mineral Ind. Expt. Sta. Penn. State Univ.*, **56:** 163–201.

———, 1953, Petrographic Variation in Oriskany "Sandstone" Complex, unpublished Ph.D. thesis, Dept. Mineralogy Pennsylvania State University, University Park, Pa., 220 pp.

——— and ———, 1953, An Experimental Test of Visual Comparison Technique in Estimating Two-dimensional Sphericity and Roundness of Quartz Grains, *Am. J. Sci.*, **251:** 553–585.

———, L. JACOBSEN, and J. C. FERM, 1953, A Comparison of Sieve and Thin Section Technique for Size Analysis, *J. Geol.*, **61:** 114–132.

———, 1954, A Modification of the Chayes' Point Counter Stage, *Am. Mineralogist*, **39:** 834–836.

ROSIWAL, A., 1898, Ueber geometrische Gesteinanalysen usw., *Verhandel. König. Geol. Reichsanstalt Wien*, pp. 143–175.

ROSS, F. W. J., 1950, The Development and Some Practical Applications of a Statistical Value Distribution Theory for the Witwatersrand Auriferous Deposits, M.Sc. thesis, Dept. Mineralogy and Metallurgy University of Witwatersrand, Johannesburg, 117 pp.

ROWLAND, R. A., 1946, Grain Shape Fabrics of Clastic Quartz, *Bull. Geol. Soc. Am.*, **59:** 547–564.

ROY, S. N., 1957, "Some Aspects of Multivariate Analysis," John Wiley & Sons, Inc., New York, 214 pp.

RUBEY, W. W., 1933, The Size Distribution of Heavy Minerals within a Water Laid Sandstone, *J. Sediment. Petrol.*, **3:** 3–29.

RUSNAK, G. A., 1957a, The Orientation of Sand Grains under Conditions of "Unidirectional" Fluid Flow, 1. Theory and Experiment, *J. Geol.*, **65:** 384–409.

———, 1957b, A Fabric and Petrologic Study of the Pleasantview Sandstone, *J. Sediment. Petrol.*, **27:** 41–55.

RUSSELL, B., 1948, "Human Knowledge: Its Scope and Limits," George Allen & Unwin, Ltd., London, 538 pp.

———, 1954, "The Analysis of Matter," Dover Publications, Inc., New York, 408 pp.

RUSSELL, R. D., 1936, The Size Distribution of Minerals in Mississippi River Sand, *J. Sediment. Petrol.*, **6:** 125–142.

———, 1937, Mineral Composition of Mississippi River Sands, *Bull. Geol. Soc. Am.*, **48:** 1307–1348.

——— and R. E. TAYLOR, 1937, Roundness and Shape of Mississippi River Sands, *J. Geol.*, **45:** 225–267.

———, 1942, Tables for the Determination of Detrital Minerals, *Natl. Res. Council Rept. Comm. Sedimentation* 1940–1941, pp. 6–8, tables I–III.

RUTHERFORD, E., and H. GEIGER, 1910, The Probability Variation in the Distribution of α Particles, *Phil. Mag.*, (6) **20:** 698.

RYDER, H. M., 1948, Permeability, Absolute, Effective, Measured, *World Oil*, **125:** 173–177.

SALMOJARGHI, F., 1904, Sullo studio mineralogico delle sabbie e sopra un modo di rappresentarne i resultati, *Atti Soc. Ital. Sci. Nat.*, **63:** 54–89.

SAVAGE, L. J., 1951, The Theory of Statistical Decisions, *J. Am. Statist. Assoc.*, **46:** 55–67.

SCATES, D. E., 1937, How Science Measures, *J. Exptl. Educ.*, **5:** 296–312.

SCHEFFE, H., 1959, "The Analysis of Variance," John Wiley & Sons, Inc., New York, 477 pp.

SCHWARZACHER, W., 1951, "Grain Orientation in Sands and Sandstones, *J. Sediment. Petrol.*, **21:** 162–172.

SEARCY, J. K., 1960, Graphical Correlation of Gaging-station Records, Manual of Hydrology, pt. 1, General Surface Water Techniques, *U.S. Geol. Surv. Water Supply Paper* 1541–C, 100 pp.

SHADLE, H. W., and J. C. GRIFFITHS, 1955, An Attempt to Establish Oil-reservoir Favorability Criteria based on Quantitative Petrographic Analysis, *Bull. Mineral Ind. Expt. Sta. Penn. State Univ.*, **68:** 61–66.

SHAND, S. J., 1916, A Recording Micrometer for Rock Analysis, *J. Geol.*, **24:** 394–403.

SHANNON, C. E., 1948, A Mathematical Theory of Communication, *Bell System Tech. J.*, **27:** 379–423, 623–656.

SHEWART, A., 1931, "Economic Control of Quality of Manufactured Product," D. Van Nostrand Company, Inc., New York, 501 pp.

———, 1939, "Statistical Method from the Viewpoint of Quality Control," U.S. Department of Agriculture, Washington, 135 pp.

SHROCK, R. R., 1948, "Sequence in Layered Rocks," McGraw-Hill Book Company, New York, 507 pp.

SHUKRI, N. M., and R. A. HIGAZY, 1944, Mechanical Analysis of Some Bottom Deposits of the Northern Red Sea, *J. Sediment. Petrol.*, **14:** 43–69.

SICHEL, H. S., 1947, An Experimental and Theoretical Investigation of Bias Error in Mine Sampling with Special Reference to Narrow Gold Reefs, *Trans. Inst. Mining Met.*, **56:** 403–474.

———, 1949, Mine Valuation and Maximum Likelihood, M.Sc. thesis, Dept. Mining and Metallurgy, University of Witwatersrand, Johannesburg, 52 pp.

———, 1952, New Methods in the Statistical Evaluation of Mine Sampling Data, *Trans. Inst. Mining Met.*, **61:** 261–288.

SIEGEL, S., 1956, "Nonparametric Statistics for the Behavioral Sciences, McGraw-Hill Book Company, New York, 312 pp.

SIEGRIST, H. G., JR., 1961, Multivariate Statistical Study of Two Sandstones, unpublished Ph.D. thesis, Dept. Mineralogy, Pennsylvania State University, 158 pp.

SLACK, H. H., and W. C. KRUMBEIN, 1955, Measurement and Statistical Evaluation of Low-level Radioactivity in Rocks, *Trans. Am. Geophys. Union*, **36:** 460–464.

SLICHTER, C. S., 1899, Theoretical Investigation of the Motion of Ground Water, *U.S. Geol. Surv. 19th, Ann. Rept.*, pt. 11, p. 305.

SLICHTER, L. B., 1960, The Need of a New Philosophy of Prospecting, *Mining Eng.*, **12:** 570–576.

SMITH, C. M., JR., 1959, Distribution of Alpha Tracks in Autoradiographs of Some Uraniferous Base Metal Sulfide Minerals, unpublished M.S. thesis, Dept. Mineralogy, Pennsylvania State University, University Park, Pa., 58 pp.

———, 1964, Quantitative Petrographic Comparison of the Bradford Third and Lewis Run Sands, unpublished Ph.D. thesis, Dept. Geochemistry and Mineralogy, Pennsylvania State University, University Park, Pa., 111 pp.

SMITH, C. S., 1948, Grains, Phases, and Interfaces: An Interpretation of Microstructure, *Trans. AIME*, **175:** 15–51.

———, 1952, Interphase Interfaces, pp. 377–401 in W. Shockley (ed.), "Imperfections in Nearly Perfect Crystals," John Wiley & Sons, Inc., New York, 403 pp.

SMITH, J. G., and A. J. DUNCAN, 1945, "Sampling Statistics and Applications," McGraw-Hill Book Company, New York, 498 pp.

SMITHSON, F., 1931, The Triassic Sandstones of Yorkshire and Durham," *Proc. Geol. Assoc. Engl.*, **42:** 125–156.

SNEDECOR, G. W., 1956, "Statistical Methods," 5th ed., The Iowa State University Press, Ames, Iowa, 534 pp.

STANONIS, F. L., 1956, The Petrology of the Chipmunk Sandstone and Its Relationship to Reservoir Properties, unpublished Ph.D. thesis, Dept Mineralogy, Pennsylvania State University, University Park, Pa., 151 pp.

STEINMETZ, R., 1956, Size and Shape of Quartzose Pebbles from Three New Jersey Gravels, unpublished M.S. thesis, Dept. Mineralogy, Pennsylvania State University, University Park, Pa., 116 pp.

———, 1962, Sampling and Size Distribution of Quartzose Pebbles from Three New Jersey Gravels, *J. Geol.*, **70:** 56–73.

STEVENS, S. S., 1946, On the Theory of Scales of Measurement, *Science*, **103:** 677–680.

———, 1955, The Measurement of Loudness, *J. Acoust. Soc. Am.*, **27:** 815–829.

———, 1956, The Direct Estimates of Sensory Magnitudes: Loudness, *Am. J. Psychol.*, **69:** 1–25.

———, 1959, Measurement, Psychophysics and Utility, pp. 18–63, in Churchman and Ratoosh, 1959.

STUART, A., 1953, The Estimates and Comparison of Strengths of Association in Contingency Tables, *Biometrika*, **40:** 105–110.

"Student," 1907, On the Error of Counting with a Hemacytometer, *Biometrika*, **5:** 351.

———, 1908, The Probable Error of a Mean, *Biometrika*, **6:** 1–34.

———, 1919, An Explanation of Deviations from Poisson's Law in Practice, *Biometrika*, **12:** 211.

SULLIVAN, J. D., 1927, Heavy Liquids for Mineralogical Analyses, *U.S. Bur. Mines Tech. Paper* 381, pp. 1–26.

SUTER, H. H., 1951, 1952, The General and Economic Geology of Trinidad, B.W.I., *Colonial Geol. Mineral Resources*, **2:** 177–217, 271–307, **3:** 3–51.

SWINEFORD, A., and F. SWINEFORD, 1946, A Comparison of Three Sieve-shakers, *J. Sediment. Petrol.*, **16:** 3–13.

TANNER, W. F., 1958, The Zig-zag Nature of Type I and Type IV Curves, *J. Sediment. Petrol.*, **28:** 372–375.

TAYLOR, J. M., 1950, Pore-space Reduction in Sandstones, *Bull. Am. Assoc. Petrol. Geologists*, **34:** 701–716.

THIEL, G. A., 1940, The Relative Resistance to Abrasion of Mineral Grains of Sand Size, *J. Sediment. Petrol.*, **10:** 103–124.

THOMAS, H. H., 1909, A Contribution to the Petrography of the New Red Sandstone in the West of England, *Quart. J. Geol. Soc. London*, **65:** 229–244.

THOULET, J., 1881, Étude minéralogique d'un sable du Sahara, *Bull. Soc. Minéral. France,* **4:** 262–268.

———, 1913, Notes de lithologie sous-marine, *Ann. Inst. Oceanogr. Paris,* **5:** 1–14.

THURSTONE, L. L., 1927, A Law of Comparative Judgment, *Psych. Rev.,* **34:** 273–286.

TIPPETT, L. H. C., 1952, "The Methods of Statistics, "4th ed., John Wiley & Sons, Inc., New York, 395 pp.

TRASK, P. D., 1932, Origin and Environment of Source Sediments of Petroleum, Gulf Publishing Company, Houston.

———, 1942, Suggested Classification of Detrital Particles, *Natl. Res. Council Rept. Comm. Sedimentation* 1941–1942, appendix G, p. 3.

UDDEN, J. A., 1898, The Mechanical Composition of Wind Deposits, *Augustana Library Publ.* 1.

VAN ORSTRAND, C. E., 1925, Note on the Representation of the Distribution of Grains in Sands, *Natl. Res. Council Res. Sedimentation* 1924, pp. 63–67.

VISTELIUS, A. B., 1960, The Skew Frequency Distributions and the Fundamental Law of Geochemical Processes, *J. Geol.,* **68:** 1–22.

——— and O. V. SARMANOV, 1961, The Correlation between Percentage Values: Major Component Correlation in Ferromagnesium Micas, *J. Geol.,* **69:** 145–153.

WADELL, H., 1932, Volume, Shape and Roundness of Rock-particles, *J. Geol.,* **40:** 443–451.

———, 1933, Sphericity and Roundness of Rock-particles, *J. Geol.,* **41:** 310–331.

———, 1936, Volume, Shape and Shape Position of Rock-fragments in Openwork Gravel, *Geograph. Ann.,* **1:** 74–92.

WALD, A., 1950, "Statistical Decision Functions," John Wiley & Sons, Inc., New York, 172 pp.

WALKER, H. M., 1931, "Studies in the History of Statistical Method," The Williams & Wilkins Company, Baltimore, 229 pp.

WALKER, S., 1941, Reproducibility of Tests of Foundry Sand, *Am. Foundrymen's Assoc.,* **49:** 1–31.

WATSON, G. S., and E. IRVING, 1957, Statistical Methods in Rock Magnetism, *Monthly Notices Roy. Astron. Soc. Geophys. Suppl.,* **7:** 289–300.

WEATHERBURN, C. E., 1957, "A First Course in Mathematical Statistics," 2d ed., Cambridge University Press, New York, 277 pp.

WELSH, B. L., 1938, The Significance of the Difference between Two Means when the Population Variances Are Unequal, *Biometrika,* **29:** 350–362.

WENTWORTH, C. K., 1922, A Scale of Grade and Class Terms for Clastic Sediments, *J. Geol.,* **30:** 377–392.

———, 1923, An Improved Recording Micrometer for Rock Analysis, *J. Geol.,* **31:** 228–232.

———, 1929, Method of Computing Mechanical Composition Types of Sediments, *Bull. Geol. Soc. Am.,* **40:** 771–790.

———, W. L. WILGUS, and H. L. KOCH, 1934, A Rotary Type of Sample Splitter, *J. Sediment. Petrol.,* **4:** 127–138.

WESTMAN, A. E. R., 1926, The Mercury Balance: An Apparatus for Measuring the Bulk Volume of Brick, *J. Am. Ceram. Soc.,* **9:** 311–318.

WHITAKER, L., 1914, On the Poisson Law of Small Numbers, *Biometrika,* **10:** 36–71.

WHITTAKER, E., and G. ROBINSON, 1949, "The Calculus of Observations," 4th ed., Blackie & Son, Ltd., Glasgow, 397 pp.

WHITTEN, E. H. T., 1961, Quantitative Areal Modal Analysis of Granitic Complexes, *Bull. Geol. Soc. Am.,* **72:** 1333–1360.

WILLIAMS, E. J., 1959, "Regression Analysis," John Wiley & Sons, Inc., New York, 214 pp.

WINCHELL, A. N., and H. WINCHELL, 1951, "Elements of Optical Mineralogy," pts. II and III, John Wiley & Sons, Inc., New York.

WINSAUER, W. O., H. M. SHEARIN, JR., P. H. MASSON, and M. WILLIAMS, 1952, Resistivity of Brine-saturated Sands in Relation to Pore Geometry, *Bull. Am. Assoc. Petrol. Geologists,* **36:** 253–277.

WOOD, G. V., 1960, A Comparison of Three Quartzites, unpublished Ph.D. thesis, Dept. Mineralogy Pennsylvania State University, University Park, Pa., 159 pp.
——— and J. C. GRIFFITHS, 1963, Modal Analysis of Three Quartzites and Their Textural Implications, *J. Geol.,* **71:** 405–421.
YATES, F., 1934, Contingency Tables Involving Small Numbers and the Chi-square Test, *J. Roy. Statist. Soc. Suppl.,* **1:** 217–235.
———, 1949, "Sampling Methods for Censuses and Surveys," Hafner Publishing Company, Inc., New York, 318 pp. (2d ed., 1953, 481 pp.).
YOUDEN, J., 1951, "Statistical Methods for Chemists," John Wiley & Sons, Inc., New York, 126 pp.
——— and W. S. CONNOR, 1953, The Chain Block Design, *Biometrics,* **9:** 127–140.
YULE, G. U., 1911, The Influence of Bias and of Personal Equation on Statistics of Ill-defined Qualities, *J. Anthropological Inst. Gt. Brit. Ireland,* **36:** 325–381.
——— AND M. G. KENDALL, 1948, "An Introduction to the Theory of Statistics," 13th ed., Charles Griffin & Company, Ltd., London, 570 pp.
YUSTER, S. T., 1937, Determining the Porosity of Oil Sands, *Producer's Monthly,* **1:** 21–25.

Index